Recent Progress in Plant Biology

Recent Progress in Plant Biology

Edited by Geoffrey Watkins

Syrawood
PUBLISHING HOUSE

New York

Published by Syrawood Publishing House,
750 Third Avenue, 9th Floor,
New York, NY 10017, USA
www.syrawoodpublishinghouse.com

Recent Progress in Plant Biology
Edited by Geoffrey Watkins

International Standard Book Number: 978-1-68286-745-7 (Hardback)

Cataloging-in-Publication Data

Recent progress in plant biology / edited by Geoffrey Watkins.
    p. cm.
Includes bibliographical references and index.
ISBN 978-1-68286-745-7
1. Botany. 2. Plants. I. Watkins, Geoffrey.
QK45.2 .R43 2019
580--dc23

TABLE OF CONTENTS

PREFACE

This book aims to highlight the current researches and provides a platform to further the scope of innovations in this area. This book is a product of the combined efforts of many researchers and scientists, after going through thorough studies and analysis from different parts of the world. The objective of this book is to provide the readers with the latest information of the field.

Plants are multicellular organisms of the kingdom Plantae. They are primarily photosynthetic eukaryotic organisms and include flowering plants, conifers, ferns, gymnosperms, etc. The branch of plant biology is concerned with the scientific study of plants. Modern botany integrates aspects of molecular genetics and epigenetics in plant studies. The provision of staple foods and materials such as oil, rubber and drugs is achieved by the application of botanical studies. Plants play a crucial role in maintaining the balance of carbon cycle and water cycle. Research in this field focuses on the identification of plant species and commercial plant varieties by DNA barcoding. This book presents the complex subject of plant biology in the most comprehensible and easy to understand language. It provides significant information of this discipline to help develop a good understanding of botany and related fields. The book is appropriate for students seeking detailed information in this area as well as for experts.

I would like to express my sincere thanks to the authors for their dedicated efforts in the completion of this book. I acknowledge the efforts of the publisher for providing constant support. Lastly, I would like to thank my family for their support in all academic endeavors.

Editor

Morphological, cellular and molecular evidences of chromosome random elimination *in vivo* upon haploid induction in maize

Fazhan Qiu*, Yanli Liang, Yan Li, Yongzhong Liu, Liming Wang, Yonglian Zheng

National Key Laboratory of Crop Genetic Improvement, Huazhong Agricultural University, Wuhan 430070, China

ARTICLE INFO

Keywords:
Maize
Chromosome elimination
Inducer
Haploid
SSR
Aneuploidy

ABSTRACT

The mechanism of maternal *in vivo* haploid induction is not fully understood. In this study, the young embryos were identified by morphology, cytology and simple sequence repeat (SSR) markers at different developmental stages in the cross HZ514 (sweet corn) × HZI1 (inducer). The results indicated that the low seed setting rate was determined by the inducer pollen during the process of fertilization. The mosaic endosperm kernels and the different percentages of aneuploidy, mixploidy, lagged chromosome, micronuclei, chromosomal bridge and ring chromosome were found in the cross; 7.37% of the haploid embryos carried chromosome segments from HZI1. About 1% twin seedlings resulted from the cross and were analyzed by cytology and SSR markers. Four pairs of twin seedlings had different chromosome numbers ($2n = 20$ and $2n = 10–20$) and there were some chromosome fragments from HZI1. Aneuploidy, mixploidy and the abnormal chromosomes occurred in the *in vivo* haploid induction by HZI1, which is the cytological basis for haploid induction and indicates that the inducer's chromosomes are prone to be lost during mitotic and meiotic divisions. Morphological, cellular and molecular evidences reveal that complete or partial chromosome elimination from inducer HZI1 controls the maize *in vivo* haploid induction.

DATA: The link refers to the raw data from: Morphological, cellular and molecular evidences of chromosome random elimination in vivo upon haploid induction in maize. Current Plant Biology.

1. Introduction

The term haploid sporophyte is generally used to refer to sporophytes having the gametic chromosome number [1]. The first haploids in flowering plants were identified by Blakeslee in 1922 [2], and the doubled-haploid (DH) technology can shorten the breeding time significantly [3]. Haploids generated from a heterozygous individual and doubled to instant homozygous lines can greatly accelerate plant breeding [4–6]. For these reasons, the potential of haploids in plant breeding is recognized and considered in crop genetic improvement.

Two methods are generally used to produce haploids in plants: cells and tissues culture (*in vitro*) and genetic induction (*in vivo*). Maize haploid can also be derived through the two methods. However, tissue culture in maize is complex and greatly limited by genetic background [7,8]. Thus the method of induction-haploid *in vivo* by inducer lines, which achieves a high haploid induction

frequency and is relatively simple to use, is important and widely used in maize breeding.

Several haploid-inducing lines have been developed in maize [9,10]. Stock6, with the induction rate of 0.5–3%, is one of the haploid-inducing lines discovered by Coe [11] and Sarkar and Coe [12]. However, the low induction rate could not meet the needs of breeders. When both maternal and paternal effects were detected in the process of haploid induction and the haploid-inducing character was found to be a heritable trait [9,12–15], a number of new inducers with much higher haploid-induction rate have been created by cross method among stock6, w23ig or other germplasm, such as KMS [16], WS14, ZMS [10], RWS [17], MHI [18] and HZI1 [19]. Unfortunately, the mechanism underlying *in vivo* haploid-inducing capacity in maize is not fully understood.

Researchers have focused on two possible mechanisms: parthenogenesis and chromosome elimination. Firstly, parthenogenesis was caused by the irregularities of microsporogenesis and fertilization [20–25]. All of these findings indicate that various irregularities appearing between microsporogenesis and fertilization may prevent double fertilization and stimulate division of the egg cell without fertilization. As a result of this process, a haploid

* Corresponding author.
  E-mail address: qiufazhan@gmail.com (F. Qiu).

embryo can be formed from an unfertilized egg cell. Secondly, a set of chromosome that is randomly eliminated after fertilization might be a major mechanism underlying *in vivo* haploid-induction in maize. Wedzony et al. [17] observed that 10% of the resulting embryos exhibit micronuclei of variable size after the inducer line RWS self-pollinated. Such micronuclei are characteristic of chromosome fragments being eliminated from the cell in subsequent divisions. Gernand et al. [26] found that the inducer chromosomes degenerate and fragment a few days after fertilization in interspecific crosses. Then the fragments coalesce to form the micronuclei and become eliminated from the cells within three weeks. Fischer et al. [27], Zhang et al. [19], Li et al. [28] and Zhao et al. [52] observed that a small proportion (1–3%) of haploids obtained from the cross between inducer lines and breeding materials carried several paternal chromosome segments *via* SSR markers analysis and cytogenetic makers to trace chromosomes from inducers. The results showed that some minor fragments of the inducer genome were introgressed into maternal genome of the haploids. However, Zhao et al. [52] found haploid formation with rare inducer fragment introgression. Furthermore, the aberrant fertilization mechanisms leading to haploidy may be related to mechanisms leading to hetero-fertilization [29]. Thus whether the formation of female haploid embryos results from single fertilization or from chromosome elimination remains unclear; whether haploid occur is determined by inducer or maternal materials is also unclear.

In the present study, the inducer line HZI1 derived from Stock 6 was used to induce haploids from the sweet maize pure line HZ514. The main objectives were (1) to study the characteristic of the inducer line HZI1 and identify its pollen and ear fertility; (2) to monitor the chromosome number during development of the haploid seeds after fertilization, upon induction of *in vivo* maize haploid production by HZI1. The genotype of haploid embryo was identified *via* SSR markers; and (3) to discuss the possible fundamental biological mechanisms underlying *in vivo* maternal haploid induction as well as implications of the results on improving high induction rate in maize.

2. Materials and methods

2.1. Plant materials and pollination

HZI1, a stock-6-derived haploid-induction maize line was used as the male parent, which carried the R-navajo gene that is responsible for the anthocyanin pigmentation of the endosperm and embryo. HZ514, a super sweet corn inbred line, and NA and 248, two normal corn inbred lines, were used as female parents (Table S1). The three inbred lines with colorless aleurone layer and colorless scutellum were developed by Huazhong Agricultural University (Hubei Province, China). The crosses between the inducer HZI1 and above three inbred lines were performed at Huazhong Agricultural University in 2009, and all F$_1$ kernels were harvested by single ear and analyzed separately. A total of 30,000 kernels were harvested from 150 ears for each cross. At the same time, HZ514 and HZI1 plants of normal development were selected for self-pollination at shedding pollens and emerging silks. The reciprocal crosses were also done between HZ514 and HZI1. HZ514 plants as receiver were crossed with NA and 248 respectively (HZ514 × NA and HZ514 × 248).

2.2. Sampling methods and cytology

2.2.1. Sample, fixation and isolation

Maize immature kernels after pollination were harvested from each ears of HZ514 × HZI1 and stored in the alcohol following a procedure similar to that used by Yang et al. [30]. From 25 to 65 h after pollination, the ears were collected every 5 h and fixed in a solution of 3:1 alcohol: glacial acetic acid for 24 h, rinsed one time every 30 min in 95% ethanol, 85% ethanol, 70% ethanol, and then stored in 70% alcohol at room temperature for further use. All the collected kernels were used for cytological observations and molecular marker analysis.

According to the method used by Herr [31] and Stelly et al. [32], the whole stain-clearing technique was used to detect the ovules development status. The ovaries were dissected in 70% ethanol, and hydrated sequentially in 50% ethanol, 30% ethanol, 15% ethanol and distilled water. After that, the ovaries were stained with diluted Enrlich's haemaloxylin dyeing liquor (primary Enrlich's haemaloxylin dyeing liquor:50% ethanol:glacial acetic acid = 1:1:1). The ovaries were rinsed 24 h with distilled water and agitated for 4–5 times in that duration. The ovaries were dehydrated one time with 15%, 30%, 50%, 70%, 85%, 95% of ethanol solutions, and then with 100% ethanol three times (dehydrated for 1 h at each step). Finally, the samples were stored in wintergreen oil for further use.

2.2.2. Microscopic examination

The cleared ovaries were put on Glass slides and observed with OLYMPUS IX71 microscope. The dissected embryos were stained with Carbol fuchsin solution for 10 min and then squashed. The samples were placed under the OLYMPUS IX71 microscope to image cell division phases and record the numbers of the chromosome present.

2.2.3. Fertility investigation

The pollen fertility of HZI1, HZ514, NA and 248 were determined as the percentage of pollen grains stained with 1% KI/I$_2$. The ear fertility was determined by the seed setting rate in the reciprocal cross between them.

2.3. SSR analysis

The haploid kernels and diploid kernels from the cross HZ514 × HZI1 were judged according to cytology analysis. The DNA will be extracted from the accurate haploid embryos for SSR analysis.

Genomic DNA was isolated individually from immature embryos according to a procedure similar to that used by Saghai-Maroof et al. [33]. The sequence of all SSR markers was obtained from the MaizeGDB database (www.maizegdb.org/ssr.php).

3. Results

3.1. Identification of fertility and morphology

The pollen fertility of HZI1, HZ514, NA and 248 were all normal, with over 90% regarded as fertile (Fig. S1). The pollen fertility of the haploid plants from the cross HZ514 × HZI1 were 0–38%; and the doubled plants from the haploid individual had a similar morphology and the same phenotype and genotype as the female HZ514.

In addition, 0.3% kernels with mosaic endosperm of purple aleurone and yellow shrunken without purple aleurone were found in the F$_1$ mature kernels from HZ514 × HZI1 (Fig. S2). The same results were also found by Zhang et al. [19].

3.2. Seed setting rate from reciprocal-cross and self-fertilization

In this study, the seed setting rate from self-fertilization or crosses among HZ514, HZI1, NA and 248 were significantly different. The seed setting rate of HZ514 self-fertilization, HZ514 × NA, HZ514 × 248 and HZI1 × HZ514 were normal. However, when HZI1 was used as the male parent for either cross or self-cross, the seed

Fig. 1. The various mating types at the developmental and mature stages. (A) The ears of HZ514 × NA, HZ514 × 248, HZ514 and HZ514 × HZI1 at 40 d after pollination; from left to right. (B) The ears of HZI1 and HZI1 × HZ514 at 40 d after pollination; from left to right. (C) The ears of HZ514 × HZI1, HZ514, HZ514 × HZI1 and HZ514 × HZI1 at 18 d after pollination; from left to right. (D) The ears of HZ514 × NA, HZ514 × HZI1, HZ514 × 248 and HZ514 at 18 d after pollination; from left to right.

setting rate was very poor and some kernels were abnormal in morphology (Fig. 1).

3.3. Fertilization status, embryo and endosperm development at nine stages after pollination in the cross HZ514 × HZI1

Three types of ovules at different sample stages (25, 30, 35, 40, 45, 50, 55, 60, 65 h) were identified in the cross HZ514 × HZI1 upon ovules whole stain-clearing: in the first type (i), the embryo and endosperm were developed (Fig. 2A); in the second type (ii), the embryo has not divided, but the endosperm was developed (Fig. 2B); and in the third type (iii), the embryo developed, but the central cell has not divided (Fig. 2C-2D). Out of 1681 ovules, the proportion of the ovules types from 25 h to 65 h after pollination was as follows: the first type was 51.45% to 57.55%, with an average of 54.19% (911); the second type was from 3.09% to 7.17%, with an average of 5.29% (89); the third type was from 37.74% to 43.57%, with an average of 40.51% (681) (Table 1);

3.4. Cytological analysis of the pollen mother cell of the HZI1, HZ514, 248 and HZ514 × HZI1

In the pollen mother cell of HZI1, only 82.3% cells had $2n = 20$ and 17.7% had $2n < 20$. In addition, 11.3% had the number of haploid cells $2n = 10$. The $2n = 10$ occupied about 10.2% of the progeny from the HZ514 × HZI1 while $2n = 20$ occupied about 86.39%. Compared with HZI1 and the cross HZ514 × HZI1, the pollen mother cell of HZ514 had 98.4% of the progeny with $2n = 20$ and only 1.6% had $2n < 20$. The pollen mother cell of 248 had 99.44% of the progeny with $2n = 20$ and only 0.56% had $2n < 20$ (Table S2).

The abnormal chromosomes were found in the pollen mother cell of the inducer HZI1, HZ514 and HZ514 × HZI1 included lagged chromosome, micronuclei and chromosome bridges. The percentages of lagged chromosomes were 1.47%, 6.83% in HZ514,

HZI1 and HZ514 × HZI1, respectively, during meiosis; the unsynchronized chromosome condensation and division were 2.01%, 7.24% and 8.33%, respectively; Chromosome bridges and fragments occupied about 0.34%, 1.69% and 3.42%, respectively (Table 2, Fig. 3).

3.5. Cytological observation and marker analysis at 12 d after pollination in the cross HZ514 × HZI1

The frequency of cells with $2n = 10$ was 6.31%, while those with $2n = 20$ was 72.13%. More than half of the embryos of F_1 kernels had variable chromosome numbers ($2n = 9$–21), as described by Zhang et al. [19], and aneuploid embryos with $2n = 19$ and 21 were about 0.5%–1%. The kernel's endosperms with $2n = 40$, 30 or 60 were present at 1.85% and 98.15%, respectively. Most of the cells in endosperms had $2n = 30$, but there were some cells that randomly doubled to $2n = 60$ in every endosperm (Fig. 4). In addition, ring chromosomes were found in the F_1 kernels of HZ514 × HZI1 cross. The percentage of the ring chromosome was 2.13% of the total 1232 embryos (Fig. 5).

The total of 286 haploid embryos identified by cytology analysis ($2n = 10$) from HZ514 × HZI1 were analyzed by 100 SSR markers, with the diploid embryos and two parents embryos serving as the control. 92.63% haploids shared the same genomic compositions as the female HZ514 and 7.37% haploids had male chromosome fragments (Fig. 6). For example, 32 $2n = 10$ embryos were analyzed by the SSR marker umc1747 and revealed that three haploid embryos had the chromosome fragments from HZI1 (Fig. 6A). No. 21, 562, 1032 and 1632 with $2n = 10$ were found to carry male chromosome fragments identified to be bnlg1909, umc1241, bnlg1600 and umc1241, respectively (Fig. 6B–D). In addition, about 12 pair SSR markers located on chromosome 5 were detected as a heterozygous band in the $2n = 10$ embryos, which indicates a possible hot region for chromosome elimination on chromosome 5.

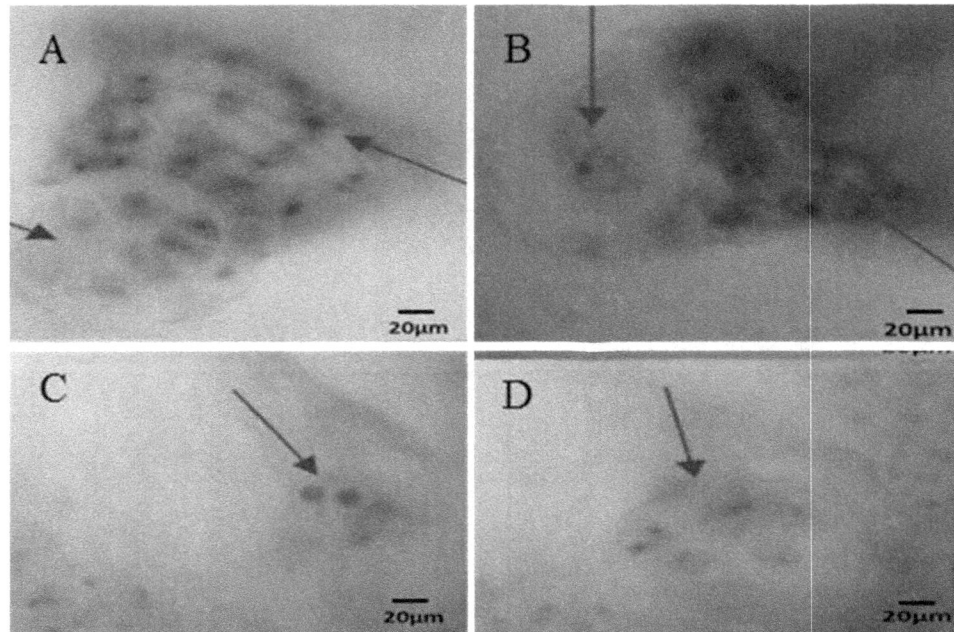

Fig. 2. Ovule whole stain-clearing at 50 h after pollination in the cross HZ514 × HZI1. (A) Arrow means dividing embryo and dividing endosperm; (B) arrow means division endosperm and no division embryo; (C) no division central cell; (D) dividing embryo.

Table 1
Three types ovules after pollination in the cross HZ514 × HZI1 at different periods.

		I	II	III	Total
25 h	Ovary no.	129	13	101	243
	%	53.09	5.35	41.56	
30 h	Ovary no.	119	16	88	223
	%	53.36	7.17	39.46	
35 h	Ovary no.	124	12	105	241
	%	51.45	4.98	43.57	
40 h	Ovary no.	85	9	70	164
	%	51.83	5.49	42.68	
45 h	Ovary no.	122	9	81	212
	%	57.55	4.25	38.21	
50 h	Ovary no.	97	10	67	174
	%	55.75	5.75	38.51	
55 h	Ovary no.	108	6	80	194
	%	55.67	3.09	41.24	
60 h	Ovary no.	68	7	49	124
	%	54.84	5.65	39.52	
65 h	Ovary no.	59	7	40	106
	%	55.66	6.6	37.74	
Total	Ovary no.	911	89	681	1681
Average	%	54.19	5.29	40.51	

Note: I, developed embryo and endosperm; II, developed embryo, undeveloped central cell; III, undeveloped egg cell, developed endosperm.

3.6. Cytology and marker analysis of twin seedlings from the progeny of the cross HZ514 × HZI1

About 1% twin seedlings were found in the progeny of the cross HZ514 × HZI1, and this percentage is higher than that in spontaneous generation [51]. There are two possible phenotypes that can be observed from the twin seedlings: (1) the two seedlings have the same phenotype and chromosomes numbers; (2) the two seedlings have different phenotypes and chromosome numbers (Fig. 7A1 and A2). Among the 20 pairs of twin seedlings, 4 pairs had different chromosome numbers ($2n = 20$ and $2n = 10$–20), and 5 twin seedlings containing $2n = 10$ and $2n = 20$. Cells with $2n = 20$ were 11 pairs (Fig. 7B1–B4). The results mean that haploid seedling and diploid seedling coexist in some twin seedlings; diploid seedlings and aneuploidy or mixoploid coexist in some twin seedlings; the rest twin seedlings are diploid.

Table 2
Frequency of the abnormal chromosomes in HZ514, HZI1 and HZ514 × HZI1.

Material and cell numbers	Total (%)	Lagged chromosome (%)	Unsynchronized chromosome condensation and division (%)	Chromosome bridge (%)
HZ514 (2682)	3.48	1.47	2.01	0.34
HZI1 (2265)	14.06	6.83	7.24	1.69
HZ514 × HZI1 (2570)	16.52	8.19	8.33	3.42

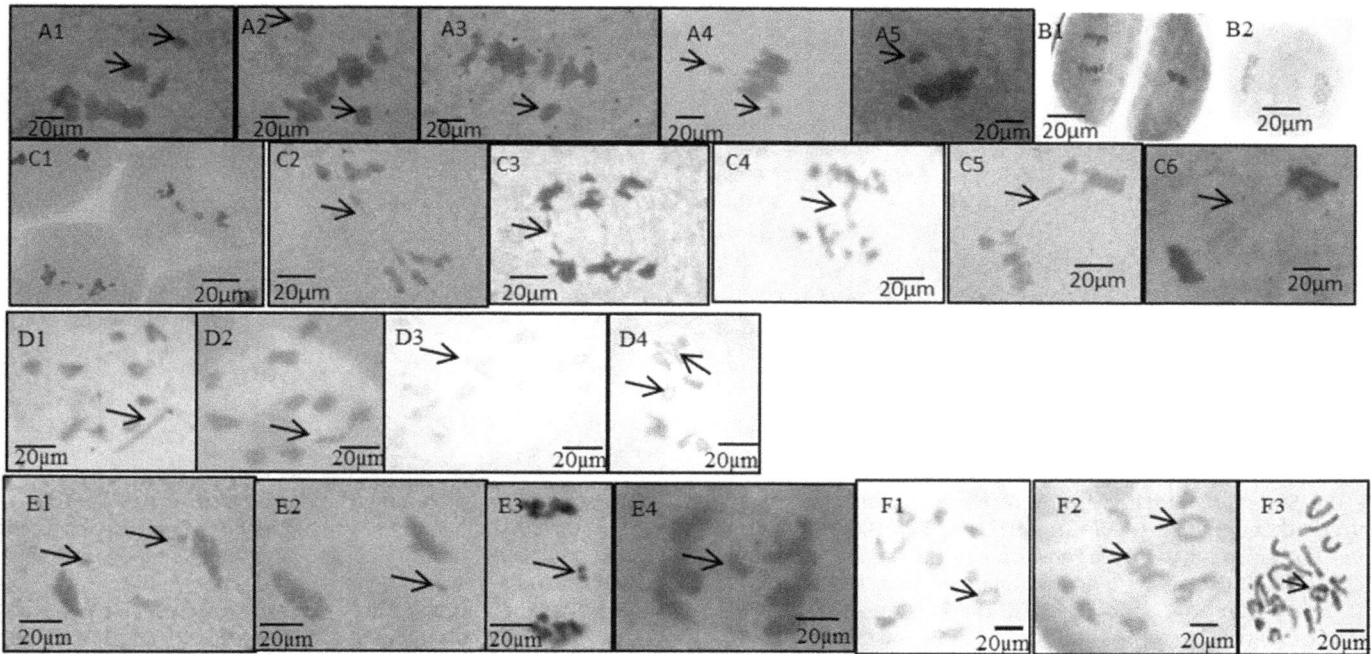

Fig. 3. Various chromosomes morphology after pollination in the cross HZ514 × HZI1. (A1–A5) Lagged chromosome; (B1–B2) unsynchronized division; (C1–C6) chromosome bridge in late mitosis; (D1–D4) unsynchronous chromosome; (E1–E4) micronuclei at telophase; (F1–F3) cycle chromosome.

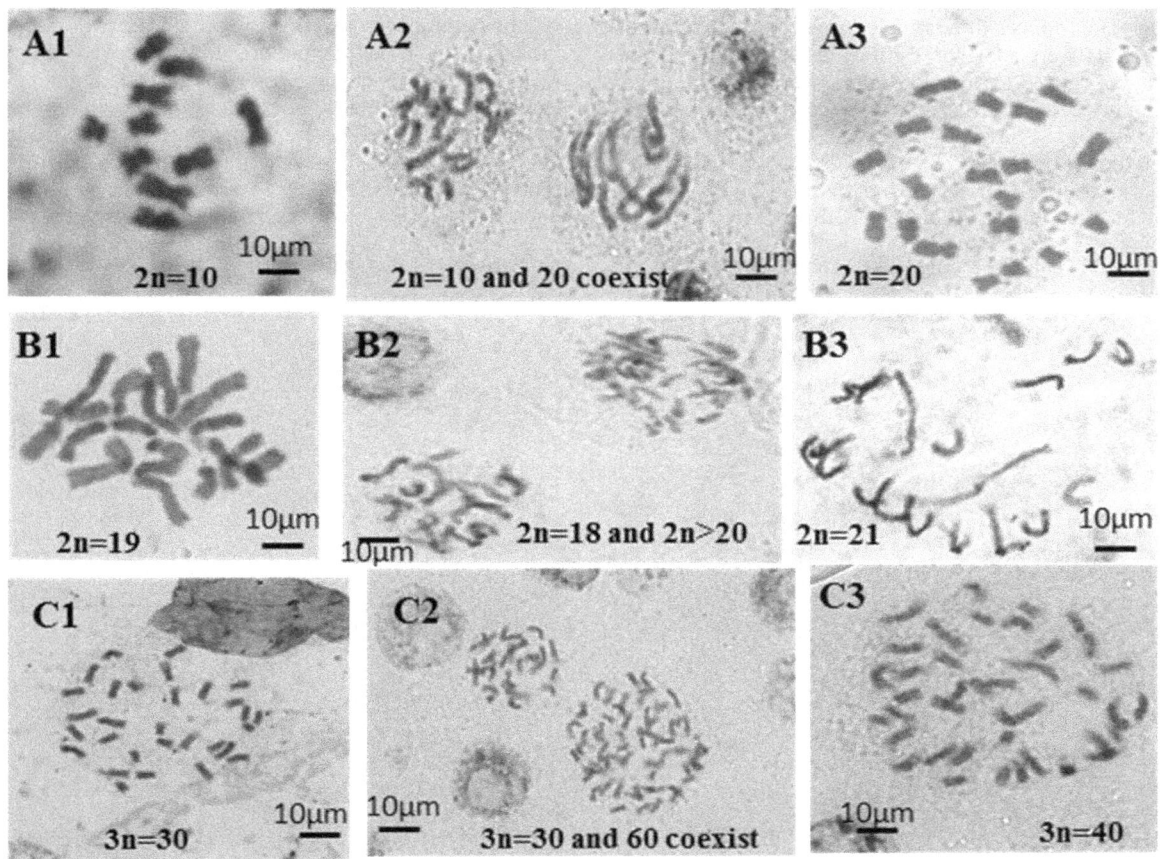

Fig. 4. Cytology of F_1 kernels from HZ514 × HZI1 cross at 12 days after pollination. (A1–A3) the embryo of the F_1 kernels with $2n = 10$ (A1), 10 and 20 coexisting (A2), 20 (A3); *Bar* 10 μm. (B1–B3) Aneuploid embryo of the F_1 kernels with $2n = 19$(B1), $2n = 18$ and $2n > 20$ coexisting (B2), $2n = 21$(B3); *Bar* 10 μm. (C1–C3) the endosperm of the F_1 kernels, with $3n = 30$ (B1), 30 and 60 coexist (B2), 40 (B3); *Bar* 10 μm.

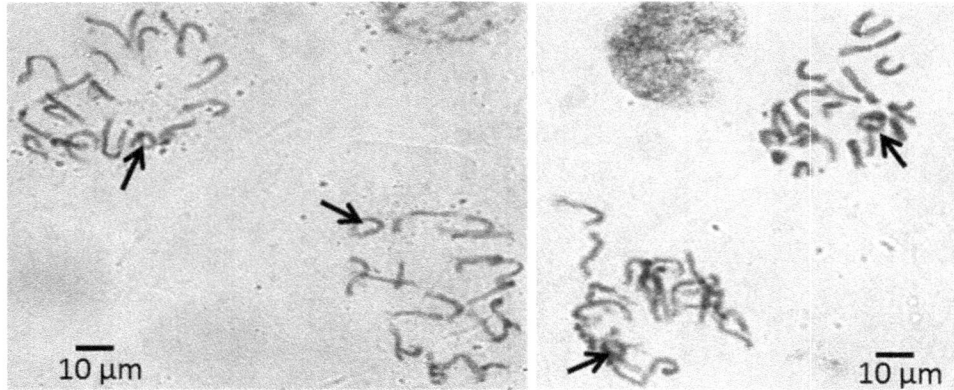

Fig. 5. Cycle chromosome of F$_1$ kernels from HZ514 × HZI1 cross at 12 days after pollination.

The DNA of the 20 twin seedlings were extracted and analyzed by 100 SSR markers located on ten maize chromosomes. Some chromosome fragments were from HZI1 in $2n = 10$ seedlings and $2n = 11$–19 seedlings, whereas no chromosome fragment was from HZI1 in $2n = 20$ seedlings. For example, five twin seedlings were analyzed by umc1447 and umc2030 markers and heterozygous bands were found in three seedlings (Fig. 7C1 and C2)

4. Discussion

Measurement of the seed setting rate from reciprocal-cross and self-fertilization among HZI1, HZ514, NA and 248 at different development stages revealed that the low seed setting rate is only detected when the inducer line is male. About 36%–50% kernels from HZ514 × HZI1 were abortive (Fig. 1), indicating that the phenomenon may be caused by inducer pollen. However, the pollen fertility of the inducer and other materials were normal and over 90% were fertile (Fig. S1). Lin et al. [34] speculated that abortive grain was related with the excrescent polar nuclei in megaspore. According to the results in this study, the main reason may be that the process of fertilization was abnormal, in that the two sperms of

inducer were not normal; and the low setting rate was not caused by the sperm number, but may be caused instead by alterations in the chromosomal structure or cell organelles in sperms during the second mitosis in microspores. The haploid induction should also be determined during this abnormal process.

$2n = 10$ kernels occupied about 10.2% in the cross HZ514 × HZI1, with the final haploid induction frequency at 6.31%, lower than that in the development after pollination. This result indicates that some kernels which are aneuploid and mixploid were sterile during the development process and the chromosomes from HZI1 may be completely lost during the development process of embryos. Thus we should take into account that the seed setting rate of the inducer is lower whether self-crossed or used as male [35]. This information will be helpful to breeders during the process of higher frequency inducer selection or haploid breeding [36].

The F$_1$ kernels of the cross HZ514 × HZI1 should be flint corn. However, over 0.3% of the mosaic endosperm kernels were found in the progenies of the cross, and these kernels exhibited different proportions of the sweet shrunken endosperm in some ears (Fig. S2). The same results were also found by Zhang et al. [19], and Zhao et al. [52] and Xu et al. [53]. These observations could

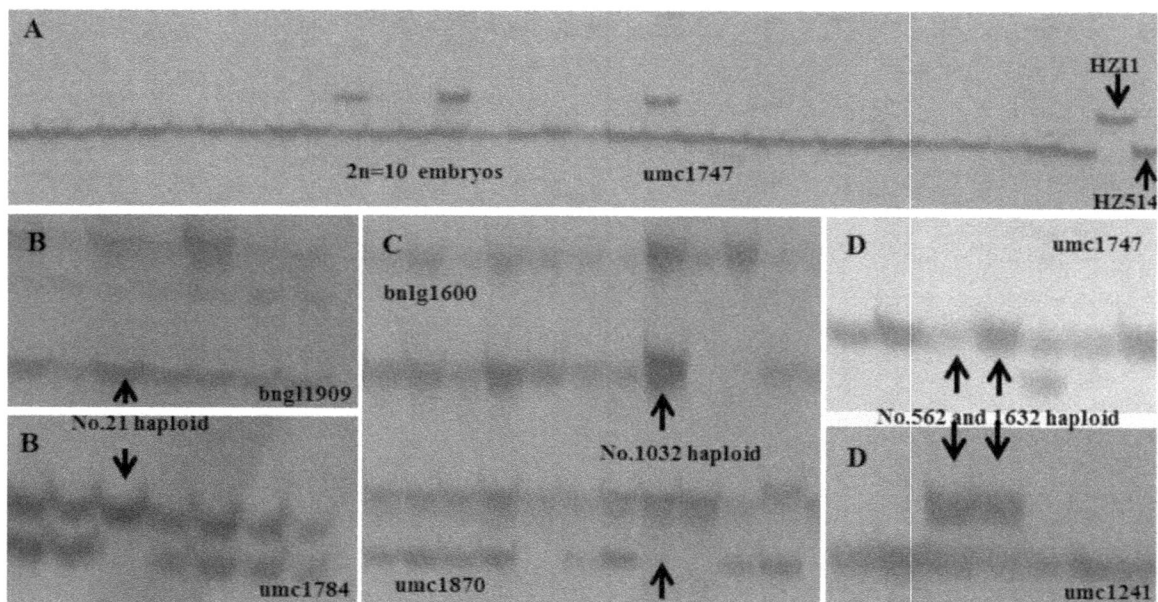

Fig. 6. SSR analysis of haploid embryos in the cross HZ514 × HZI1 at 12 days after pollination. (A) 32 $2n = 10$ embryos were analyzed by the SSR marker umc1747 and three haploid embryos with inducer's chromosome fragments; (B) arrow is No. 21 with haploid embryo, was analyzed by umc1784 and bnlg1909, the inducer's chromosome fragments was found in No. 21 by bnlg1909; (C) arrow is No. 1032 with haploid embryo, was analyzed by umc1870 and bnlg1600, the inducer's chromosome fragments was found in No. 1032 by bnlg1600; (D) arrows are Nos. 562 and 1632 with haploid embryos, which carried male fragments identified by marker umc1241 and no male fragments by marker umc1747.

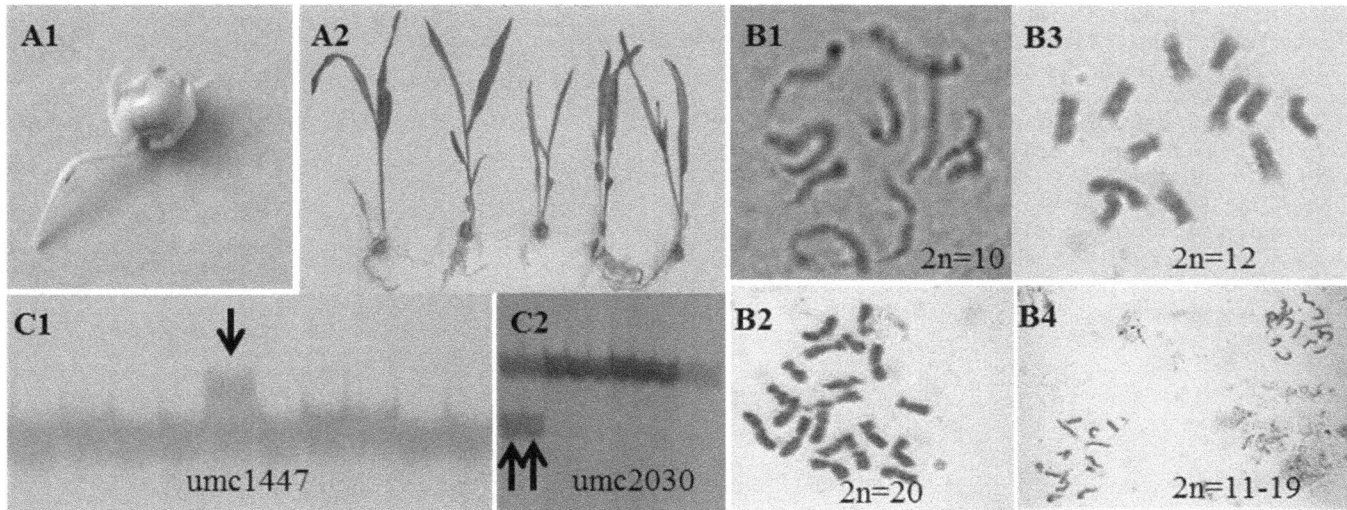

Fig. 7. The genotype and phenotype of twin seedlings in the HZ514 × HZI1 cross. (A1 and A2) twin seedlings; (B1) the small seedlings with 2n = 10; (B2) the large seedlings with 2n = 20; (B3 and B4) the seedlings are aneuploidy or mixoploidy with 2n = 12 and 2n = 11–19; (C1 and C2) The seedlings with 2n = 10 carried male fragments identified by marker umc1447 and umc2030.

be accounted for by the double recessive sweet endosperm and yellow aleurone layer organization. The mosaic mixoploid traits were caused by the complete elimination of the paternal chromosome in one daughter cell from the first mitosis of the primary endosperm nuclei during the process of cell division. In addition, the whole ears phenotypes of the induced progeny showed that the different ratios of mosaic kernels were caused by the different elimination times of paternal chromosomes during the endosperm developmental process after cross fertilization.

We observed three types of ovules with different embryos and endosperms after pollination at different development periods in the cross HZ514 × HZI1 (nine stages, from 25 to 65 h). Kernels obtained from the cross-pollinated ears also were classified into three categories: (i) normal kernels with normal embryos and endosperms, normal diploids and haploids were included; (ii) endosperm abortion, with shrunken endosperms or defective kernels; (iii) embryo abortion, with normal endosperms but without embryos. Based on the fertilization status of the embryos and endosperms in the cross, we speculate that the process of fertilization and the different developmental stages must be abnormal and that this was the main reason for the haploidy (Table 1 and 2; Fig. 2). Similar results were found by Bylich et al. [21] and Coe et al. [37], but they did not provide possible mechanisms; Rotarenco and Eder [25] detected a much higher rate of heterofertilization when the haploid inducer MHI was used instead of the normal line. Xu et al. [53] reported that the different kernel phenotypes are most probably caused by the sed1 locus, they speculated that the defective kernels and haploids are caused by the same genetic mechanism. However, we could not speculate what caused the change in the number of sperm in synergid in this study. Is it because the sperm stayed in synergid and did not make it to the egg cell, or that fertilization was finished and could not proceed to egg cell division? Did the egg cell stay somewhere at the megaspore stage? We could not confirm any of these possibilities by whole stain-cleaning method and optical microscope alone. Whether parthenogenesis occurred or not need further study, but our results support that the abnormal process of fertilization must be one reason for the maternal haploid induction.

The paternal haploid could be found in the progeny of the inducers [38]. The percentage of inducer pollen cells with aneuploidy was significantly higher than other normal inbred lines. Similar results were also identified for microsporocytes of the inducer MHI [22]

and radicle cells of the inducer [19]. Although the pollen fertility of the inducer was normal, the chromosome ploidy variation was the main reason that caused high frequency of heterofertilization. Lagged chromosomes, the micronuclei and chromosome bridges were often used as direct evidences of chromosome elimination and haploid production in inter-, intra-specific hybridizations in crops [17,39,40] also found micronuclei in the maize haploid inducer RWS. In this study, the abnormal chromosomes, such as lagged chromosome, micronuclei and chromosome bridges, indicate that chromosome elimination occurred in the primordium of the inducer after self-pollination and that chromosome segregations were not synchronous and equal during cell meiosis. According to the results, chromosome random elimination should occur during haploid induction. The aneuploidy should be induced by partial elimination of chromosomes from HZI1. About 1.85% endosperms had 2n = 40 which may be caused by chromosome randomly partial elimination. Remarkably, about 2.13% ring chromosome was observed in 1232 embryos (Fig. 5). The ring chromosome is formed after chromosome deletion or elimination. Thus our results indicate that random elimination of chromosome occurs during the in vivo haploid induction by inducers in maize.

Khokhlov et al. [41] pointed out that diploid cells exist in any tissues of haploid and Wei et al. [42] observed that 80% plants had doubled haploid cells. There were some cells that always randomly doubled to 2n = 60 in endosperms (Fig. 4 C2), which was similarly observed with haploid embryos. However, the doubling frequency of haploid embryos is lower than in endosperm. In this study, 80% of the identified megaspores had fertilized polar nucleus and normally divided endosperm. These also contained multi nucleoli which may have triggered the random doubling of the endosperm.

Our results from the genome-wide SSR markers demonstrated that the introgression of genetic element from the inducer HZI1 was occurred during the process of chromosome elimination. The results of cytological analysis and genotype analysis based on twin seedlings also indicate that chromosome introgression or elimination occurs during the haploid induction by the inducer. Similar results were found by Fischer et al. [27], Zhang et al. [19] and Li et al. [28]; but Zhao et al. [53] found haploid formation with rare inducer fragment introgression, the discrepancy might be due to different genetic background, different markers and marker numbers. One major QTLs controlled haploid induction rate were identified on chromosome 5 [43], and haploid embryos with male chromosome fragment could be identified by 12 pairs SSR markers on chromosome 5. This indicates the presence of one hot region of chromosome introgression or exchange which may control the haploid induction rate.

The chromosome elimination mechanism is similar to that in genome wide hybridizations and probably controlled by some genetic factors in the inducers [39,40,43–50,52,53]. Aneuploidy, mixploidy and the abnormal chromosomes observed upon *in vivo* haploid induction by inducer were found in the pollen cells, microsporocytes and radicle cells of inducers. This is the cytological basis for haploid induction, suggesting that the inducer's chromosomes were prone to be lost during mitotic and meiotic divisions after the inducers were crossed with other lines. The morphological, cellular and molecular evidences suggest that maize *in vivo* haploid induction is controlled by complete or partial chromosome elimination from inducer HZI1.

Acknowledgements

This research was supported by the National Natural Science Foundation of China (No. 30900899) and partly supported by the open funds of the National Key Laboratory of Crop Genetic Improvement.

References

[1] C.E. Palmer, W.A. Keller, Overview of haploidy Haploids in Crop, Improvement II, vol. 56, Springer, Heidelberg, Germany, 2005, pp. 3–9.
[2] A.F. Blakeslee, J. Belling, M.E. Farnham, A.D. Bergner, A haploid mutant in *Datura stramonium*, Science 55 (1922) 646–647.
[3] S. Guha, S. Maheshwari, In vitro production of embryos from anthers of *Datura*, Nature 204 (1964) 497.
[4] B.P. Forster, W.T.B. Thomas, Doubled haploids in genetics and plant breeding, Plant Breed Rev. 25 (2005) 57–88.
[5] B.P. Forster, E. Heberle-Bors, K.J. Kasha, A. Touraev, The resurgence of haploids in higher plants, Trends Plant Sci. 12 (8) (2007) 368–375.
[6] M. Ravi, S.W. Chan, Haploid plants produced by centromere-mediated genome elimination, Nature 464 (2010), http://dx.doi.org/10.1038/nature08842.
[7] M. Beckert, in: Y.P.S. Bajaj (Ed.), Advantages and disadvantages of the use of in vitro/in situ produced DH maize plants, Biotechnol. Agric. For. 25 (1994) 201–213.
[8] O.A. Shatskaya, E.R. Zabirova, V.S. Shcherbak, M.V. Chumak, Mass induction of maternal haploids in corn, Maize Genet. Coop. Newsl. 68 (1994) 51.
[9] K.R. Sarkar, S. Panke, J.K.S. Sachan, Development of maternal-haploidy-inducer lines in maize (*Zea mays* L.), Indian J. Anim. Sci. 4 (1972) 781.
[10] S.T. Chalyk, Properties of maternal haploid maize plants and potential application to maize breeding, Euphytica 79 (1994) 13–18.
[11] E.H. Coe, A line of maize with high haploid frequency, Am. Nat. 93 (1959) 381–382.
[12] K.R. Sarkar, E.H. Coe, A genetic analysis of the origin of maternal haploids in maize, Genetics 54 (1966) 453–464.
[13] P. Lashermes, M. Beckert, Genetic control of maternal haploidy in maize (*Zea mays* L.) and selection of haploid inducing lines, Theor. Appl. Genet. 76 (1988) 405–410.
[14] M.A. Aman, K.R. Sarkar, Selection for haploidy inducing potential in maize, Indian J. Genet. 38 (1978) 452–457.
[15] E.H. Coe, K.R. Sarkar, The detection of haploids in maize, J. Hered. 55 (1964) 231–233.
[16] V.S. Tyrnov, A.N. Zavalishina, Inducing high frequency of matroclinal haploids in maize, Dokl. Akad. Nauk SSSR 276 (1984) 735–738.
[17] M. Wedzony, F.K. Rober, H.H. Geiger, Chromosome elimination observed in selfed progenies of maize inducer line RWS, in: XVIIth International Congress on Sex Plant Rep, Maria Curie-Sklodowska University Press, Lublin, 2002, p. 173.
[18] J. Eder, S. Chalyk, In vivo haploid induction in maize, Theor. Appl. Genet. 104 (2002) 703–708.
[19] Z.L. Zhang, F.Z. Qiu, Y.Z. Liu, K.J. Ma, Z.Y. Li, S.Z. Xu, Chromosome elimination and in vivo haploid induction by stock 6-derived inducer line in maize (*Zea mays* L.), Plant Cell Rep. 27 (2008) 1851–1860.
[20] S. Hu, Male germ unit and sperm heteromorphism: the current status, Acta Bot. Sin. 32 (1990) 230–240.
[21] V.G. Bylich, S.T. Chalyk, Existence of pollen grains with a pair of morphologically different sperm nuclei as a possible cause of the haploid-inducing capacity in ZMS line, Maize Genet. Coop. Newsl. 70 (1996) 33.
[22] S.T. Chalyk, A. Baumann, G. Daniel, J. Eder, Aneuploidy as a possible cause of haploid-induction in maize, Maize Genet. Coop. Newsl. 77 (2003) 29.
[23] M.T. Chang, Preferential fertilization induced from Stock 6, Maize Genet. Coop. Newsl. 66 (1992) 99–100.
[24] N.K. Enaleeva, V.S. Tyrnov, L.P. Selivanova, A.N. Zavalishina, Single fertilization and the problem of haploidy induction in maize, Dokl. Biol. Sci. 353 (1996) 225–226.
[25] V.A. Rotarenco, J. Eder, Possible effect of heterofertilization on the induction of maternal haploids in maize, Maize Genet. Coop. Newsl. 77 (2003) 30.
[26] D. Gernand, T. Rutten, A. Varshney, C. Bruss, R. Pickering, F. Matzk, A. Houben, Selective chromosome elimination in embryos of interspecific crosses, in: 15th International Chromosome Conference (poster), September 5–10, Brunel University, London, 2004.
[27] E. Fischer, Molekulargenetische untersuchungen zum vorkommen paternaler DNA-übertragung bei der in-vivo-haploideninduktion beimais (*Zea mays* L.) (PhD dissertation), University of Hohenheim, Grauer Verlag, Stuttgart, 2004.
[28] L. Li, X. Xu, W. Jin, S. Chen, Morphological and molecular evidences for DNA introgression in haploid induction via a high oil inducer CAUHOI in maize, Planta 230 (2009) 367–376.
[29] A. Kato, Induced single fertilization in maize, Sex. Plant Reprod. 10 (1997) 96–100.
[30] X. Yang, X. Dai, Observation of mitosis and meiosis in rice cells by simple squash method, Agric. Sci. Technol. 10 (5) (2009) 96–99.
[31] J.M. Herr, A new clearing-squash technique for the study of ovule in angiosperms, Am. J. Bot. 58 (1971) 785–790.
[32] D.M. Stelly, S. Peloquin, R.G. Palmer, Mayer's hemalummthyl salicylate: a stain-clearing technique for observations within whole ovules, Stain Technol. 59 (1984) 155–161.
[33] M.A. Saghai-Maroof, K.M. Soliman, R.A. Jorgensen, R.W. Allard, Ribosomal DNA spacer-length polymorphisms in barley: mendelian inheritance, chromosomal location, and population dynamics, Proc. Natl. Acad. Sci. 81 (1984) 8014–8018.
[34] B.Y. Lin, Ploidy barrier to endosperm development in maize, Genetics 107 (1984) 103–115.
[35] J.L. Kermicle, Pleiotropic effects on seed development of the indeterminate gametophyte gene in maize, Am. J. Bot. 58 (1971) 1–7.
[36] V.A. Rotarenco, G. Dicu, D. State, S. Fuia, New inducers of maternal haploids in maize, Maize Genet. Coop. Newsl. 84 (2010), http://www.agron.missouri.edu/mnl/84/PDF/15rotarenco.pdf
[37] E.H. Coe, M.G. Neuffer, Darker orange endosperm color associated with haploid embryos: Y1 dosage and the mechanism of haploid induction, Maize Genet. Coop. Newsl. 79 (2005) 7.
[38] J.L. Kermicle, Androgenesis conditioned by a mutation in maize, Science 166 (1969) 1422.
[39] K.J. Kasha, K.N. Kao, High frequency haploid production in barley (*Hordeum vulgare* L.), Nature 225 (1970) 874–876.
[40] D.A. Laurie, M.D. Bennett, The timing of chromosome elimination in hexaploid wheat and maize crosses, Genome 32 (1989) 953–961.
[41] S.S. Khokhlov, V.S. Tyrnov, E.V. Grishina, N.I. Davoyan, Haploidy and Breeding, Nauka, Moscow, 1976, pp. 221 (in Russian).
[42] J.J. Wei, H.M. Li, Z.Z. Liu, Cytology study of maize Haploid, J. Baoding Teach. Coll. 16 (4) (2003) 21–23 (in Chinese).
[43] P. Vanessa, S. Wolfgang, M. George, A.N. Gary, E.M. Albrecht, Development of in vivo haploid inducers for tropical maize breeding programs, Euphytica 185 (2012) 481–490.
[44] D.A. Laurie, The frequency of fertilization in wheat × pearl millet crosses, Genome 32 (1989) 1063–1067.
[45] P. Barret, M. Brinkmann, M. Beckert, A major locus expressed in the male gametophyte with incomplete penetrance is responsible for in situ gynogenesis in maize, Theor. Appl. Genet. 117 (2008) 581–594.
[46] S. Deimling, F.K. Rober, H.H. Geiger, Methodik und genetic der in-vivo- haploideninduktion bei mais, Vortr. PXanzenzuchtg. 38 (1997) 203–224.
[47] D. Gernand, T. Rutten, A. Varshney, M. Rubtsova, S. Prodanovic, C. Bruß, J. Kumlehn, F. Matzk, A. Houben, Uniparental chromosome elimination at mitosis and interphase in wheat and pearl millet crosses involves micronucleus formation, progressive heterochromatinization, and DNA fragmentation, Plant Cell 17 (2005) 2431–2438.
[48] D. Gernand, T. Rutten, R. Pickering, A. Houben, Elimination of chromosomes in *Hordeum vulgare* 9H, bulbosum crosses at mitosis and interphase involves micronucleus formation and progressive heterochromatinization, Cytogenet. Genome Res. 114 (2006) 169–174.
[49] K. Mochida, H. Tsujimoto, T. Sasakuma, Confocal analysis of chromosome behavior in wheat × maize zygotes, Genome 47 (2004) 199–205.
[50] O. Riera-Lizarazu, H.W. Rines, R.L. Phillips, Cytological and molecular characterization of oat × maize partial hybrids, Theor. Appl. Genet. 93 (1996) 123–135.
[51] L. Li, X. Dong, X. Xu, L. Liu, C. Liu, S. Chen, Observation of twin seedling in maternal haploid induction in maize, J. China Agric. Univ. 17 (5) (2012) 1–6.
[52] X. Zhao, X. Xu, H. Xie, S. Chen, W. Jin, Fertilization and uniparental chromosome elimination during crosses with maize haploid inducers, Plant Physiol. 163 (2013) 72–731.
[53] X. Xu, L. Li, X. Dong, W. Jin, A.E. Melchinger, S. Chen, Gametophytic and zygotic selection leads to segregation distortion through in vivo induction of a maternal haploid in maize, J. Exp. Bot. 64 (4) (2013) 1083–1096.

2

Application of iPBS in high-throughput sequencing for the development of retrotransposon-based molecular markers

Yuki Monden, Kentaro Yamaguchi, Makoto Tahara*

Graduate School of Environmental and Life Science, Okayama University, 1-1-1 Tsushimanaka Kitaku, Okayama, Okayama 700-8530, Japan

A R T I C L E   I N F O

Keywords:
Retrotransposon
Next-generation sequencing
Polymorphism
Molecular marker
Strawberry

A B S T R A C T

Retrotransposons are major components of higher plant genomes, and long terminal repeat (LTR) retrotransposons are especially predominant. Thus, numerous LTR retrotransposon families with high copy numbers exist in most plant genomes. As the integrated copies of these retrotransposons are genetically inherited, their insertion polymorphisms among crop cultivars have been used as functional molecular markers such as inter-retrotransposon amplification polymorphism (IRAP), retrotransposon microsatellite amplification polymorphism (REMAP), retrotransposon-based insertion polymorphism (RBIP) and sequence-specific amplification polymorphism (S-SAP). However, the effective use of these methods requires suitable LTR sequences showing high insertion polymorphism among crop cultivars. Recently, we conducted an efficient screening of LTR retrotransposon families that showed high insertion polymorphism among closely related strawberry cultivars using a next-generation sequencing platform. This method focuses on the primer binding site (PBS), which is adjacent to the 5′ LTR sequence and is conserved among different LTR retrotransposon families. Construction of a sequencing library using the PBS motif allowed us to identify a large number of LTR sequences and their insertion sites throughout the genome. The LTR sequences identified by our method showed high insertion polymorphism among closely related strawberry cultivars, and these families should thus be useful in the development of molecular markers for phylogenetic and genetic diversity studies. This article briefly describes the general aspects of retrotransposon-based molecular markers and also outlines our method for screening LTR sequences suitable for genetic analyses.

1. Introduction

Retrotransposons are a type of transposable element (TE) consisting of mobile DNA sequences within eukaryotic genomes [1–3]. In higher plant genomes, retrotransposons are major components [1,4]. For example, over 75% of the maize genome is derived from retrotransposon sequences, and long terminal repeat (LTR) retrotransposons are especially predominant [5]. LTR retrotransposons contain LTR sequences at both ends and are divided into Ty1-*copia* and Ty3-*gypsy* classes based on the structures of their protein-coding domains, which contain capsid protein (CP), protease (PR), integrase (INT), reverse transcriptase (RT) and RNaseH [1,2,6]. In addition, several nonautonomous elements have

been characterized, including terminal repeat retrotransposons in miniature (TRIMs) and large retrotransposon derivatives (LARDs) [7–9]. These elements amplify their copy numbers using a "copy and paste" mechanism involving the reverse transcription and integration of cDNA fragments, which leads to the substantial accumulation of retrotransposon families with high copy numbers within the genome [1,4].

2. Retrotransposon-based molecular markers

Several molecular markers based on retrotransposon insertion polymorphisms (IRAP, REMAP, S-SAP and RBIP) have previously been developed [10–21]. The multiple insertion sites of retrotransposons are distributed throughout the entire genome and are genetically inherited without excision, which makes insertion polymorphisms highly useful as molecular markers [22–24] (Fig. 1). The IRAP method amplifies the genomic regions between two different LTR sequences using PCR primers that are specific to the LTR sequences [9,12,16,19,21], whereas the REMAP method

* Corresponding author.
E-mail addresses: y_monden@cc.okayama-u.ac.jp (Y. Monden),
gag423134@s.okayama-u.ac.jp (K. Yamaguchi), tahara@cc.okayama-u.ac.jp,
makoto.tahara@gmail.com (M. Tahara).

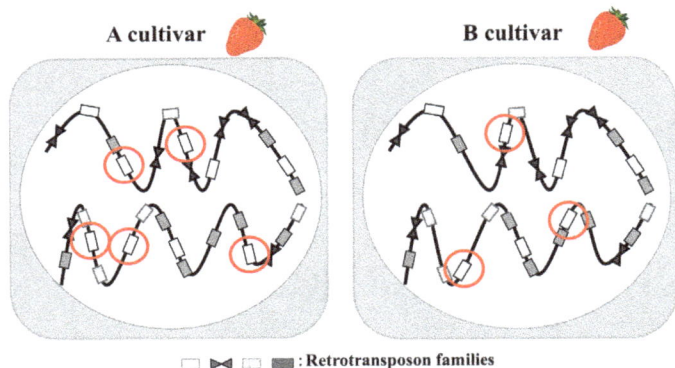

Fig. 1. Retrotransposon insertion polymorphism between strawberry cultivars. Numerous retrotransposon families exist in the nuclear genome, and their insertion sites are distributed throughout the genome. Several of these insertion sites (indicated by circles) differ among modern crop cultivars.

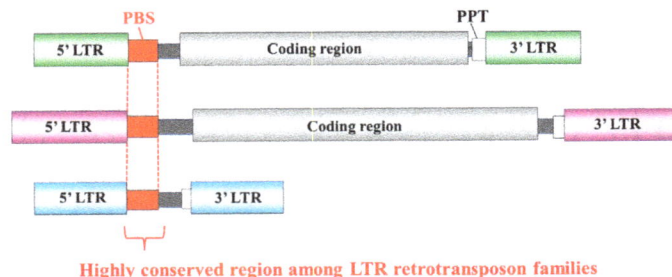

Fig. 3. Conserved region among LTR retrotransposon families. The PBS is adjacent to the 5′ LTR sequence. Reverse transcription of LTR retrotransposons is initiated by the binding of tRNA to the PBS region. Thus, while LTR sequences differ among different LTR retrotransposon families, the PBS sequence is highly conserved.

amplifies the regions between LTR sequences and simple sequence repeats (SSRs) using LTR-specific and SSR-specific primer combinations [9,12,13,21] (Fig. 2(a) and (b)). The S-SAP method detects insertion polymorphisms by amplifying the DNA fragments from LTR sequences to the nearest restriction enzyme cutting sites, following a procedure of restriction enzyme digestion, adapter ligation to the digested DNA fragment and PCR amplification with LTR-specific and adapter-specific primer sets [10,14,15,17,18,20] (Fig. 2(c)). These methods can visualize multiple and polymorphic DNA fragments through a single reaction using agarose/acrylamide electrophoresis. In contrast, the RBIP method targets one insertion site and detects polymorphism based on the presence/absence of this insertion [11]. PCR amplification is performed using LTR-specific and flanking region-specific primer sets to provide codominant markers [11] (Fig. 2(d)). These retrotransposon-based molecular markers have been widely applied in phylogenetic and genetic diversity analyses [22–24]. However, the availability of these methods depends largely on the presence of suitable LTR sequences.

iPBS, developed by Kalendar et al. (2010), is a method for identifying diverse LTR sequences and directly visualizing their polymorphism among cultivars [25]. This method focuses on the PBS region, which is adjacent to the 5′ LTR and is conserved among different LTR retrotransposon families [25] (Fig. 3). Because tRNA binds to the PBS region to initiate reverse transcription, the PBS sequence is complementary to the 3′ terminal sequence of tRNA and is conserved across nearly all LTR retrotransposon families, with several exceptions [26–29]. Thus, designing PCR primers at this region produces DNA fragments that contain diverse LTR sequences, including nonautonomous elements such as TRIMs and LARDs that lack protein-coding regions [25] (Fig. 4). Earlier methods for cloning LTR sequences relied on conserved protein-coding domain, such as reverse transcriptase and integrase, and required genomic walking to the LTRs, which limited the screening of autonomous elements. Therefore, the iPBS method has several advantages for screening diverse LTR sequences and conducting DNA fingerprinting [25]. Additionally, next-generation sequencing technologies have accelerated genetic and genomic studies in recent years by allowing larger volumes of sequence data to be acquired in shorter times and with lower costs. Thus, the combination of the iPBS method with a high-throughput sequencing platform was hypothesized to be capable of screening a large number of LTR retrotransposon families on a genome-wide scale.

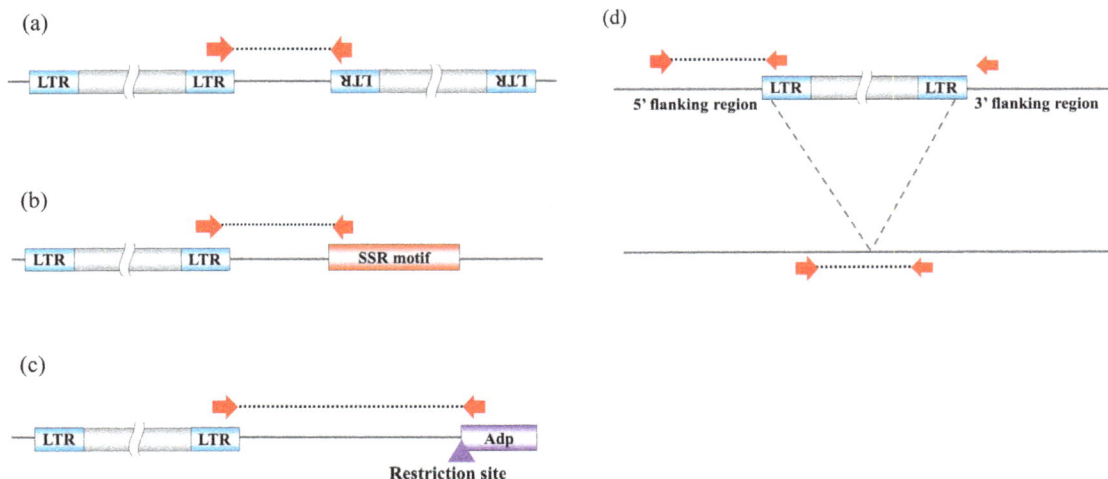

Fig. 2. Retrotransposon-based molecular markers. (a) Inter-retrotransposon amplification polymorphism (IRAP). IRAP amplifies the intervening region between two LTR sequences. (b) Retrotransposon microsatellite amplification polymorphism (REMAP). The DNA fragment between an LTR sequence and an SSR motif is amplified. (c) Sequence-specific amplification polymorphism (S-SAP). Genomic DNAs are digested by a restriction enzyme, and adapters are ligated to the digested DNA fragments. PCR amplification is performed with LTR-specific and adapter-specific primer sets. (d) Retrotransposon-based insertion polymorphism (RBIP). RBIP targets one retrotransposon insertion site in the presence/absence allelic states. To detect the presence of the LTR retrotransposon insertion, primers specific to the flanking regions of the insertion site are used with an LTR-specific primer. In contrast, primers specific to the flanking regions are used for the empty site. The arrows represent PCR primer, and the dotted lines represent the amplified region. (For interpretation of the references to color in this figure legend, the reader is referred to the web version of this article.)

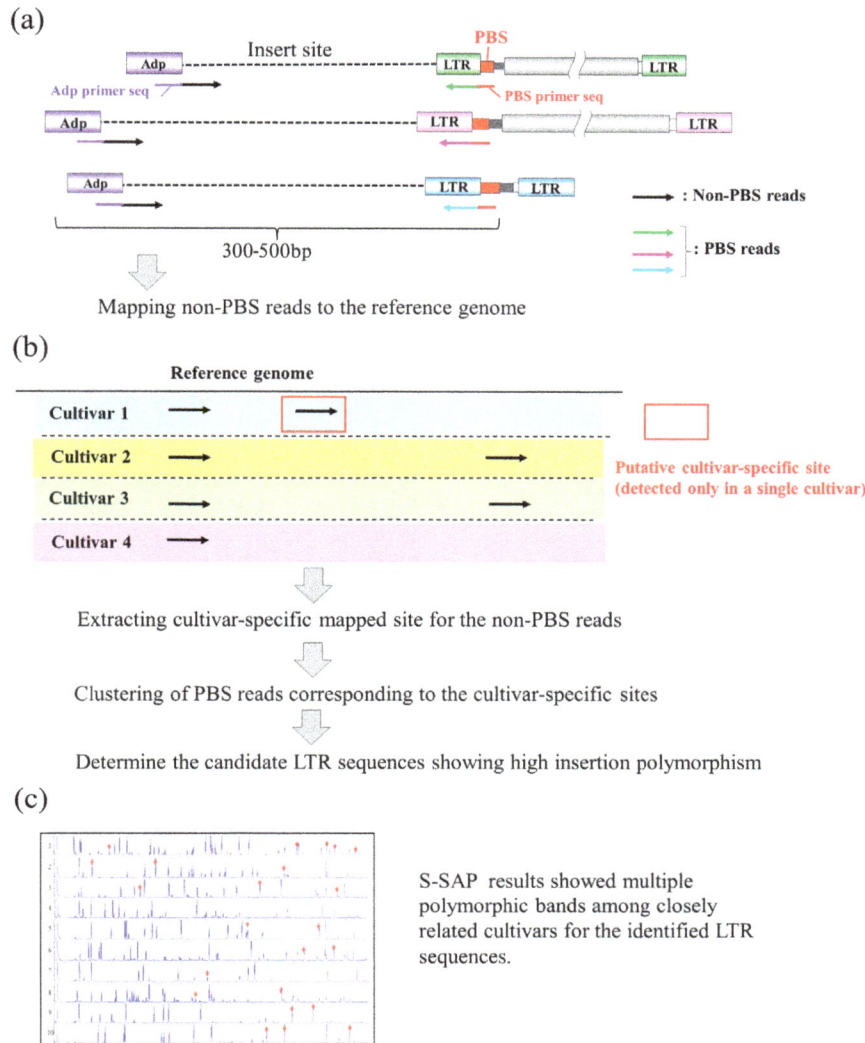

Fig. 4. The procedure of our method. (a) Output of paired-end sequencing. One set of reads, from the PBS-specific primer (PBS reads), contain diverse LTR sequences, while the other set of reads, from the adapter-specific primer (non-PBS reads), contain insertion site sequences. Thus, the non-PBS reads of several cultivars were mapped to the reference genome to identify the insertion sites of LTR sequences. (b) Comparison of mapped sites among several cultivars. Cultivar-specific sites were extracted from these sites, as this type of insertion was hypothesized to have occurred quite recently. Pooling and clustering analyses were conducted using PBS reads corresponding to the cultivar-specific sites, which led to the candidate LTR sequences. (c) Experimental confirmation of insertion polymorphism by the S-SAP method. The S-SAP method visualized DNA fragments by specifically amplifying the insertion sites for the identified LTR sequences.

3. High-throughput sequencing for screening LTR sequences showing high insertion polymorphism

In our research, we employed an Illumina HiSeq 2000 sequencing platform; this platform is widely used and can produce huge amounts of sequencing data. We constructed a sequencing library through PCR amplification using PBS-specific primers with a multiplex barcoding system for eight strawberry cultivars (Fig. 4(a)) [30]. The procedure of our method was briefly described. At first, genomic DNA from eight cultivars was fragmented using g-TUBE (Covaris). These fragmented DNA samples were end repaired, modified to add 3′ A overhangs and ligated to the forked adapters. The forked adapters were prepared by annealing two DNA oligomers: one of the Forked_Type1-4 and Forked_Com (Table 1). We designed PBS-specific primers based on the two types of PBS sequences that matched part of the sequence of the iMET and Asp tRNA genes (indicated as Fr_Mal_iMET_* and Fr_Mal_Asp_* in Table 1). These PBS-specific (iMET and Asp) primers contained 7–8 nt barcode sequence for multiplexing (Table 1). We performed primary PCR with adapter-specific (AP2_Type*) and PBS-specific (Fr_Mal_iMET_*

and Fr_Mal_Asp_*) primer combinations using the ligated products as the template. Then, secondary PCR was performed with adapter-specific (AP3_Type*) and PBS-specific (Fr_Mal_iMET_* and Fr_Mal_Asp_*) primer combinations using the initial PCR products as the template. The PCR cycling program consisted of an initial denaturation at 94 °C for 4 min, 30 cycles of 94 °C for 30 s, 80 °C for 30 s, 58 °C for 30 s and 72 °C for 60 s and a final extension at 72 °C for 15 min. These PCR products were size-selected (300–500 bp) on agarose gels and purified with QIAquick Gel Extraction Kit (QIA-GEN). Finally, these libraries were pooled into one sequencing sample. Paired-end reads of 101 bp were generated on an Illumina HiSeq 2000 platform. In this library, the set of reads amplified from the PBS-specific primer (designated as PBS-reads) should represent the LTR sequences, whereas the set of reads from the adapter-specific primer (designated as non-PBS reads) should contain insertion site sequences (Fig. 4(a)). After filtering those reads based on the primer sequence of the cultivar barcode and PBS with the Q scores of all base calls being ≥30, the non-PBS reads were mapped to the *Fragaria vesca* reference genome [31] to identify the insertion sites of these LTR sequences for each cultivar (Fig. 4(b)).

Table 1
Sequences of the adapters and primers used in our method.

Primer name	Sequence (5'→3')
Forked_Type1	AATAGGGCTCGAGCGGCAGCTATTAATAGTACT
Forked_Type2	AATAGGGCAGCTGCGGCAGCTATTAATAGTACT
Forked_Type3	AATAGGGCGATGGCGGCAGCTATTAATAGTACT
Forked_Type4	AATAGGGCCTACGCGGCAGCTATTAATAGTACT
Forked_Com	GTACTATTAATAGCATCTTCGTTCGTCGAT
AP2_Type1	AATAGGGCTCGAGCGGC
AP2_Type2	AATAGGGCAGCTGCGGC
AP2_Type3	AATAGGGCGATGGCGGC
AP2_Type4	AATAGGGCCTACGCGGC
AP3_Type1	TCGAGCGGCAGCTATTAATAGTACT
AP3_Type2	AGCTGCGGCAGCTATTAATAGTACT
AP3_Type3	GATGGCGGCAGCTATTAATAGTACT
AP3_Type4	CTACGCGGCAGCTATTAATAGTACT
Fr_Mal_iMET_1	AGACTGCNNGCTCTGATACCA
Fr_Mal_iMET_2	ATGATCGCNNGCTCTGATACCA
Fr_Mal_iMET_3	CGTCCAANNGCTCTGATACCA
Fr_Mal_iMET_4	CTTGACCNNGCTCTGATACCA
Fr_Mal_iMET_5	GACTAGTCNNGCTCTGATACCA
Fr_Mal_iMET_6	GAGTGTGNNGCTCTGATACCA
Fr_Mal_iMET_7	TCAGCTAGNNGCTCTGATACCA
Fr_Mal_iMET_8	TCCAGATGNNGCTCTGATACCA
Fr_Mal_Asp_1	AGACTGCAGACGGCGCCA
Fr_Mal_Asp_2	ATGATCGAGACGGCGCCA
Fr_Mal_Asp_3	CGTCCAAAGACGGCGCCA
Fr_Mal_Asp_4	CTTGACCAGACGGCGCCA
Fr_Mal_Asp_5	GACTAGTAGACGGCGCCA
Fr_Mal_Asp_6	GAGTGTGAGACGGCGCCA
Fr_Mal_Asp_7	TCAGCTAAGACGGCGCCA
Fr_Mal_Asp_8	TCCAGATAGACGGCGCCA

Out of these mapped sites, we extracted uniquely mapped loci for the non-PBS reads. Because the lengths of the LTR sequences varied from a few hundred to over a thousand bases, several of the non-PBS reads may have contained LTR sequences, but these reads could be mapped to the reference genome at multiple loci. After comparing the uniquely mapped loci among several cultivars, we focused on the cultivar specific insertion loci which were detected only in one cultivar, but not in the others (Fig. 4(b)). These insertion sites were considered to have occurred relatively recently, presumably after cultivar divergence. Thus, we pooled these insertion sites and extracted the corresponding PBS reads which represented LTR sequences (Fig. 4(b)). By clustering these extracted PBS reads, we obtained 24 candidate LTR sequences that contained more than two reliable cultivar specific insertion sites. In contrast, we identified a total of 20,918 LTR retrotransposon family candidates by clustering all PBS reads in this study [unpublished data]. This number may represent the majority of LTR retrotransposon families present in the strawberry genome, because we selected the two most frequently used tRNA sequences (iMET and Asp) to construct the sequencing library. Finally, the insertion polymorphism in these identified LTR sequences was experimentally investigated by S-SAP analysis, indicating high insertion polymorphism even among Japanese strawberry cultivars known to be closely related genetically (Fig. 4(c)). Thus, these LTR sequences should be useful for the development of retrotransposon-based molecular markers.

4. Concluding remarks

Due to the prevalence and progress of high-throughput sequencing technologies, a number of methods and techniques for sequence analysis have been developed, including genome-wide association studies (GWAS), restriction-site associated DNA (RAD) sequencing and MutMap [32–40]. High-throughput sequencing platforms and multiplex barcoding systems allow us to investigate a large number of genomic loci, including retrotransposon insertion sites, for multiple samples in a single sequencing run [41–46]. For example, our research showed that 76,912 genotypic

data points (2024 retrotransposon insertion loci for 38 cultivars) were obtained in one Illumina MiSeq sequencing run for sweet potato (*Ipomoea batatas*) [47]. We constructed a MiSeq sequencing library through the PCR amplification of the insertion loci of target retrotransposon families, namely *Rtsp-1* [48] and *Llb* [49]. Because these insertion loci were highly polymorphic among cultivars, we acquired a number of molecular markers for cultivar screening and also revealed the genetic relationships among the cultivars without using the reference genome. In addition, our recent research has indicated that retrotransposon insertion sites are quite suitable for linkage map construction aimed at the isolation of agronomically important genes in outcrossing polyploid plant species [unpublished results]. Thus, the targeted sequencing of these polymorphic retrotransposon insertion sites is highly effective for extensive DNA genotyping and marker development. However, retrotransposon families that show little insertion polymorphism among crop cultivars are obviously impractical for these genetic analyses. Although retrotransposon-based molecular markers techniques such as IRAP, REMAP and S-SAP have previously been applied for a number of genetic analyses, these methods are limited by their need for suitable LTR sequences and ability to detect, at most, several dozen sites through agarose/acrylamide electrophoresis. Thus, we believe that our method for screening LTR sequences with high insertion polymorphism and conducting extensive genotyping based on these insertion polymorphisms via high-throughput sequencing platforms should accelerate the genetic analyses such as phylogenetic and genetic diversity studies and promote plant breeding through the map-based cloning of genes in a wide range of plant species.

Acknowledgements

We thank Nobuyuki Fujii and Kazuho Ikeo of the National Institute of Genetics for their support of the data analyses, as well as Yoshiko Nakazawa and Takamitsu Waki of the Tochigi Prefectural Agricultural Experiment Station and Keita Hirashima and Yosuke Uchimura of the Fukuoka Agricultural Research Center for their assistance in the S-SAP experiment. This work was supported by a Research and Development Projects for Application in Promoting New Policy of Agriculture, Forestry and Fisheries grant from the Ministry of Agriculture, Forestry and Fisheries of Japan, as well as by funding from the Program to Disseminate Tenure Tracking System of the Ministry of Education, Culture, Sports, Science and Technology (MEXT), Japan (to Y.M.).

References

[1] A. Kumar, J.L. Bennetzen, Plant retrotransposons, Annu. Rev. Genet. 33 (1999) 479–532, http://dx.doi.org/10.1146/annurev.genet.33.1.479.
[2] C. Feschotte, N. Jiang, S.R. Wessler, Plant transposable elements: where genetics meets genomics, Nat. Rev. Genet. 3 (2002) 329–341, http://dx.doi.org/10.1038/nrg793.
[3] S.R. Wessler, Transposable elements and the evolution of eukaryotic genomes, Proc. Natl. Acad. Sci. U. S. A. 103 (2006) 17600–17601, http://dx.doi.org/10.1073/pnas.0607612103.
[4] M. Bento, D. Tomas, W. Viegas, M. Silva, Retrotransposons represent the most labile fraction for genomic rearrangements in polyploid plant species, Cytogenet. Genome Res. 140 (2013) 286–294, http://dx.doi.org/10.1159/000353308.
[5] P.S. Schnable, D. Ware, R.S. Fulton, J.C. Stein, F. Wei, S. Pasternak, C. Liang, J. Zhang, L. Fulton, T.A. Graves, et al., The B73 maize genome: complexity, diversity, and dynamics, Science 326 (2009) 1112–1115, http://dx.doi.org/10.1126/science.1178534.
[6] E.R. Havecker, X. Gao, D.F. Voytas, The diversity of LTR retrotransposons, Genome Biol. 5 (2004) 225, http://dx.doi.org/10.1186/gb-2004-5-6-225.
[7] C. Witte, Q.H. Le, T. Bureau, A. Kumar, Terminal-repeat retrotransposons in miniature (TRIM) are involved in restructuring plant genomes, Proc. Natl. Acad. Sci. U. S. A. 98 (2001) 13778–13783, http://dx.doi.org/10.1073/pnas.241341898.
[8] R. Kalendar, C.M. Vicient, O. Peleg, K. Anamthawat-Jonsson, A. Bolshoy, A.H. Schulman, Large retrotransposon derivatives: abundant, conserved but nonautonomous retroelements of barley and related genomes, Genetics 166 (2004) 1437–1450, http://dx.doi.org/10.1534/genetics.166.3.1437.

[9] K. Antonius-Klemola, R. Kalendar, A.H. Schulman, TRIM retrotransposons occur in apple and are polymorphic between varieties but not sports, Theor. Appl. Genet. 112 (2006) 999–1008, http://dx.doi.org/10.1007/s00122-005-0203-0.

[10] R. Waugh, K. McLean, A.J. Flavell, S.R. Pearce, A. Kumar, B.B. Thomas, W. Powell, Genetic distribution of Bare-1-like retrotransposable elements in the barley genome revealed by sequence-specific amplification polymorphisms (S-SAP), Mol. Gen. Genet. 253 (1997) 687–694 http://www.ncbi.nlm.nih.gov/pubmed/9079879

[11] A.J. Flavell, M.R. Knox, S.R. Pearce, T.H. Ellis, Retrotransposon-based insertion polymorphisms (RBIP) for high throughput marker analysis, Plant J. 16 (1998) 643–650 http://www.ncbi.nlm.nih.gov/pubmed/10036780

[12] R. Kalendar, T. Grob, M. Regina, A. Suoniemi, A. Schulman, IRAP and REMAP: two new retrotransposon-based DNA fingerprinting techniques, Theor. Appl. Genet. 98 (1998) 704–711, http://dx.doi.org/10.1007/s001220051124.

[13] R. Kalendar, J. Tanskanen, S. Immonen, E. Nevo, A.H. Schulman, Genome evolution of wild barley (Hordeum spontaneum) by BARE-1 retrotransposon dynamics in response to sharp microclimatic divergence, Proc. Natl. Acad. Sci. U. S. A. 97 (2000) 6603–6607, http://dx.doi.org/10.1073/pnas.110587497.

[14] N.H. Syed, S. Sureshsundar, M.J. Wilkinson, B.S. Bhau, J.J.V. Cavalcanti, A.J. Flavell, Ty1-copia retrotransposon-based SSAP marker development in cashew (Anacardium occidentale L.), Theor. Appl. Genet. 110 (2005) 1195–1202, http://dx.doi.org/10.1007/s00122-005-1948-1.

[15] Q. Lou, J. Chen, Ty1-copia retrotransposon-based SSAP marker development and its potential in the genetic study of cucurbits, Genome 810 (2007) 802–810, http://dx.doi.org/10.1139/G07-067.

[16] A. Belyayev, R. Kalendar, L. Brodsky, E. Nevo, A. Schulman, O. Raskina, Transposable elements in a marginal plant population: temporal fluctuations provide new insights into genome evolution of wild diploid wheat, Mob. DNA 1 (2010) 1–6, http://dx.doi.org/10.1186/1759-8753-1-6.

[17] M. Petit, C. Guidat, J. Daniel, E. Denis, E. Montoriol, Q.T. Bui, K.Y. Lim, A. Kovarik, A.R. Leitch, M. Grandbastien, et al., Mobilization of retrotransposons in synthetic allotetraploid tobacco, N. Phytol. 186 (2010) 135–147, http://dx.doi.org/10.1111/j.1469-8137.2009.03140.x.

[18] F.A. Konovalov, N.P. Goncharov, S. Goryunova, A. Shaturova, T. Proshlyakova, A. Kudryavtsev, Molecular markers based on LTR retrotransposons BARE-1 and Jeli uncover different strata of evolutionary relationships in diploid wheats, Mol. Genet. Genomics 283 (2010) 551–553, http://dx.doi.org/10.1007/s00438-010-0539-2.

[19] P. Smýkal, N. Bačová-Kerteszová, R. Kalendar, J. Corander, A.H. Schulman, M. Pavelek, Genetic diversity of cultivated flax (Linum usitatissimum L.) germplasm assessed by retrotransposon-based markers, Theor. Appl. Genet. 122 (2011) 1385–1397, http://dx.doi.org/10.1007/s00122-011-1539-2.

[20] N.V. Melnikova, A.V. Kudryavtseva, A.S. Speranskaya, A.A. Krinitsina, A.A. Dmitriev, M.S. Belenikin, V.P. Upelniek, E.R. Batrak, I.S. Kovaleva, A.M. Kudryavtsev, The FaRE1 LTR-retrotransposon based SSAP markers reveal genetic polymorphism of strawberry (Fragaria x ananassa) cultivars, J. Agric. Sci. 4 (2012) 111–118, http://dx.doi.org/10.5539/jas.v4n11p111.

[21] S. Nasri, B. Abdollahi Mandoulakani, R. Darvishzadeh, I. Bernousi, Retrotransposon insertional polymorphism in Iranian bread wheat cultivars and breeding lines revealed by IRAP and REMAP markers, Biochem. Genet. 51 (2013) 927–943, http://dx.doi.org/10.1007/s10528-013-9618-5.

[22] A. Kumar, H. Hirochika, Applications of retrotransposons as genetic tools in plant biology, Trends Plant Sci. 6 (2001) 127–134 http://www.ncbi.nlm.nih.gov/pubmed/11239612

[23] A.H. Schulman, A.H. Flavell, T.H.N. Ellis, The application of LTR retrotransposons as molecular markers in plants, Methods Mol. Biol. 260 (2004) 145–173, http://dx.doi.org/10.1385/1-59259-755-6.145.

[24] P. Poczai, I. Varga, M. Laos, A. Cseh, N. Bell, J.P. Valkonen, J. Hyvönen, Advances in plant gene-targeted and functional markers: a review, Plant Methods 9 (2013) 6, http://dx.doi.org/10.1186/1746-4811-9-6.

[25] R. Kalendar, K. Antonius, P. Smýkal, A.H. Schulman, iPBS: a universal method for DNA fingerprinting and retrotransposon isolation, Theor. Appl. Genet. 121 (2010) 1419–1430, http://dx.doi.org/10.1007/s00122-010-1398-2.

[26] R. Marquet, C. Isel, C. Ehresmann, B. Ehresmann, tRNAs as primer of reverse transcriptases, Biochimie 77 (1995) 113–124 http://www.ncbi.nlm.nih.gov/pubmed/7541250

[27] J. Mak, L. Kleiman, Primer tRNAs for reverse transcription, J. Virol. 71 (1997) 8087–8095.

[28] N.J. Kelly, M.T. Palmer, C.D. Morrow, Selection of retroviral reverse transcription primer is coordinated with tRNA biogenesis, J. Virol. 77 (2003) 8695–8701, http://dx.doi.org/10.1128/jvi.77.16.8695.

[29] A. Hizi, The reverse transcriptase of the Tf1 retrotransposon has a specific novel activity for generating the RNA self-primer that is functional in cDNA synthesis, J. Virol. 82 (2008) 10906–10910, http://dx.doi.org/10.1128/jvi.01370-08.

[30] Y. Monden, N. Fujii, K. Yamaguchi, K. Ikeo, Y. Nakazawa, T. Waki, K. Hirashima, Y. Uchimura, M. Tahara, Efficient screening of the long terminal repeat (LTR) retrotransposons that show high insertion polymorphism via high-throughput sequencing of the primer binding site, Genome 57 (2014) 1–8, http://dx.doi.org/10.1139/gen-2014-0031.

[31] Fragaria vesca Genome v1.1 Assembly. http://www.rosaceae.org/projects/strawberry_genome/v1.1/assembly

[32] X. Huang, X. Wei, T. Sang, Q. Zhao, Q. Feng, Y. Zhao, C. Li, C. Zhu, T. Lu, Z. Zhang, et al., Genome-wide association studies of 14 agronomic traits in rice landraces, Nat. Genet. 42 (2010) 961–967, http://dx.doi.org/10.1038/ng.695.

[33] S.W. Baxter, J.W. Davey, J.S. Johnston, A.M. Shelton, D.G. Heckel, C.D. Jiggins, M.L. Blaxter, Linkage mapping and comparative genomics using next-generation RAD sequencing of a non-model organism, PLoS ONE 6 (2011) e19315, http://dx.doi.org/10.1371/journal.pone.0019315.

[34] R.L. Elshire, J.C. Glaubitz, Q. Sun, J.A. Poland, K. Kawamoto, E.S. Buckler, S.E. Mitchell, A robust, simple genotyping-by-sequencing (GBS) approach for high diversity species, PLoS ONE 6 (2011) e19379, http://dx.doi.org/10.1371/journal.pone.0019379.

[35] F. Tian, P.J. Bradbury, P.J. Brown, H. Hung, Q. Sun, S. Flint-Garcia, T.R. Rocheford, M.D. McMullen, J.B. Holland, E.S. Buckler, Genome-wide association study of leaf architecture in the maize nested association mapping population, Nat. Genet. 43 (2011) 6–11, http://dx.doi.org/10.1038/ng.746.

[36] A. Abe, S. Kosugi, K. Yoshida, S. Natsume, H. Takagi, H. Kanzaki, H. Matsumura, K. Yoshida, C. Mitsuoka, M. Tamiru, et al., Genome sequencing reveals agronomically important loci in rice using MutMap, Nat. Biotechnol. 30 (2012) 174–178, http://dx.doi.org/10.1038/nbt.2095.

[37] A.L. Harper, M. Trick, J. Higgins, F. Fraser, L. Clissold, R. Wells, C. Hattori, P. Werner, I. Bancroft, Associative transcriptomics of traits in the polyploid crop species Brassica napus, Nat. Biotechnol. 30 (2012) 798–802, http://dx.doi.org/10.1038/nbt.2302.

[38] R. Fekih, H. Takagi, M. Tamiru, A. Abe, S. Natsume, H. Yaegashi, S. Sharma, H. Kanzaki, H. Matsumura, M. Saitoh, et al., MutMap+: genetic mapping and mutant identification without crossing in rice, PLoS ONE 8 (2013) e68529, http://dx.doi.org/10.1371/journal.pone.0068529.

[39] D.T. Morishige, P.E. Klein, J.L. Hilley, S.M.E. Sahraeian, A. Sharma, J.E. Mullet, Digital genotyping of sorghum – a diverse plant species with a large repeat-rich genome, BMC Genomics 14 (2013) 448, http://dx.doi.org/10.1186/1471-2164-14-448.

[40] H. Takagi, A. Uemura, H. Yaegashi, M. Tamiru, A. Abe, C. Mitsuoka, H. Utsushi, S. Natsume, H. Kanzaki, H. Matsumura, et al., Methods MutMap-Gap: whole-genome resequencing of mutant F_2 progeny bulk combined with de novo assembly of gap regions identifies the rice blast resistance gene Pii, N. Phytol. 200 (2013) 276–283, http://dx.doi.org/10.1111/nph.12369.

[41] R.C. Iskow, M.T. McCabe, R.E. Mills, S. Torene, W.S. Pittard, A.F. Neuwald, E.G. Van Meir, P.M. Vertino, S.E. Devine, Natural mutagenesis of human genomes by endogenous retrotransposons, Cell 141 (2010) 1253–1261, http://dx.doi.org/10.1016/j.cell.2010.05.020.

[42] A.D. Ewing, H.H. Kazazian, Whole-genome resequencing allows detection of many rare LINE-1 insertion alleles in humans, Genome Res. 21 (2011) 985–990, http://dx.doi.org/10.1101/gr.114777.110.

[43] F. Hormozdiari, C. Alkan, M. Ventura, I. Hajirasouliha, M. Malig, F. Hach, D. Yorukoglu, P. Dao, M. Bakhshi, S.C. Sahinalp, et al., Alu repeat discovery and characterization within human genomes, Genome Res. 21 (2011) 840–849, http://dx.doi.org/10.1101/gr.115956.110.

[44] D.F. Urbański, A. Małolepszy, J. Stougaard, S.U. Andersen, Genome-wide LORE1 retrotransposon mutagenesis and high-throughput insertion detection in Lotus japonicas, Plant J. 69 (2012) 731–741, http://dx.doi.org/10.1111/j.1365-313X.2011.04827.x.

[45] M. David, H. Mustafa, M. Brudno, Detecting Alu insertions from high-throughput sequencing data, Nucleic Acids Res. 41 (2013) e169, http://dx.doi.org/10.1093/nar/gkt612.

[46] J. Xing, D.J. Witherspoon, L.B. Jorde, Mobile element biology: new possibilities with high-throughput sequencing, Trends Genet. 29 (2013) 280–289, http://dx.doi.org/10.1016/j.tig.2012.12.002.

[47] Y. Monden, A. Yamamoto, A. Shindo, M. Tahara, Efficient DNA fingerprinting based on the targeted sequencing of active retrotransposon insertion sites using a bench-top high-throughput sequencing platform, DNA Res. (2014), http://dx.doi.org/10.1093/dnares/dsu015.

[48] M. Tahara, T. Aoki, S. Suzuka, H. Yamashita, M. Tanaka, S. Matsunaga, S. Kokumai, Isolation of an active element from a high-copy-number family of retrotransposons in the sweetpotato genome, Mol. Genet. Genomics 272 (2004) 116–127, http://dx.doi.org/10.1007/s00438-004-1044-2.

[49] H. Yamashita, M. Tahara, A LINE-type retrotransposon active in meristem stem cells causes heritable transpositions in the sweetpotato genome, Plant Mol. Biol. 61 (2006) 79–94, http://dx.doi.org/10.1007/s11103-005-6002-9.

3

Exploring genomic databases for *in silico* discovery of Pht1 genes in high syntenic close related grass species with focus in sugarcane (*Saccharum* spp.)☆

Arthur Tavares de Oliveira Melo*

University of New Hampshire, College of Life Science and Agriculture, Department of Biological Sciences, Durham, NH, USA

ARTICLE INFO

Keywords:
Comparative genomics
Pht1 genes
Sugarcane
Illumina reads

ABSTRACT

The genomic sequences available at several public databases and the local alignments allow a fast and accurate search for nucleotide homology among different species, revealing the structure, composition and function of nucleotides and amino acids. The Phosphate Transporter 1 gene family (Pht1) has crucial role in several metabolic processes from plants and has been identified in different crops. Here, the CDS sequence of 28 Pht1 genes described for *Oryza sativa*, *Zea mays* and *Arabidopsis thaliana* were used to identify nine homologous sequences in *Sorghum bicolor* and *Setaria italica* and the *S. bicolor* sequences were used as reference to guide the identification of those homologous sequences in *Saccharum* spp. Using five different libraries of paired-end Illumina reads, nine sugarcane genes (ScPht1) were assembled and the rate of gene recover, the average of gene length, the sequencing depth and the GC content were estimated in 93.72%, 2053 bp, 16.08X and 58.70%, respectively. Putative SNPs and microsatellites loci were found for those ScPht1 genes and three of them seem to be under positive selection pressure. A phylogenetic tree corroborates with orthology relationship among *Saccharum* spp., *O. sativa* and *Z. mays* genes, emphasizing the levels of synteny described for grass species. This study discuss the potential use of homology in genomic studies as one of the most useful tools for gene discovery and annotation, illustrating how useful is the exploitation of public databases for gene identification and characterization, allowing further gene manipulation in crop improvements.

1. Introduction

The actual genomic-scale studies provide information over the entire genome and as consequence, an enormous amount of genomic sequences for model and non-model species are currently available at several public databases. The Phytozome database is the Plant Comparative Genomics portal that provides access to fifty-eight sequenced and annotated green plant genomes [1]. Along with databases like GenBank NCBI and Gramene [2], they store an incredible amount of genomic sequences, being indispensable for any comparative genome project. Comparative and functional genomics seeks characterize each species' gene content and chromosomal arrangements, explaining the observed similarities and differences within a molecular evolutionary context aiming to assess functional significance of distinct genetic traits [3].

Phylogenetic closed related species, as have been reported to Poaceae family [4], generally shows high genomic similarity such gene localization and orientation (synteny). Sharing a common ancestor about 8–9 million years ago [5], the *Sorghum bicolor* and *Saccharum* spp. are very close relatives, classified at the same subtribe (Saccharinae), being a well-known example of syntenic species. Even showing a large (~10 Gb) and complex genome (2n = 8–12X = 80–128 chromosomes) as consequence of multiple polyploidization events and interspecific hybridization occurred during *Saccharum* spp. genome evolution [6,7], the level of genic conservation between sugarcane and *S. bicolor* is so extensive that is characterized as a macrosynteny [8] and the completion of the sorghum genome sequence offered unprecedented opportunities for sugarcane genomic research [9].

Species belongs to Andropogoneae tribe as maize, sorghum and sugarcane stands out due the efficiency in convert CO_2 molecules into biomass, through the C_4 photosynthetic mechanism [10,11] and the sugarcane in particular, is considered an important crop species for global supply of sugar and energy [12].

☆ This article is part of a special issue entitled "Plants and global climate change: a need for sustainable agriculture", published in the journal Current Plant Biology 6, 2016.

* Corresponding author at: University of New Hampshire, Department of Biological Sciences, Rudman Hall, Room 391, 46 College Road, Durham, NH 03824, USA.

E-mail address: arthurmelobio@gmail.com

The phosphorus (Pi) is an important inorganic chemical element as primary macronutrient for plant development, being component of macromolecules like DNAs and RNAs [13]. In recent years, significant progress has been made to isolate and characterize molecular determinants of Pi acquisition, transportation and storage by plants. The gene family described as responsible for this metabolic process in high affinity system (activated under conditions of early stress with low Pi availability) is the Pht1 transporter genes [14]. Those genes has been identified for several important plant species such *Arabidopsis thaliana* [15], *Medicago truncatula* [16], *Solanum tuberosum* [17], tobacco [18], *Oryza sativa* [19], *Zea mays* [20], wheat [21] and *Populus trichocarpa* [22].

So, the present study explored the conservation of gene arrangements and syntenic genomic blocks among closely related grass species and by performing the nucleotide sequence similarity analysis, revealed and characterized Pht1 genes in high-affinity systems for *Saccharum* spp.

2. Material and methods

2.1. *Phytozome cDNA sequences*

The Phytozome v.9.1 database was used to access the nucleotide CDS sequences from Pht1 transporter genes in *A. thaliana* (nine genes), *O. sativa* (13 genes) and *Z. mays* (six genes). These 28 Pht1 genes were aligned in both genomes of *S. bicolor* v.2.1 [4] and *Setaria italica* v.2.1 [23] using tBLASTx v.2.2.30 procedure [24] in order to identify homologous sequences in these last two species. The best-hit sequence with e-valor <10^{-6} and that maximize the coverage of query sequence was chosen as the homologous copies. The *S. bicolor* Pht1 gene sequences (SbPht1) were used as a reference to identify homologous sugarcane genomic nucleotide reads.

2.2. In silico *identification of sugarcane Pht1 genes*

Five libraries of whole genome sequence from the *Saccharum* spp. hybrid cultivar SP80-3280 was sampled from NCBI SRA database (SRX853962–SRX853966). The fragments were sequenced in an Illumina HiSeq2000 system using paired-end chemistry, yielding, respectively 41.9, 42.1, 33.7, 34.6 and 55.6 billion of nucleotides. The raw sequence data were initially parsed to remove low-quality bases (Phred score Q $\leq$ 30), Illumina common adapter and reads shorter than 50 bp. These procedures were performed using the Trimmomatic software [25]. The high quality paired-end reads were separately mapped in each SbPht1 gene previously identified. The BWA aligner [26] was used for mapping procedures while the SAMTools [27] was used to retain appropriate reads and assembly the consensus sequences from each Pht1 genes from *Saccharum* spp. (ScPht1). The gaps and the non-assembled nucleotides were filled using the ORFeome database composed by full-length enriched cDNA libraries from different sugarcane genotypes [28]. The alignment dotplot for each pair of reference *S. bicolor* sequences and the sugarcane genes assembled was evaluated by using Gepard software [29].

2.3. *ScPht1 genes characterization and annotation*

Five molecular-genetic parameters were estimated for those ScPht1 genes identified: (1) the putative number of SNPs (using *S. bicolor* as reference); the average of (2) transition rates (Ts) and (3) transversion rates (Tv); (4) the GC content and (5) the gene length. The putative SNPs were called through SAMTools mpileup algorithm. Open Read Frames (ORFs) was identified using ORF Finder tool available in NCBI platform for all ScPht1 genes. The MEGA5 software [30] was used to estimate GC content and identify polymorphic nucleotide sites for each gene, using Nei and Li [31] gene diversity measure (π). In order to evaluate whether some ScPht1 gene are under positive selection pressure, the ratio (ω) of non-synonymous (Ka) *versus* synonymous (Ks) nucleotide substitution was estimated using KaKs Calculator software [32].

The iMEx software [33] was used to sample perfect and imperfect microsatellites (SSR) regions in *Saccharum* spp. and *S. bicolor*. Besides original length of genes, 2 Kb of nucleotide sequences upstream and downstream were also used. Specific ESTs (Expressed Sequenced Tags) derived microsatellites primers were designed using the Primer3Plus software [34] and primers with 15–25 bases and with a GC content of approximately 55%, annealing temperature of approximately 60 °C and PCR product size between 100 and 400 bp were sampled. Mononucleotide SSR regions were considered when a motif repeated more than ten times. For di- and tri- nucleotides, at least six times of repeated motif was required while for tetra-, penta- and hexa- nucleotides at least four times of repeated motif was required. A maximum of 10% was allowed for imperfect microsatellites.

The functional annotation analysis of those putative sugarcane genes (ScPht1) was carried out through the BLAST2GO platform [35], using the Gene Ontology terms. The InterProScan [36] and Pfam [37] databases were also used to reveal the functionality of the protein family.

2.4. *Phylogenetic reconstruction for Pht1 genes from Poaceae family species*

A phylogenetic tree was estimated using the 28 Pht1 genes sampled from Phytozome database (*A. thaliana*, *O. sativa* and *Z. mays*) and all genes identified for *S. bicolor*, *S. italica* and *Saccharum* spp., using a Bayesian method available at Mr. Bayes [38]. 100,000 bootstraps were used to evaluate the tree nodes (monophyletic groups) and an evolutionary model was previously defined based on Bayesian informative criterion (BIC) using jModelTest2 software [39]. The phylogenetic tree was constructed using nucleotide sequences aligned by MUSCLE [40] and edited by Jalview software [41].

3. Results

3.1. *The Pht1 genes from* A. thaliana, O. sativa, Z. mays, S. bicolor *and* S. italica

The 28 Pht1 genes from *A. thaliana*, *O. sativa* and *Z. mays* revealed in both *S. bicolor* (Table 1) and *S. italica* (Table S1), nine homologous gene copies. However, the 13 genes from *O. sativa* (OsPht1) showed the highest level of sequencing similarity with *S. bicolor* (80%) and *S. italica* (86%) being enough to sample the nine homologous gene copies in these last two species. The sequence similarity analysis also suggested different number of genes between these two species and *O. sativa*, i.e., some genes showed homology with two different copies from *O. sativa* (Table 1) and in this particular case, genes with highest query coverage were chosen for further analysis.

3.2. *Assembling Pht1 genes for* Saccharum *spp.*

Nearly 488 million of high-quality paired-end sugarcane reads (average of 80 bp) were mapped separately in each *S. bicolor* gene sequence previously identified, allowing to sample sugarcane reads homologous to *S. bicolor* genes. Using 16.08X of reads coverage, nine homologous copies from the sugarcane Pht1 genes (ScPht1) were assembled, recovering, on average, 84.62% of the sorghum reference sequences. Furthermore, after use 48 different transcripts deposit on ORFeome database to fill the gaps and non-assembled sites, the recovering rate of the sugarcane assembled genes (ScPht1)

Table 1

Pht1 gene family homology between *O. sativa* (query sequence) and *S. bicolor* (subject sequence) exposed through sequence similarity analysis. Nine Pht1 genes in *S. bicolor* homologous to those from *O. sativa* were found. The genes SbPht1-1, SbPht1-5, SbPht1-6 and SbPht1-9 had two homologous in *O. sativa*.

Oryza sativa					*Sorghum bicolor*				tBLASTx parameters[a]	
n.	Gene name	Phytozome ID	Chr	Length (bp)	Gene name	Phytozome ID	Chr	Length (bp)	Score[b]	Query coverage[c]
1	OsPht1-1	Os03g05620	3	2124	SbPht1-6	Sobic.001G234900	1	2309	518	71%
2	OsPht1-2	Os03g05640	3	1930	SbPht1-7	Sobic.001G502000	1	2112	928	81%
3	OsPht1-3	Os10g30770	10	1581	SbPht1-6*	Sobic.001G234900	1	2309	2066	98%
4	OsPht1-4	Os04g10750	4	3105	SbPht1-1	Sobic.001G234800	1	2630	514	50%
5	OsPht1-5	Os04g10690	4	1760	SbPht1-8	Sobic.006G027300	6	2033	548	82%
6	OsPht1-6	Os08g45000	8	1827	SbPht1-4	Sobic.007G164400	7	2243	510	82%
7	OsPht1-7	Os03g04360	3	2546	SbPht1-9	Sobic.001G502100	1	2179	406	58%
8	OsPht1-8	Os10g30790	10	2829	SbPht1-1*	Sobic.001G234800	1	2630	923	53%
9	OsPht1-9	Os06g21920	6	1214	SbPht1-5	Sobic.010G133300	10	2037	199	72%
10	OsPht1-10	Os06g21950	6	1989	SbPht1-5*	Sobic.010G133300	10	2037	193	79%
11	OsPht1-11	Os01g46860	1	1668	SbPht1-2	Sobic.003G243400	3	2522	335	96%
12	OsPht1-12	Os03g05610	3	1803	SbPht1-9*	Sobic.001G502100	1	2179	921	87%
13	OsPht1-13	Os04g10800	4	1527	SbPht1-3	Sobic.006G026900	6	1771	317	97%

[a] The e-values were less than the threshold of 10^{-6}, showing a low probability to found some blast hit by chance.
[b] Score is value that describes the overall quality of an alignment. Higher numbers correspond to higher similarity.
[c] Query coverage means the percent of the query sequence that overlaps with subject sequence.
* The SbPht1 genes chosen for further analysis from the genes that showed two homology copies with *O. sativa*.

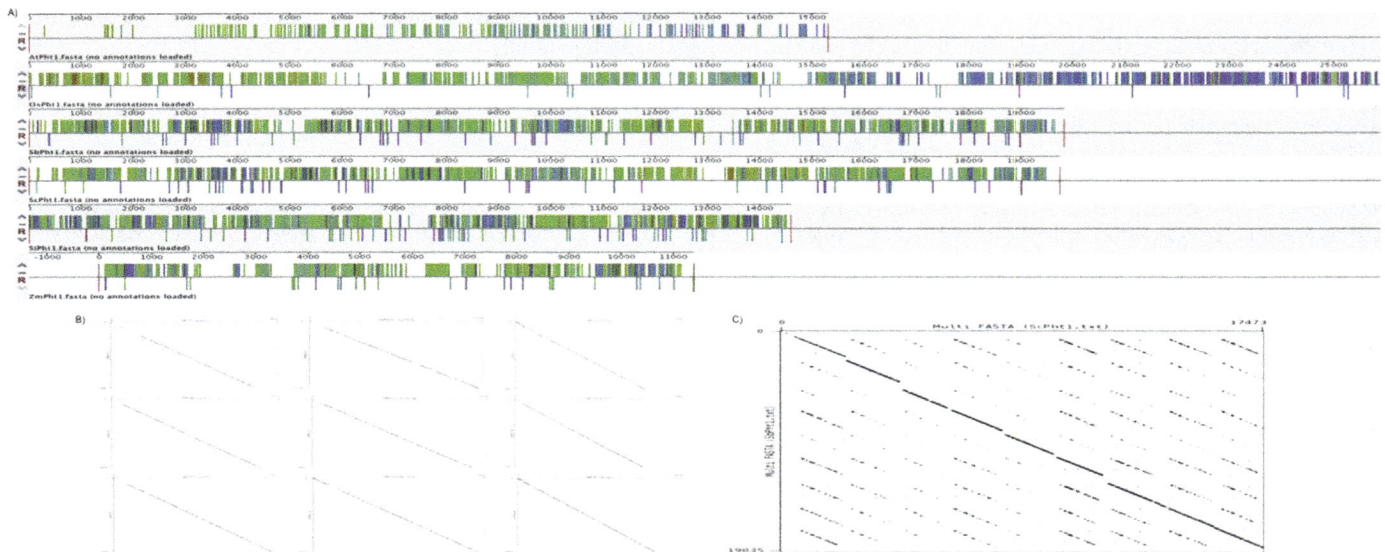

Fig. 1. The Pht1 gene sequence similarity. (A) The collor bar graph represent the nucleotide composition similarity across the all species evaluated. From top to botton, *A. thaliana*, *O. sativa*, *S. bicolor*, *Saccharum* spp., *S. italica* and *Z. mays*. (B and C) The dotplot alignment showing the concordance between the reference *S. bicolor* genes sequences and the sugarcane assembled genes.

Table 2

The nine copies of SbPht1 genes from *S. bicolor* and the correspondent homologous genes identified for *Saccharum officinarum* (ScPht1). Those nine SbPht1 genes was used as reference to found the sugarcane homologous copies and for them, six molecular-genetic parameters were estimated to characterize the assembled sequences.

Sorgum bicolor					*Saccharum* spp.						
n.	Gene name	Phytozome ID	Chr	Length (bp)	Gene name	Assembly[a]	Coverage[b]	Length (bp)[c]	SNPs[d]	ω[e]	GC Content[f]
1	SbPht1-1	Sobic.001G234800	1	2630	ScPht1-1	81.69	13.62	2175	11	1.135	52.51
2	SbPht1-2	Sobic.003G243400	3	2522	ScPht1-2	86.24	14.12	2164	5	0.830	59.23
3	SbPht1-3	Sobic.006G026900	6	1771	ScPht1-3	100.00	12.23	1771	8	0.741	50.30
4	SbPht1-4	Sobic.007G164400	7	2243	ScPht1-4	96.61	25.10	2167	13	0.740	61.32
5	SbPht1-5	Sobic.010G133300	10	2037	ScPht1-5	89.24	10.33	1818	4	0.745	64.63
6	SbPht1-6	Sobic.001G234900	1	2309	ScPht1-6	92.85	19.41	2144	23	0.932	61.14
7	SbPht1-7	Sobic.001G502000	1	2112	ScPht1-7	100.00	18.78	2112	18	1.017	58.31
8	SbPht1-8	Sobic.006G027300	6	2033	ScPht1-8	100.00	18.93	2033	39	0.591	58.08
9	SbPht1-9	Sobic.001G502100	1	2179	ScPht1-9	95.91	12.26	2090	11	1.003	62.79
Average	–	–	–	2204		93.72	16.08	2053	14.66	–	58.70
Total	–	–	–	–		–	–	18,485	132	–	–

[a] The percentage of total base pair recovered from Pht1 *S. bicolor* genes in *Saccharum* spp.
[b] Average depth of reads used to assembled Pht1 genes in sugarcane.
[c] Estimated gene size (bp) for each gene assembled.
[d] Number of putative SNPs called using SAMTools mpileup algorithm based on *S. bicolor* reference alignment.
[e] The ratio (ω) of non-synonymous (Ka) *versus* synonymous (Ks) nucleotide substitution rate.
[f] Percentage of GC content estimated by MEGA5.0 software.

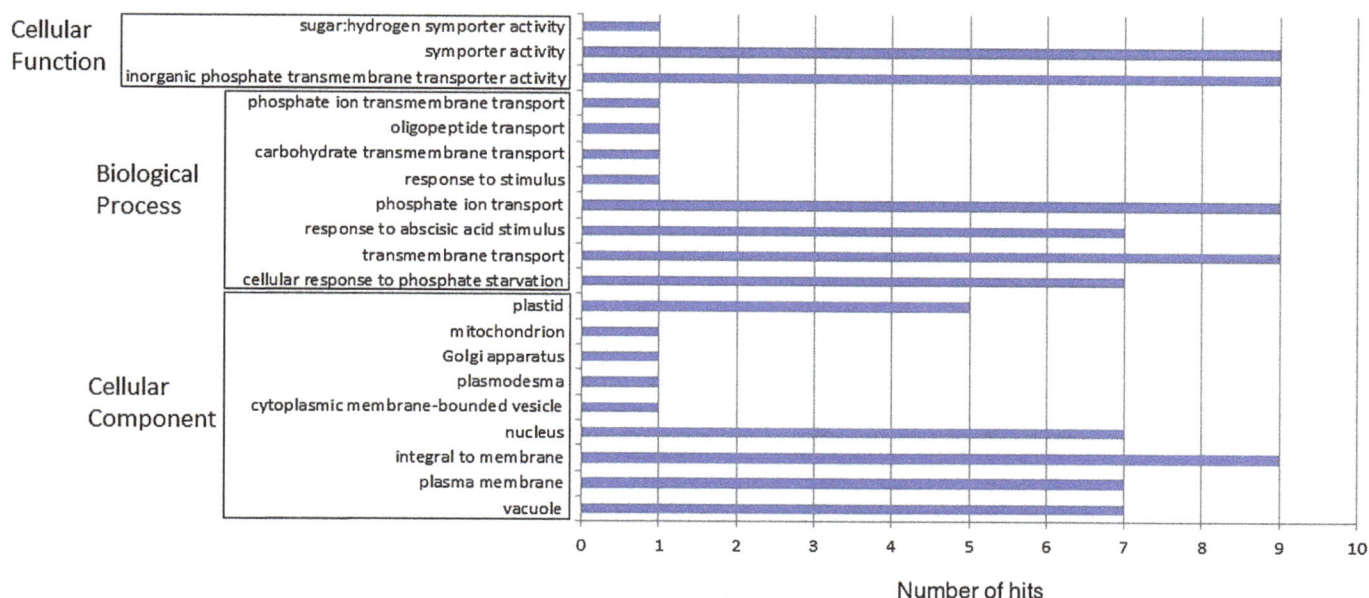

Fig. 2. ScPht1 genes functional annotation showing most of gene hits were assigned to Biological Process and Cellular Component GO terms.

increased to 93.72%. The Fig. 1A shows the base pair similarity of the Pht1 genes evaluated for all six grass species and *A. thaliana*, while Fig. 1B and C shows the recover rate of sugarcane genes compared with *S. bicolor* reference sequence. The estimated average length and CG content of those sugarcane genes were 2053 bp and 58.70%, respectively (Table 2). The number of reads used to assemble such genes was 35,242, ranges from 112 (ScPht1-3) to 15,154 (ScPht1-1), with an average of 5118 reads mapped for each ScPht1 gene assembled. The complete ORFs of those genes were deposited on NCBI GenBank database with follow accession numbers: KU048769-KU048777.

3.3. Characterization of ScPht1 putative genes

One hundred thirty-two putative SNPs and eleven Indels (insertions/deletions) were identified for those ScPht1 genes and 74.72% of these nucleotide substitutions are transversion (Tv). 78% of putative SNPs called showed a high Phred scored (>Q30), revealing a low probability of has been miscalled due sequencing errors. The ratio between Ka/Ks revealed three genes under positive selection pressure ($\omega > 1$; p-value < 0.01) (Table 2).

Most perfect microsatellites loci identified in sugarcane are di-nucleotide and were found mainly for genes ScPht1-7, ScPht1-8 and ScPht1-9. In addition, 11 perfect SSR loci were also identified for *S. bicolor*, four on 3' downstream position, four on 5' upstream position and three within CDS genes region (intragene). Similarly with sugarcane, most of *S. bicolor* perfect loci are di-nucleotides. These analysis also revealed 27 imperfect SSR regions in *S. bicolor* and seven in *Saccharum* spp. EST-derived primers of 22 bp in length were drawn for all these perfect and imperfect loci in both species, providing molecular markers for further population polymorphism investigation (Table S2).

The gene annotation confirmed that those nine ScPht1 genes assembled have functions related with absorption, transport and storage of Pi. Cellular Component and Biological Process GO terms represent almost 80% of the total annotations, being characterized mainly like cellular response to phosphorus starvation, ions phosphate transportation and response to abscisic acid (ABA) stimulus, known as part of the response to abiotic stress (Fig. 2). The protein family annotation performed by both Pfam and Panther databases revealed that all ScPht1 genes are associated with a class of mem-

brane transport proteins that facilitate movement of small solutes across cell membranes in response to chemiosmotic gradients, with accession ID PF00083 and PTHR24064, respectively.

3.4. Phylogenetic reconstruction

After remove the non-assembled positions and gaps, 526 remained sites from 55 homologous Pht1 genes from five grass species and *A. thaliana* (Table S3) were used for phylogenetic reconstruction. Respectively, 83% and 70% of used nucleotides sites were considered polymorphic and parsimoniously informative. The BIC criterion suggested the General Time Reversible + Gamma distribution (GTR + G) as the appropriated model to explain the molecular evolution of gene sequences, being used for Bayesian phylogenetic analysis. The bootstraps values of the tree nodes had on average 88.69%, supporting monophyletic groups of orthologous Pht1 genes among different Poaceae species. Eight monophyletic groups may be identified showing gene homology between ScPht1 and *O. sativa* homologous copies. The only exception is the ScPht1-1 gene that showed the orthology with *Z. mays* ZmPht1-2 and ZmPht1-4 genes (Fig. 3).

4. Discussion

4.1. Orthology among Pht1 genes in close related Poaceae family species

The physical position of Pht1 genes on genomes of the evaluated species are concentrated on chromosomes 3 and 4 for *O. sativa* and in chromosome 01 for *S. bicolor* and *Z. mays*. High genetic similarity (>97%) between *S. bicolor* and *Saccharum* spp. Pht1 genes was found (Fig. 1B and C), agreeing with the existence of synteny and collinearity between the two species, highlight the possibility to use the diploid *S. bicolor* genome as a model for genomic studies in polyploid sugarcane genome [42].

The use of gene arrangements among close related species to support the orthology identification of genes related with response to abiotic stress like water scarcity has been reported for grass species and an important example are the identification and characterization of orthology PSY3 genes between *S. bicolor* and *O. sativa*. This gene family affects directly the carotenoid biosynthe-

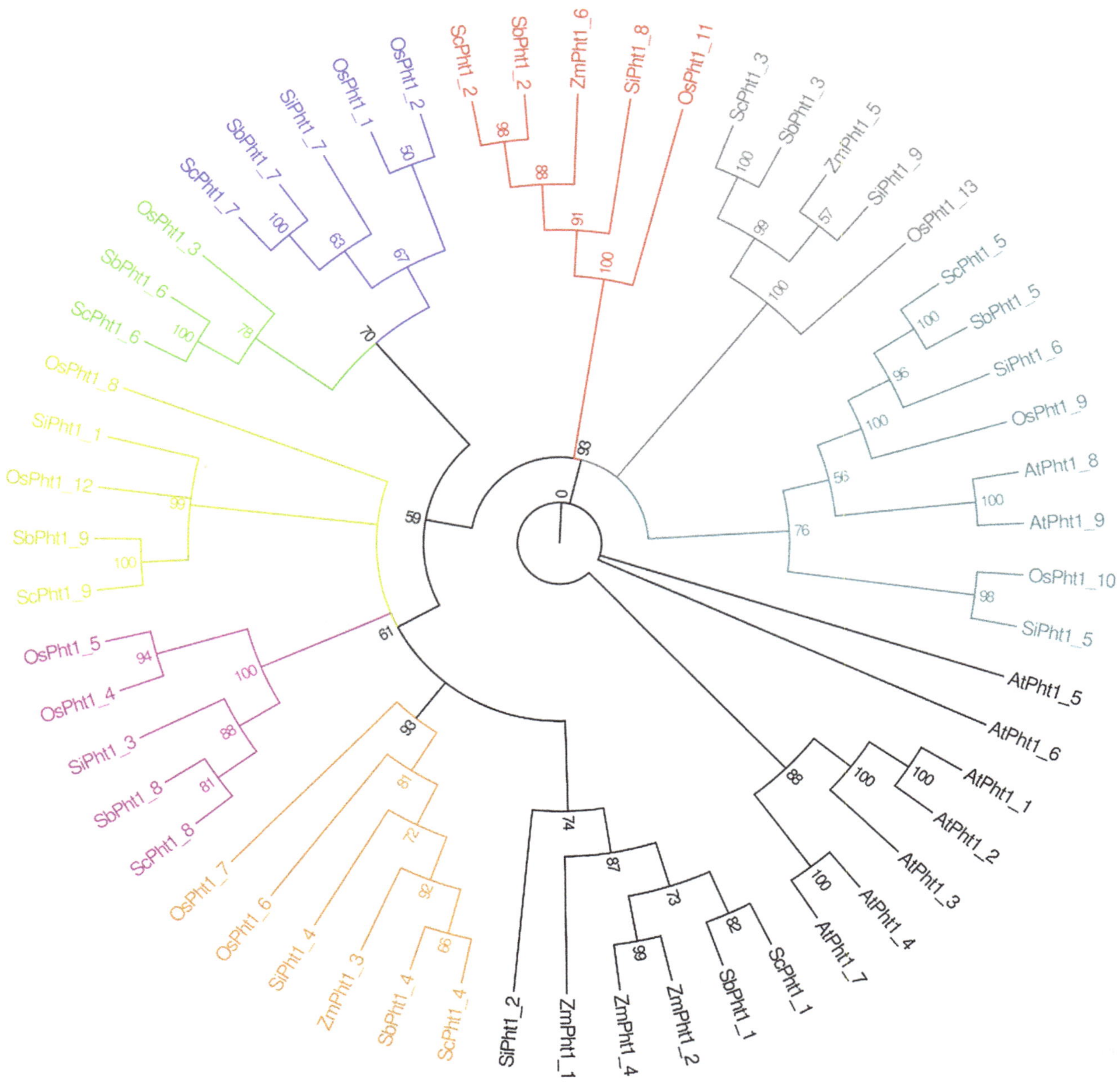

Fig. 3. The phylogenetic tree estimated for all 55 Pht1 genes evaluated. Except for the ScPht1-1 gene that had orthology with *Z. mays* (ZmPht1-2/ZmPht1-4) genes, all others eight genes identified for sugarcane shows orthology with *O. sativa* Pht1 genes. The average of bootstrap values was 89.33%, supporting the monophyletic groups.

sis in roots under drought stress in both species [43]. Moreover, two genes responsible for abscisic acid (ABA) metabolism have been identified in maize using the orthology with rice genes [44]. The genes controlling the (ABA) production are important genes for further investigation to detail the genetic based of drought stress metabolism in plants. This biochemical compound is a phytohormone critical for plant growth and development and plays an important role in integrating various stress signals and controlling downstream stress responses [45].

In maize, both genes ZmPht1-2 and ZmPht1-4 had similar pattern of gene expression in response to Pi scarcity in root tissues sampled from plants growth in hydroponic solution [20]. The sugarcane gene ScPht1-1 is orthologous to both ZmPht1-2 and ZmPht1-4 genes. This sugarcane gene was assigned to a Biological Process GO term during the annotation process and more specifically it was annotated in response to abscisic acid (ABA) stimulus (Fig. 4). Blast-

ing this particular gene sequence on Plant Reactome database [46] was found a homology with *O. sativa* gene also assigned to Biological Process GO term characterized as a series of molecular signals generated by the binding of the plant hormone abscisic acid (ABA) to a receptor, and ending with modulation of a cellular process, *e.g.* transcription.

The Pht1 gene family has also been reported as response to drought stress in non-grass species like bean. A study comparing the transcript profiles in roots of *Phaseolus acutifolius* and *P. vulgaris* under water deficit conditions demonstrated that the drought resistance phenotype of *P. acutifolius* is related with polymorphism in two putative phosphate transporter 1 genes [47]. A very helpful database to evaluate the orthology of genes involved in drought stress response is available on DroughtDB. It includes information about the originally identified gene, its physiological and/or molecular function and mutant phenotypes and provides detailed

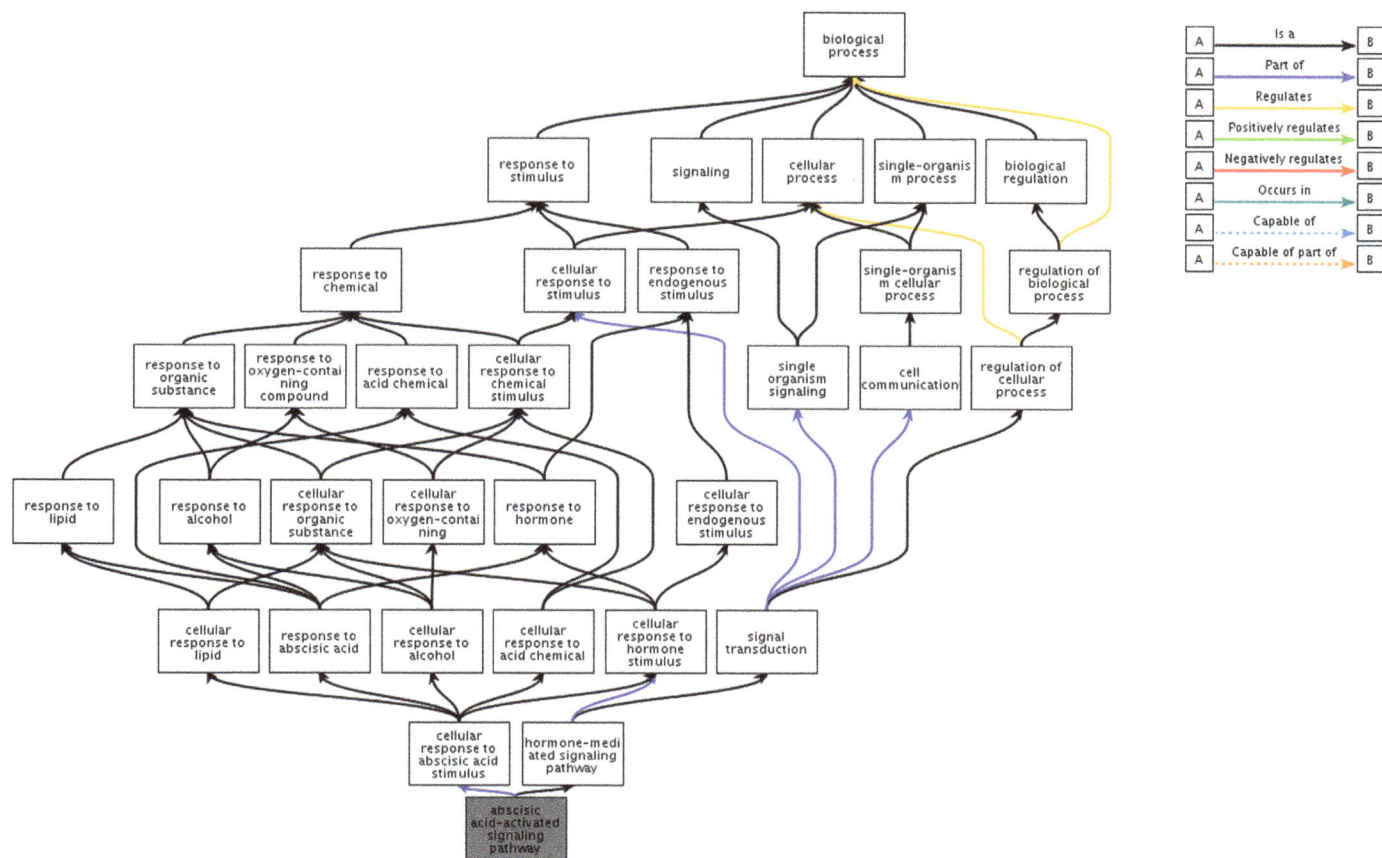

Fig. 4. Abscisic acid activated signaling patway revealed by annotation of ScPht1-1 gene, performed by QuickGO database.

information about computed orthologous genes in nine model and crop plant species including *A. thaliana*, rice, maize and barley [48].

The ω ratio found for ScPht1-1 gene can suggest this gene may be under a positive selection pressure (ω = 1.135). One hypothesis to explain why some genes such ScPht1-1, ScPht1-7 and ScPht1-9 had significant ω > 1 may be those genes are first activated as response to low concentrations of Pi on soils. These genes may act quickly in response to changes in Ca^{2+} caused by abiotic stress [49]. Whether this hypothesis may be confirmed, these genes could be better investigated in attempt to provide basis for the development superior genotypes tolerant to phosphorus scarcity in soil. The SUCEST contig SCEQRT1028B07.g was previously reported as responsible for the phosphorus uptake in high affinity systems in sugarcane [50] and high levels of genetic similarity between ScPht1-1 and this contig was found (Fig. S2).

The sugarcane gene ScPht1-2 shows orthology with two genes from *O. sativa* (OsPht1-11) and *Z. mays* (ZmPht1-6). High level of gene expression was found for OsPht1-11 when roots of *O. sativa* are colonized by *Glomus intraradices* fungus in symbiosis system, which increase the ability of the plant uptake phosphates ions from soil, increase grain yielding [19]. The same pattern of gene activation in conditions of symbiosis with the same fungus was found in maize for gene ZmPht1-6 [20]. Probably, the sugarcane gene orthologous to these rice and maize genes (see Fig. 3) might be also considered a potential candidate gene for further investigations in order to associate the increase of gene expression in symbiosis systems. The protein ID classification of these orthologous genes on UniProt database is A1C9B0. A gene annotation analysis performed by KEEG database [51], revealed the genes OsPht1-11 and ZmPht1-6 have a metabolic pathway (K08176) related with a protein superfamily responsible for Pi and organophosphate transportation

(PHS – Phosphate Pht1 + symporter). This phosphate transporter family can be found in Transporter Classification Data Base (TCDB) with accession number 2.A.1.9, where more than ten ratings and functional characterization were done for *O. sativa* [52].

High levels of gene expression were also reported in pollen and seed tissue for ZmPht1-3 gene. Both pollen and seed plant tissues are involved with phytate storage, a rich inorganic phosphate compound. This maize gene is orthologous to ScPht1-4 and the functional annotation performed by this particular gene confirmed the association of them with the transport and storage of inorganic phosphate, mainly during pollen tube growth, which corroborates with the assignment of ZmPht1-3 and ScPht1-4 in a particular monophyletic group (see Fig. 3).

The gene duplication events have already been reported for Oryzoideae sub-group (*Oryza* genus), providing evolutionary divergence for this group with Panicoideae subfamily (*S. bicolor, S. italica, Saccharum* spp. and *Z. mays*) [53]. Therefore, there is no consensus if the common ancestor of these two taxonomic groups had nine copies of homologous Pht1 genes, and some duplication events favored the emergence of four exclusive copies in Oryzoideae subgroup, suggesting a complex genomic evolution by this taxonomic group [54]. However, within the Panicoideae subfamily, *Z. mays* had only six copies of Pht1 genes, which can be a sign of gene loss during *Z. mays* particular evolutionary history and this event have been reported for tetraploid nature of maize genome [55].

4.2. Candidate loci to polymorphic molecular markers

The microsatellite and SNPs loci may be considered as the two major types of genetic variation in eukaryotic genomes that are widely converted into reliable molecular markers providing molecular bases to explore the genetic polymorphism within or between populations [56,57].

Even with low identification of EST-derived loci from sugarcane Pht1 genes (see Table S2), the transferability of microsatellite markers is a viable and wild-used alternative to use primers described to particular specie in other closely related specie. This procedure has been described for several different species from Poaceae family [58–60]. Thus, the primers identified in *S. bicolor* certain could be checked transferability to sugarcane as an attempt to identify polymorphisms associated with Pi metabolism and drought stress response in both species.

5. Conclusion and perspectives

This study illustrates the potential use of genomic databases associated with sequence similarity analysis for gene discovery and characterization in a comparative genomic approach. The existence of synteny among species from the Pannicoideae subfamily was also supported. Homologous sequences of three sugarcane genes (ScPht1-1, ScPht1-2 and ScPht1-4) were described acting in important plant metabolic process, encouraging the use of modern biotechnological techniques for further investigation, supporting the sugarcane breeding programs that aims develop superior genotypes tolerant to Pi scarcity on agricultural lands.

According to Varmus [61], over the last decades, several technical tools revolutionized the genomic field allowing scientists to generate extraordinarily useful information, including the nucleotide-by-nucleotide description of the genetic blueprint of many of the organisms we care about. Furthermore, the amount of factual knowledge has expanded so precipitously that all modern biologists using genomic methods have become dependent on computer science to store, organize, search, manipulate and retrieve the information.

Once the sequencing costs continue decreasing, many other plant genomes are been sequenced each year, supporting initiatives like the 1000 Plant Genome Projects [62]. The advent of third-generation sequencing and the development of bioinformatics tools able to analyze those data, allow that huge and complex plant genomes like the sugarcane genome can be sequenced, assembled and annotated. Clearly, it still requiring that public databases increases its storage capacity following the increasing number of genome projects around the world.

Acknowledgments

I would like to thank the institutions that support the execution of this study. CAPES (Coordenação de Aperfeiçoamento de Pessoal de Nível Superior) and Federal University of Goiás (UFG) for the PhD scholarship granted. I also thank Professor ASG Coelho, and the Ph.D. candidates IDB Oliveira and IP Souza for the critical suggestions to this study.

References

[1] D.M. Goodstein, S. Shu, R. Howson, R. Neupane, R.D. Hayes, J. Fazo, T. Mitros, W. Dirks, U. Hellsten, N. Putnam, D.S. Rokhsar, Phytozome: a comparative platform for green plant genomics, Nucleic Acids Res. 40 (2011) 1178–1186.

[2] M.K. Monaco, J. Stein, S. Naithani, S. Wei, P. Dharmawardhana, S. Kumari, V. Amarasinghe, K. Youens-Clark, J. Thomason, J. Preece, S. Pasternak, Gramene 2013: comparative plant genomics resources, Nucleic Acids Res. 42 (2013), http://dx.doi.org/10.1093/nar/gkt1110.

[3] Q. Dong, S.D. Schlueter, V. Brendel, PlantGDB, plant genome database and analysis tools, Nucleic Acids Res. 32 (2004), http://dx.doi.org/10.1093/nar/gkh046.

[4] A.H. Paterson, J.E. Bowers, R. Bruggmann, I. Bubchak, J. Grimwood, H. Gundlach, G. Haberer, U. Hellsten, T. Mitros, A. Poliakov, The *Sorghum bicolor* genome and the diversification of grasses, Nature 457 (2009) 551–556.

[5] N. Jannoo, L. Grivet, N. Chantret, O. Garsmeur, J.C. Glaszmann, P. Arruda, A. D'Hont, Orthologous comparison in a gene-rich region among grasses reveals stability in the sugarcane polyploid genome, Plant J. 50 (2007) 574–585.

[6] L. Grivet, P. Arruda, Sugarcane genomics: depicting the complex genome of an important tropical crop, Curr. Opin. Plant Biol. 5 (2002) 122–127.

[7] N. Setta, G.B.M. Vitorello, J.C. Metcalfe, G.M.Q. Cruz, L.E.D. Bem, R. Vicentini, F.T.S. Nogueira, R.A. Campos, S.L. Nunes, P.C.G. Turrini, Building the sugarcane genome for biotechnology and identifying evolutionary trends, BMC Genom. 15 (2014), http://dx.doi.org/10.1186/1471-2164-15-540.

[8] T.R. Figueira, V. Okura, R.F. Silva, J.M. Silva, D. Kudrna, J.S.S. Ammiraju, P. Arruda, A BAC library of the SP 80-3280 sugarcane variety (*Saccharum* spp.) and its inferred microsynteny with the sorghum genome, BMC Res. Notes 5 (2012), http://dx.doi.org/10.1186/1756-0500-5-185.

[9] K.M. Devos, M.D. Gale, Genome relationship: the grass model in current research, Plant Cell 1 (2000) 636–646.

[10] P.D.R. Van-Heerden, R.A. Donaldson, D.A. Watt, A. Singels, Biomass accumulation in sugarcane: unravelling the factors underpinning reduced growth phenomena, J. Exp. Bot. 61 (2010) 2877–2887.

[11] A.J. Waclawovsky, P.M. Sato, C.G. Lembke, P.H. Moore, G.M. Souza, Sugarcane for bioenergy production: an assessment of yield and regulation of sucrose content, Plant Biotechnol. J. 8 (2010) 263–276.

[12] S.S. Ferreira, M.Y. Nishiyama, A.H. Paterson, G.M. Souza, Biofuel and energy crops: high-yield Saccharinae take center stage in the post-genomics era, Genome Biol. 14 (2013), http://dx.doi.org/10.1186/gb-2013-14-6-210.

[13] M. Hawkesford, W. Horst, T. Kichey, H. Lambers, J. Schjoerring, I.S. Moller, P. White, Functions of macronutrients, in: P. Marschner (Ed.), Mineral Nutrition of Higher Plants, ScienceDirect: Elsevier, 2012, pp. 135–189.

[14] L. Nussaume, S. Kanno, H. Javot, E. Marin, N. Pochon, A. Ayadi, T.M. Nakanishi, M.C. Thibaud, Phosphate import in plants: focus on the PHT1 transporters, Front. Plant Sci. 2 (2011), http://dx.doi.org/10.3389/fpls.2011.00083.

[15] S.M. Mudge, A.L. Rae, E. Diatloff, F.W. Smith, Expression analysis suggests novel roles for members of the Pht1 family of the phosphate transporters in Arabidopsis, Plant J. 31 (2002) 341–353.

[16] J. Liu, W.K. Versaw, N. Pumplin, S.K. Gomez, L.A. Blaylock, M.J. Harrison, Closely related members of the *Medicago truncatula* Pht1 phosphate transporter gene family encode phosphate transporters with distinct biochemical activities, J. Biol. Chem. 283 (2008) 24673–24681.

[17] G. Leggewie, L. Willmitze, J.W. Riesmeier, Two cDNAs from potato are able to complement a phosphate uptake-deficient yeast mutant: identification of phosphate transporters from higher plants, Plant Cell 9 (1997) 381–392.

[18] S.H. Baek, I.M. Chung, S.J. Yun, Molecular cloning and characterization of a tobacco leaf cDNA encoding a phosphate transporter, Mol. Cells 11 (2001) 1–6.

[19] U. Paszkowski, S. Kroken, C. Roux, S.P. Briggs, Rice phosphate transporters include an evolutionarily divergent gene specifically activated in arbuscular mycorrhizal symbiosis, Proc. Natl. Acad. Sci. U. S. A. 99 (2002) 13324–13329.

[20] R. Nagy, M.J.V. Vasconcelos, S. Zhao, J. McElver, W. Bruce, N. Amrhein, K.G. Raghothama, M. Bucher, Differential regulation of five Pht1 phosphate transporters from maize (*Zea mays* L.), Plant Biol. 8 (2006) 186–197.

[21] A. Tittarelli, L. Milla, F. Vargas, A. Morales, C. Neupert, L.A. Meisel, G.H. Salvo, E. Penaloza, G. Munoz, L.J. Corcuera, H. Silva, Isolation and comparative analysis of the wheat TaPT2 promoter: identification *in silico* of new putative regulatory motifs conserved between monocots and dicots, J. Exp. Bot. 58 (2007) 2573–2582.

[22] V. Loth-Pereda, E. Orsini, P.E. Courty, F. Lota, A. Kohler, L. Diss, D. Blaudez, M. Chalot, U. Nehls, M. Bucher, F. Martin, Structure and expression profile of the phosphate Pht1 transporter gene family in mycorrhizal *Populus trichocarpa*, Plant Physiol. 156 (2011) 2141–2154.

[23] G. Zhang, X. Liu, Z. Quan, S. Cheng, X. Xu, S. Pan, M. Xie, P. Zeng, Z. Yue, W. Wang, Genome sequence of foxtail millet (*Setaria italica*) provides insights into grass evolution and biofuel potential, Nat. Biotechnol. 30 (2012) 549–556.

[24] S.F. Altschul, T.L. Madden, A.A. Schäffer, J. Zhang, Z. Zhang, W. Miller, D.J. Lipman, Gapped BLAST and PSI-BLAST: a new generation of protein database search programs, Nucleic Acids Res. 25 (1997) 3389–3402.

[25] A.M. Bolger, M. Lohse, B. Usadel, Trimmomatic: a flexible trimmer for illumina sequence data, Bioinformatics 30 (2014) 2114–2120.

[26] H. Li, R. Durbin, Fast and accurate short read alignment with Burrows-Wheeler Transform, Bioinformatics 25 (2009) 1754–1760.

[27] H. Li, B. Handsaker, A. Wysoker, T. Fennell, J. Ruan, J. Homer, G. Marth, G. Abecasis, R. Durbin, The sequence alignment/map format and SAMtools, Bioinformatics 25 (2009) 2078–2079.

[28] M.Y. Nishiyama, S.S. Ferreira, P.Z. Tang, S. Becker, A.P. Taliana, G.M. Souza, Full-length enriched cDNA libraries and ORFeome analysis of sugarcane hybrid and ancestor genotypes, PLoS One 9 (2014), http://dx.doi.org/10.1371/journal.pone.0107351.

[29] J. Krumsiek, R. Arnold, T. Rattei, Gepard: a rapid and sensitive tool for creating dotplots on genome scale, Bioinformatics 23 (2007) 1026–1028.

[30] K. Tamura, D. Peterson, G. Stecher, M. Nei, S. Kumar, MEGA5: molecular evolutionary genetics analysis using maximum likelihood, evolutionary distance, and maximum parsimony method, Mol. Biol. Evol. 28 (2011) 2731–2739.

[31] M. Nei, W.H. Li, Mathematical model for studying genetic variation in terms of restriction endonucleases, Proc. Natl. Acad. Sci. U. S. A. 76 (1979) 5269–5273.

[32] Z. Zhang, J. Li, X.Q. Zhao, J. Wang, G.K. Wong, J. Yu, KaKs Calculator: calculating Ka and Ks through model selection and model averaging, Genom. Proteom. Bioinform. 4 (2006) 259–263.

[33] S.B. Mudunuri, H.A. Nagarajaram, IMEx: imperfect microsatellite extractor, Bioinformatics 23 (2007) 1181–1187.

[34] A. Untergasser, H. Nijveen, X. Rao, T. Bisseling, R. Geurts, J.A.M. Leunissen, Primer3Plus, an enhanced web interface to Primer3, Nucleic Acids Res. 35 (2007) 71–74.

[35] A. Conesa, S. Gotz, Blast2GO: a comprehensive suite for functional analysis in plant genomics, Int. J. Plant Genom. (2008) 1–12, 619832.

[36] E. Quevillon, V. Silventoinen, S. Pillai, N. Harte, N. Mulder, R. Apweiler, R. Lopez, InterProScan: protein domains identifier, Nucleic Acids Res. 33 (2005) 116–120.

[37] R.D. Finn, A. Bateman, J. Clements, P. Coggill, R.Y. Eberhardt, S.R. Eddy, A. Heger, K. Hetherington, L. Holm, J. Mistry, Pfam: the protein families database, Nucleic Acids Res. 42 (2014) 222–230.

[38] J.P. Huelsenbeck, F. Ronquist, MrBayes: bayesian inference of phylogenetic trees, Bioinformatics 17 (2001) 754–755.

[39] D. Darriba, G.L. Taboada, R. Doallo, D. Posada, jModelTest 2: more models, new heuristics and parallel computing, Nat. Methods 9 (2012), http://dx.doi.org/10.1038/nmeth.2109.

[40] R.C. Edgar, MUSCLE: multiple sequence alignment with high accuracy and high throughput, Nucleic Acids Res. 32 (2004) 1792–1797.

[41] A.M. Waterhouse, J.B. Procter, D.M.A. Martin, M. Clamp, G.J. Barton, Jalview: version 2-a multiple sequence alignment editor and analysis workbench, Bioinformatics 25 (2009) 1189–1191.

[42] J. Wang, B. Roe, S. Macmil, Q. Yu, J.E. Murray, H. Tang, C. Chen, F. Najar, G. Wiley, J. Bowers, Microcollinearity between autopolyploid sugarcane and diploid sorghum genomes, BMC Genom. 11 (2010) 261, http://dx.doi.org/10.1186/1471-2164-11-261.

[43] F. Li, V. Ratnakar, E.T. Wurtzel, PSY3, a new member of the phytoene synthase gene family conserved in the Poaceae and regulator of abiotic stress-induced root carotenogenesis, Plant Physiol. 146 (2008) 1333–1345.

[44] H. Saika, M. Okamoto, K. Miyoshi, T. Kushiro, S. Shinoda, Y. Jikumaru, M. Fujimoto, T. Arikawa, H. Takahashi, E. Ando, Ethylene promotes submergence-induced expression of OsABA8ox1, a gene that encodes ABA 8'-hydroxylase in rice, Plant Cell Physiol. 48 (2007) 287–298.

[45] N. Tuteka, Abscisic Acid and abiotic stress signaling, Plant Signal. Behav. 3 (2007) 135–138.

[46] M.K. Tello-Ruiz, J. Stein, S. Wei, J. Preece, A. Olson, S. Naithani, V. Amarasinghe, P. Dharmawardhana, Y. Jiao, J. Mulvaney, Gramene 2016: comparative plant genomics and pathway resources, Nucleic Acids Res. (2015), http://dx.doi.org/10.1093/nar/gkv1179.

[47] S. Micheletto, L. Rodriguez-Uribe, R. Hernandez, R.D. Richins, J. Curry, M.A. O'Connell, Comparative transcript profiling in roots of Phaseolus acutifolius and P. vulgaris under water deficit stress, Plant Sci. 173 (2007) 510–520.

[48] S. Alter, K.C. Bader, M. Spannagl, Y. Wang, E. Bauer, C.C. Schön, K.F.X. Mayer, DroughtDB: an expert-curated compilation of plant drought stress genes and their homologs in nine species, Database 2015 (2015), http://dx.doi.org/10.1093/database/bav046

[49] S. Fraire-Velazquez, R. Rodriguez-Guerra, L. Sanches-Colderon, Abiotic and biotic stress response crosstalk in plants, in: A.K. Shanker, B. Venkateswarlu (Eds.), Abiotic stress response in plants—Physiological, biochemical and genetic perspectives, Rijeka, 2010, pp. 3–26.

[50] R.S. Almeida, Identificação e caracterização de genes transportadores de fosfato em cana-de-acúcar, Master degree Thesis, University of São Paulo, 2002.

[51] M. Kanehisa, S. Goto, KEGG: Kyoto encyclopedia of genes and genomes, Nucleic Acids Res. 28 (2000) 27–30.

[52] P. Ai, S. Sun, J. Zhao, X. Fan, W. Xin, Q. Guo, L. Yu, Q. Shen, P. Wu, A.J. Miller, G. Xu, Two rice phosphate transporters OsPht1;2 and OsPht1;6, have different functions and kinetic properties in uptake and translocation, Plant J. 57 (2009) 798–809.

[53] O. Morrone, L. Aagesen, M.A. Scataglini, D.L. Salariato, S.S. Denham, M.A. Chemisquy, S.M. Sede, L.M. Giussani, E.A. Kellogg, F.O. Zuloaga, Phylogeny of the Paniceae (Poaceae: Panicoideae): integrating plastid DNA sequences and morphology into a new classification, Cladistic 38 (2012) 333–356.

[54] K. Illic, P.J. Sanmiguel, J.L. Bennetzen, A complex history of rearrangement in an orthologous region of the maize, sorghum and rice genomes, Proc. Natl. Acad. Sci. U. S. A. 100 (2003) 12265–12270.

[55] J. Lai, J. Ma, Z. Swigonová, E. Linton, V. Llaca, B. Tanyolac, Y.J. Park, O.Y. Jeong, J.L. Bennetzen, J. Messing, Gene loss and movement in the maize genome, Genome Res. 14 (2004) 1924–1931.

[56] J. Mammadov, R. Aggarwal, R. Buyyarapu, S. Kumpatla, SNP markers and their impact on plant breeding, Int. J. Plant Genom. (2012) 1–12, 728398.

[57] A.M. Pérez-de-Castro, S. Vilanova, J. Cañizares, L. Pascual, L.M. Blanca, M.J. Díez, J. Prohens, B. Picó, Application of genomic tools in plant breeding, Current Genom. 13 (2012) 179–195.

[58] V. Decroocq, M.G. Favé, L. Hagen, L. Bordenave, S. Decroocq, Development and transferability of apricot and grape EST microsatellite markers across taxa, Theor. Appl. Genet. 106 (2003) 912–922.

[59] M.C. Sasha, M.A.R. Mian, I. Eujavl, J.C. Zwonitzer, L. Wang, G.D. May, Tall fescue EST-SSR markers with transferability across several grass species, Theor. Appl. Genet. 109 (2004) 783–791.

[60] O.P. Yadav, S.E. Mitchell, T.M. Fulton, S. Kresovich, Transferring molecular markers from sorghum, rice and other cereals to pearl millet and identifying polymorphic markers, J. SAT Agric. Res. 6 (2008) 1–4.

[61] H. Varmus, Genomic empowerment: the importance of public databases, Nat. Genet. 32 (2002), http://dx.doi.org/10.1038/ng963.

[62] N. Matasci, L.H. Hung, Z. Yan, E.J. Carpenter, N.J. Wickett, S. Mirarab, N. Nguyen, T. Warnow, S. Ayyampalayam, M. Barker, Data access for the 1,000 plants (1KP) project, Gigascience 3 (2014), http://dx.doi.org/10.1186/2047-217x-3-17.

4

Manually Curated Database of Rice Proteins (MCDRP), a database of digitized experimental data on rice ☆

Saurabh Raghuvanshi*, Pratibha Gour, Shaji V. Joseph

Department of Plant Molecular Biology, University of Delhi South Campus, Benito Juarez Road, New Delhi, 110021, India

ARTICLE INFO

Keywords:
Rice
Manual curation
Data digitization

ABSTRACT

MCDRP or 'Manually Curated Database of Rice Proteins' is a database of digitized experimental datasets on rice proteins. Every aspect of the experimental data published in peer-reviewed research articles on rice biology has been digitized with the help of novel data curation models. These models use a semantic and structured arrangement of alpha-numeric notation, including several well known ontologies, to represent various aspect of the data. As a result data from more than 15,000 different experiments pertaining to about 2400 rice proteins has been digitized from over 540 published and peer-reviewed research articles. The database portal provides access to the digitized experimental data via search or browse functions. In essence, one can instantly access data from even a single data-point from a collection of thousands of the experimental datasets. On the other hand, one can easily access the digitized experimental data from multiple research articles on a rice protein. Based on the analysis and integration of the digitized experimental data, more than 800 different traits (molecular, biochemical or phenotypic) have been precisely mapped onto the rice proteins along with the underlying experimental evidences. Similarly, over 4370 associations, based on experimental evidence, have been established between the rice proteins and various gene ontology terms. The database is being continuously updated and is freely available at www.genomeindia.org.in/biocuration.

1. Introduction

Semantic integration and fast access of the experimental data in biological sciences is essential in order to understand the multi-dimensional nature of most biological processes. Rice is a model monocot system as well as one of the most important food grain. Consequently, extensive amount of research has been conducted over the years with a view to better understand its biology. Several database have also been developed that have greatly facilitated the research on rice biology [1–14]. While the databases ensure that most high throughput datasets like genome/gene sequences, microarray of RNA-seq expression data etc. are catalogued and made available to the users, however, a huge amount of invaluable high quality experimental data remains buried in published literature and cannot be searched or integrated computationally. This is because most published experimental data is presented in pictorial format either as an image or graph and thus not amenable to computerized search, let alone seamless integration. The only way to access such data is via reading the entire publication. The ever increasing number of publications on rice biology has resulted in a massive accumulation of such high quality experimental data. In order to gain a 'systems level' perspective of rice biology, it is imperative that such experimental data is rendered computer indexible so that it can be rapidly searched and integrated. 'Manually Curated Database of Rice Proteins' address this aspect and provides the user with digitized published experimental data on rice biology. It is a manually curated database which utilizes in-house developed data curation models that enable digitization every aspect of the experimental data.

2. Database description

The current release of the database consists of digitized experimental data for over 2400 rice proteins spread over more than 540 published research articles. More than 15,000 individual experiments containing over 90,000 data-points have been digitized with the help of novel data curation models developed earlier [15]. The entire curation or digitization is based on manual curation and thus leads to a very high quality of digitized and validated experimental data. Data from a wide variety of experimental techniques (>150 different types) such as gene expression measurements, enzymatic activities, interaction studies, trait analysis (phenotypic or

☆ This article is part of a special issue entitled "Genomic resources and databases", published in the journal Current Plant Biology 7–8, 2016.
* Corresponding author.
  E-mail address: Saurabh@genomeindia.org (S. Raghuvanshi).

metabolic) etc. have been digitized. The detailed digitization process ensures that information for a single data-point from any of the experiments can be individually retrieved, if required. A data-point, for example, would typically corresponds to the underlying data of single 'bar' of a 'bar graph' depicting a RT-PCR expression data. Information contained within every data-point is represented by a collection of alpha-numeric terms which include various ontology terms such as plant ontology, environmental ontology or trait ontology. Thus, the data of the entire experiment can be represented by a semantic collection of these alpha-numeric terms. The digitization of the data has been done by utilizing >600 plant ontology, >350 environment ontology, >800 trait ontology and >350 gene ontology terms. In other words, every experiment has been extensively annotated with the help of various ontology terms. The plant ontology terms have been used to represent the growth/developmental stage and the tissue of the plant that has been analyzed in the experiment. Similarly, environmental ontology terms have been used to record the growth conditions such as temperature, light or water status as well as treatment with any chemical. The trait ontology terms have been used to represent any trait (molecular, physiological or phenotypic) of the proteins whereas gene ontology terms have been used to encode the functional details of the protein.

Analysis and integration of the digitized data unravels several interesting aspects of the data. As a result of detailed digitization it was possible to map more than 800 different traits to rice proteins. In summary, 831 traits have been mapped to 398 different rice proteins. The database contains a wide range of traits such as 'anatomy and morphology related', 'biochemical profile and physiology related', 'enzymatic activity related', 'growth and development related', 'stress related' and 'yield and biomass related' traits. The top 5 most frequently related traits are 'survival rate', 'plant height', 'root length' 'seedling height' and 'seedling vigour'. The association of the traits to the genes is of very high confidence since it is based on high quality experimental data. Similarly, more than 4300 associations have been made between the rice proteins to various molecular functions as well as biological process and cellular component based on the digitized experimental data. This data is represented with the help of gene ontology terms. Further, in order to facilitate better usage of the database, metabolic pathways have also been mapped on the proteins that have been curated in the database. Consequently, more than 200 metabolic pathways are represented by the rice proteins that have been curated in the database. More than half of these pathways are related to one or the other biosynthetic processes.

3. How to use the database?

Since every aspect of the experimental data has been digitized, the data can be easily and rapidly searched. In other words one can easily search or retrieve any experiment from over 500 published research articles in a matter of seconds and without even opening the research publication. In general the database can be either browsed or searched with a specific query.

3.1. Browsing the database

The contents of the database can be browsed from eight different perspectives. As a result of digitization the experimental data can be retrieved based on the either the PubMed id, gene locus id, growth conditions (Environmental Ontology), plant tissue/developmental stage (Plant Ontology term), phenotypic or biochemical trait (Trait Ontology) and gene function/localization (Gene Ontology term). Ultimately all the aspects are inter-related and one can start from any end and access the information from any of the perspectives.

For example, one can browse the database based on any of the known molecular functions of the rice proteins or on the basis of the biological processes wherein they have been implicated. Such data can be accessed via the page summarizing all the mapper GO terms (Fig. 1a). On selecting any one of the molecular functions such as 'DNA binding' or GO:0003677, the database portal would provide a list of all the rice proteins that have been curated in the database, in this case 32 rice gene (Fig. 1b). The data is arranged as per individual experimental dataset. This information has been compiled from 44 digitized experimental data sets from 21 different research articles. This information is primarily acquired by digitizing experimental techniques such as EMSA (Electrophoretic Mobility Shift Assay), CHIP assay or transient co-expression assays. Further, on selecting any of the listed rice protein ids a 'Rice gene details' is shown (Fig. 2). Fig. 2 summarizes information of one of an 'AP2 domain containing protein' that has been assembled via digitization of experimental data from 10 published articles [16–26]. .The information in the 'Rice gene details' page is divided into several sections. The 'Basic information' section gives the information of the protein domain present in the rice protein as per the Pfam database. The 'Functional details' section lists all the molecular function/s or biological process/s that the protein is known to possess or involved in based on published experimental data. 'Clicking' on any of the term id provides the details of the digitized experimental dataset. Similarly, the 'Plant developmental stage/tissue details' section lists all the developmental stages and plant tissue wherein the selected rice gene had been studied. The '(+)' or '(−)' signs indicate the presence or absence of expression/protein activity in a particular tissue/developmental stage. Thus, '(−)' means that while expression/activity of the gene has been analyzed in that particular tissue but no detectable expression or activity was found. The 'Rice gene details' page also summarizes information about the environmental conditions under which the expression/activity of the gene has been analyzed. This information is presented as a list of Environmental Ontology terms in the 'Environment details' section. The impact of abiotic stress conditions such as salinity, drought, heat, or hormones/other chemicals is also recorded with the help of the EO terms. Similarly the 'Trait details' section summarizes all the phenotypic or biochemical traits that have been associated with the selected gene based on experimental data. The page also summarizes the physical interaction data (protein–protein and protein-DNA) for the rice protein as well as any metabolic pathway or QTL associated with the rice protein.

3.2. Searching the database

Specific searches can also be done on the digitized experimental data in the database. One can retrieve any experimental datasets on the basis of gene id, plant developmental stage/tissue, growth conditions including environmental parameters or any chemical treatment. The experimental data can also be retrieved based on any molecular function, biological process, cellular localization or any trait associated with the protein that has been studied. Further one can also use a combination of terms to formulate a detailed query. For example one can ask the question 'Is there a "protein kinase" gene related to the 'plant height' trait in rice. In order to formulate this query one can use the GO term 'protein kinase activity' and the trait ontology term TO: 'Plant height'….for initiating the search. This will give a list of genes that have protein kinase activity and have been found regulating the trait plant height. The user can then select any one or all the rice gene loci to access the experimental details. The output of the search is a list of rice gene ids as well as the link to the exact experiment (PubMed id and experiment no.) where the search term has been used. Similarly, one can use any other combination of search terms to access the digitized experimental data. These searches can be done by specify-

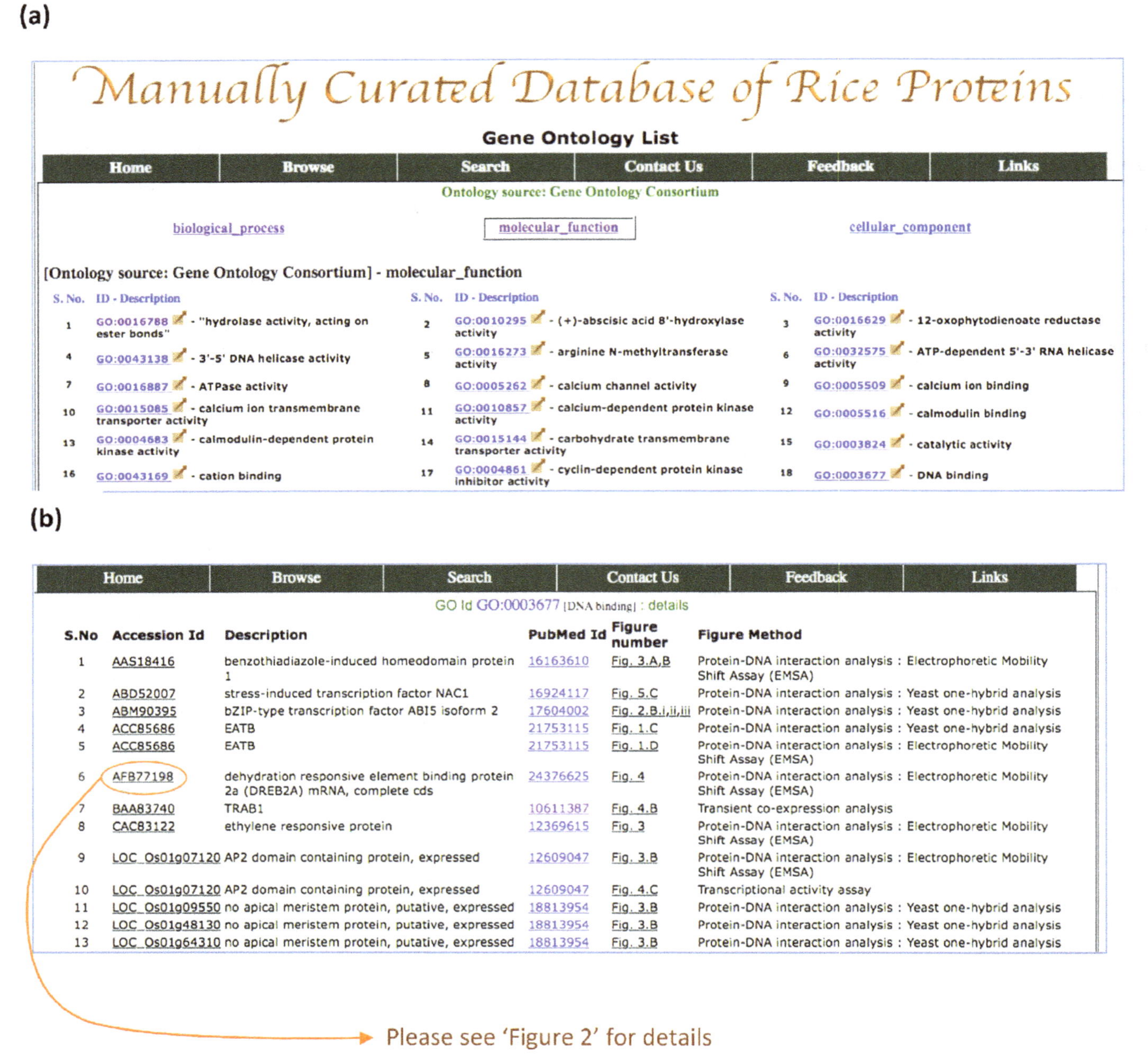

(a)

(b)

> Please see 'Figure 2' for details

Fig 1. Screenshots showing browse results. (a) When browse by 'Gene Ontology' is selected a list of all the GO terms that have been associated on the basis of digitized experimental data is shown. (b) On selecting any one of the GO terms (in this case GO:0003677) a list of all the genes associated with that term are shown along with links to the relevant publication and the digitized experimental data that was used to associate the GO term to the gene. Details of the gene id encircled with a 'red circle' are shown in Fig. 2. In order to accommodate within limited space the figures show only partial list.

ing either the exact 'term id' (ontology term) or any keyword. The basic search function outputs the results in two tiers. If an exact 'term id' is defined (such as protein id, ontology term id) the relevant results are shown immediately. In case a keyword is provided; then in the first stage a list of all the related terms would be displayed. User can then select one or more of these terms and then proceed to the second stage where data relevant to the selected term would be shown.

4. Discussion

Digitization of experimental data is essential due to a phenomenal increase in the bulk of the data as well as the need for seam-less integration of diverse experimental datasets in order to understand complex biological traits. Thus, efforts are being made globally to address the issue. Several repositories like DRYAD (https://datadryad.org) and FIGSHARE (https://figshare.com/) facilitate submission of diverse experimental data. 'Scientific Data' (http://www.nature.com/sdata/) is a peer-reviewed journal that accepts experimental datasets. The datasets need to be deposited in one of the repositories (http://www.nature.com/sdata/policies/

Searching for LOC_Os01g07120 ☑ [AP2 domain containing protein, expressed]

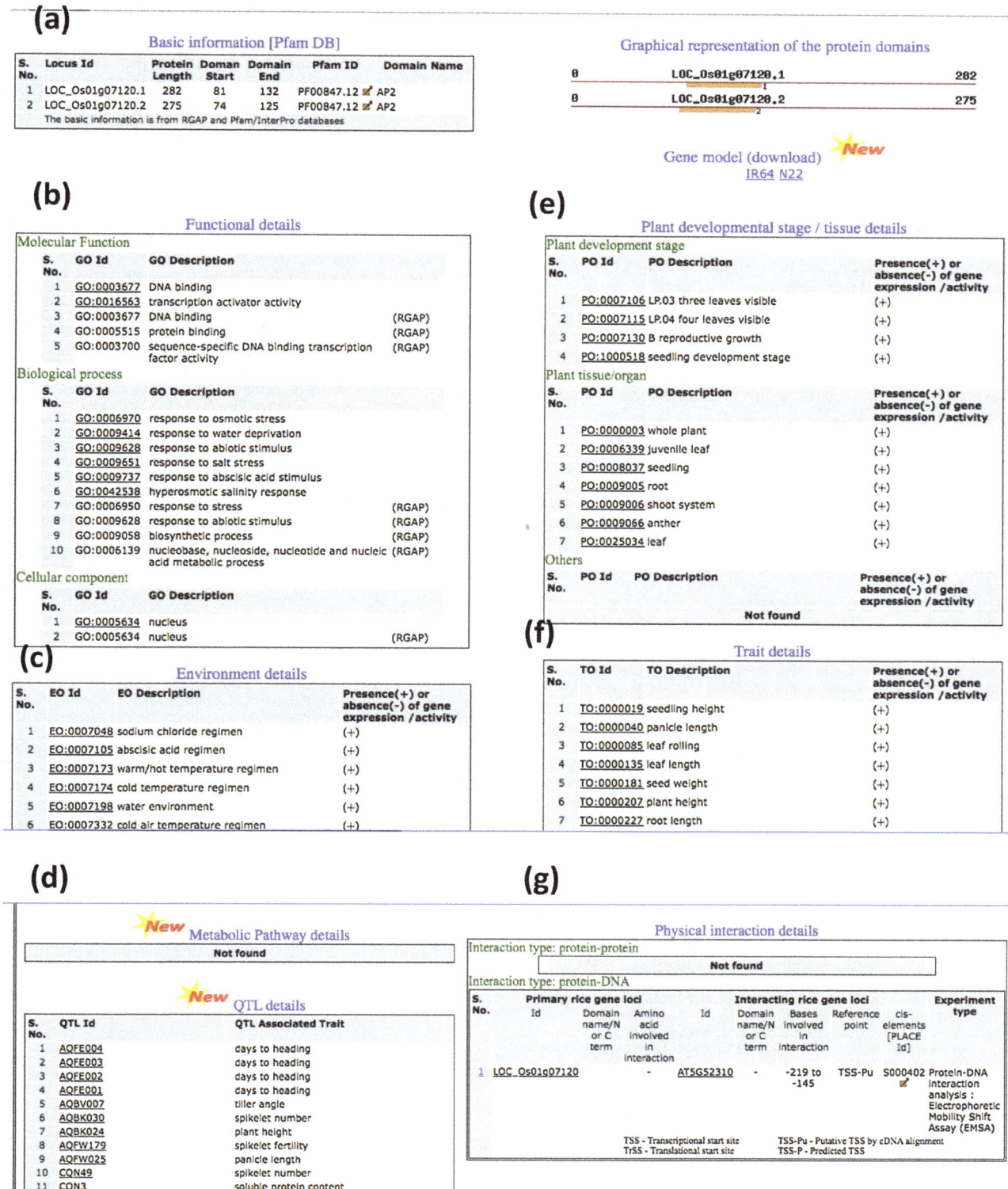

(a)

Basic information [Pfam DB]

S. No.	Locus Id	Protein Length	Domain Start	Domain End	Pfam ID	Domain Name
1	LOC_Os01g07120.1	282	81	132	PF00847.12 ☑	AP2
2	LOC_Os01g07120.2	275	74	125	PF00847.12 ☑	AP2

The basic information is from RGAP and Pfam/InterPro databases

Graphical representation of the protein domains

```
0        LOC_Os01g07120.1         282
                  1
0        LOC_Os01g07120.2         275
                  2
```

Gene model (download) ★New
IR64 N22

(b)

Functional details

Molecular Function

S. No.	GO Id	GO Description	
1	GO:0003677	DNA binding	
2	GO:0016563	transcription activator activity	
3	GO:0003677	DNA binding	(RGAP)
4	GO:0005515	protein binding	(RGAP)
5	GO:0003700	sequence-specific DNA binding transcription factor activity	(RGAP)

Biological process

S. No.	GO Id	GO Description	
1	GO:0006970	response to osmotic stress	
2	GO:0009414	response to water deprivation	
3	GO:0009628	response to abiotic stimulus	
4	GO:0009651	response to salt stress	
5	GO:0009737	response to abscisic acid stimulus	
6	GO:0042538	hyperosmotic salinity response	
7	GO:0006950	response to stress	(RGAP)
8	GO:0009628	response to abiotic stimulus	(RGAP)
9	GO:0009058	biosynthetic process	(RGAP)
10	GO:0006139	nucleobase, nucleoside, nucleotide and nucleic acid metabolic process	(RGAP)

Cellular component

S. No.	GO Id	GO Description	
1	GO:0005634	nucleus	
2	GO:0005634	nucleus	(RGAP)

(c)

Environment details

S. No.	EO Id	EO Description	Presence(+) or absence(-) of gene expression /activity
1	EO:0007048	sodium chloride regimen	(+)
2	EO:0007105	abscisic acid regimen	(+)
3	EO:0007173	warm/hot temperature regimen	(+)
4	EO:0007174	cold temperature regimen	(+)
5	EO:0007198	water environment	(+)
6	EO:0007332	cold air temperature regimen	(+)

(e)

Plant developmental stage / tissue details

Plant development stage

S. No.	PO Id	PO Description	Presence(+) or absence(-) of gene expression /activity
1	PO:0007106	LP.03 three leaves visible	(+)
2	PO:0007115	LP.04 four leaves visible	(+)
3	PO:0007130	B reproductive growth	(+)
4	PO:1000518	seedling development stage	(+)

Plant tissue/organ

S. No.	PO Id	PO Description	Presence(+) or absence(-) of gene expression /activity
1	PO:0000003	whole plant	(+)
2	PO:0006339	juvenile leaf	(+)
3	PO:0008037	seedling	(+)
4	PO:0009005	root	(+)
5	PO:0009006	shoot system	(+)
6	PO:0009066	anther	(+)
7	PO:0025034	leaf	(+)

Others

S. No.	PO Id	PO Description	Presence(+) or absence(-) of gene expression /activity
		Not found	

(f)

Trait details

S. No.	TO Id	TO Description	Presence(+) or absence(-) of gene expression /activity
1	TO:0000019	seedling height	(+)
2	TO:0000040	panicle length	(+)
3	TO:0000085	leaf rolling	(+)
4	TO:0000135	leaf length	(+)
5	TO:0000181	seed weight	(+)
6	TO:0000207	plant height	(+)
7	TO:0000227	root length	(+)

(d)

★New Metabolic Pathway details

Not found

★New QTL details

S. No.	QTL Id	QTL Associated Trait
1	AQFE004	days to heading
2	AQFE003	days to heading
3	AQFE002	days to heading
4	AQFE001	days to heading
5	AQBV007	tiller angle
6	AQBK030	spikelet number
7	AQBK024	plant height
8	AQFW179	spikelet fertility
9	AQFW025	panicle length
10	CQN49	spikelet number
11	CQN3	soluble protein content

(g)

Physical interaction details

Interaction type: protein-protein

Not found

Interaction type: protein-DNA

S. No.	Primary rice gene loci Id	Domain name/N or C term	Amino acid involved in interaction	Interacting rice gene loci Id	Domain name/N or C term	Bases involved in interaction	Reference point	cis-elements [PLACE Id]	Experiment type
1	LOC_Os01g07120	-		AT5G52310	-	-219 to -145	TSS-Pu	S000402	Protein-DNA interaction analysis : Electrophoretic Mobility Shift Assay (EMSA) ☑

TSS - Transcriptional start site   TSS-Pu - Putative TSS by cDNA alignment
TrSS - Translational start site   TSS-P - Predicted TSS

Fig. 2. A snapshot of the 'Rice gene Details' page. The page summarizes information for a particular rice gene from digitized experimental data across all the curated publications. Parts of the output have been truncated to accommodate the whole information. The page has several sections (a-g) which are divided based on the type information. (a) Basic information regarding the domains present in the protein as well as link to the database 'Indica Rice Genome Database' (IRDB) which contains gene models sequences from related indica rice varieties Nagina 22 and IR 64. (b) Mapping of gene ontology terms based on the digitized experimental data. Ontology terms taken from the RGAP database have been indicated with the suffix '(RGAP)' while the others are based on the current curation exercise. (c) Summarizes all environmental conditions or treatments under which the gene has been studied. Plus (+) or (−) signs indicate whether the gene had expression under the condition or not. (d) Lists all the metabolic pathways and QTLs wherein the gene is a constituent., (e) Summarizes all the rice tissues and developmental stages wherein the gene has been studied. (f, g) Details of the molecular/biochemical/phenotypic traits and the physical interaction (DNA-protein or protein-protein) associated with the gene.

repositories) accompanied by a 'Data Descriptor' describing the dataset. However, despite these efforts no data resource provides digitized experimental data on rice proteins. 'Manually Curated Database of Rice Proteins' (MCDRP) was established to address this issue and provide the user with digitized experimental data on rice proteins. The manual data curation/digitization models implemented in MCDRP ensures that all the aspects of every data-point of the experimental data are digitized. This is in contrast to several other formats that add *meta*-data over the entire experimental dataset.

One very important aspect of the curation/digitization process is that it uses very similar elements and fundamentals to digitize data from a wide variety of experimental techniques thus enables efficient integration of the data. The search function is able to retrieve data from even a single data point instantly from a collection of over thousands of experiments. Further, the data can be retrieved from several different perspectives. For instance one can search experimental data based on a particular tissue of growth conditions from across all the curated research articles. In summary, MCDRP, on one end acts an important resource for rice biologist while on the other acts as proof-of-concept regarding the possibility and efficacy of digitizing experimental data.

Funding

The authors acknowledge funding from Department of Biotechnology, Govt. of India and the Delhi University-UGC R&D grant.

References

[1] H. Gu, P. Zhu, Y. Jiao, Y. Meng, M. Chen, PRIN: a predicted rice interactome network, BMC Bioinf. 12 (2011) 161, http://dx.doi.org/10.1186/1471-2105-12-161.
[2] D. Wang, Y. Xia, X. Li, L. Hou, J. Yu, The Rice Genome Knowledgebase (RGKbase): An annotation database for rice comparative genomics and evolutionary biology, Nucleic Acids Res. 41 (2013) 1199–1205, http://dx.doi.org/10.1093/nar/gks1225.
[3] B. Pan, J. Sheng, W. Sun, Y. Zhao, P. Hao, X. Li, OrysPSSP: a comparative Platform for Small Secreted Proteins from rice and other plants, Nucleic Acids Res. 41 (2013) 1192–1198, http://dx.doi.org/10.1093/nar/gks1090.
[4] Y. Sato, N. Namiki, H. Takehisa, K. Kamatsuki, H. Minami, H. Ikawa, et al., RiceFREND: a platform for retrieving coexpressed gene networks in rice, Nucleic Acids Res. 41 (2013) 1214–1221, http://dx.doi.org/10.1093/nar/gks1122.
[5] Y. Sato, H. Takehisa, K. Kamatsuki, H. Minami, N. Namiki, H. Ikawa, et al., RiceXPro Version 3.0: expanding the informatics resource for rice transcriptome, Nucleic Acids Res. 41 (2013) 1206–1213, http://dx.doi.org/10.1093/nar/gks1125.
[6] T. Lu, X. Huang, C. Zhu, T. Huang, Q. Zhao, K. Xie, et al., RICD: a rice indica cDNA database resource for rice functional genomics, BMC Plant Biol. 8 (2008) 118, http://dx.doi.org/10.1186/1471-2229-8-118.
[7] T. Sakurai, Y. Kondou, K. Akiyama, A. Kurotani, M. Higuchi, T. Ichikawa, et al., RiceFOX: a database of arabidopsis mutant lines overexpressing rice full-length cDNA that contains a wide range of trait information to facilitate analysis of gene function, Plant Cell Physiol. 52 (2011) 265–273, http://dx.doi.org/10.1093/pcp/pcq190.
[8] N. Kurata, Y. Yamazaki, Oryzabase. An integrated biological and genome information database for rice, Plant Physiol. 140 (2006) 12–17, http://dx.doi.org/10.1104/pp.105.063008.
[9] H. Sakai, S.S. Lee, T. Tanaka, H. Numa, J. Kim, Y. Kawahara, et al., Rice annotation project database (RAP-DB): An integrative and interactive database for rice genomics, Plant Cell Physiol. 54 (2013), http://dx.doi.org/10.1093/pcp/pcs183.
[10] Z. Zhang, J. Sang, L. Ma, G. Wu, H. Wu, D. Huang, et al., RiceWiki: a wiki-based database for community curation of rice genes, Nucleic Acids Res. 42 (2014) 1222–1228, http://dx.doi.org/10.1093/nar/gkt926.
[11] G. Droc, M. Ruiz, P. Larmande, A. Pereira, P. Piffanelli, J.B. Morel, et al., OryGenesDB: a database for rice reverse genetics, Nucleic Acids Res. 34 (2006) D736–D740, http://dx.doi.org/10.1093/nar/gkj012.
[12] P. Larmande, C. Gay, M. Lorieux, C. Périn, M. Bouniol, G. Droc, et al., Oryza Tag Line, a phenotypic mutant database for the Génoplante rice insertion line library, Nucleic Acids Res. 36 (2008) 1022–1027, http://dx.doi.org/10.1093/nar/gkm762.
[13] R. Narsai, J. Devenish, I. Castleden, K. Narsai, L. Xu, H. Shou, et al., Rice DB: an Oryza Information Portal linking annotation, subcellular location, function, expression, regulation, and evolutionary information for rice and Arabidopsis, Plant J. 76 (2013) 1057–1073, http://dx.doi.org/10.1111/tpj.12357.
[14] P. Jaiswal, D. Ware, J. Ni, K. Chang, W. Zhao, S. Schmidt, et al., Gramene: development and integration of trait and gene ontologies for rice, Comp. Funct. Genomics. 3 (2002) 132–136, http://dx.doi.org/10.1002/cfg.156.
[15] P. Gour, P. Garg, R. Jain, S.V. Joseph, A.K. Tyagi, S. Raghuvanshi, Manually curated database of rice proteins, Nucleic Acids Res. 42 (2014).
[16] J.G. Dubouzet, Y. Sakuma, Y. Ito, M. Kasuga, E.G. Dubouzet, S. Miura, et al., OsDREB genes in rice, Oryza sativa L., encode transcription activators that function in drought-, high-salt- and cold-responsive gene expression, Plant J. 33 (2003) 751–763, http://www.ncbi.nlm.nih.gov/pubmed/12609047 (Accessed October 27, 2016).
[17] W. Yang, Z. Kong, E. Omo-Ikerodah, W. Xu, Q. Li, Y. Xue, Calcineurin B-like interacting protein kinase OsCIPK23 functions in pollination and drought stress responses in rice (Oryza sativa L.), J. Genet. Genom. 35 (531-543) (2008) S1–S2, http://dx.doi.org/10.1016/S1673-8527(08)60073-9.
[18] L. Zhang, L.-H. Tian, J.-F. Zhao, Y. Song, C.-J. Zhang, Y. Guo, Identification of an apoplastic protein involved in the initial phase of salt stress response in rice root by two-dimensional electrophoresis, Plant Physiol. 149 (2009) 916–928, http://dx.doi.org/10.1104/pp.108.131144.
[19] S.-J. Sun, S.-Q. Guo, X. Yang, Y.-M. Bao, H.-J. Tang, H. Sun, et al., Functional analysis of a novel Cys2/His2-type zinc finger protein involved in salt tolerance in rice, J. Exp. Bot. 61 (2010) 2807–2818, http://dx.doi.org/10.1093/jxb/erq120.
[20] S. Matsukura, J. Mizoi, T. Yoshida, D. Todaka, Y. Ito, K. Maruyama, et al., Comprehensive analysis of rice DREB2-type genes that encode transcription factors involved in the expression of abiotic stress-responsive genes, Mol. Genet. Genom. 283 (2010) 185–196, http://dx.doi.org/10.1007/s00438-009-0506-y.
[21] M. Cui, W. Zhang, Q. Zhang, Z. Xu, Z. Zhu, F. Duan, et al., Induced over-expression of the transcription factor OsDREB2A improves drought tolerance in rice, Plant Physiol. Biochem. PPB 49 (2011) 1384–1391, http://dx.doi.org/10.1016/j.plaphy.2011.09.012.
[22] Y. Ning, C. Jantasuriyarat, Q. Zhao, H. Zhang, S. Chen, J. Liu, et al., The SINA E3 ligase OsDIS1 negatively regulates drought response in rice, Plant Physiol. 157 (2011) 242–255, http://dx.doi.org/10.1104/pp.111.180893.
[23] G. Mallikarjuna, K. Mallikarjuna, M.K. Reddy, T. Kaul, Expression of OsDREB2A transcription factor confers enhanced dehydration and salt stress tolerance in rice (Oryza sativa L.), Biotechnol. Lett. 33 (2011) 1689–1697, http://dx.doi.org/10.1007/s10529-011-0620-x.
[24] J. You, H. Hu, L. Xiong, An ornithine δ-aminotransferase gene OsOAT confers drought and oxidative stress tolerance in rice, Plant Sci. 197 (2012) 59–69, http://dx.doi.org/10.1016/j.plantsci.2012.09.002.
[25] A. Yang, X. Dai, W.-H. Zhang, A R2R3-type MYB gene, OsMYB2, is involved in salt, cold, and dehydration tolerance in rice, J. Exp. Bot. 63 (2012) 2541–2556, http://dx.doi.org/10.1093/jxb/err431.
[26] Z. Lang, S. Xie, J.-K. Zhu, X.-J. He, M.W. Horton, et al., The 1001 arabidopsis DNA methylomes: an important resource for studying natural genetic, epigenetic, and phenotypic variation, Trends Plant Sci. 21 (2016) 906–908, http://dx.doi.org/10.1016/j.tplants.2016.09.001.

5

The Bio-Analytic Resource: Data visualization and analytic tools for multiple levels of plant biology☆

Jamie Waese, Nicholas J. Provart*

Cell and Systems Biology, University of Toronto, Toronto, Ontario M5S 3B2, Canada

ARTICLE INFO

Keywords:
Arabidopsis thaliana
Data visualization
Transcriptomics
Protein-protein interactions
Protein-DNA interactions
cis-regulatory elements
Natural variation
Agriculture

ABSTRACT

The Bio-Analytic Resource for Plant Biology (BAR) is a portal for accessing large data sets from approximately 15 different plant species, with a focus on transcriptomic, protein-protein interaction, and promoter data. It consists of numerous databases for which its curators have added useful metadata, data visualization tools to display the query results from these databases, and visual analytic tools to identify *e.g.* gene expression patterns of interest based on publicly-available data. We briefly cover some of these tools and scenarios in which they might be useful for plant researchers.

1. Introduction

Researchers are collecting unprecedented quantities of data from virtually every level of plant biology. Many journals and granting agencies now require that all data reported in published works be made available publicly. These open data have transformed biology research workflows because they make it possible to generate and test hypotheses *in silico* before validating them in the lab, as a complement or alternative to traditional genetic screens or molecular approaches. Typically, large data sets are generated and curated by the research labs that create them, and are subsequently deposited upon publication in infrastructure-tier databases like those run by the National Center for Biotechnology Information, NCBI, or the European Bioinformatics Institute, EBI. They are easily downloadable from these repositories, but some expertise in bioinformatics and data science is needed to manipulate, interpret and analyse them. Web-based data visualization tools that rely on publicly-available data sets can help researchers answer many questions faster than if they had to acquire and analyse the raw data themselves.

The Bio-Analytic Resource for Plant Biology (BAR; http://bar.utoronto.ca) offers a combination of data visualization tools (53 in total) and database web services (15 in total) for exploring a wide variety of large data sets covering multiple levels of plant biology. It aggregates published data from diverse sources and provides user friendly tools for accessing and analysing them without having to download anything or write custom scripts. Many of the tools combine data from various sources into the same view, such as the ePlant Molecule Viewer [1], which paints Pfam [2] domains, CDD [3] protein motifs, and polymorphisms causing non-synonymous changes from the 1001 Proteomes [4] site directly onto Phyre2- predicted [5] tertiary structures. Fig. 1 shows a CDD "DNA binding site" motif mapped onto the partial structure of the ABI3 transcription factor, along with nearby non-synonymous changes, which may potentially affect DNA binding in the ecotypes where those polymorphisms occur. The BAR has been online since 2005 [6] and had an average of 60,000 uses per month in 2015. Roughly 2300 published papers have cited one or more of our visual analytic tools. The following sections describe some of the more popular tools on the BAR and how they may apply to one's own research.

2. Data and web services

The BAR currently consists of 145.2 million gene expression measurements from several agronomically-important plant species (poplar, maize, soybean, tomato, potato, grape, peanut, *Medicago truncatula*, *Camelina sativa*, rice, barley, and triticale) and

☆ This article is part of a special issue entitled "Genomic resources and databases", published in the journal Current Plant Biology 7–8, 2016.
* Corresponding author. Present address: Department of Cell and Systems Biology, Centre for the Analysis of Genome Evolution and Function, University of Toronto, 25 Willcocks St., Toronto, Ontario M5S 3B2, Canada.
E-mail address: nicholas.provart@utoronto.ca (N.J. Provart).

Molecule Viewer: AT3G24650 / ABI3, SIS10

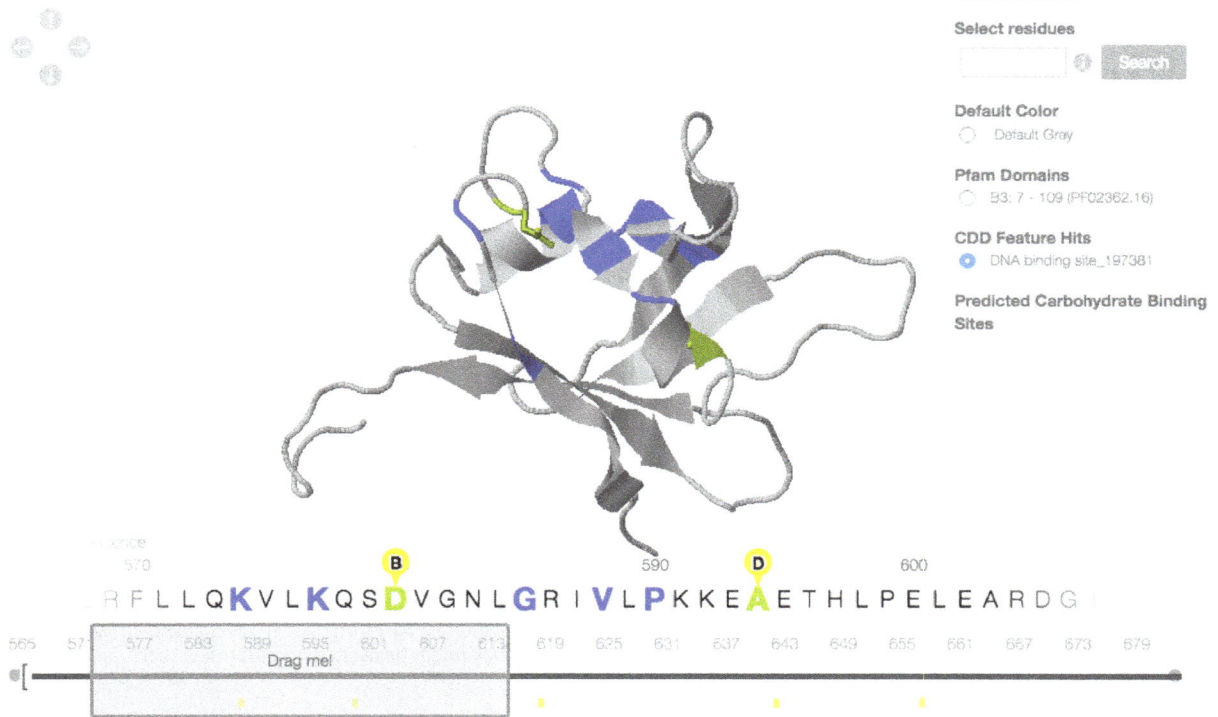

Fig. 1. BAR ePlant Molecule Viewer showing the transcription factor ABI3's partial tertiary structure with its DNA binding site highlighted in blue (identified using motifs from the CDD database), and two non-synonymous changes caused by polymorphisms (based on data from the 1001 Proteomes site) highlighted in green.

the model plants *Arabidopsis thaliana* and moss, across tens of thousands of genes; documented subcellular localizations for more than 9300 proteins in Arabidopsis from the SUBA3 database [7]; 70,944 predicted [8] and 36,306 documented Arabidopsis protein-protein interactions (from many publications); 123,484 SNP polymorphisms across 96 Arabidopsis accessions (in addition to millions available *via* web services from the 1001 Proteomes [4] site); and 29,180 Arabidopsis Phyre2-predicted [5] protein tertiary structures covering ~84% of the Arabidopsis proteome, along with 885 experimentally-determined protein tertiary structures from the Protein Data Bank. Many of these data sets are accessible *via* web services that return JSON (Javascript Object Notation)-formatted results. See http://bar.utoronto.ca/webservices/ for details on using the BAR's web services.

3. Data visualization and analytic tools

3.1. ePlant

ePlant [1] helps biologists visualize the natural connections between DNA sequences, natural variation (polymorphisms), molecular structures, protein-protein interactions, and gene expression patterns by combining several data visualization tools in a "zoomable user interface". ePlant connects to several publicly-available web services to download the latest genome, interactome, and transcriptome data for any number of genes or gene products of interest. Data are displayed with a set of visualization tools that are presented as a conceptual hierarchy from "big" to "small". Links between the different views help underscore connections between multiple levels of analysis. It is based on an earlier version of ePlant by Fucile et al. [9] and was developed using an agile software development process that included several rounds of user testing to ensure user-centered design. Fig. 2 shows the ePlant Heat Map

Viewer, which displays all the expression levels for all the samples of all the genes that are loaded (in the figure, it's all genes that exhibit the same expression pattern as *ABI3*). From here, users may select which gene they wish to look at and zoom in and out from kilometer level to nanometer level data (*i.e.*, natural variation to molecular structure).

3.2. eFP browsers

3.2.1. Arabidopsis eFP browser [10]

Create 17 "electronic fluorescent pictographic" representations of a gene of interest's expression patterns based on open data from 64 publications and dozens of experiments from the AtGenExpress initiative. These data sets encompass around 75 million expression measurements. Some views, such as the Lateral Root Initiation view, have been supported by the community and have taken on a life of their own, with contributors working with the BAR curators to add new data sets as they are published.

3.2.2. Other eFP browsers

We have worked with numerous plant laboratories to create eFP Browsers (based on array or RNA-seq expression atlases) for poplar [11,12], maize [13], soybean [14], tomato [15], potato [16], grape [17], *Medicago truncatula* [18], *Camelina sativa* [19], rice [20], barley [21], triticale [22], the moss *Physcomitrella patens* [23], and most recently for peanut [24].

The BAR has recently released an "eFP-Seq Browser" at http://bar.utoronto.ca/~dev/eFP-Seq_Browser/ for exploring gene expression data generated using the RNA-seq method. Here it is possible to see both the readmap patterns and expression level summaries (as eFP images) in an easily sortable table. The current version taps into more than 100GB of RNA-seq data used re-annotate the Arabidopsis genome for the Araport 11 release [25].

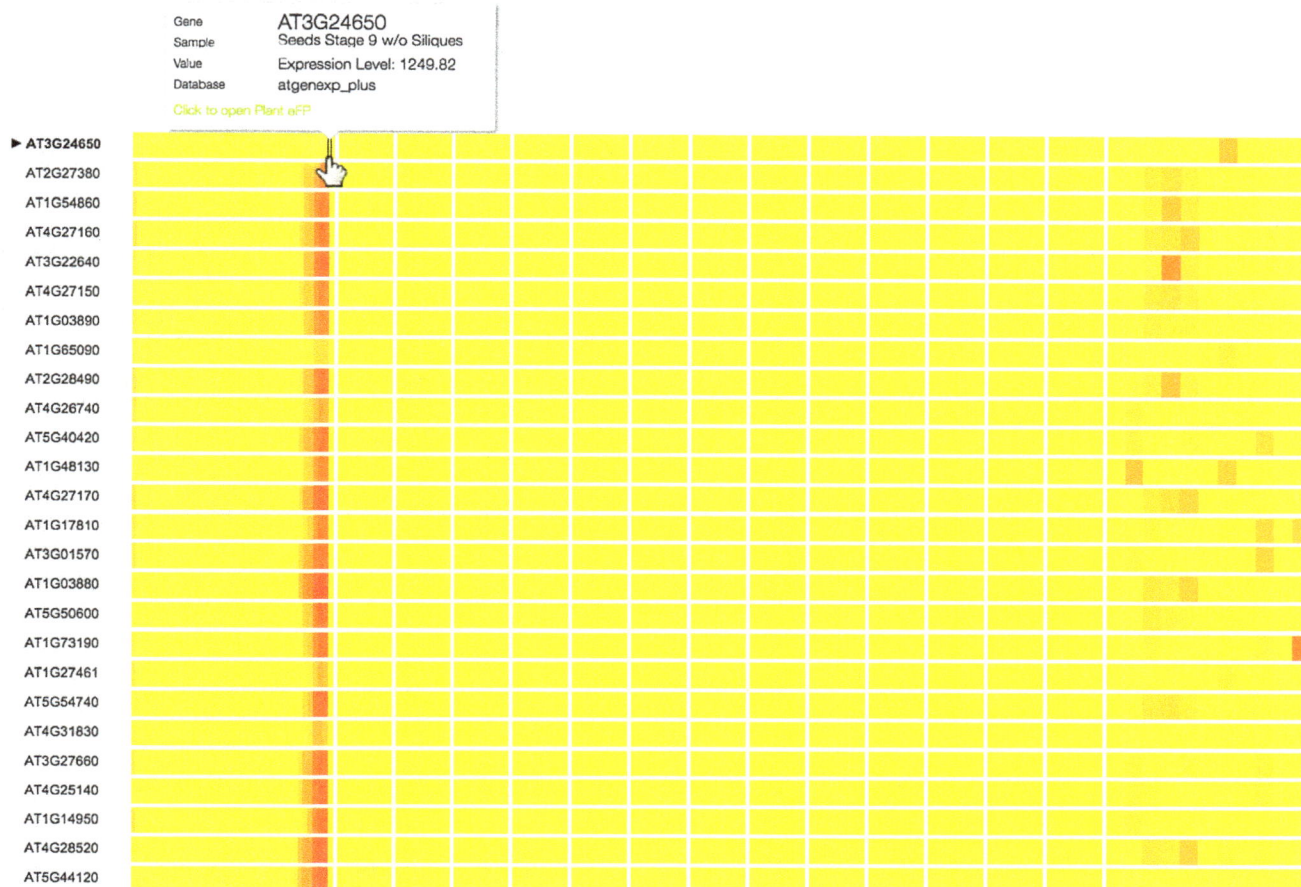

Fig. 2. BAR ePlant Heat Map Viewer showing 350+ expression level samples for twenty-five genes identified with the BAR's Expression Angler tool as having expression patterns similar to *ABI3*'s. The "global" colour gradient has been selected, making it easy to see the similarity in the expression levels of the various genes.

3.3. Other gene expression & protein tools

3.3.1. Arabidopsis interactions viewer [8]

The BAR's Arabidopsis Interactions Viewer provides access to a database of 70,944 predicted [8] and 36,352 experimentally-confirmed Arabidopsis protein-protein interactions. It has been continuously improved and updated since its release in 2007. The most recent innovations include the addition of around 40,000 of protein-DNA interactions, determined either with the yeast one hybrid system [26,27] or predicted using the mapping program FIMO [28] together with hundreds of transcription factor binding specificities from JASPAR [29] and Weirauch et al. [30] Additionally, it is now possible using the AIV to query other protein-protein interaction databases, such as BioGRID [31] and IntAct [32], *via* PSICQUIC [33] web services. There is also a Rice Interactions Viewer, which permits the exploration of 37,472 predicted and 430 experimentally-confirmed rice interacting proteins [34].

3.3.2. Cistome [35]

Cistome is a multi-purpose tool that can predict novel *cis*-elements in the promoters of co-expressed genes using several published *cis*-element prediction programs, or a new version of Promomer [6] called Promomer2. Or, it can be used to map known *cis*-elements from PLACE [36], JASPAR [29], or Weirauch et al. [30]. Cistome can also be used to map transcription factor binding sites based on position-specific scoring matrices or consensus sequences onto the promoters of a defined list of AGI (Arabidopsis Genome

Initiative) identifiers. The output includes sequence logo representations of the elements identified by the mapping.

3.3.3. Gene slider [37]

Gene Slider helps examine the conservation across aligned DNA or protein sequences by presenting one long "sequence logo" [38] that can be zoomed in and out of, from an overview across the entire sequence length down to a few residues. In addition to displaying user-supplied FASTA alignment files, the version of Gene Slider on the BAR loads and displays data for 90,000+ conserved non-coding regions across the Brassicaceae [39] indexed to the TAIR10 Col-0 Arabidopsis sequence. It also displays transcription factor binding sites from JASPAR [29] and Weirauch et al. [30], enabling easy identification of regions that are both conserved across multiple species and may contain transcription factor binding sites.

3.3.4. Expression angler [6,35]

This tool (or an earlier legacy version) identifies Arabidopsis genes with similar expression patterns by calculating the Pearson correlation coefficients for all gene expression vectors as compared to an expression pattern that one defines with a graphical input tool, or to an expression pattern associated with an AGI ID or gene name that one specifies. It outputs a series of eFP images depicting the expression data for each gene that meets one's correlational cut-off criterion or encompassing a specified number of hits.

4. Future prospects

The BAR is a collaborative effort that demonstrates the power of a web service approach in integrating and visualizing data from multiple sources. It is significantly integrated with Araport [40], both as a provider of visualization "apps" and as a user of their web services for several BAR hosted tools. More integration with other plant portals is underway—eFP images are already served up within MaizeGDB [41] gene pages and on SoyKB [42]. Additional ePlants have been funded through a Genome Canada grant, so researchers may look forward to exploring multiple levels of data about any gene of interest across several species. This work is cross disciplinary, combining bioinformatics with software engineering, user experience design and data visualization best practices. We have adopted an agile development process that includes several rounds of user testing to support our commitment to user-centered design.

Funding

This work was supported by the National Sciences and Engineering Research Council of Canada and Genome Canada. Funding for open access charge: NSERC.

Acknowledgements

We wish to thank Asher Pasha, Jim Fan, Geoff Fucile, and Hans Yu for their contributions to the work described above.

References

[1] Waese, J. et al. ePlant: visualizing and exploring multiple levels of data for hypothesis generation in plant biology. (in preparation).
[2] A. Bateman, et al., The Pfam protein families database, Nucleic Acids Res. 30 (2002) 276–280.
[3] A. Marchler-Bauer, et al., CDD: a database of conserved domain alignments with links to domain three-dimensional structure, Nucleic Acids Res. 30 (2002) 281–283.
[4] H.J. Joshi, et al., Proteomes: a functional proteomics portal for the analysis of Arabidopsis thaliana accessions, Bioinf. Oxf. Engl. 28 (2012) 1303–1306.
[5] L.A. Kelley, S. Mezulis, C.M. Yates, M.N. Wass, M.J.E. Sternberg, The Phyre2 web portal for protein modeling, prediction and analysis, Nat. Protoc. 10 (2015) 845–858.
[6] K. Toufighi, S.M. Brady, R. Austin, E. Ly, N.J. Provart, The botany array resource: e-Northerns, expression angling, and promoter analyses, Plant J. 43 (2005) 153–163.
[7] J.L. Heazlewood, R.E. Verboom, J. Tonti-Filippini, I. Small, A.H. Millar, SUBA: the arabidopsis subcellular database, Nucleic Acids Res. 35 (2007) D213–D218.
[8] J. Geisler-Lee, et al., A predicted interactome for Arabidopsis, Plant Physiol. 145 (2007) 317–329.
[9] G. Fucile, et al., ePlant and the 3D data display initiative: integrative systems biology on the world wide web, PLoS One 6 (2011), e15237.
[10] D. Winter, et al., An 'electronic fluorescent pictograph' browser for exploring and analyzing large-scale biological data sets, PLoS One 2 (2007), e718.
[11] O. Wilkins, H. Nahal, J. Foong, N.J. Provart, M.M. Campbell, Expansion and diversification of the Populus R2R3-MYB family of transcription factors, Plant Physiol. 149 (2009) 981–993.
[12] O. Wilkins, L. Waldron, H. Nahal, N.J. Provart, M.M. Campbell, Genotype and time of day shape the Populus drought response, Plant J. 60 (2009) 703–715.
[13] L. Wang, et al., Comparative analyses of C4 and C3 photosynthesis in developing leaves of maize and rice, Nat. Biotechnol. 32 (2014) 1158–1165.
[14] A.J. Severin, et al., RNA-Seq atlas of glycine max: a guide to the soybean transcriptome, BMC Plant Biol. 10 (2010) 160.
[15] T.T.G. Consortium, The tomato genome sequence provides insights into fleshy fruit evolution, Nature 485 (2012) 635–641.
[16] A.N. Massa, et al., The transcriptome of the reference potato genome Solanum tuberosum group phureja clone DM1-3 516R44, PLoS One 6 (2011), e26801.
[17] M. Fasoli, et al., The grapevine expression atlas reveals a deep transcriptome shift driving the entire plant into a maturation program[W][OA], Plant Cell 24 (2012) 3489–3505.
[18] V.A. Benedito, et al., A gene expression atlas of the model legume Medicago truncatula, Plant J. 55 (2008) 504–513.
[19] S. Kagale, et al., The emerging biofuel crop Camelina sativa retains a highly undifferentiated hexaploid genome structure, Nat. Commun. 5 (2014).
[20] M. Jain, et al., F-Box proteins in rice. genome-wide analysis, classification, temporal and spatial gene expression during panicle and seed development, and regulation by light and abiotic stress, Plant Physiol. 143 (2007) 1467–1483.
[21] A. Druka, et al., An atlas of gene expression from seed to seed through barley development, Funct. Integr. Genomics 6 (2006) 202–211.
[22] F. Tran, et al., Developmental transcriptional profiling reveals key insights into Triticeae reproductive development, Plant J. 74 (2013) 971–988.
[23] C. Ortiz-Ramírez, et al., A transcriptome atlas of Physcomitrella patens provides insights into the evolution and development of land plants, Mol. Plant (2016), http://dx.doi.org/10.1016/j.molp.2015.12.002.
[24] J. Clevenger, Y. Chu, B. Scheffler, P.A. Ozias-Akins, Developmental transcriptome map for allotetraploid Arachis hypogaea, Plant Genet. Genomics (2016) 1446, http://dx.doi.org/10.3389/fpls.2016.01446.
[25] C.-Y. Cheng, V. Krishnakumar, A. Chan, S. Schobel, C.D. Town, Araport11: a complete reannotation of the Arabidopsis thaliana reference genome, BioRxiv (2016) 047308, http://dx.doi.org/10.1101/047308.
[26] A. Gaudinier, et al., Enhanced Y1H assays for Arabidopsis, Nat. Methods 8 (2011) 1053–1055.
[27] M. Taylor-Teeples, et al., An Arabidopsis gene regulatory network for secondary cell wall synthesis, Nature 517 (2015) 571–575.
[28] C.E. Grant, T.L. Bailey, W.S. Noble, FIMO: scanning for occurrences of a given motif, Bioinformatics 27 (2011) 1017–1018.
[29] A. Mathelier, et al., JASPAR 2014: an extensively expanded and updated open-access database of transcription factor binding profiles, Nucleic Acids Res. (2013), http://dx.doi.org/10.1093/nar/gkt997, gkt997.
[30] M.T. Weirauch, et al., Determination and inference of eukaryotic transcription factor sequence specificity, Cell 158 (2014) 1431–1443.
[31] A. Chatr-Aryamontri, et al., The BioGRID interaction database: 2015 update, Nucleic Acids Res. 43 (2015) D470–D478.
[32] S. Kerrien, et al., The IntAct molecular interaction database in 2012, Nucleic Acids Res. 40 (2012) D841–D846.
[33] B. Aranda, et al., PSICQUIC and PSISCORE: accessing and scoring molecular interactions, Nat. Methods 8 (2011) 528–529.
[34] C.-L. Ho, Y. Wu, H. Shen, N.J. Provart, M. Geisler, A predicted protein interactome for, Rice 5 (2012) 15.
[35] R.S. Austin, et al., New BAR tools for mining expression data and exploring Cis-elements in Arabidopsis thaliana, Plant J. Cell Mol. Biol. (2016), http://dx.doi.org/10.1111/tpj.13261.
[36] K. Higo, Y. Ugawa, M. Iwamoto, H. Higo, PLACE: A database of plant cis-acting regulatory DNA elements, Nucleic Acids Res. 26 (1998) 358–359.
[37] J. Waese, et al., Gene slider: sequence logo interactive data-visualization for education and research, Bioinf. Oxf. Engl. (2016), http://dx.doi.org/10.1093/bioinformatics/btw525.
[38] T.D. Schneider, R.M. Stephens, Sequence logos: a new way to display consensus sequences, Nucleic Acids Res. 18 (1990) 6097–6100.
[39] A. Haudry, et al., An atlas of over 90,000 conserved noncoding sequences provides insight into crucifer regulatory regions, Nat. Genet. 45 (2013) 891–898.
[40] V. Krishnakumar, et al., Araport: the Arabidopsis information portal, Nucleic Acids Res. (2014), http://dx.doi.org/10.1093/nar/gku1200.
[41] C.M. Andorf, et al., MaizeGDB update: new tools, data and interface for the maize model organism database, Nucleic Acids Res. (2015), http://dx.doi.org/10.1093/nar/gkv1007, gkv1007.
[42] T. Joshi, et al., Soybean knowledge base (SoyKB): a web resource for integration of soybean translational genomics and molecular breeding, Nucleic Acids Res. 42 (2014) D1245–D1252.

6

Introgression of the high grain protein gene *Gpc-B1* in an elite wheat variety of Indo-Gangetic Plains through marker assisted backcross breeding

Manish K. Vishwakarma[a], V.K. Mishra[a], P.K. Gupta[b], P.S. Yadav[a], H. Kumar[a], Arun K. Joshi[a,c,*]

[a] *Department of Genetics and Plant Breeding, Institute of Agricultural Sciences, Banaras Hindu University, Varanasi, India*
[b] *CCS University, Meerut, India*
[c] *CIMMYT, South Asia Regional Office, P.O. Box 5186, Kathmandu, Nepal*

ARTICLE INFO

Keywords:
Grain protein content
Triticum aestivum
Marker assisted selection
Foreground selection
Background selection

ABSTRACT

Grain protein content (GPC) in wheat has been a major trait of interest for breeders since it has enormous end use potential. In the present study, marker-assisted backcrossing (MABC) was successfully used to improve GPC in wheat cultivar HUW468. The genotype Glu269 was used as the donor parent for introgression of the gene *Gpc-B1* that confers high GPC. In a segregating population, SSR marker Xucw108, with its locus linked to *Gpc-B1* was used for foreground selection to select plants carrying *Gpc-B1*. Background selection, involving 86 polymorphic SSR markers dispersed throughout the genome, was exercised to recover the genome of HUW468. For eliminating linkage drag, markers spanning a 10 cM region around the gene *Gpc-B1* were employed to select lines with a donor segment of the minimum size carrying the gene of interest. Improved lines had significantly higher GPC and displayed 88.4–92.3 per cent of the recurrent parent genome (RPG). For grain yield, selected lines were at par with the recurrent parent HUW468, suggesting that there was no yield penalty. The whole exercise of transfer of *Gpc-B1* and reconstitution of the genome of HUW468 was completed within a period of two and half years (five crop cycles) demonstrating practical utility of MABC for developing high GPC lines in the background of any elite and popular wheat cultivar with relatively higher speed and precision.

1. Introduction

Wheat (*Triticum aestivum* L. em. Thell.) is one of the most important food crops in the world with global yield over 700 million tons annually, and providing 20% of the total calorie intake for the world population [1,2]. Wheat is grown in ~128 countries involving all the continents of the world, the top five leading producers being China, India, United States of America, France and Russia [3]. Among cereals involved in cross-border trading, wheat has the highest tonnage, with an estimate value of 135 Mt in 2012/13, with major importing countries being in Asia and Africa [3]. In India, wheat is a staple food for more than 65% of the population with annual production of around 94 Mt [4]. The demand of wheat is expected to keep growing due to steady population increase. The demand for better quality wheat grain will also increase due to increased urbanization [5]. Although, many Indian wheat varieties have been characterized for various end products, these varieties and their traits are spillovers of the routine breeding program for high yield and disease resistance, rather than the product of a systematic quality breeding program [6]. The current challenges for wheat breeding programs around the world are to maintain or improve agronomic performance along with improvement in wheat quality, thus maintaining competitiveness in the increasingly discriminating international market [7].

Wheat is a crop with several end-use products such as pasta, macaroni, biscuit, chapatti and bread. These end-use products differ for their requirements of GPC and the type of wheat. In general, GPC in Indian wheat cultivars is relatively lower than the standards of international market [7]. Under these conditions, either we need to accept wheat with a lower GPC or apply more nitrogen to achieve the desired level of GPC, since some increase in GPC through increased nitrogen application has been documented [8].

* Corresponding author.
   E-mail address: a.k.joshi@cgiar.org (A.K. Joshi).

GPC is also an important for bread-making quality, which is known to depend upon both, the content and composition of grain protein [9,10].

According to an estimate of World Health Organization [11], over 3 billion people were deficient in key micronutrients Zn and Fe, and about 160 million children below the age of 5 lack adequate protein, amounting to malnutrition [11–13]; this suggests that not only GPC, but also the content of micronutrients like Zinc (Zn) and Iron (Fe) need to be improved for improving the grain quality of wheat. Progress in breeding for high GPC wheat has been rather slow, because GPC is controlled by a complex genetic system and is also influenced by the environment, thus making it difficult to select effectively for this trait [8,14,15]. However, GPC and grain yield are reported to be negatively correlated [8,14], making it difficult to breed for high GPC without a yield penalty. Although the theoretical basis for this inverse correlation has been debated [16], high GPC cereals are unlikely to be commercially successful without a financial incentive to growers.

Yield is an essential trait for commercial success of a variety, hence developing wheat varieties combining improved grain quality with high grain yield is an important goal in wheat breeding. However specific quality parameters such as protein %, grain hardness, bread loaf volume and biscuit spread factor are getting increased attention due to growing demand for industrial end-products such as bread, biscuit, cake, pasta, etc. Wheat varieties with high GPC (>12%) and micronutrients (Zinc and Iron) are also important for providing nutritionally improved wheat based diet and for enhancing export potential of wheat. In addition, high yielding wheat with superior internal (protein %) and external (grain weight, luster) traits is easy to market and may provide extra cash to poor farmers. In India, although wheat is overwhelmingly consumed in the form of chapatti [7], the demand for other end-products like bread, biscuit, pasta and cakes is growing with expanding urbanization (estimated urban population in 2020 = 550 million) and growing industrialization [5]. Therefore, it is important to combine the high grain yield with better grain quality to meet the twin challenges of nutritionally superior and high quality wheat products [6].

In the recent past, the introgression and pyramiding of major genes/QTL for different traits through marker-assisted selection (MAS) has proved successful in wheat [17–24]. Several RFLP, SSR and CAPS markers were reported to be closely linked with high GPC locus (*Gpc-B1*) on the short arm of chromosome 6B [25–28]. Among these markers, a tightly linked marker at a narrow distance of 0.1 cM within a physical location of a 250 kb, was the SSR marker Xuhw89 for the locus *Gpc-B1* [29]. Since *Gpc B1* has been cloned and characterized, a "gene-specific" marker is also available for this locus [30]. The introgression of *Gpc-B1* has been achieved for improving GPC without yield penalty mostly in the developed countries [8,31], although a report of successful introgression of *Gpc-B1* in 10 elite varieties of India is also available [23].

Conventional breeding program, if supplemented with MAS, can become cost and time-effective [20]. For the last more than 20 years, MAS is being used on a large-scale in several countries including USA, Australia, Canada, and Mexico (CIMMYT). In majority of these MAS programs in wheat, MABC involving backcrossing has been deployed to ensure maximum recovery of the genome and particularly, the carrier chromosome [32]. According to a recent report, more than 60 genes/markers are being deployed for wheat improvement through MAS [33], of which more than 20 traits/genes belong to grain quality like gain hardness, dough strength and swelling volume [34]. Molecular markers for quality traits (protein content, pre-harvest sprouting tolerance, gluten strength and grain weight) are also being increasingly used in Indian wheat breeding program successfully [19,22–24].

The present study was planned to improve GPC through MABC coupled with stringent phenotypic selection into the genetic background of wheat cultivar HUW468, which is a very promising cultivar with good performance under conventional and zero-tillage conditions in the North Eastern Plains Zone (NEPZ) of India [6].

2. Materials and methods

2.1. Plant materials

Plant materials used in the present work included recipient parent HUW468 and a donor parent Glu269 with high GPC procured from Punjab Agriculture University, Ludhiana, Punjab. HUW468 (CPAN-1962//TONI/LIRA'S'/PRL'S'). Glu269 is a high yielding, disease resistant and double dwarf wheat variety released in 1999 for timely sown high fertility irrigated conditions of NEPZ and since then has maintained resistance and popularity among farmers due to its superior agronomic performance. The donor parent Glu269 (DBW16/GluPro//2*DBW16) is a wheat breeding line that is resistant to yellow and brown rusts and is also amenable to late sowings. It also carries higher level of resistance to spot botch relative to all the existing varieties, and has been identified for cultivation in the North Western Plains Zone (NWPZ) of India. Since, it had GluPro as one the parents in its pedigree, it carries *Gpc-B1* providing higher level of GPC (>14%) relative to the recurrent parent HUW468 with only 10% GPC.

2.2. Molecular marker used

Seven GPC linked markers (Xuhw89, QGpc.ccsu-2D.1/2DL, CAPS/ASA/XNor-B2, Xwmc415, Xucw108 and Xucw109) were validated based on published results [26,29,30,35,36]. Out of seven, only one (Xucw108) [30] was selected for foreground selection. Primers were synthesized from Eurofins Genomics India Pvt Ltd., Bangaluru, India. To analyze the recovery of RPG of the segregating backcross progeny during background selection, a total of 744 SSR (simple sequence repeats) markers covering all the 21 chromosomes of wheat were selected to detect polymorphism between HUW468 and Glu269. The primer sequences were obtained from http://wheat.pw.usda.gov/GG2/index.shtml (Grain-Genes 2.0: A database for Triticeae and Avena), Somers et al. [37] and Röder et al. [38]. Primers that were polymorphic between parents were used for background selection.

2.3. MABC breeding

2.3.1. DNA isolation, PCR conditions and electrophoresis

DNA isolation of parental genotypes and backcross progenies was carried out from one-month-old plants using a modified CTAB method [39]. The PCR amplification was carried out in a reaction mixture of 20 μL containing 200 μM dNTPs (MBI; Fermentas, Lithuania, USA), 0.75 U Taq DNA polymerase (MBI; Fermentas, Lithuania, USA), 5 pmole of each primer, 20–30 ng template DNA and 10 X PCR buffer (10 mM Tris, pH 8.4, 50 mM KCl, 1.8 mM MgCl$_2$). PCR cycle consisted of an initial denaturation for 5 min at 94 °C, followed by 40 cycles each with 1 min at 94 °C, 1 min at annealing temperature (which differs for different primers), with a final extension of 7 min at 72 °C. The amplified products were resolved on 2.5% agarose gel for the foreground selection (involving use of gene specific marker Xucw108), and on 10% PAGE (followed by silver staining for visualization) for the background selection (used for RPG recovery).

2.3.2. Breeding scheme

MABC scheme [32] was followed to transfer the *Gpc-B1* gene from Glu269 into the genetic background of HUW468. Recurrent

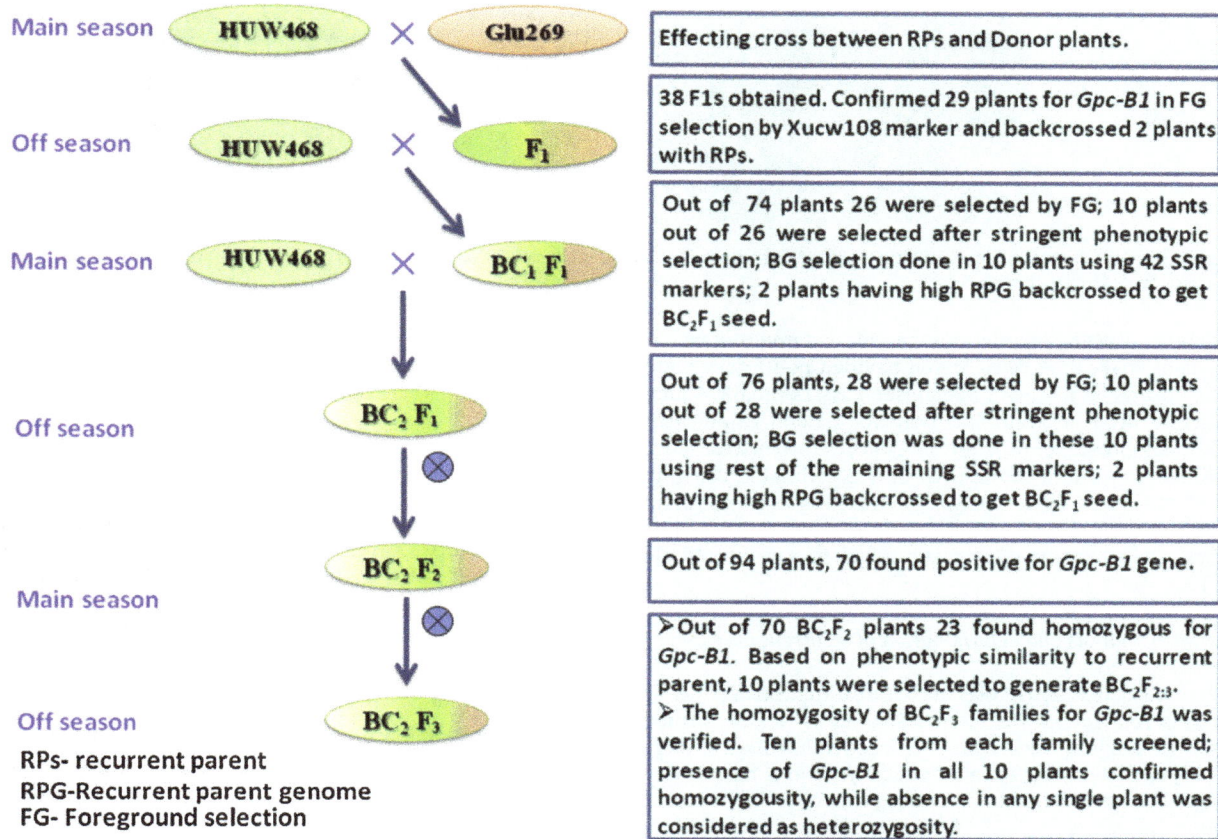

Fig. 1. Development of HUW468 for protein content with details of markers used for foreground and background selection and numbers of plants selected in each generation.

parent HUW468 was used as female and crossed with Glu269 as male to generate the F_1 seeds. The cross was designated HUW468-09. Detailed flowchart of MABC approach deployed to transfer Gpc-B1 is given in Fig. 1. Following testing hybridity of F_1 plants using gene-linked marker, two true F_1 plants were crossed with the recurrent parent to obtain BC_1F_1 seeds. Both foreground and background selections were deployed. Foreground selection for a trait using molecular marker facilitates identification of positive plants for the gene of interest at early plant stage and thus enables a breeder to reduce the population size by around 50% in a backcross breeding program [40,41].

Foreground selection for desirable BC_1F_1 plants with Gpc-B1 gene was exercised using gene-linked marker. The plants possessing Gpc-B1 were subjected to further phenotypic selection to identify top ten plant with desirable recurrent parent phenotype (RPP) for analyzing RPG recovery. One selected plant with high RPG recovery was then backcrossed to produce BC_2F_1 seed. Like BC_1F_1, foreground selection (for Gpc-B1), phenotypic selection (for plants having agronomic similarity to recurrent parent) and background selections (for RPG recovery) were again exercised to identify suitable plants for obtaining BC_2F_2 seeds. The selected BC_2F_2 plants with high RPG recovery were selfed and advanced up to BC_2F_3 using marker-assisted pedigree method of selection. The homozygosity of BC2F3 families for Gpc-B1 gene was further confirmed by screening of randomly selected 10 plants from each family using the marker Xucw108 associated with Gpc-B1.

2.3.3. Estimation of linkage drag

Analysis of the size of introgressed donor segment on carrier chromosome was conducted to eliminate linkage drag [42]; this was possible through the use of six additional markers (Xgwm132,

Xcfd190, Xgwm193, Xgwm361, Xgwm219 and Xcfd2110) from the 10 cM genomic region on either side of Gpc-B1 marker Xucw108 given in http://wheat.pw.usda.gov/GG2/index.shtml (GrainGenes 2.0: A database for Triticeae and Avena).

2.3.4. Evaluation of MABC lines for agronomic performance

Field trials to evaluate MABC lines were conducted during main Rabi season (2008–2009 to 2011–2012) at Agricultural Research Farm, Institute of Agricultural Sciences, Banaras Hindu University (BHU), Varanasi, Uttar Pradesh, India and during off-season (2009–2011) at IARI, Regional Station, Wellington, Tamil Nadu during 2008–2009 to 2011–2012. In the different backcross generations, plants were hand sown in three rows of 2 meter with row to row spacing of 22.5 and plant to plant distance of 20 cm. After every third progeny row, recurrent parent HUW468 was planted as a check for facilitating morphological evaluation. Recommended agronomic and fertilization (120 kg N: 60 kg P_2O_5 : 40 kg K_2O per ha) practices were followed at both locations and seasons. Zinc was not applied. Full doses of K_2O and P_2O_5 were applied at sowing; nitrogen was supplied in split applications, with 60 kg N per ha at sowing, 30 kg N per ha at the first irrigation (21 days after sowing), and 30 kg N per ha at the second irrigation (45 days after sowing) [56]. Data on phenotypic traits in BC_1F_1, BC_2F_1 and BC_2F_2 generations was recorded on single selected plants that were positive for Gpc-B1 gene, as determined through associated marker. In BC_2F_3 generation, data were taken from ten randomly selected plants from each homozygous family for Gpc-B1 identified in BC_2F_2. Following agronomic traits were used for recording data on phenotypes: days to maturity (DM), plant height (PH), number of effective tillers/plant (TP), spike length (SL), spikelet number (SN), thousand grain weight (TGW) and grain yield per plant (GY).

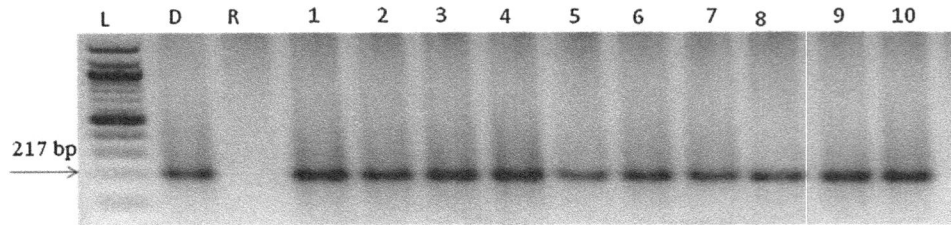

Fig. 2. Molecular profile of BC_2F_3 families of HUW468 × Glu269 cross for confirming homozygosity for *Gpc-B1* by marker Xucw108. L-100 bp ladder, D-Glu269, R-HUW468, lane 1–10 for each BC_2F_3 families.

2.3.5. Evaluation of MABC lines for protein, zinc and iron content

The seeds of BC_2F_3 lines of HUW468 having *Gpc-B1* gene were analyzed by Infratec™ 1241 Grain Analyser, Foss, Denmark. A total of 5–7 g sample of clean grain was used for Zn and Fe analysis, based on X-ray fluorescence (X-Supreme8000, Oxford Instruments, Oxford, UK). Grain protein content (GPC) was estimated using an Infratec 1241 Grain Analyser (Foss, Hilleroed, Denmark). Protein data was recorded at 12% of seed moisture level in (%) unit, while data on Zn and Fe content were taken as particle per molecule (ppm) [43].

2.3.6. Statistical analysis and determination of recurrent parent genome recovery

The data recorded as above was used for estimation of progeny means for each replication. The statistical analysis was performed on the basis of progeny means in each replication by the PAST software [44] and Microsoft excel. The extent of RPG recovery in backcross generations was calculated using the following formula.

$$RPG\% = \frac{2(R) + (H)}{2N} \times 100$$

where, R = number of marker loci homozygous for recurrent parent allele; H = number of marker loci still remaining heterozygous and N = total number of polymorphic markers used for background analysis. In BC_2F_3, the genetic similarity between the recurrent parent HUW468 and the high GPC introgressed lines was determined through data on morphological features. To check the robustness of the clustering, boot-strap analysis was carried out. Further graphical genotyping was done by using GGT software [45].

3. Results

3.1. MABC for GPC

Of seven markers for *Gpc-B1*, only two (Xucw108 and Xucw109) were found polymorphic between parents. The pipeline (crossing program) followed for MABC to incorporate *Gpc-B1* into HUW468 is presented in Fig. 1. Details of 10 phenotypically superior plants each among 26 BC_1F_1 and 28 BC_2F_1 plants carrying *Gpc-B1* are available in Supplementary Tables S1 and S2, respectively. The homozygosity of BC_2F_3 families for *Gpc-B1* by screening of randomly selected 10 plants from each family using the marker Xucw108 is given in Fig. 2.

3.2. Grain quality in BC₂F₃

A summary of data on grain quality of 23 BC_2F_3 lines is presented in Table 1. The GPC of the BC_2F_3 lines ranged from 13 to 17.2% compared to 10% in HUW468 and 14.3% in donor parent Glu269. All these 23 lines had a significant increase in the mean value of GPC relative to the recurrent parent HUW468. Similarly, in these 23 BC_2F_3 lines, Fe content ranged from 39.3 to 53.8 ppm, while Zn content ranged from 35 to 54.2 ppm, the Zn and Fe in the recurrent parent HUW468 being 39 and 30 ppm respectively. For Zn and Fe

content, all the 23 improved lines except HUW468-09-233 for Fe were superior to the recurrent parent HUW468, and only one line (HUW468-09-131) exceeded the level in the donor.

3.3. Evaluation of MABC lines for agronomic performances

The 23 BC_2F_3 MABC lines showed variable expression of agronomic traits. The plant height among the families ranged from 80-92 cm compared to 85.7 cm of HUW468 (Table 1). Six lines showed significantly higher tillers/plant relative to both parents HUW468 and Glu269. The number of spikelet per spike ranged from 41 to 51, while only one line HUW468-09-6 (12 cm) showed better spike length relative to HUW468 (11.2 cm). The TGW among selected families ranged from 37.8 to 42.4 g as against 35.10 g in HUW468. For TGW, all the 23 lines were better than the recurrent parent HUW468, but none was better than the donor Glu269. For yield per plant, 14 families showed significant improvement over HUW468 (7.85 g), while only one (HUW468-09-95) was better than the donor parent. On the basis of overall performance, three lines (HUW468-09-131, HUW468-09-132 and HUW468-09-59) were considered to be superior to the recurrent parent HUW468.

3.4. Recurrent parent genome recovery

3.4.1. Recovery for whole genome

Out of 744 SSR markers, 106 (14.0%) were polymorphic between both the parents. However, on the basis of the difference in product size (eliminated ≤10 bp difference), 86 SSR (11.5%) were selected for background analysis of 23 BC_2F_3 lines (Table 1); these 86 markers covered all the 21 chromosomes. The per cent RPG recovery among the 23 selected lines ranged from 90.7% (HUW468-09-3 and HUW468-09-6) to 95.4% (HUW468-09-244) (Table 1).

3.4.2. Carrier chromosome recovery with the minimum size of donor segment

As expected, donor segments were present in all the 23 improved lines, and a segment carrying the gene *Gpc-B1* was available in all the lines; this segment was not the minimum possible, so that some linkage drag was unavoidable. A screening for recombinants carrying the minimum size of donor fragment was undertaken using six markers, which were the only polymorphic markers among the 48 flanking markers that were tested for the *Gpc-B1* region. These six markers included Xgwm132, Xcfd190, Xgwm193, Xgwm361, Xgwm219 and Xcfd2110 (Fig. 3). Based on carrier chromosome analysis of 23 lines using these six markers, four (HUW468-09-96, HUW468-09-131, HUW468-09-132 and HUW468-09-244) were such, which did not carry any segment other than the small segment carrying *Gpc-B1*; all other lines carried additional segments, away from *Gpc-B1*.

4. Discussion

Breeding for agronomic and nutritional traits of wheat continues to be important for food security and human health especially in developing countries of south Asia, where demand of wheat is

Table 1

Quality characters and agronomic performance of BC_2F_3 families of the cross HUW 468 × Glu269 with high *GPC-B1* in the background of recurrent parent HUW468.

Grain quality				Agronomic traits							%RPG
Lines/traits	GPC (%)	Zn (ppm)	Fe (ppm)	DM	PH (cm)	SPN	SPL	TLN	TGW	YPP	
HUW468-09-3	14.5	45.8	47.5	107	86	47	10	13	38.80	7.56	90.70
HUW468-09-5	13.5	43.5	53.0	102	92	41	9	12	37.84	8.23	91.86
HUW468-09-6	13.0	43.0	52.3	109	89	45	12	16	41.44	8.45	90.70
HUW468-09-29	13.8	44.3	41.0	105	82	46	10	16	40.04	8.41	91.86
HUW468-09-30	13.8	44.5	43.0	104	80	45	9	16	41.94	8.50	91.86
HUW468-09-56	14.8	50.2	49.4	104	89	50	9	9	41.56	7.56	91.86
HUW468-09-57	13.6	48.4	47.4	102	89	45	9	10	42.43	8.23	93.02
HUW468-09-59	13.3	42.6	52.5	102	84	48	10	10	41.48	8.45	93.02
HUW468-09-88	13.1	48.0	52.0	107	87	51	11	12	38.96	7.90	91.86
HUW468-09-89	13.2	42.2	35.1	102	82	50	10	13	37.08	7.55	91.86
HUW468-09-95	13.2	47.3	44.8	108	85	47	11	15	41.20	9.24	93.02
HUW468-09-96	13.1	43.5	47.2	110	89	46	10	15	41.76	8.40	94.19
HUW468-09-128	13.8	47.2	46.9	104	84	46	10	9	40.40	8.60	93.02
HUW468-09-131	16.4	53.8	54.2	107	85	47	11	12	41.60	7.90	94.19
HUW468-09-132	14.1	44.6	47.0	107	87	48	10	13	40.84	7.85	94.19
HUW468-09-142	17.2	44.0	49.8	103	86	48	9	11	40.44	8.41	93.02
HUW468-09-144	13.3	46.5	43.1	104	87	50	10	12	39.04	8.50	93.02
HUW468-09-233	13.6	39.3	45.4	106	91	45	11	10	41.94	7.56	93.02
HUW468-09-235	13.6	43.7	46.8	104	91	47	10	12	41.56	8.23	91.86
HUW468-09-236	13.7	41.4	49.2	103	83	48	9	13	42.44	7.56	91.86
HUW468-09-241	13.8	43.3	51.1	104	88	45	10	13	41.44	8.23	91.86
HUW468-09-242	14.1	48.4	52.6	102	90	45	11	12	40.04	8.45	93.02
HUW468-09-244	14.2	47.2	43.3	111	89	47	10	14	41.94	7.90	95.35
HUW468	10.0	39.0	30.0	103	85.7	46.5	11.2	9	35.10	7.85	
Glu 269	14.3	45.2	49.0	120	82.6	53.6	11	11	42.24	8.54	
CD (0.05)	0.43	1.38	1.97	1.17	1.39	0.96	0.36	3.93	0.64	0.19	

GPC (%), grain protein content (per cent); Fe (ppm), iron content; Zn (ppm), zinc content, DM, days to maturity; PH, plant height; SPN, spikelet number; SPL, spikelet length; TLN, number of tillers; TGW, 1000-grain weight; % RPG, per cent recurrent parent genome recovery.

growing due to increasing population and income levels. For a long time, conventional breeding methods have given desired success in wheat improvement in the absence of alternative approaches. However conventional methods are time consuming, needing up to 12 years for the development leading to release of a new variety [46]. It has been suggested that conventional breeding methods, if supplemented with MAS, could help reduce time for development of a new cultivar and offer an approach for reliable improvement of elite breeding material [20]. During the last few years, use of molecular tools has grown substantially due to development of high-throughput markers that can be used in a cost-effective manner [33]. There are several examples of successful use of MAS for introgression or pyramiding of major genes/QTL for different traits in wheat [18,21,22,33]. There are also examples of introgression of *GpcB1* through MAS for improving GPC without yield penalty [8,31], but most of these examples are from developed countries. There are at least two reports of its introgression in elite wheat lines of India, one of them involving pyramiding of eight genes including *Gpc-B1* [23,24]. In the present study, *Gpc-B1* was incorporated into elite variety HUW468 using MABC combined with stringent phenotypic selection within a short period. HUW468 is a popular cultivar of NEPZ of India but has low protein content. The derived lines HUW468 (*Gpc-B1*) showed significant improvement in GPC, Fe and Zn as compared to the recipient parent. In addition, the improved lines were superior in TGW, TLN and YPP, while at par for other agronomic traits. Distelfeld et al. [29] reported that GPC was significantly ($P < 0.001$) correlated with grain Zn and Fe content, since the locus *Gpc-B1* (imparting high GPC) on the short arm of chromosome 6B was also effective in increasing Zn and Fe in the grain. In our study, some MABC lines were found to show higher Zn content compared to Fe. Additive main effects and multiplicative interactions (AMMI) analysis of genotype × environment interactions for grain Fe and Zn content also revealed that Fe content in a large measure is genetically controlled, whereas Zn was almost totally dependent on location effects [47].

4.1. Marker assisted background selection with stringent phenotypic selection

The expected average RPG recovery in the BC_2F_3 generation was 93.75%. However, RPG in the final product (HUW468-09-244) in BC_2F_3 was 95.35% as revealed by background analysis with polymorphic SSR markers. Eighty six (86) genome-wide polymorphic SSR markers distributed at an average interval of 5 Mb (~20 cM) were employed to analyze the recovery of RPG in 23 improved lines in the background of HUW468. The higher recovery of RPG was achieved mainly through stringent phenotypic selection followed by background selection using molecular markers. Background analysis exercised through phenotypic evaluation is reported to be useful in efficient recovery of the RPG [48]. It has been suggested that stringent phenotypic selection for the recovery of recurrent parent phenotype is a good substitute for background selection through molecular markers in terms of judicious use of resources [42,49–51]. Recently, the problem of linkage drag has also been reported in rice when stringent phenotypic selection with MAS was used [51]. In the present study, the progenies selected through foreground markers were subjected to phenotypic selection for background traits followed by background analysis using molecular markers. This approach reduces the cost of MAS and also helps to retain useful interactive loci of both the parents [42]. Cost of molecular genotyping in routine marker-assisted breeding programs is an important factor especially in developing countries.

4.2. Marker assisted step-wise background selection (MASBS)

Most of the MABC studies conducted earlier utilized all the background polymorphic markers in early generation for the recovery of RPG [23]. The use of a large number of markers in a single generation is a tedious job and keeps molecular breeders engaged for a longer period of time. In the present study, we utilized step-wise background selection for RPG recovery. We judicially used two markers per chromosomes and six markers on carrier chromosome

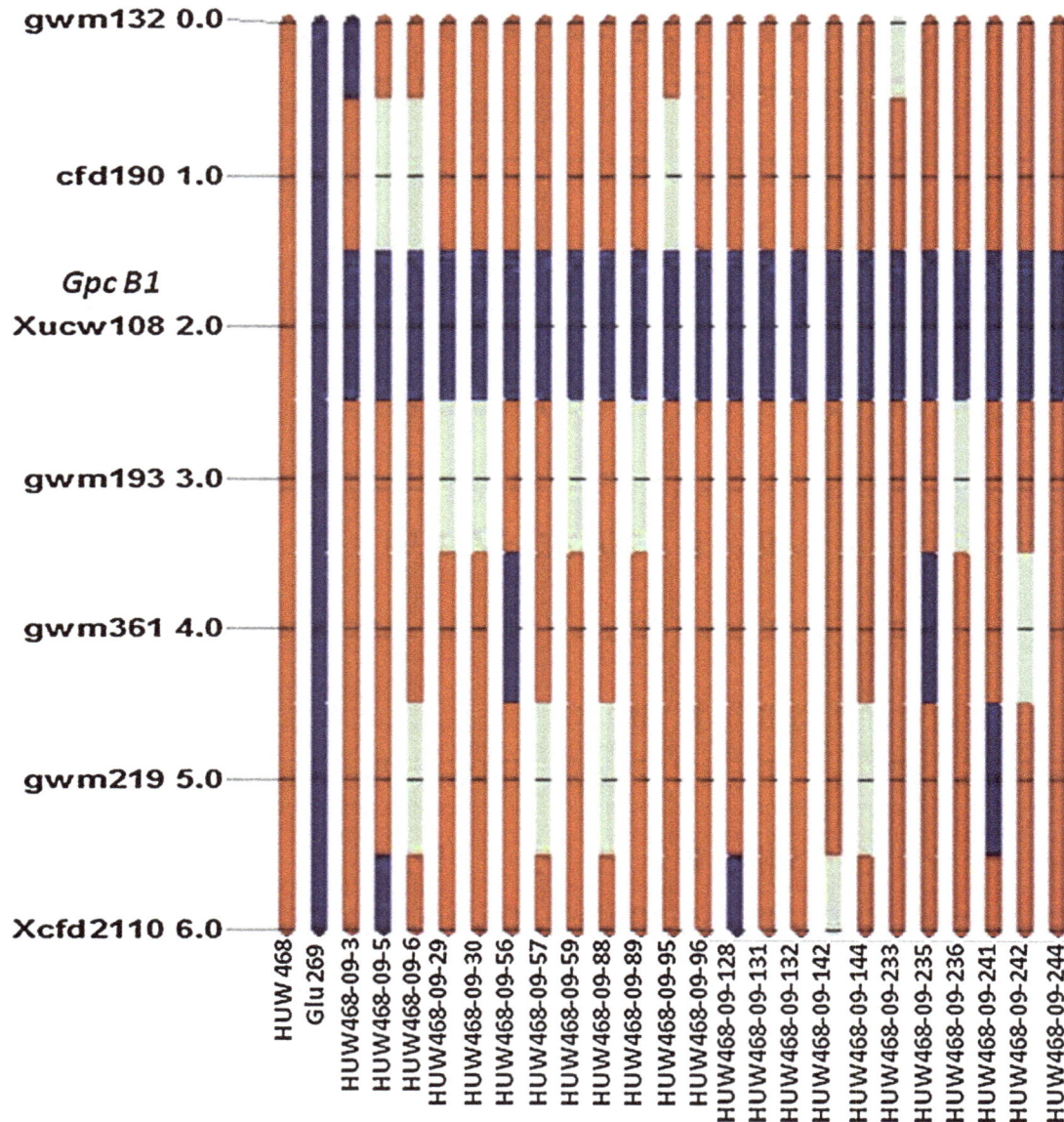

Fig. 3. Analysis of improved lines of HUW468 for introgressed genomic regions of chromosome 6B, including segments carrying *Gpc-B1* (blue segments from donor; red segments from recurrent parent, and gray segments showing heterozygosity). (For interpretation of the references to color in this figure legend, the reader is referred to the web version of the article.)

in the first backcross generation after stringent phenotypic selection. While the remaining markers out of a total of 86 were used in the second backcross. The markers that were heterozygous in BC_2F_1 were again used in BC_2F_2. In this manner, RPG recovery could be fastened in a cost-effective manner. The application of this step-wise background selection has also been demonstrated in rice [52].

4.3. Linkage drag and its elimination

During introgression of desirable/target gene(s) in the backcross-breeding program the probability of other closely linked genes getting introgressed is quite high, which affects the performance of the end product [53]. The probability of undesirable end product due to linkage drag is high, if the donor genotype is of unadapted/wild type. This can be significantly reduced by selecting the genomic region of recurrent parent flanking the desirable gene(s) [42,54].

In the present study, the donor parent Glu269 (DBW16/GluPro//2*DBW16) carrying the gene *Gpc-B1* was a non-adaptive genotype for NEPZ and had poor grain texture and chapatti making quality. This could be because the introgressed gene *Gpc-B1* for high GPC was originally identified in wild emmer wheat *Triticum turgidum* ssp. dicoccoides accession FA15-3 (referred hereafter as DIC; [55]). Hexaploid cultivar Glu-pro was developed from a three-way cross between two bread wheat cultivars and the *dicoccoides* accession earlier used to develop the substitution line LDN(DIC6B) [26]. Therefore, the only option to get rid of undesirable effects, particularly poor grain texture and chapatti quality, was to minimize linkage drag of the donor parent Glu269 used for introgressing the gene *Gpc B1*. For this, we used flanking markers (*Xcfd190* and *Xgwm193*) on the carrier chromosome 6B. These markers were approximately 12 cM apart from each other covering the gene *Gpc-B1*. In a previous study by Kumar et al. [23] flanking markers *XNor-B2* (*CAPS*) and Xgwm193 were used for selection of *Gpc-B1* gene to eliminate the risk of losing the gene segment due to lack of closely linked markers. Later, more precise (Xucw108 *and* Xucw109) markers (within 500 bp region in gene) became available for the transfer of *Gpc-B1* segment [30]. This facilitates the flanking PCR markers XNor-B2, Xucw66, Xgwm508 and Xgwm193 to minimize the linkage drag during MAS [27]. One of the BC_2F_3 line HUW468-09-244 had a single small donor segment (no other segment in the vicinity of Gpc-B1) in the carrier chromosome 6B in the region flanking the Gpc-B1 gene. In this study we utilized carrier chromosome genome recovery in the plants selected through foreground selected in each of the backcross generations. After foreground selection in BC_2F_1, we covered genome of all 21 chromosomes by utilizing single marker for each chromosome along with 6 markers for carrier chromosomes, covering 12 cM region of recurrent parent flanking the donor gene segment. Previously, two backcrosses along with maximum recovery of carrier chromosomes were reported in wheat [32].

4.4. Agronomic performance in the backcross derived lines

Marker-assisted backcrossing has been used for introgression of the gene Gpc-B1 in a few landmark wheat cultivars in India [23,24]. However, only seven desirable selected progenies showed high GPC without yield penalty. The RPG recovery varied from 72.0 to 95.71%. However in the present study, the improved HUW468 derivatives showed high GPC as well as agronomic superiority. Due to directional phenotypic selection for yield and yield related traits, we obtained superiority in four HUW468 derived lines (HUW468-09-95, HUW468-09-131, HUW468-09-132 and HUW468-09-59) in the background of HUW468. Thus, the present work is a successful example of an integrated approach of combining phenotypic selection with marker assisted backcross breeding in wheat for introgression of Gpc-B1, high grain weight and leaf rust resistance in the background of HUW468. The results indicate that the marker-assisted selection with stringent phenotypic selection is effective in rapid recovery of the RPG in a time and cost-effective manner.

Author's contribution

All authors contributed significantly in different ways. In the planning, coordination, conduct of the field experiments, data collection, tabulation, field and molecular analysis, interpretation and writing of the manuscript.

Acknowledgement

The off-season facility extended by Indian Agriculture Research Institute, Regional Station, Wellington, Tamil Nadu is gratefully acknowledged.

References

[1] F.J.P.H. Brouns, V.J. van Buul, P.R. Shewry, Does wheat make us fat and sick, J. Cereal Sci. 58 (2013) 209–215.
[2] A. Rasheed, X. Xia, Y. Yan, R. Appels, T. Mahmood, Z. He, Wheat seed storage proteins: advances in molecular genetics, diversity and breeding applications, J. Cereal Sci. (2014), http://dx.doi.org/10.1016/j.jcs.2014.01.020.
[3] FAOSTAT (2012) faostat.fao.org/site/339/default.aspx and www.fao.org/worldfoodsituation/wfs-home/csdben
[4] M.J. Hawkesford, J. Araus, R. Park, D. Calderini, D. Miralles, T. Shen, J. Zhang, M.A.J. Parry, Prospects of doubling global wheat yields, Food Energy Secur. 2 (1) (2013) 34–48.
[5] R.J. Pena, Current and future trends of wheat quality needs, in: H.T. Buck, J.E. Nisi, N. Salomon (Eds.), Wheat Production in Stressed Environments, Springer, Berlin, 2007, pp. 411–424.
[6] A.K. Joshi, B. Mishra, R. Chatrath, G.O. Ferrara, R.P. Singh, Wheat improvement in India: present status, emerging challenges and future prospects, Euphytica 157 (2007) 431–446.
[7] P.K. Gupta, Quality of Indian wheat and infrastructure for analysis, in: A.K. Joshi, R. Chand, B. Arun, G. Singh (Eds.), A Compendium of the Training Program on Wheat Improvement in Eastern and Warmer Regions of India: Conventional and Non-Conventional Approaches, NATP Project, (ICAR), BHU, Varanasi, India, 2004.
[8] J.C. Brevis, J. Dubcovsky, Effect of chromosome region including the Gpc-B1 locus on wheat protein and grain yield, Crop Sci. 50 (2010) 93–104.
[9] P.R. Shewry, The HEALTHGRAIN programme opens new opportunities for improving wheat for nutrition and health, Nutr. Bull. 34 (2009) 225–231.
[10] R. Avni, R. Zhao, S. Pearce, Y. Jun, C. Uauy, F. Tabbita, T. Fahima, A. Slade, J. Dubcovsky, A. Distelfeld, Functional characterization of GPC-1 genes in hexaploid wheat, Planta 239 (2014) 313–324.
[11] P.J. White, M.R. Broadley, Biofortification of crops with seven mineral elements often lacking in human diets – iron, zinc, copper, calcium, magnesium, selenium and iodine, New Phytol. 182 (2009) 49–84.
[12] R.M. Welch, R.D. Graham, Breeding for micronutrients in staple food crops from a human nutrition perspective, J. Exp. Bot. 55 (2004) 353–364.
[13] T.D. Thacher, P.R. Fischer, M.A. Strand, J.M. Pettifor, Nutritional rickets around the world: causes and future directions, Annu. Trop. Paediatr. 26 (2006) 1–16.
[14] N.W. Simmonds, The relation between yield and protein in cereal grain, J. Sci. Food Agric. 67 (1995) 309–315.
[15] D.W. Lawlor, Carbon and nitrogen assimilation in relation to yield: mechanisms are the key to understanding production systems, J. Exp. Bot. 53 (2002) 773–787.
[16] B.J. Miflin, Nitrogen metabolism and amino acid biosynthesis in crop plants, in: PS Carlson (Ed.), The Biology of Crop Productivity, Academic Press, London, 1980, pp. 255–296.
[17] A.L. Gao, H.G. He, Q.Z. Chen, Pyramiding wheat powdery mildew resistance genes Pm2, Pm4a and Pm21 by molecular marker-assisted selection, Acta Agron. Sin. 31 (2005) 1400–1405.
[18] T. Miedaner, F. Wilde, V. Korzun, E. Ebmeyer, M. Schmolke, L. Hart, C.C. Schon, Marker selection for Fusarium head blight resistance based on quantitative trait loci (QTL) from two European sources compared to phenotypic selection in winter wheat, Euphytica 166 (2009) 219–227.
[19] P.K. Gupta, H.S. Balyan, J. Kumar, P.K. Kulwal, N. Kumar, R.R. Mir, A. Kumar, K.V. Prabhu, QTL analysis and marker assisted selection for improvement in grain protein content and pre-harvest sprouting tolerance in bread wheat, in: 11th Wheat Genetics Symposium (IWGS), Brisbane, Australia, 2008, pp. 1–3.
[20] P.K. Gupta, J. Kumar, R.R. Mir, A. Kumar, Marker-assisted selection as a component of conventional plant breeding, Plant Breed. Rev. 33 (2010) 145–217.
[21] D. Barloy, J. Lemoine, P. Abelard, A.M. Tanguy, R. Rivoal, J. Jahier, Marker assisted pyramiding of two cereal cyst nematode resistance genes from Aegilops variabilis in wheat, Mol. Breed. 20 (2007) 31–40.
[22] J. Kumar, R.R. Mir, N. Kumar, A. Kumar, A. Mohan, K.V. Prabhu, H.S. Balyan, P.K. Gupta, Marker-assisted selection for pre-harvest sprouting tolerance and leaf rust resistance in bread wheat, Plant Breed. 129 (2010) 617–621.
[23] J. Kumar, V. Jaiswal, A. Kumar, N. Kumar, R.R. Mir, S. Kumar, R. Dhariwal, S. Tyagi, M. Khandelwal, K.V. Prabhu, R. Prasad, H.S. Balyan, P.K. Gupta, Introgression of a major gene for high grain protein content in some Indian bread wheat cultivars, Field Crop Res. 123 (2011) 226–233.
[24] S. Tyagi, R.R. Mi, P. Chhuneja, B. Ramesh, H.S. Balyan, P.K. Gupta, Marker-assisted pyramiding of eight QTLs/genes for seven different traits in common wheat (Triticum aestivum L.), Mol. Breed. 34 (2014) 167–175.
[25] A. Mesfin, R.C. Frohberg, J.A. Anderson, RFLP markers associated with high grain protein from Triticum turgidum L. var. dicoccoides introgressed into hard red spring wheat, Crop Sci. 39 (1999) 508–513.
[26] I.A. Khan, J.D. Procunier, D.G. Humphreys, G. Tranquilli, A.R. Schlatter, S. Marcucci-Poltri, R. Frohberg, J. Dubcovsky, Development of PCR-based markers for a high grain protein content gene from Triticum turgidum ssp. dicoccoides transferred to bread wheat, Crop Sci. 40 (2000) 518–524.
[27] S. Olmos, A. Distelfeld, O. Chicaiza, A.R. Schlatter, T. Fahima, V. Echenique, J. Dubcovsky, Precise mapping of a locus affecting grain protein content in durum wheat, Theor. Appl. Genet. 107 (2003) 1243–1251.
[28] A. Distelfeld, C. Uauy, S. Olmos, A.R. Schlatter, J. Dubcovsky, T. Fahima, Microcolinearity between a 2-cM region encompassing the grain protein content locus GPC-6B1 on wheat chromosome 6B and a 350-kb region on rice chromosome 2, Funct. Integr. Genomics 4 (2004) 59–66.
[29] A. Distelfeld, C. Uauy, T. Fahima, J. Dubcovsky, Physical map of the wheat high-grain protein content gene Gpc-B1 and development of a high-throughput molecular marker, New Phytol. 169 (2006) 753–763.
[30] C. Uauy, A. Distelfeld, T. Fahima, A. Blechl, J. Dubcovsky, A NAC gene regulating senescence improves grain protein, zinc, and iron content in wheat, Science 314 (2006) 1298–1301.
[31] M. Kade, A.J. Barneix, S. Olmos, J. Dubcovsky, Nitrogen uptake and remobilization in tetraploid 'Langdon' durum wheat and a recombinant substitution line with the high grain protein gene Gpc-B1, Plant Breed. 124 (2005) 343–349.
[32] H.S. Randhawa, J.S. Mutti, K. Kidwell, C.F. Morris, X. Chen, K.S. Gill, Rapid and targeted introgression of genes into popular wheat cultivars using marker-assisted background selection, PLoS ONE 4 (6) (2009) e5752, http://dx.doi.org/10.1371/journal.pone.0005752.
[33] P.K. Gupta, P. Langridge, R.R. Mir, Marker-assisted wheat breeding: present status and future possibilities, Mol. Breed. 26 (2010) 145–161.
[34] H.M. William, R.M. Trethowan, E.M. Crosby-Galvan, Wheat breeding assisted by markers: CIMMYT's experience, Euphytica 157 (2007) 307–319.
[35] M. Prasad, R.K. Varshney, A. Kumar, H.S. Balyan, P.C. Sharma, K.J. Edwards, H. Singh, H.S. Dhaliwal, J.K. Roy, P.K. Gupta, A microsatellite marker associated with a QTL for grain protein content on chromosome arm 2DL of bread wheat, Theor. Appl. Genet. 99 (1999) 341–345.

[36] H. Singh, M. Prasad, R.K. Varshney, J.K. Roy, H.S. Balyan, H.S. Dhaliwal, P.K. Gupta, STMS markers for grain protein content and their validation using near-isogenic lines in bread wheat, Plant Breed. 120 (2001) 273–278.

[37] D.J. Somers, P. Isaac, K. Edwards, A high-density microsatellite consensus map for bread wheat (*Triticum aestivum* L.), Theor. Appl. Genet. 109 (2004) 1105–1114.

[38] M.S. Röder, V. Korzun, K. Wendehake, J. Plaschke, M.H. Tixier, P. Leroy, M.W. Ganal, A microsatellite map of wheat, Genetics 149 (1998) 2007–2023.

[39] M.A. Saghai-Maroof, K.M. Soliman, R.A. Jorgensen, R.W. Allard, Ribosomal DNA spacer length polymorphisms in barley: Mendelian inheritance, chromosomal location, and population dynamics, Proc. Natl. Acad. Sci. U.S.A. 81 (1984) 8014–8018.

[40] B.C.Y. Collard, D.J. Mackill, Marker-assisted selection: an approach for precision plant breeding in the twenty-first century, Philos. Trans. R. Soc. B 363 (2008) 557–572.

[41] F. Hospital, A. Charcosset, Marker-assisted introgression of quantitative trait loci, Genetics 147 (1997) 1469–1485.

[42] S. Gopalakrishnan, R.K. Sharma, Rajkumar, A. Kumar, M. Joseph, V.P. Singh, A.K. Singh, K.V. Bhat, N.K. Singh, T. Mohapatra, Integrating marker assisted background analysis with foreground selection for identification of superior bacterial blight resistant recombinants in Basmati rice, Plant Breed. 127 (2008) 131–139.

[43] N.G. Paltridge, L.J. Palmer, P.J. Milham, G.E. Guild, J.C.R. Stangoulis, Energy-dispersive X-ray fluorescence analysis of zinc and iron concentration in rice and pearl millet grains, Plant Soil 361 (2012) 251–260.

[44] Ø. Hammer, D.A.T. Harper, P.D. Ryan, PAST: paleontological statistics software package for education and data analysis, Palaeontol. Electron. 4 (2001) 9.

[45] R. Van Berloo, GGT: software for display of graphical genotypes, J. Hered. 90 (1999) 328–329.

[46] R.P. Singh, J. Huerta-Espino, R. Sharma, A.K. Joshi, R. Trethowan, High yielding spring bread wheat germplasm for global irrigated and rainfed production systems, Euphytica 157 (2007) 351–363.

[47] A. Morgounov, H.F. Go'mez-Becerra, A. Abugalieva, M. Dzhunusova, M. Yessimbekova, H. Muminjanov, Y. Zelenskiy, L. Ozturk, I. Cakmak, Iron and zinc grain density in common wheat grown in Central Asia, Euphytica 155 (2007) 193–203.

[48] V.K. Singh, A. Singh, S.P. Singh, R.K. Ellur, D. Singh, S. Gopalakrishnan, P.K. Bhowmick, M. Nagarajan, K.K. Vinod, U.D. Singh, T. Mohapatra, K.V. Prabhu, A.K. Singh, Marker-assisted simultaneous but stepwise backcross breeding for pyramiding blast resistance genes Piz5 and Pi54 into an elite Basmati rice restorer line 'PRR78', Plant Breed. 132 (2013) 486–495.

[49] M. Joseph, S. Gopalakrishnan, R.K. Sharma, Combining bacterial blight resistance and basmati quality characteristics by phenotypic and molecular marker-assisted selection in rice, Mol. Breed. 13 (2004) 377–387.

[50] S.H. Basavaraj, V.K. Singh, A. Singh, A. Singh, R.K. Ellur, D. Singh, S. Gopalakrishnan, M. Nagarajan, T. Mohapatra, K.V. Prabhu, A.K. Singh, Marker-assisted improvement of bacterial blight resistance in parental lines of Pusa RH10, a superfine grain aromatic rice hybrid, Mol. Breed. 26 (2010) 293–305.

[51] V.K. Singh, A. Singh, S.P. Singh, R.K. Ellur, D. Singh, S. Gopalakrishnan, M. Nagarajan, K.K. Vinod, U.D. Singh, R. Rathore, S.K. Prasanthi, P.K. Agrawal, J.C. Bhatt, T. Mohapatra, K.V. Prabhu, A.K. Singh, Incorporation of blast resistance gene in elite Basmati rice restorer line PRR78, using marker assisted selection, Field Crop Res. 128 (2012) 8–16.

[52] S.H. Basavaraj, V.K. Singh, A. Singh, D. Singh, M. Nagarajan, T. Mohapatra, K.V. Prabhu, A.K. Singh, Marker aided improvement of Pusa6B, the maintainer parent of hybrid Pusa RH10, for resistance to bacterial blight, Indian J. Genet. Plant Breed. 69 (2009) 10–16.

[53] P. Stam, A.C. Zeven, The theoretical proportion of the donor genome in near isogenic lines of self-fertilizers bred by backcrossing, Euphytica 30 (1981) 227–237.

[54] C.N. Neeraja, R.R. Maghirang, A. Pamplona, S. Heuer, B.C.Y. Collard, E.M. Eptinigsih, G. Vergara, D. Sanchez, K. Xu, A.M. Ismail, D.J. Mackill, A marker assisted backcross approach for developing submergence tolerant rice cultivars, Theor. Appl. Genet. 115 (2007) 767–776.

[55] L. Avivi, High grain protein content in wild tetraploid wheat Korn, in: S. Ramanujam (Ed.), 5th Wheat Genetics Symposium, Indian Society of Genetics and Plant Breeding, New Delhi, 1978, pp. 372–380.

[56] A.K. Joshi, J. Crossab, B. Arun, R. Chand, R. Trethowan, M. Vargas, I. Ortiz-Monasterio, Genotype × environment interaction for zinc and iron concentration of wheat grain in eastern Gangetic plains of India, Field Crop Res. 116 (2010) 268–277.

7

Time-dependent, glucose-regulated *Arabidopsis* Regulator of G-protein Signaling 1 network☆

Dinesh Kumar Jaiswal[a], Emily G. Werth[b], Evan W. McConnell[b], Leslie M. Hicks[b,**], Alan M. Jones[a,c,*]

[a] *Department of Biology, University of North Carolina, Chapel Hill, NC 27599, USA*
[b] *Department of Chemistry, University of North Carolina, Chapel Hill, NC 27599, USA*
[c] *Department of Pharmacology, University of North Carolina, Chapel Hill, NC 27599, USA*

ARTICLE INFO

Keywords:
Heterotrimeric G-protein
Membrane protein complexes
Tandem affinity purification
Mass spectrometry
Glucose signaling

ABSTRACT

Plants lack 7-transmembrane, G-protein coupled receptors (GPCRs) because the G alpha subunit of the heterotrimeric G protein complex is "self-activating"—meaning that it spontaneously exchanges bound GDP for GTP without the need of a GPCR. In lieu of GPCRs, most plants have a seven transmembrane receptor-like regulator of G-protein signaling (RGS) protein, a component of the complex that keeps G-protein signaling in its non-activated state. The addition of glucose physically uncouples AtRGS1 from the complex through specific endocytosis leaving the activated G protein at the plasma membrane. The complement of proteins in the AtRGS1/G-protein complex over time from glucose-induced endocytosis was profiled by immunoprecipitation coupled to mass spectrometry (IP-MS). A total of 119 proteins in the AtRGS1 complex were identified. Several known interactors of the complex were identified, thus validating the approach, but the vast majority (93/119) were not known previously. AtRGS1 protein interactions were dynamically modulated by D-glucose. At low glucose levels, the AtRGS1 complex is comprised of proteins involved in transport, stress and metabolism. After glucose application, the AtRGS1 complex rapidly sheds many of these proteins and recruits other proteins involved in vesicular trafficking and signal transduction. The profile of the AtRGS1 components answers several questions about the type of coat protein and vesicular trafficking GTPases used in AtRGS1 endocytosis and the function of endocytic AtRGS1.

1. Introduction

The *Arabidopsis* G-proteins, despite having a simpler repertoire than that in metazoans, impart many physiological and biochemical responses and affect growth and development in plants [1]. In metazoans, the duration of G-protein signal termination is dependent on (1) the residence time of the GPCR agonist, (2) the number of activating interactions between the cognate GPCR coupled to the G protein complex over time and (3) the rate of deactivation through intrinsic GTPase activity of the Gα protein. The latter is accelerated by interaction with a group of regulator of G protein signaling (RGS) proteins, accelerating hydrolysis of $G_{\alpha GTP}$ into $G_{\alpha GDP}$ and returning the G protein complex to the resting state [2]. It is now well established that plants use a distinct mechanism to regulate G-protein signaling from metazoans and differ in many aspects. First, plant cells lack G-protein coupled receptors (GPCR) that stimulate guanine nucleotide exchange. Second, plant Gα proteins exchange guanine nucleotides spontaneously *in vitro*. Third, all plants, except the cereals, contain a receptor-like RGS protein that deactivates until decoupled from the G protein complex. In *Arabidopsis*, the RGS protein (AtRGS1) has a seven transmembrane (7TM) domain at its N-terminus and a catalytic RGS box at its C-terminal domain [3–7].

The mechanism of glucose-induced G protein activation is known. Urano et al. demonstrated that D-glucose recruits WITH NO LYSINE (WNK) kinases to phosphorylate AtRGS1 and that this phosphorylation is necessary and sufficient for endocytosis [7,8]. AtRGS1 endocytosis leads to physical uncoupling of AtRGS1 from the *Arabidopsis* G protein α subunit (AtGPA1) and thus a

☆ This article is part of a special issue entitled "Protein networks – a driving force for discovery in plant science".
  * Corresponding author at: Department of Biology, CB# 3280 Coker Hall, University of North Carolina, Chapel Hill, USA.
 ** Corresponding author at: Department of Chemistry, CB# 3290 Kenan A221, University of North Carolina, Chapel Hill, USA.
   E-mail addresses: lmhicks@unc.edu (L.M. Hicks), alan.jones@unc.edu (A.M. Jones).

release of the GAP activity and concomitant sustained activation of G-protein signaling. One of the most intriguing aspects of glucose-regulated, AtRGS1-mediated G-protein activation is that the response is receptive to both signal concentration and timing information, unlike G-protein signaling in animals which is triggered by a threshold of signal [8]. This emergent property has dose and duration reciprocity such that an acute dose of glucose (e.g. 6%) induces complete AtRGS1 endocytosis in 30 min while a low dose induces endocytosis over many hours [8]. While AtRGS1 is considered an inhibitor of G-protein signaling as a GTPase Activating Protein (GAP), the effect of genetic ablation of AtRGS1 suggests that AtRGS1 is a positive regulator of G-protein signaling [9]. Given that trafficking of AtRGS1 is an important part of plant G protein signaling, we hypothesized that signaling through AtRGS1 is both time and sub-cellular location contingent. Specifically, AtRGS1 signaling from the endosome may be an obligatory aspect of signaling output as has been recently shown for β2-adrenoceptor-mediated signaling [10].

Studies of plant G-proteins in the last decade revealed associations with fundamental biological processes such as sugar perception [11,7], organ development [12], hormone signaling [13,14], light responsiveness [15], biotic and abiotic stress [16–19,6], among others. Sugar-induced signal transduction pathways play significant roles in many physiological processes in plants such as photosynthetic efficiency [20], cell wall hexose composition [21], and pathogen defense [22,23]. AtRGS1 plays an important role not only in sugar-mediated seedling development [5], but also in responses to environmental cues [18].

While the heterotrimeric plant G-protein complex is similar in atomic structure to that from animal cells [24], the mechanism of activation of G-protein signaling is dramatically different [1]. Activation involves unknown proteins that operate on the core complex. Targets of the activated G protein complex as defined for animal cells are mostly lacking in plant cells. Therefore, a yeast complementation was previously conducted to assemble a set of candidate targets of G protein complex proteins, but that analysis clearly indicated that the screen was not saturated [25]. Moreover, yeast complementation only detects direct interactions between bait and prey although indirect interactions can be deduced from an *in silico* construction of the network. Finally, while post-translational modifications are often critical for protein–protein interactions, complementation screens in yeast are insensitive to modifications such as the phosphorylation state [26].

For these reasons, it was necessary to initiate an *ab initio* approach to discover the missing elements to G protein signaling in plant cells. Here we used tandem affinity purification that leads to highly enriched samples for AtRGS1-containing complexes. Since AtRGS1 regulates the activation state of G-proteins in a time-dependent manner, we sought *in planta* interacting proteins of AtRGS1 over time after induction with glucose. Mass spectrometry analyses of purified protein complexes led to the identification of 119 interacting AtRGS1 complex proteins associated with diverse biological functions such as response to stimulus, cell organization, transport, and metabolism.

2. Materials and methods

2.1. Plant materials, seedling growth and sugar treatment

All the wild type plants and mutant alleles used in this study were in the Col-0 ecotype of *Arabidopsis thaliana*. The ORF of AtRGS1 and AtRGS1(E320K) were cloned into pEarleyGate 205 vector. The 35S:RGS1-TAP and 35S:RGS1(E320K)-TAP overexpressing lines in the rgs1-2 background were generated by the floral dip method [27]. ACD2-TAP transgenic lines were described by Sakuraba et al.

[28]. Seeds of *Arabidopsis* were surface sterilized and sugar treatments given as described earlier [7,9]. Sterilized and stratified *Arabidopsis* seeds were grown for seven days in flasks containing quarter-strength MS liquid media, subjected to two days sugar starvation in the dark and then treated with 6% glucose for 0, 10 and 30 min. Two hours before sugar treatment, seedlings were treated with 70 μM cycloheximide to block protein translation. The flasks were grown hydroponically in a constant low light (50 μmol s^{-1} m^{-2}) growth chamber at 23 °C with gentle rotation (120 rpm).

2.2. Confocal microscopy and AtRGS1 internalization

All confocal microscopy analyses were performed using a Zeiss LSM710 confocal laser scanning microscope equipped with a C-Apochromat × 40 NA = 1.20 water immersion objective. The imaging and fluorescence internalization were analyzed using ImageJ software as previously described [7].

2.3. Isolation of total membranes and tandem affinity purification

Approximately 10 gm of seedlings (non-treated and 6% glucose-treated for 10 min and 30 min) were ground into powder in liquid nitrogen. The tissue powder was homogenized in a Waring Blender homogenizer with small size 30-mL bowl (Waring, USA) around 18000 rpm speed for 10–15 s with homogenizing buffer [50 mM Tris–Cl (pH 8.0), 150 mM NaCl, 10 mM EDTA (ethylenediaminetetraacetic acid), 1 mM PMSF (phenylmethylsulfonyl fluoride), 2 mM DTT (dithiothreitol), 0.1% Protease inhibitor cocktail Sigma]. The cell debris was filtered through one-layer of Miracloth (Calbiochem) and the filtrate was centrifuged at 12,000 × g for 15 min at 4 °C. The supernatant was recovered and centrifuged at 100,000 × g for 1 h at 4 °C. The resulting pellet containing the membrane fraction was suspended in detergent containing membrane suspension buffer [50 mM Tris–Cl (pH 8.0), 150 mM NaCl, 10 mM EDTA, 1 mM PMSF, 2 mM DTT, 0.5–1% detergent (as indicated), 0.1% Protease inhibitor cocktail Sigma]. The suspension was rotated overnight at 4 °C [29], and then centrifuged for 15 min to separate the insoluble debris from solubilized protein. The protein concentration of the solubilized fraction was estimated using an ESL (Exact, Sensitive, Low Interference) protein assay kit (Roche Molecular Biochemicals) and diluted to 1 mg/ml. The IgG-agarose beads were washed three times with membrane suspension buffer and then mixed with the diluted protein fraction (100 μl slurry for 1 mL solubilized protein) and incubated for 4 h at 4 °C on gentle rotation. After incubation, samples were briefly spun at 2000 × g for 1 min and supernatants were collected. The IgG-agarose beads were washed three times with membrane suspension buffer and then incubated with Tobacco Etch Virus (TEV) protease in the same buffer (0.5% detergent) for 2 h at room temperature (22 °C). His6-tagged TEV protease was purified as described previously [30]. After incubation, samples were briefly spun at 2000 × g for 1 min and the supernatants were recovered. Before proceeding to the next step of purification, the EDTA present in eluates was equilibrated by adding CaCl$_2$. Calmodulin-sepharose beads were washed three times with calmodulin-binding buffer [50 mM Tris–Cl (pH 8.0), 150 mM NaCl, 1 mM Mg-acetate, 2 mM CaCl$_2$, 2 mM DTT, 0.5% ASB-14] and then eluate was mixed and incubated for 2 h at 4 °C on gentle rotation. After incubation, samples were briefly spun at 2000 × g for 1 min and the supernatants were collected and calmodulin-sepharose beads were washed three times with calmodulin-binding buffer. Bound complexes were eluted with 200 μl of calmodulin elution buffer [50 mM Tris–Cl (pH 8.0), 150 mM NaCl, 1 mM magnesium acetate, 2 mM CaCl$_2$, 2 mM DTT, 2 mM EGTA (ethylene glycol tetra acetic acid), 1% ASB-14].

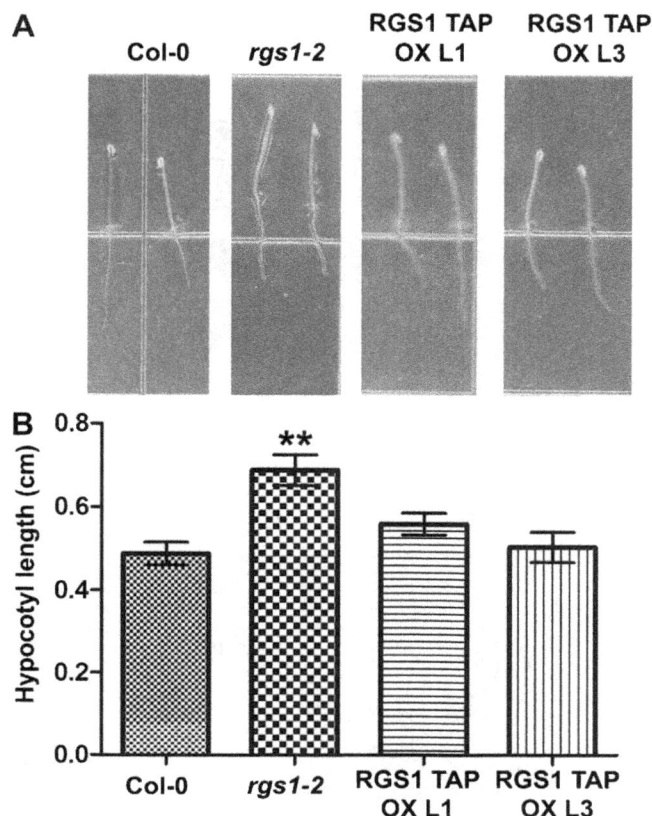

Fig. 1. Genetic complementation of the *rgs1-2* null mutant with AtRGS1-TAP. (A) Representative images of hypocotyl growth (mm) of plants for each genotype on ¼ X-strength MS with 1% (w/v) sucrose at 3 days. (B) Quantitation of hypocotyl growth measured at 3 days. ANOVA was used to determine statistical difference ($p < 0.05$, $n \geq 10$). RGS1-TAP OX L1 and RGS1-TAP OX L3 represents *rgs1-2* complemented line 1 and line 3 by ectopically expressing 35S:AtRGS1–TAP construct.

2.4. SDS-PAGE and Western blotting

The purified protein complexes were pooled and precipitated by chloroform/methanol as previously described [31]. The pellet was redissolved in buffer (7 M urea, 2 M thiourea and 1% ASB-14), diluted 1:1 in Laemmli buffer containing 4% β-mercaptoethanol and separated by 12% SDS PAGE. The proteins were detected by silver staining with the Pierce™ Silver Stain Kit per the manufacturer's instructions. For mass spectrometry analyses, protein complexes were separated on 12% SDS PAGE precast gels (Bio-Rad), stained with Sypro Ruby, and imaged using a Typhoon scanner (GE Healthcare). The apparent molecular weight was estimated using Precision Plus protein standards (Bio-Rad). Protein concentrations were determined with the Protein Assay ESL (Roche). For immunoblot analysis, proteins were blotted onto PVDF membranes (Bio-Rad) and blocked in 3–5% (v/v) milk powder in TBS buffer for at least 1 h at room temperature and then incubated overnight at 4 °C with different primary antibodies. Protein bands were detected with horseradish peroxidase-conjugated IgG diluted 1/10,000 (Amersham Biosciences) as per the manufacturer's instruction (SuperSignal Western Blotting Kits, Thermo Scientific).

2.5. In-gel tryptic digestion, mass spectrometry, and protein identification

SYPRO Ruby stained protein bands were excised manually for downstream processing. The entire gel lane was processed accordingly. Gel slices were destained using 50 mM ammonium

bicarbonate/50% acetonitrile (ACN) solution, reduced with 10 mM dithiothreitol (30 min, RT), alkylated with 55 mM iodoacetamide (30 min, RT, dark) and an in-gel trypsin digestion (25 ng trypsin in 50 mM NH₄HCO₃) was performed overnight at 37 °C as previously described [32]. Following digestion, peptides were extracted first with 1% formic acid in 2% ACN, and second with 60% ACN. Peptide extracts were dried by vacuum centrifugation and resuspended in 10 μl of 5% ACN/0.1% trifluoroacetic acid prior to separation using a nanoACQUITY UPLC (Waters, Milford, MA, USA) coupled to a TripleTOF 5600 MS/MS (AB Sciex, Framingham, MA, USA). Samples (5 μL) were injected onto a trap column (nanoACQUITY UPLC 2G-W/M Trap 5 μm Symmetry C18, 180 μm × 20 mm) at a flow rate of 5 μL/min for 5 min. Peptides were separated using a C18 column (nanoACQUITY UPLC 1.8 μm BEH, 75 μm × 250 mm) at a flow rate of 300 nL/min. Mobile phase A consisted of 0.1% formic acid in H₂O and mobile phase B was 0.1% formic acid in ACN. Peptides were separated using a 30-min linear gradient from 5% to 30% mobile phase B. MS data acquisition was performed as previously described [33]. Raw mass spectral files were converted to mascot generic format (*.mgf) using the ProteinPilot algorithm (AB Sciex). All LC-MS/MS files for bands from the same gel lane were merged into a single peak list and protein identification was performed using a Mascot Server (Matrix Science, London, UK; v2.5.1) against the *A. thaliana* UniProtKB database (Proteome ID: UP000006548, 31,527 entries; accessed July 15, 2015) appended with sequences for common laboratory contaminants (http://www.thegpm.org/cRAP/, 116 entries). Searches of MS/MS data used trypsin protease specificity with the possibility of up to 2 missed cleavages, peptide mass tolerance of 20 ppm, and MS/MS ion mass tolerance of 0.8 Da. Acetylation of the protein N-terminus, carbamidomethylation of cysteine, deamidation of asparagine/glutamine, and oxidation of methionine were selected as variable modifications. Significant peptide identifications above the identity or homology threshold were adjusted in Mascot to ≤1% peptide FDR and resulting matches were exported for data processing. The raw mass spectrometry data were deposited into the ProteomeXchange Consortium *via* the PRIDE partner repository with the dataset identifier PXD003103 and 10.6019/PXD003103.

2.6. SAINT and bioinformatic analysis

ProHits Lite VM v3.0.3 [34] was used to parse result files for proteins with ≥2 unique peptides and format input files for protein–protein interaction analysis. A significance analysis of interactome algorithm (SAINTexpress v3.6.1) [35] was run under default settings to identify proteins that were statistically enriched. Data sets from the AtRGS1 purifications were analyzed using appropriate negative controls, *i.e.*, ACD2-TAP and Col-0 plants without expression of TAP tag. Negative control replicates were treated as different baits to improve statistical performance of the SAINT algorithm [36]. Known contaminants (http://www.thegpm.org/crap/index.html) were removed before SAINT analysis (see Section 3.4). We only considered prey proteins detected with probability avgP ≥ 0.3 for further inspection. Network visualization was performed using Cytoscape v3.2.1 [37] where edge thickness is proportional to the average probability (avgP) for each bait-prey interaction detected.

To define the functional annotation of the identified candidate proteins obtained after SAINT analysis, gene ontology (GO) analysis was performed using PlantGSEA (Plant GeneSet Enrichment Analysis) [38]. The GO terms and gene families enrichments were detected using Fisher's test with Yekutieli correction (false discovery rate cutoff of 0.05). *Arabidopsis* whole genome annotation was used as the background. The data were visualized using REVIGO [39].

2.7. Mating-based split ubiquitin system

Mating-based split ubiquitin assays were performed as previously described [40]. Entry clones of selected candidate proteins, namely, ras-related protein RABB1c (At4g17170), 14-3-3-like protein GF14 phi (At1g35160), aquaporin PIP2-1 (At3g53420), guanylate-binding protein (GBP, At1g03830), ADP-ribosylation factor A1E (At3G62290), mitochondrial outer membrane protein porin 1 (VDAC1, At3g01280), and 3 (VDAC3, At5g15090) were obtained from ABRC. They were subsequently mobilized by LR recombination [Gateway LR Clonase Enzyme Mix (Invitrogen, USA)] into Nub destination vectors (pNX32_GW). The AtRGS1 ORF was subsequently mobilized by LR recombination into pMetYC-GW, to generate the C-terminal Cub fusions of AtRGS1. Empty vectors were used as negative controls. The Cub and Nub clones were transformed into *Saccharomyces cerevisiae* haploid strains THY.AP4 and THY.AP5, respectively [41]. Clones from each THY.AP5 and THY.AP4 transformation were mixed and used for subsequent interaction assays. The protein interactions were detected by assessment of growth of the diploid cells on selective medium lacking leucine, tryptophan, histidine and adenine supplemented with various concentrate of methionine (200 μM and 1 mM).

3. Results

3.1. Detergent screening and tandem affinity purification of AtRGS1-associated proteins

We first investigated whether the AtRGS1–TAP protein was functional. To test this genetically, we ectopically expressed AtRGS1–TAP lines in the *rgs1*-2 null mutant and found that *rgs1*-2 phenotypes were rescued to wild type (Fig. 1). This indicates that the AtRGS1–TAP protein was folded properly and functional *in vivo*. Our approach involved isolation of proteins from a lipid environment, therefore it was necessary to optimize detergent solubilization. AtRGS1 has an N-terminal seven-transmembrane (7TM) helical domain and it is known that membrane protein complexes are sensitive to the detergents used during purification. Therefore, we first needed to determine which detergents should be used for the purification of the AtRGS1 protein complex and then optimize effective tandem affinity purification procedures. We systematically evaluated 5 detergents [*n*-dodecyl-β-D-maltopyranoside (DDM), amidosulfobetaine-14 (ASB-14), Nonidet P-40 (NP-40), *n*-Octyl-*b*-D-glucopyranoside (OG) and Triton X-100] to define the minimum concentration needed to maximally solubilize AtRSG1 (Fig. 2A). ASB-14 is a sulfobetaine-type zwitterionic detergent, while DDM, NP-40, OG and Triton-X100 are all nonionic detergents. The critical micelle concentration (CMC) for ASB-14 is 8 mM, whereas DDM, NP-40, OG and Triton-X100 are 0.17, 0.29, 23 and 0.22 mM, respectively. It has been shown that DDM, NP40 and Triton X-100 are used to solubilize tagged membrane proteins in yeast [42], whereas OG and ASB-14 for solubilizing plant proteins [43,7]. For each of these treatments, we screened by Western blotting to detect the AtRGS1 protein (Fig. 2A). These detergents were used at 0.5% and 1% concentration. The tested detergents displayed different efficiency in the solubilization of AtRGS1 and we concluded that ASB-14 and DDM were the most effective detergents in solubilizing the tagged bait protein (Fig. 2A). To determine if ASB14 at 1% is overly stringent, we determined if known components of the complex were stripped by the detergent. As shown in Fig. 2B, the ASB-14 solubilized complex contains AtRGS1, the Gα subunit (AtGPA1), and the Gβ subunit (AGB1). Therefore, ASB-14 (1%) was chosen for purification in the subsequent large-scale experiments.

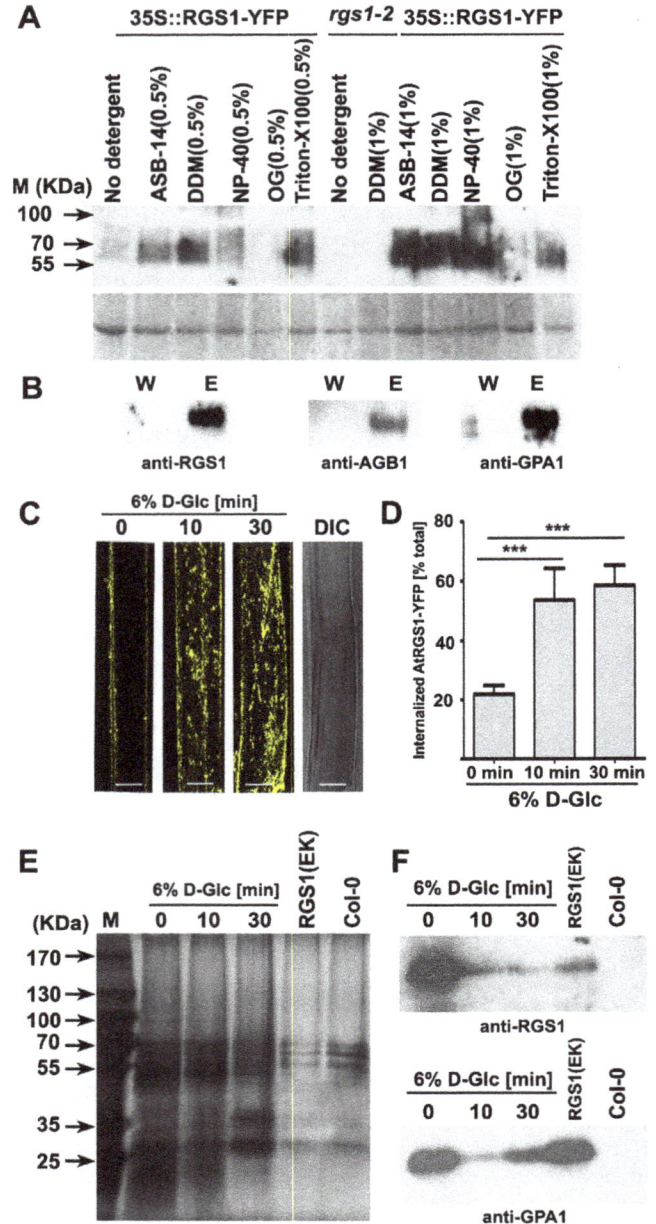

Fig. 2. Detergent screening and tandem affinity purification of AtRGS1-associated protein complex. (A) Western blots showing solubilization efficiency of AtRGS1–TAP using five different detergents at 0.5% and 1% concentrations. Approximately 20 μg of total membrane proteins were subjected to SDS PAGE and probed with anti-RGS1 antibody (upper panel). The lower panel represents the corresponding CBB stained membrane. (B) Pilot test for ABS-14 solubilization of core *Arabidopsis* G protein complex. The ABS-14-solubilized sample was interrogated with the indicated antisera to AtRGS1, AtGPA1, and AGB1. (C, D) *Arabidopsis* seedlings stably expressing AtRGS1-YFP were treated with 6% D-glucose in presence of cycloheximide (70 μM) and internalization was imaged by confocal microscopy (C) and quantified (D). (E) Total membrane samples were prepared from untreated (Col 0 and 35S:AtRGS1E320K-TAP) and treated seedlings (35S:AtRGS1-TAP, 6% D-Glucose, 0, 10 and 30 min). Protein complexes were purified using the TAP protocol described in Section 2. Eluted fractions were precipitated by chloroform/methanol, separated by gel electrophoresis, and proteins were detected by silver staining as described in Section 2 (E). Co-purification of AtRGS1 and associated complexes were monitored by immunoblot analysis using anti-RGS1 and anti-GPA1 antibodies. W, washthrough before final elution; E, final eluate.

Sustained activation of G-protein signaling in *Arabidopsis* has been quantified by endocytosis of AtRGS1 in response to sugar and salt in single seedling experiments [7,18]. For this proteomic study, it was necessary to scale up to thousands of seedlings per sample. Therefore to determine if the biological response survives scale up, we examined AtRGS1-YFP internalization in the treatment format required here. Transgenic *Arabidopsis* expressing AtRGS1-YFP were treated in bulk with 6% D-glucose and internalization was analyzed at 10 min and 30 min by confocal microscopy. Approximately 60% of AtRGS1 was internalized within 30 min of treatment (Fig. 2C and D), consistent with previous studies [7]. Therefore, we conclude that the scaled-up format used for the proteomics analysis is sufficient to recapitulate AtRGS1 endocytosis that has been extensively quantified at the single seedling level.

3.2. Pilot-scale proteome for validated partners used to set large scale filtering criteria

Establishing excellent experimental and downstream processing conditions is a prerequisite for development of affinity capture based proteome studies, especially when dealing with seven transmembrane proteins. In animal systems, different strategies have been used to identify membrane-associated complexes such as use of biotinylated ligands to purify protein complexes [44], interaction motifs [45] and tandemly-tagged full length receptor protein [29]. However, these kinds of analyses with multi transmembrane proteins are rare in plants [46]. In fact, membrane proteome data sets describing various threshold parameters associated with high confidence interactomes *via* developer software such as SAINT are not available for plants. Therefore, to build a comprehensive and robust data set with optimal sensitivity to capture the maximum positive interactions of AtRGS1, we decided to identify and validate randomly selected candidates, which can be used later as an internal positive reference for SAINT analysis (see Section 3.4).

We used optimized conditions for large-scale purification of AtRGS1–TAP complexes and subsequent LC–MS/MS analyses for identification of validated-interacting partners. Proteins present in the complex were separated by one-dimensional-SDS-PAGE, and visualized using SYPRO Ruby staining (Fig. S1). During conventional IP experiments, immunoprecipitated proteins are usually separated on a 1D-SDS PAGE and only those proteins detected selectively in the experiment (*i.e.*, proteins observed in the experiment lane and not in the control lane) are subjected to further identification by mass spectrometry. However, this approach is error prone. We therefore used a more systematic approach in which each lane was systematically divided into different sections and manually excised. Corresponding gel sections excised from three biological replicates were pooled and then subjected to in-gel trypsin digestion and LC–MS/MS for identification. Several expected interacting proteins were identified in the AtRGS1–TAP purified sample including: WNK10, SALT-INDUCIBLE ZINC FINGER 1 (SZF1) and LOW EXPRESSION OF OSMOTICALLY RESPONSIVE GENES 1 (LOS1). The complete sets of proteins identified in this pilot experiment are listed in Table S1.

In order to obtain a set of validated AtRGS1 interactors needed to determine subsequent filtering criteria for large-scale proteome profiling, 7 candidates were randomly selected for testing using mating-based split-ubiquitin yeast two hybrid system. The full length AtRGS1 was fused to the Cub domain, while the full length candidate prey proteins were fused to the NubG domain. The interaction was determined by cell growth on selective medium lacking leucine, tryptophan, histidine, and adenine. We used varying concentrations of methionine in selective medium to increase selection stringency. Of the seven tested, four were confirmed to interact with AtRGS1 as indicated by growth on selective medium (Fig. 3). These were ADP-ribosylation factor A1E

(ARFA1E, At3G62290), aquaporin PIP2-1 (At3g53420), 14-3-3-like protein GF14 phi (At1g35160), and ras-related protein RABB1c (At4g17170). Aquaporins are well known channel proteins with diverse subcellular localizations. They are involved in hydraulic regulation in response to various stimuli [47]. General regulatory factors (GRFs *i.e.*, 14-3-3) are highly conserved proteins, bind to phosphorylated proteins to modulate their function and have been implicated in diverse physiological functions in plants [48]. ARFA1E and RABB1c are members of a small GTPases superfamily and play an important role as regulators in membrane trafficking [49]. The validation rate is 57% (4/7 in Y2H, Fig. 3), which is conservatively low, it is to be noted that Y2H assay is sensitive to orientation of the split ubiquitin tags and does not detect post-translational modification-dependent interactions. These four candidates were used as internal positive standards for the subsequent full-scale experiment.

3.3. Identification of AtRGS1-associated proteins by mass spectrometry

Each of three time points sampled included 3 biological replicates with 3 technical replicates each. Biological samples were taken at 0, 10, and 30 min after glucose addition to the seedlings as described in Section 2. We also included a TAP-tagged AtRGS1 mutant having a single glutamic (E320) mutated to lysine. This glutamic acid residue is critical for AtRGS1 interaction with the Gα subunit. This mutation abolishes the GAP activity of the AtRGS protein [3], however it does not disrupt its interaction with Gα subunit at the plasma membrane [9]. AtRGS1 (E320K) mutated proteins do not leave the plasma membrane upon acute D-glucose treatment [7]. Two negative controls were included: (1) untransformed Col-0 seedlings and (2) Col-0 expressing ACCELERATED CELL DEATH 2 (ACD2) tagged with TAP (ACD2-TAP) [28]. ACD2-TAP is a good negative control because it is associated with the chloroplast membrane fraction [28] and has the same TAP backbone as in AtRGS1-TAP. To acquire sufficient material for triplicate sampling *via* LC–MS/MS analysis, protein complexes were purified from nine independent IP experiments for each biological condition [RGS1-TAP (0, 10, 30 min 6% Glc), RGS1 (E320K)-TAP, ACD2-TAP and wild type] and grouped into three separate replicates per biological condition for downstream handling. The complete list of identified proteins is presented in Table S2.

3.4. Construction of the glucose-induced AtRGS1 interactome

In order to construct a high quality interactome from the proteomics data, we performed 'significance analysis of interactome' (SAINT) analysis [50], a probabilistic scoring method to filter out non-specific interactions. Since little is known about validated AtRGS1 protein interactions, we optimized the selection of SAINT thresholds using the four validated proteins (Fig. 3) from the pilot run. The AvgP of validated candidate proteins were assessed and based on these values of internal candidates, we set a lower SAINT cut off score for all the datasets accordingly. Previous literature reports with the use of SAINT cut off score (AvgP ≥ 0.5) for interactome analysis [51–53]; however, we assessed the distribution of SAINT probability scores and accepted only those proteins showing a significantly-enriched SAINT score of 0.3 or higher (Table S3). The applied threshold is more lenient, however this might be necessary because of the penalty SAINT puts on interactions not detected in all replicates severely reduces the avgP score. This may also increase the false positive interactions. The control data sets were included from ACD2-TAP and Col-0 plants not expressing TAP-tagged AtRGS1. The SAINT scores and spectrum counts for all proteins assigned to each affinity purification and controls are provided in Table S3. The analysis revealed 119 proteins significantly

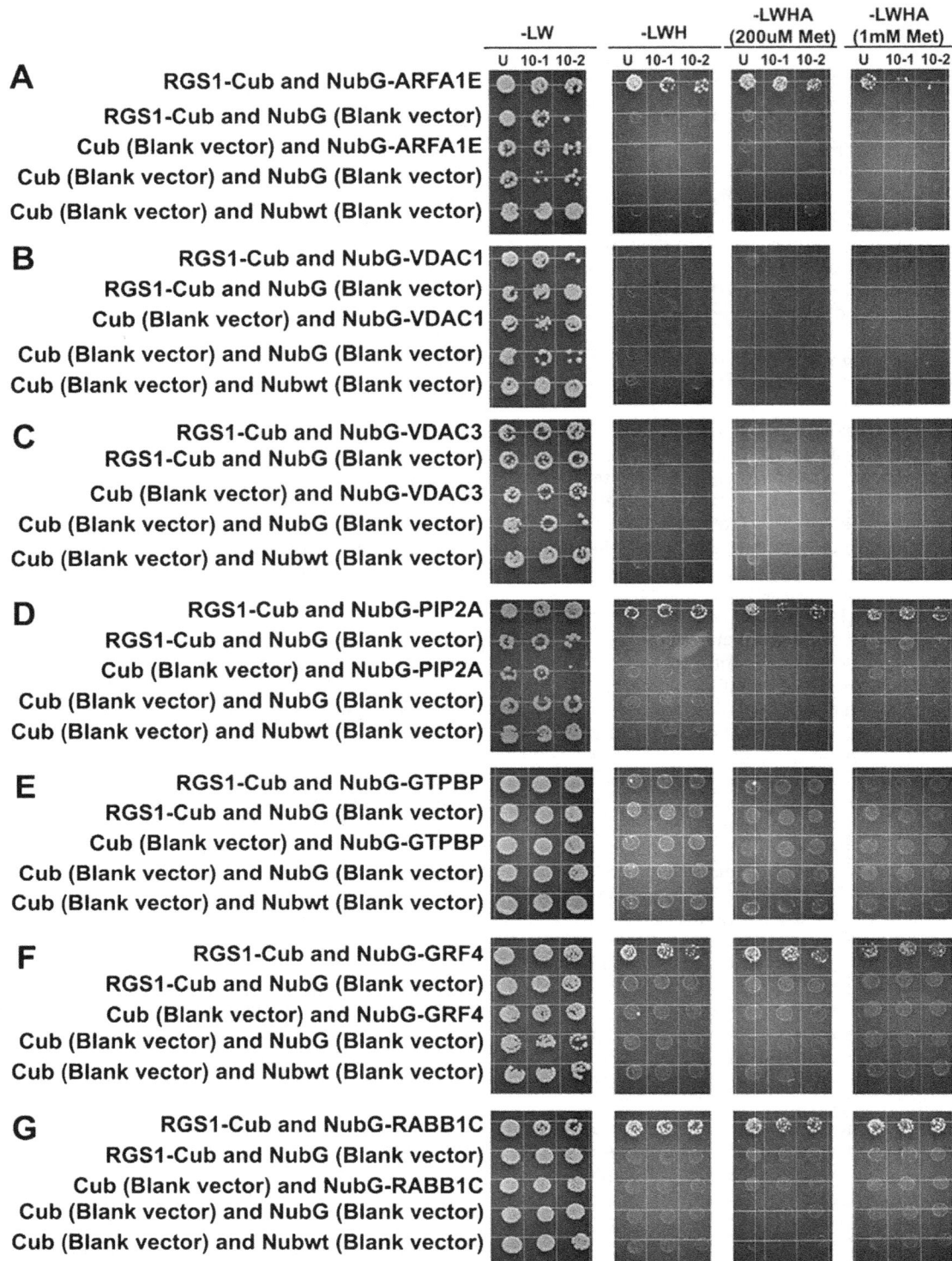

Fig. 3. Validation of candidate interactors from the pilot run. *In vivo* interaction for a random test set of complex components were tested using yeast complementation as described in Section 2. AtRGS1 physically interacts with AtARFA1 E (A), AtPIP2A (D), AtGRF4 (F), and AtRABB1 c (G) in yeast, detected using mating-based split ubiquitin system. Yeast THY.AP4 and THY.AP5 clones expressing full-length AtRGS1 and different complex proteins, respectively were mated. Yeast growth was observed in different selection media. L, leucine; W, tryptophan; H, histidine; A, adenine; Met, methionine; U, undiluted cells; 10^{-1} and 10^{-2} are sequential dilution of cells.

enriched during affinity purification; however we also observed a number of common and unique proteins (Fig. 4). Most of the proteins were identified from the untreated AtRGS1–TAP enrichment (6% Glc, 0 min) of which 74 proteins (64%) was uniquely assigned to this bait for this condition. Other conditions such as 10 and 30 min

had the smallest number of proteins assigned to their enrichments [6 proteins (5%) and 19 proteins (16%)], respectively (Table S3). The interaction of AtRGS1 with Y2H validated protein/isoforms were observed in due course of D-glucose treatment. For example, the family members of aquaporin and general regulatory factors (GRF)

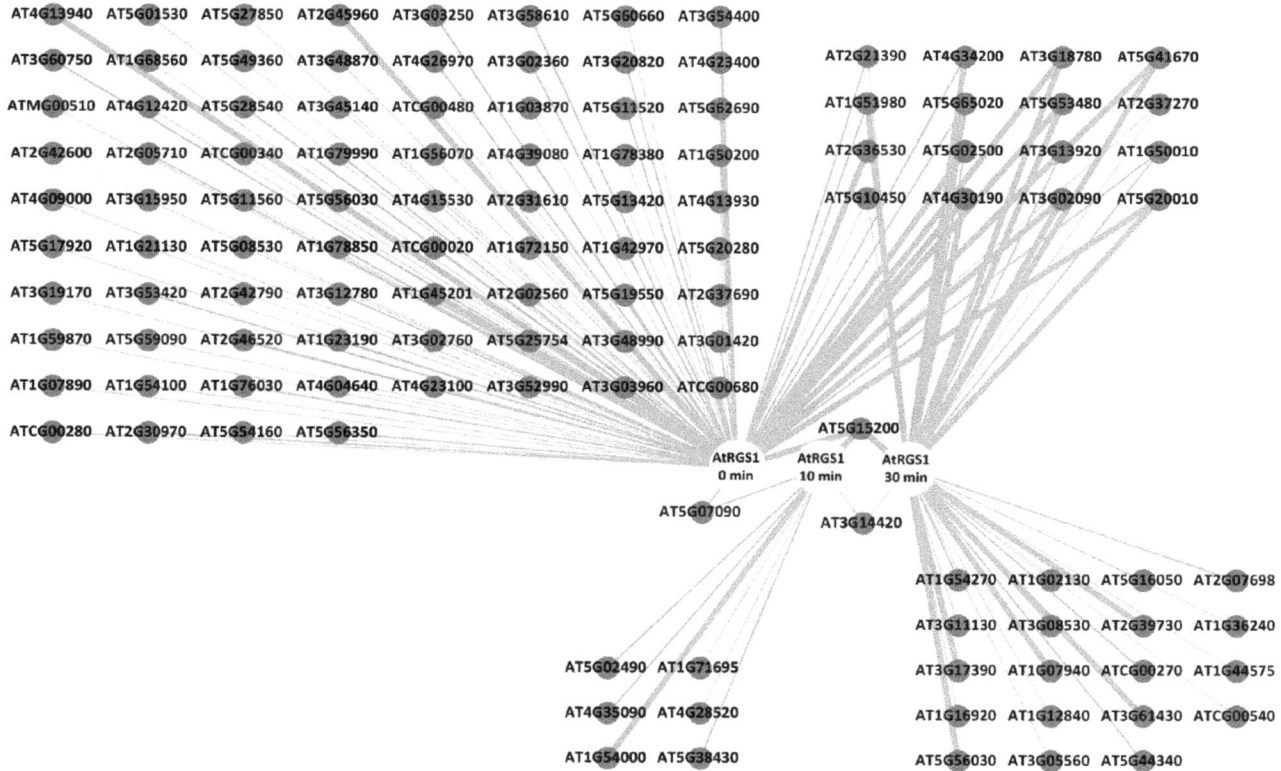

Fig. 4. Network analysis of AtRGS1 complex proteins. A total of 119 proteins obtained from three independent IP-MS experiments showing the dynamics of the AtRGS1-associated complex following treatment with glucose. Nodes represent proteins detected in IP-MS experiments whereas edges (lines connecting different nodes) indicate protein–protein interaction. Edge thickness is proportional to the average probability of interaction (avgP) determined by SAINT analysis.

interact with AtRGS1 at 0 min and 30 min time points; however the family member of small GTPase, RAB protein isoform, was found to be associated with AtRGS1 at 30 min following D-glucose treatment (Table S3). We developed a separate interactome map for the AtRGS1 (E320K) mutant (Fig. S3). SAINT analysis was performed as mentioned above using ACD2-TAP and Col-0 as the negative controls. The SAINT filtered prey data sets (Table S3) were analyzed using the *Arabidopsis* G-protein interactome database which showed that 21% of the proteins/isoforms identified as candidates of AtRGS1 complex were previously reported to be involved in the interaction with G-protein signaling components by yeast complementation assays [Y2H and Y3H, 24].

3.5. Functional annotation of AtRGS1-associated proteins

For functional categorization of candidate proteins, we applied GO analysis using the PlantGSEA search algorithm. TAIR accession numbers of the SAINT filtered list were submitted to PlantGSEA and GO terms with associated p-values were obtained (Table S4). Statistically overrepresented GO terms were subjected to REVIGO analyses for visualization of various processes [39]. The input list of proteins based on p-values was broadly grouped into biological processes (Fig. 5A), cellular component (Fig. 5B) and molecular function (Fig. 5C). In biological process, the most enriched GO terms associated with AtRGS1 complex protein were "response to metal ion", "generation of precursor metabolites and energy", "glucose metabolic pathway" and "response to stimulus" (Fig. 5A). In terms of cellular components, these proteins were distributed in different cellular and subcellular components. A broad-spectrum distribution of proteins were observed as evident from "membrane", "cytoplasm", "plasmodesma", "plastid", and "plasma membrane" (Fig. 5B). Molecular function classification revealed an overrep-

resentation of GO terms associated with "catalytic activity" and "nucleotide binding" as shown in Fig. 5C.

4. Discussion

Analysis of G protein interactome [25] showed that 149 candidates can directly interact with AtRGS1, whereas membrane-based interactome database showed 126 interactors in Y2H screen [54]. In this study, 119 proteins were identified as candidates of AtRGS1 complex protein *in vivo*. The validation rate of AtRGS1 complex protein interaction based on Y2H confirmation of our pilot screen was ~60% (Fig. 3). Twenty one percent of proteins identified from large screen (Table S3) were reported in the previous Y2H screen for AtRGS1/G-protein interacting proteins [25,54]. This clearly shows that many of them identified here (~80%, Table S3) are possibly novel candidates for AtRGS1 complex proteins [25,54]. For example, the yeast two-hybrid screen using AtRGS1 as bait did not identify 14-3-3 proteins as identified and validated as an AtRGS1 interactor here (Fig. 3). Furthermore, some proteins that associate with AtRGS1 indirectly through other AtRGS1-binding proteins or in a post-translational modification dependent manner were identified by affinity purification but were not identified by yeast two-hybrid assays. For example, while no evidence exists for a direct interaction between LOS1 with AtRGS1, LOS1 interacts with the AtRGS1 partner, AtGPA1 [25]. This suggests that novel AtRGS1-associated proteins identified in this study might interact indirectly through other AtRGS1-interacting proteins, and thus have a function in AtRGS1-regulated processes. Furthermore, the copurification of enolase 2 known to interact with AGB1 [25], suggests that the AtRGS1-signaling components exist in a multi-protein complex and possibly involve adapter proteins hitherto undiscovered. V-ATPase c subunit (VHA-c) detected here is an interesting target because

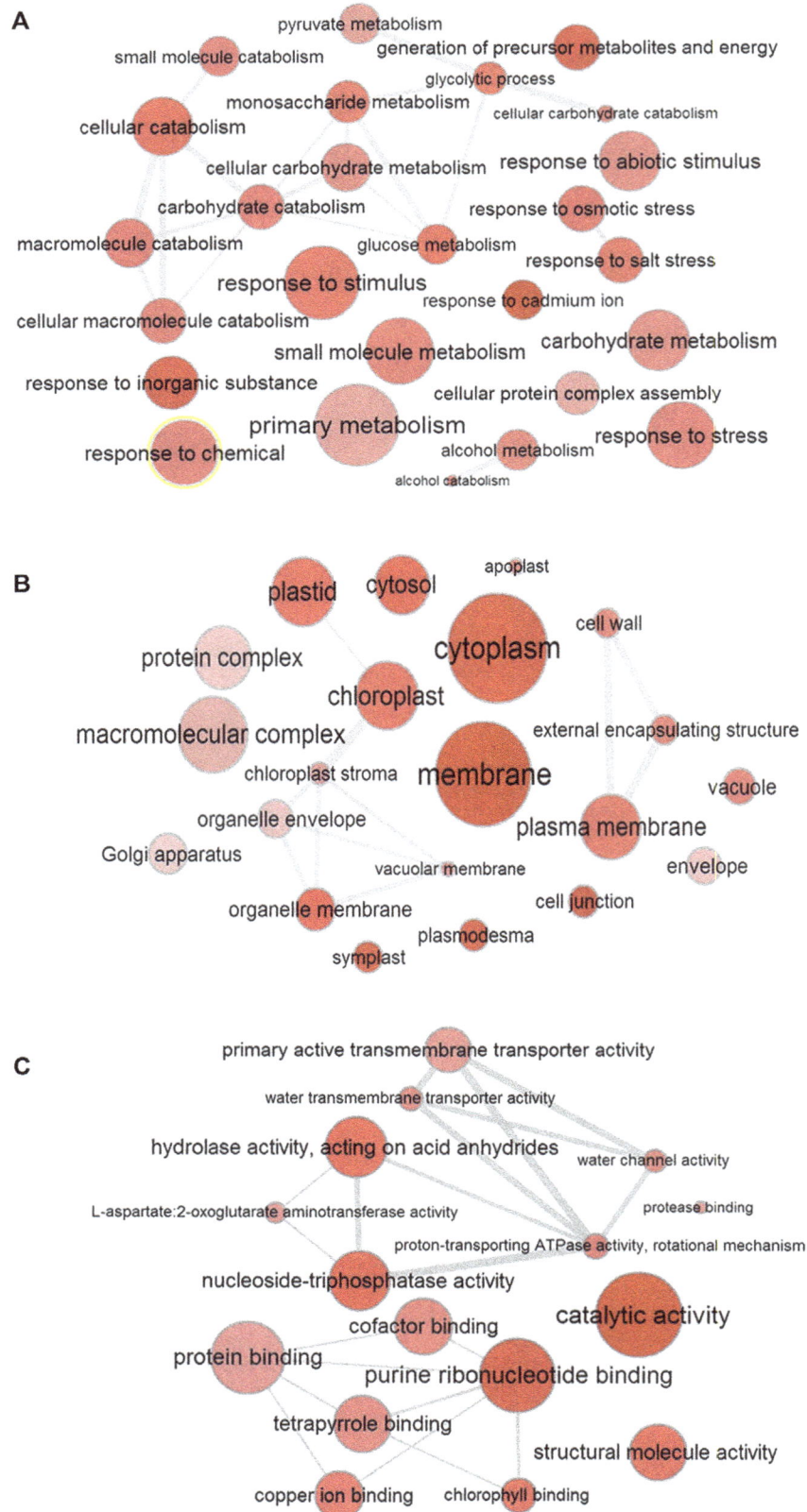

Fig. 5. Functional annotation of AtRGS1 complex proteins. GO term of SAINT analyzed AtRGS1 complex proteins were obtained using the Plant GeneSet Enrichment Analysis Toolkit. Interactive graph of over-represented GO terms were visualized by REVIGO. (A) biological process, (B) cellular component, and (C) molecular function.

functionally active VHA-c is required for proper localization of AtRGS1 on the plasma membrane [55].

Ligand-dependent endocytosis of GPCRs in animal cells may or may not require clathrin [56,57]. Clathrins are highly conserved coat proteins involved in the endocytic pathway regulating protein abundance at the plasma membrane and or the trans-Golgi network during cellular signaling [58]. While it is well known that sugar induces endocytosis of AtRGS1, it was not known whether AtRGS1 endocytosis is clathrin dependent or independent. We observed the association of AtRGS1 with the vesicle coat protein clathrin heavy chain 1 (At3g11130) and 2 (At3g08530) 30 min after D-glucose treatment (Table S3) indicating RGS1 endocytosis is clathrin dependent. Consistent with an endocytic role for AtRGS1, clathrin proteins were associated with AtRGS1 following D-glucose treatment.

In animals, agonists induce GPCR internalization in a phosphorylation-dependent manner. Phosphorylated GPCRs are recognized by a group of proteins called arrestin and arrestin-fold containing proteins designated VPS proteins [59,60]; both types recruit adaptor protein complex which then recruit clathrin triskelions to initiate endocytosis [61]. Endocytosis of animal GPCRs sequesters the GPCR away from the external stimulus, leading to signal desensitization. However, signaling also occurs at the endosome [62,63]. The possible fates of endocytosed proteins are (1) recycled back to the plasma membrane, (2) transported to the lysosome for degradation, and (3) retrograde transported to the trans-Golgi network [64]. RAB (Ras-like small GTP binding) and ARF (ADP-ribosylation factor) are small GTP-binding proteins and well characterized in vesicular trafficking [49]. During endocytosis, Ras-like GTPase family members selectively associate with recycling and sorting endosomes [49,65], whereas ARF family members bind to the target and recruit adapter protein AP-2 and consequently clathrin [66]. ARF proteins are not only involved in the endocytosis of GPCRs [67,68] but also in recycling from endosomes [69]. Since AtRGS1 interacts with members of the RAB and ARF families of small GTPases (Fig. 3), we hypothesize that interaction with these molecules play an important role in the initiation of endocytosis of AtRGS1 as well as its recycling back to the plasma membrane. We further propose that AtRGS1 promotes continued signaling from endosomes as evidenced from the interaction of RAB member with the receptor on endosomes appearing to be important for inhibition of the internalized receptor to lysosomes [70].

In animal cells, 14-3-3 proteins bind to the phosphorylated RGS protein to inhibit the GAP activity [71]. We show here that AtRGS1 interacts with a 14-3-3 protein. This 14-3-3 protein likely provides additional regulation of G-protein signaling either at the plasma membrane and/or from endosomes. Family members of this protein were identified during untreated as well as following D-glucose treatment. By analogy, once AtRGS1 is phosphorylated, 14-3-3 protein may bind to the phosphorylated form of AtRGS1 to inhibit its GAP activity, allowing the "self-activation" of AtGPA1 and consequently G-protein signaling.

Many candidate proteins were identified with other cellular functions. These proteins include water channels (e.g., aquaporin), transporters [e.g., H+-ATPase (AHA) and ABC transporter ABCG.36], chaperones (e.g., HSP90-2 and HSP70-1), and metabolic enzymes (e.g., enolase, pyruvate kinase and phosphoglycerate kinase) among others. Although the presence of metabolic enzymes and proteins from the chloroplast in the AtRGS1 interactome is surprising, many of them were found to interact with AtRGS1/G-protein signaling complex in Y2H [25]. The biological implications of the interactions between AtRGS1 and these proteins are yet to be determined.

The charge reversal mutant of AtRGS1 (E320K) disrupts the GTPase accelerating activity and glucose-induced endocytosis but not the interaction with AtGPA1 [3]. Some of the partners to AtRGS1

are shared by the AtRGS1(E320K) mutant such as aquaporin. As expected, proteins associated with endocytosis and membrane trafficking were not detected in the AtRGS1 (E320K) network. AtGPA1 was associated with the AtRGS1 (E320K) complex (Fig. 2F) indicating the presence of multiple interaction interfaces. Given the presence of the 7-transmembrane domain, it is plausible that AtRGS1 structure is unique and provides a novel interaction interface with AtGPA1 and other interactor molecules as the original yeast complementation results suggested [5].

The dynamic properties of the D-glucose-regulated AtRGS1 signaling networks are likely modulated by AtRGS1 internalization, which depend on the dose and duration of sugar applied [8]. A high dose of glucose was applied to shorten the window of time for glucose-induced endocytosis of AtRGS1. Our result shows that the AtRGS1 interactome changes in minutes of glucose application. AtRGS1 is initially associated with a group of protein involved in transport, stress and metabolism. Immediately after glucose addition, there is a recruitment of proteins that stimulate the endocytic pathways (Table S3) and therefore promote downstream signaling events. By 30 min, the AtRGS1 complex is dramatically altered and increased in size. These later complex components include annexin, 14-3-3, aquaporin, clathrin and ras-like GTPases among others (Table S3). We speculate that this change in the composition of the AtRGS1 complex leads to signaling originating from the endosome.

Acknowledgements

Work in the Jones Lab is supported by grants from the NIGMS (R01GM065989), and NSF (MCB-0723515, MCB-1158054, and MCB-0718202). The Division of Chemical Sciences, Geosciences, and Biosciences, Office of Basic Energy Sciences of the US Department of Energy provided technical support for this study (DE-FG02-05er15671). We thank Dr. Daisuke Urano for his helpful comments, Dr. Meral Tunc-Ozdemir for confocal analysis of AtRGS1 internalization and Ms. Jing Yang for her technical assistance.

References

[1] D. Urano, J.G. Chen, J.R. Botella, A.M. Jones, Heterotrimeric G protein signalling in the plant kingdom, Open Biol. 3 (2013) 120186.

[2] T.M. Cabrera-Vera, J. Vanhauwe, T.O. Thomas, M. Medkova, A. Preininger, M.R. Mazzoni, H.E. Hamm, Insights into G protein structure function, and regulation, Endocr. Rev. 24 (2003) 765–781.

[3] C.A. Johnston, J.P. Taylor, Y. Gao, A.J. Kimple, J.C. Grigston, J.G. Chen, D.P. Siderovski, A.M. Jones, F.S. Willard, GTPase acceleration as the rate-limiting step in Arabidopsis G protein-coupled sugar signaling, Proc. Natl. Acad. Sci. U. S. A. 104 (2007) 17317–17322.

[4] W. Bradford, A. Buckholz, J. Morton, C. Price, A.M. Jones, D. Urano, Eukaryotic G protein signaling evolved to require G protein-coupled receptors for activation, Sci. Signal. 6 (2013) ra37.

[5] J.G. Chen, F.S. Willard, J. Huang, J. Liang, S.A. Chasse, A.M. Jones, D.P. Siderovski, A seven-transmembrane RGS protein that modulates plant cell proliferation, Science 301 (2003) 1728–1731.

[6] D. Urano, A.M. Jones, Heterotrimeric G protein-coupled signaling in plants, Annu. Rev. Plant Biol. 65 (2014) 365–384.

[7] D. Urano, N. Phan, J.C. Jones, J. Yang, J. Huang, J. Grigston, J.P. Taylor, A.M. Jones, Endocytosis of the seven-transmembrane RGS1 protein activates G-protein-coupled signalling in Arabidopsis, Nat. Cell Biol. 14 (2012) 1079–1088.

[8] Y. Fu, S. Lim, D. Urano, M. Tunc-Ozdemir, N.G. Phan, T.C. Elston, A.M. Jones, Reciprocal encoding of signal intensity and duration in a glucose-sensing circuit, Cell 156 (2014) 1084–1095.

[9] J.C. Grigston, D. Osuna, W.R. Scheible, C. Liu, M. Stitt, A.M. Jones, D-Glucose sensing by a plasma membrane regulator of G signaling protein, AtRGS1, FEBS Lett. 582 (2008) 3577–3584.

[10] R. Irannejad, J.C. Tomshine, J.R. Tomshine, M. Chevalier, J.P. Mahoney, J. Steyaert, S.G. Rasmussen, R.K. Sunahara, H. El-Samad, B. Huang, M. von Zastrow, Conformational biosensors reveal GPCR signalling from endosomes, Nature 495 (2013) 534–538.

[11] J.G. Chen, Y. Gao, A.M. Jones, Differential roles of Arabidopsis heterotrimeric G-protein subunits in modulating cell division in roots, Plant Physiol. 141 (2006) 887–897.

[12] D. Urano, A. Colaneri, A.M. Jones, Galpha modulates salt-induced cellular senescence and cell division in rice and maize, J. Exp. Bot. 65 (2014) 6553–6561.

[13] H. Ullah, J.G. Chen, B. Temple, D.C. Boyes, J.M. Alonso, K.R. Davis, J.R. Ecker, A.M. Jones, The beta-subunit of the *Arabidopsis* G protein negatively regulates auxin-induced cell division and affects multiple developmental processes, Plant Cell 15 (2003) 393–409.

[14] Y. Fujisawa, T. Kato, S. Ohki, A. Ishikawa, H. Kitano, T. Sasaki, T. Asahi, Y. Iwasaki, Suppression of the heterotrimeric G protein causes abnormal morphology including dwarfism, in rice, Proc. Natl. Acad. Sci. U. S. A. 96 (1999) 7575–7580.

[15] J.F. Botto, S. Ibarra, A.M. Jones, The heterotrimeric G-protein complex modulates light sensitivity in *Arabidopsis thaliana* seed germination, Photochem. Photobiol. 85 (2009) 949–954.

[16] H. Zhang, Z. Gao, X. Zheng, Z. Zhang, The role of G-proteins in plant immunity, Plant Signal. Behav. 7 (2012) 1284–1288.

[17] M.N. Aranda-Sicilia, Y. Trusov, N. Maruta, D. Chakravorty, Y. Zhang, J.R. Botella, Heterotrimeric G proteins interact with defense-related receptor-like kinases in *Arabidopsis*, J. Plant Physiol. 188 (2015) 44–48.

[18] A.C. Colaneri, M. Tunc-Ozdemir, J.P. Huang, A.M. Jones, Growth attenuation under saline stress is mediated by the heterotrimeric G protein complex, BMC Plant Biol. 14 (2014), http://dx.doi.org/10.1186/1471-2229-14-129.

[19] Y. Yu, S.M. Assmann, The heterotrimeric G-protein beta subunit AGB1, plays multiple roles in the *Arabidopsis* salinity response, Plant Cell Environ. 38 (2015) 2143–2156.

[20] A. Wingler, S. Purdy, J.A. MacLean, N. Pourtau, The role of sugars in integrating environmental signals during the regulation of leaf senescence, J. Exp. Bot. 57 (2006) 391–399.

[21] W.H. Cheng, E.W. Taliercio, P.S. Chourey, Sugars modulate an unusual mode of control of the cell-wall invertase gene (Incw1) through its 3′ untranslated region in a cell suspension culture of maize, Proc. Natl. Acad. Sci. U. S. A. 96 (1999) 10512–10517.

[22] M.R. Bolouri Moghaddam, W. Van den Ende, Sugars and plant innate immunity, J. Exp. Bot. 63 (2012) 3989–3998.

[23] I. Morkunas, L. Ratajczak, The role of sugar signaling in plant defense responses against fungal pathogens, Acta Physiol. Plant. 36 (2014) 1607–1619.

[24] J.C. Jones, J.W. Duffy, M. Machius, B.R. Temple, H.G. Dohlman, A.M. Jones, The crystal structure of a self-activating G protein alpha subunit reveals its distinct mechanism of signal initiation, Sci. Signal. 4 (2011) ra8.

[25] K. Klopffleisch, N. Phan, K. Augustin, R.S. Bayne, K.S. Booker, J.R. Botella, N.C. Carpita, T. Carr, J.G. Chen, T.R. Cooke, A. Frick-Cheng, E.J. Friedman, B. Fulk, M.G. Hahn, K. Jiang, L. Jorda, L. Kruppe, C. Liu, J. Lorek, M.C. McCann, A. Molina, E.N. Moriyama, M.S. Mukhtar, Y. Mudgil, S. Pattathil, J. Schwarz, S. Seta, M. Tan, U. Temp, Y. Trusov, D. Urano, B. Welter, J. Yang, R. Panstruga, J.F. Uhrig, A.M. Jones, *Arabidopsis* G-protein interactome reveals connections to cell wall carbohydrates and morphogenesis, Mol. Syst. Biol. 7 (2011) 532.

[26] A. Bruckner, C. Polge, N. Lentze, D. Auerbach, U. Schlattner, Yeast two-hybrid, a powerful tool for systems biology, Int. J. Mol. Sci. 10 (2009) 2763–2788.

[27] S.J. Clough, A.F. Bent, Floral dip: a simplified method for *Agrobacterium*-mediated transformation of *Arabidopsis thaliana*, Plant J. 16 (1998) 735–743.

[28] Y. Sakuraba, S. Schelbert, S.Y. Park, S.H. Han, B.D. Lee, C.B. Andres, F. Kessler, S. Hortensteiner, N.C. Paek, STAY-GREEN and chlorophyll catabolic enzymes interact at light-harvesting complex II for chlorophyll detoxification during leaf senescence in *Arabidopsis*, Plant Cell 24 (2012) 507–518.

[29] A.M. Daulat, P. Maurice, C. Froment, J.L. Guillaume, C. Broussard, B. Monsarrat, P. Delagrange, R. Jockers, Purification and identification of G protein-coupled receptor protein complexes under native conditions, Mol. Cell. Proteom. 6 (2007) 835–844.

[30] J.P. Seifert, Y. Zhou, S.N. Hicks, J. Sondek, T.K. Harden, Dual activation of phospholipase C-epsilon by Rho and Ras GTPases, J. Biol. Chem. 283 (2008) 29690–29698.

[31] D. Friedman, K. Lilley, Optimizing the Difference Gel Electrophoresis (DIGE) Technology, in: A. Vlahou (Ed.), Humana Press, 2008, pp. 93–124.

[32] S. Alvarez, B.M. Berla, J. Sheffield, R.E. Cahoon, J.M. Jez, L.M. Hicks, Comprehensive analysis of the *Brassica juncea* root proteome in response to cadmium exposure by complementary proteomic approaches, Proteomics 9 (2009) 2419–2431.

[33] W.O. Slade, E.G. Werth, E.W. McConnell, S. Alvarez, L.M. Hicks, Quantifying reversible oxidation of protein thiols in photosynthetic organisms, J. Am. Soc. Mass Spectrom. 26 (2015) 631–640.

[34] G. Liu, J. Zhang, B. Larsen, C. Stark, A. Breitkreutz, Z.Y. Lin, B.J. Breitkreutz, Y. Ding, K. Colwill, A. Pasculescu, T. Pawson, J.L. Wrana, A.I. Nesvizhskii, B. Raught, M. Tyers, A.C. Gingras, ProHits: integrated software for mass spectrometry-based interaction proteomics, Nat. Biotechnol. 28 (2010) 1015–1017.

[35] G. Teo, G. Liu, J. Zhang, A.I. Nesvizhskii, A.C. Gingras, H. Choi, SAINTexpress: improvements and additional features in significance analysis of INTeractome software, J. Proteom. 100 (2014) 37–43.

[36] H. Choi, G. Liu, D. Mellacheruvu, M. Tyers, A.C. Gingras, A.I. Nesvizhskii, Analyzing protein-protein interactions from affinity purification-mass spectrometry data with SAINT, Curr. Protoc. Bioinform. (2012), Chapter 8, Unit8.15.

[37] C.T. Lopes, M. Franz, F. Kazi, S.L. Donaldson, Q. Morris, G.D. Bader, Cytoscape, Web: an interactive web-based network browser, Bioinformatics 26 (2010) 2347–2348.

[38] X. Yi, Z. Du, Z. Su, PlantGSEA: a gene set enrichment analysis toolkit for plant community, Nucleic Acids Res. 41 (2013) W98–103.

[39] F. Supek, M. Bosnjak, N. Skunca, T. Smuc, REVIGO summarizes and visualizes long lists of gene ontology terms, PLoS One 6 (2011) e21800.

[40] S. Lalonde, A. Sero, R. Pratelli, G. Pilot, J. Chen, M.I. Sardi, S.A. Parsa, D.Y. Kim, B.R. Acharya, E.V. Stein, H.C. Hu, F. Villiers, K. Takeda, Y. Yang, Y.S. Han, R. Schwacke, W. Chiang, N. Kato, D. Loque, S.M. Assmann, J.M. Kwak, J.I. Schroeder, S.Y. Rhee, W.B. Frommer, A membrane protein/signaling protein interaction network for *Arabidopsis* version AMPv2, Front. Physiol. 1 (2010) 24.

[41] P. Obrdlik, M. El-Bakkoury, T. Hamacher, C. Cappellaro, C. Vilarino, C. Fleischer, H. Ellerbrok, R. Kamuzinzi, V. Ledent, D. Blaudez, D. Sanders, J.L. Revuelta, E. Boles, B. Andre, W.B. Frommer, K+ channel interactions detected by a genetic system optimized for systematic studies of membrane protein interactions, Proc. Natl. Acad. Sci. U. S. A. 101 (2004) 12242–12247.

[42] M. Babu, J. Vlasblom, S. Pu, X. Guo, C. Graham, B.D. Bean, H.E. Burston, F.J. Vizeacoumar, J. Snider, S. Phanse, V. Fong, Y.Y. Tam, M. Davey, O. Hnatshak, N. Bajaj, S. Chandran, T. Punna, C. Christopolous, V. Wong, A. Yu, G. Zhong, J. Li, I. Stagljar, E. Conibear, S.J. Wodak, A. Emili, J.F. Greenblatt, Interaction landscape of membrane-protein complexes in *Saccharomyces cerevisiae*, Nature 489 (2012) 585–589.

[43] S. Rivas, T. Mucyn, H.A. van den Burg, J. Vervoort, J.D. Jones, An approximately 400 kDa membrane-associated complex that contains one molecule of the resistance protein Cf-4, Plant J. 29 (2002) 783–796.

[44] P.J. Brown, A. Schonbrunn, Affinity purification of a somatostatin receptor–G-protein complex demonstrates specificity in receptor–G-protein coupling, J. Biol. Chem. 268 (1993) 6668–6676.

[45] C. Becamel, S. Gavarini, B. Chanrion, G. Alonso, N. Galeotti, A. Dumuis, J. Bockaert, P. Marin, The serotonin 5-HT2A and 5-HT2C receptors interact with specific sets of PDZ proteins, J. Biol. Chem. 279 (2004) 20257–20266.

[46] R.B. Rodrigues, G. Sabat, B.B. Minkoff, H.L. Burch, T.T. Nguyen, M.R. Sussman, Expression of a translationally fused TAP-tagged plasma membrane proton pump in *Arabidopsis thaliana*, Biochemistry 53 (2014) 566–578.

[47] C. Maurel, Y. Boursiac, D.T. Luu, V. Santoni, Z. Shahzad, L. Verdoucq, Aquaporins in plants, Physiol. Rev. 95 (2015) 1321–1358.

[48] F.C. Denison, A.L. Paul, A.K. Zupanska, R.J. Ferl, 14-3-3 Proteins in plant physiology, Semin. Cell Dev. Biol. 22 (2011) 720–727.

[49] E. Nielsen, A.Y. Cheung, T. Ueda, The regulatory RAB and ARF GTPases for vesicular trafficking, Plant Physiol. 147 (2008) 1516–1526.

[50] H. Choi, B. Larsen, Z.Y. Lin, A. Breitkreutz, D. Mellacheruvu, D. Fermin, Z.S. Qin, M. Tyers, A.C. Gingras, A.I. Nesvizhskii, SAINT: probabilistic scoring of affinity purification-mass spectrometry data, Nat. Methods 8 (2011) 70–73.

[51] S. Moon, S. Han, D. Kim, Y. Jin, W.K. Ho, Y. Kim, Interactome analysis of AMP-activated protein kinase (AMPK)-alpha1 and -beta1 in INS-1 pancreatic beta-cells by affinity purification-mass spectrometry, Sci. Rep. 4 (2014) 4376.

[52] M. Taipale, G. Tucker, J. Peng, I. Krykbaeva, Z.Y. Lin, B. Larsen, H. Choi, B. Berger, A.C. Gingras, S. Lindquist, A quantitative chaperone interaction network reveals the architecture of cellular protein homeostasis pathways, Cell 158 (2014) 434–448.

[53] D.V. Skarra, M. Goudreault, H. Choi, M. Mullin, A.I. Nesvizhskii, A.C. Gingras, R.E. Honkanen, Label-free quantitative proteomics and SAINT analysis enable interactome mapping for the human Ser/Thr protein phosphatase 5, Proteomics 11 (2011) 1508–1516.

[54] A.M. Jones, Y. Xuan, M. Xu, R.S. Wang, C.H. Ho, S. Lalonde, C.H. You, M.I. Sardi, S.A. Parsa, E. Smith-Valle, T. Su, K.A. Frazer, G. Pilot, R. Pratelli, G. Grossmann, B.R. Acharya, H.C. Hu, C. Engineer, F. Villiers, C. Ju, K. Takeda, Z. Su, Q. Dong, S.M. Assmann, J. Chen, J.M. Kwak, J.I. Schroeder, R. Albert, S.Y. Rhee, W.B. Frommer, Border control—a membrane-linked interactome of *Arabidopsis*, Science 344 (2014) 711–716.

[55] A. Zhou, Y. Bu, T. Takano, X. Zhang, S. Liu, Conserved V-ATPase c subunit plays a role in plant growth by influencing V-ATPase-dependent endosomal trafficking, Plant Biotechnol. J. (2015).

[56] M. Wan, W. Zhang, Y. Tian, C. Xu, T. Xu, J. Liu, R. Zhang, Unraveling a molecular determinant for clathrin-independent internalization of the M2 muscarinic acetylcholine receptor, Sci. Rep. 5 (2015) 11408.

[57] G. Lavezzari, K.W. Roche, Constitutive endocytosis of the metabotropic glutamate receptor mGluR7 is clathrin-independent, Neuropharmacology 52 (2007) 100–107.

[58] X. Chen, N.G. Irani, J. Friml, Clathrin-mediated endocytosis: the gateway into plant cells, Curr. Opin. Plant Biol. 14 (2011) 674–682.

[59] V.V. Gurevich, E.V. Gurevich, Structural determinants of arrestin functions, Prog. Mol. Biol. Transl. Sci. 118 (2013) 57–92.

[60] L. Aubry, D. Guetta, G. Klein, The arrestin fold: variations on a theme, Curr. Genom. 10 (2009) 133–142.

[61] F. Delom, D. Fessart, Role of phosphorylation in the control of clathrin-mediated internalization of GPCR, Int. J. Cell Biol. 2011 (2011) 246954.

[62] D. Calebiro, V.O. Nikolaev, M. Persani, M.J. Lohse, Signaling by internalized G-protein-coupled receptors, Trends Pharmacol. Sci. 31 (2010) 221–228.

[63] A. Sorkin, M. von Zastrow, Endocytosis and signalling: intertwining molecular networks, Nat. Rev. Mol. Cell Biol. 10 (2009) 609–622.

[64] V.W. Hsu, M. Bai, J. Li, Getting active: protein sorting in endocytic recycling, Nat. Rev. Mol. Cell Biol. 13 (2012) 323–328.

[65] J.L. Seachrist, S.S. Ferguson, Regulation of G protein-coupled receptor endocytosis and trafficking by Rab GTPases, Life Sci. 74 (2003) 225–235.

[66] M.T. Drake, S.K. Shenoy, R.J. Lefkowitz, Trafficking of G protein-coupled receptors, Circ. Res. 99 (2006) 570–582.

[67] E. Macia, M. Partisani, O. Paleotti, F. Luton, M. Franco, Arf6 negatively controls the rapid recycling of the beta2 adrenergic receptor, J. Cell. Sci. 125 (2012) 4026–4035.

[68] C. Reiner, N.M. Nathanson, The internalization of the M2 and M4 muscarinic acetylcholine receptors involves distinct subsets of small G-proteins, Life Sci. 82 (2008) 718–727.

[69] M. Prigent, T. Dubois, G. Raposo, V. Derrien, D. Tenza, C. Rosse, J. Camonis, P. Chavrier, ARF6 controls post-endocytic recycling through its downstream exocyst complex effector, J. Cell Biol. 163 (2003) 1111–1121.

[70] L.B. Dale, J.L. Seachrist, A.V. Babwah, S.S. Ferguson, Regulation of angiotensin II type 1A receptor intracellular retention degradation, and recycling by Rab5, Rab7, and Rab11 GTPases, J. Biol. Chem. 279 (2004) 13110–13118.

[71] M. Abramow-Newerly, H. Ming, P. Chidiac, Modulation of subfamily B/R4 RGS protein function by 14-3-3 proteins, Cell. Signal. 18 (2006) 2209–2222.

8

CressInt: A user-friendly web resource for genome-scale exploration of gene regulation in *Arabidopsis thaliana*

Xiaoting Chen [a,b,1], Kevin Ernst [a,b,1], Frances Soman [a], Mike Borowczak [a,b,2], Matthew T. Weirauch [b,c,*]

[a] Department of Electrical Engineering and Computing Systems, College of Engineering and Applied Sciences, University of Cincinnati, Cincinnati, OH 45221, United States
[b] Center for Autoimmune Genomics and Etiology, Cincinnati Children's Hospital Medical Center, Department of Pediatrics, College of Medicine, University of Cincinnati, Cincinnati, OH 45229, United States
[c] Division of Biomedical Informatics and Division of Developmental Biology, Cincinnati Children's Hospital Medical Center, Department of Pediatrics, College of Medicine, University of Cincinnati, Cincinnati, OH 45229, United States

ARTICLE INFO

Keywords:
Arabidopsis
Functional genomics
Transcription factors
Gene regulation
Systems biology
Web server
Computational tools

ABSTRACT

The thale cress *Arabidopsis thaliana* is a powerful model organism for studying a wide variety of biological processes. Recent advances in sequencing technology have resulted in a wealth of information describing numerous aspects of *A. thaliana* genome function. However, there is a relative paucity of computational systems for efficiently and effectively using these data to create testable hypotheses. We present *CressInt*, a user-friendly web resource for exploring gene regulatory mechanisms in *A. thaliana* on a genomic scale. The *CressInt* system incorporates a variety of genome-wide data types relevant to gene regulation, including transcription factor (TF) binding site models, ChIP-seq, DNase-seq, eQTLs, and GWAS. We demonstrate the utility of *CressInt* by showing how the system can be used to (1) Identify TFs binding to the promoter of a gene of interest; (2) identify genetic variants that are likely to impact TF binding based on a ChIP-seq dataset; and (3) identify specific TFs whose binding might be impacted by phenotype-associated variants.

1. Introduction

The sequencing of the *Arabidopsis thaliana* genome over 15 years ago [1] enabled a new era of scientific exploration of this versatile model organism. As "next generation" sequencing technologies continue to mature, datasets capable of measuring function on a genome-wide scale continue to become more prevalent. Despite an exponential increase in our ability to generate data probing function on a genome-scale, there remains a lag in our analytical

capability to effectively analyze these data to attain new biological insights.

Several useful bioinformatics tools are currently in widespread use in the *Arabidopsis* community (see de Lucas et al. [2], Bassel et al. [3], and Brady and Provart [4] for reviews). However, as more complex and higher resolution data types become available, there is an increasing need for the development of user-friendly computational tools for their analysis. In the past five years alone, *Arabidopsis* data have been released describing genetic variants associated with particular traits [5] or altered gene expression levels [6], open chromatin regions in multiple tissue types and conditions [7], and DNA binding specificities for hundreds of transcription factors (TFs) [8]. Collectively, these data offer new opportunities to probe gene regulation and genome function. However, access to the wide range of analytical capabilities afforded by these data remains largely limited to bioinformaticians.

We present CressInt (thale cress data intersector), a user-friendly, freely accessible web server for integrating and analyzing genome-scale *A. thaliana gene* datasets. Conceptually, CressInt is similar to visually analyzing data in a genome browser such as those

Abbreviations: TF, transcription factor; ChIP-seq, Chromatin immunoprecipitation followed by sequencing; DNase-seq, sequencing of DNase I hypersensitive sites; eQTL, expression quantitative trait locus; GWAS, genome-wide association study; PBM, protein binding microarray.
* Corresponding author at: Center for Autoimmune Genomics and Etiology, Cincinnati Children's Hospital Medical Center, Department of Pediatrics, College of Medicine, University of Cincinnati, Cincinnati, OH 45229, United States.
E-mail address: Matthew.Weirauch@cchmc.org (M.T. Weirauch).
[1] These authors contributed equally.
[2] Current affiliation: Erebus Labs, Laramie, WY 82073, United States.

provided by UC Santa Cruz [9] or Ensembl [10], with the key differences that (1) up to thousands of loci of interest can be queried at once; (2) quality-controlled data specific to *A. thaliana* are preloaded into the CressInt system; and (3) results are downloadable in formats easily amenable to further downstream analysis. CressInt combines data from a wide variety of sources, including TF genomic binding regions (from ChIP-seq), TF DNA binding specificities (from Protein Binding Microarrays (PBMs) [11]), chromatin accessibility (DNAse-seq), and genetic variants associated with specific phenotypes (from GWAS) or genotype-dependent gene expression levels (*i.e.*, expression quantitative trait loci or eQTLs). The CressInt system enables a wide range of queries, from simple (*e.g.*, identifying all datasets intersecting genomic regions of interest), to complex (*e.g.*, identifying genetic variants of interest likely to affect the binding of specific TFs). To our knowledge, there is currently no web server capable of performing these operations on *A. thaliana* datasets that are already integrated into the system. This not only enables easy access to these data for non-computational experts, but also saves hours of time that would otherwise be spent identifying, obtaining, quality checking, and re-formatting the various data sets.

CressInt's intuitive graphical user interface is designed to be easy to use for non-bioinformaticians, while maintaining sufficient power and capabilities to enable downstream computational analysis. Using three case studies, we demonstrate the ability of CressInt to effectively use functional genomics data to generate testable hypotheses involving genes or phenotypes of interest. The CressInt web server is freely available at https://cressint.cchmc.org.

2. Materials and methods

2.1. Data and code availability

All source code developed for the web server is available on Bitbucket (https://bitbucket.org/weirauchlab/tf-tools-cressint). All datasets are available from the authors upon request.

2.2. Data collection and quality control

We obtained data from a variety of sources (Table 1). All genome-based datasets are organized by plant tissue type (*e.g.*, seedling, leaf, inflorescence, *etc.*), and stored as UC Santa Cruz BED6 files [9]. DNase-seq data indicating open chromatin regions in *A. thaliana* seedlings exposed to heatshock, darkness, and light were taken from Sullivan et al. [7]. 4,355,790 naturally occurring genetic variants and eQTLs derived from seedlings were obtained from Gan et al. [6]. The eQTL set was filtered to only include SNPs with *P*-values <0.001. GWAS data were obtained from Atwell et al. [5], and genetic variants were included in our set of phenotype-associated variants if they either (1) have associations exceeding genome-wide significance ($P < 2.75 \times 10^{-7}$, which corresponds to the Bonferroni-corrected $P < 0.05$ cutoff used in the original study; 178 SNPs in total) or (2) are among the top 10 most strongly associated variants for each phenotype, regardless of *P*-value (943 SNPs in total). TF binding specificity models were taken from build 1.01 of the CisBP database [8] (http://cisbp.ccbr.utoronto.ca/).

We obtained ChIP-seq data from the gene expression omnibus (GEO) [12]. Beginning with all 26 *A. thaliana* ChIP-seq datasets available in GEO in March of 2015, we used a three-step quality control procedure to ensure that only high-quality datasets are included in the CressInt system. First, we removed any datasets whose peak regions cover >5% of the *A. thaliana* genome, deeming them too non-specific (with the exception of ChIP-seq for histone marks, which mark general regulatory regions and tend to have wider peaks). Next, we removed any datasets where the number of peaks obtained from the GEO dataset did not match the number

of peaks reported in the publication associated with the data—this step is necessary because both GEO datasets and methods sections of manuscripts are often insufficiently documented to reproduce the reported peak calls. Finally, we ran all peak sets through the TF DNA binding motif enrichment algorithm used by HOMER [13], and only included datasets where the ChIP'ed TF's motif ranks in the top three of enriched motifs. A total of 16 ChIP-seq datasets, taken from 13 different studies, passed our QC process (Table 1).

2.3. Differential binding of transcription factors to genetic variants

We used PBM data describing the DNA binding specificities of 575 *A. thaliana* TFs taken from Weirauch et al. [8], and a similar procedure used in that study and another recent study [14] to identify TFs whose binding might be affected by the alleles of 4,355,790 naturally occurring *A. thaliana* genetic variants [6]. One type of data produced by a PBM experiment is the E-score, which ranges from −0.50 to +0.50, and quantifies the relative preference of the binding of the tested TF to each of the 32,896 possible 8 base sequences [11]. We constructed a matrix containing the PBM 8-mer E-scores for 534 PBM experiments (267 constructs, each assayed on two independent array designs). 466 of these experiments directly assay the DNA binding specificity an *A. thaliana* TF. 68 of them measure a related TF in another organism that has a similar DNA binding domain (DBD) to at least one *A. thaliana* TF (68 experiments). Each PBM experiment was mapped to its "closest" *A. thaliana* TF by either (1) assigning it to the *A. thaliana* TF that was directly measured (trivial); or (2) (for PBMs measuring non-*A. thaliana* TFs) assigning it to the *A. thaliana* TF with the most similar DBD (based on percent amino acid identity in DBD alignments—see Weirauch et al. [8] for details of how thresholds for these inferred binding specificities are established).

We then scored the alleles of each genetic variant using the resulting 8-mer E-score matrix. For a given variant, we first determined all 8-mers in the reference genome sequence overlapping each allele—for example, a SNP will overlap eight 8-mers, plus their reverse complements, for each allele. For each PBM experiment, we then identified the highest scoring 8-mer E-score attained by any of the reference allele sequences (E_{ref}), and the highest attained by any non-reference allele ($E_{non-ref}$). We then identified all PBM experiments where only one of E_{ref} and $E_{non-ref}$ has an E-score value exceeding 0.45 (values above this threshold will likely be strongly bound by the given TF [15]). All experiments meeting this criterion were then assigned a final score E_{final}, which is the maximum value of (E_{ref} and $E_{non-ref}$). Finally, we also calculated the predicted difference in binding strength between the two alleles as $E_{delta} = |E_{ref} - E_{non-ref}|$. We then created a final ranked list of TFs (sorted by E_{final}) whose binding is likely to be affected by the alleles of a given SNP (*e.g.*, strongly binding to one allele, but not binding to the other).

2.4. Web server implementation

The user interface to the CressInt analysis pipeline is served by a GNU/Linux virtual machine running CentOS 6 and the Apache 2.2 web server. The web front-end is implemented primarily as HTML "templates" rendered through the use of a PHP library (http://twig.sensiolabs.org/), which maintains a separation of concerns between interface and application logic. Client-side JavaScript manages interaction among input form elements in the web front-end, and the form submission is done asynchronously (*via* Ajax), allowing certain types of validation errors such as missing inputs or malformed BED files to be detected and reported without a page reload. Input data for analysis is received and processed by a Perl CGI (Common Gateway Interface) script, which in turn inter-

Table 1
Datasets incorporated into the *CressInt* system.

Data type	Source	Description
TF DNA binding specificity models	Weirauch et al. [8]	CisBP database, which contains thousands of TF binding models across eukaryotes
DNase-seq	Sullivan et al. [7]	Genome-wide mapping of DNase hypersensitive sites in *A. thaliana* seedlings
Genetic variants and eQTLs	Gan et al. [6]	Multiple reference genomes and transcriptomes for 19 *A. thaliana* strains
GWAS	Atwell et al. [5]	Genome-wide association study of 107 phenotypes in *A. thaliana* inbred lines
ChIP-seq	Chica et al. [24]	H3K4me3 and H3K27me3 marks in leaves
ChIP-seq	Willing et al. [25]	H3K27me1 marks in leaves
ChIP-seq	Ómaoiléidigh et al. [26]	AG binding in inflorescences
ChIP-seq	Pajoro et al. [27]	AP1 and SEP3 binding in inflorescences
ChIP-seq	Wuest et al. [28]	AP3 and PI binding in flowers
ChIP-seq	Oh et al. [29]	ARF6 binding in seedlings
ChIP-seq	Heyman et al. [30]	ERF115 binding in dark growing cells
ChIP-seq	Fan et al. [31]	HBI1 binding in seedlings
ChIP-seq	Zhiponova et al. [32]	IBL1 binding in seedlings
ChIP-seq	Moyroud et al. [33]	LFY binding in seedlings
ChIP-seq	Pfeiffer et al. [34]	PIF1 binding in seedlings
ChIP-seq	Zhang et al. [35]	PIF3 binding in seedlings
ChIP-seq	Brandt et al. [36]	REV binding in seedlings
ChIP-seq	Huang et al. [37]	TOC1 binding in seedlings

faces with an in-house high-performance computing (HPC) cluster (currently containing over 700 processing cores) through a set of locally-developed Perl modules, generating shell scripts for batch processing. These Perl modules abstract away the implementation details of the batch facility (IBM's load sharing facility [LSF]) and allow interfaces to be written for other local or remote HPC load-sharing systems without impacting the front-end web service. Intersection analyses are performed using the BedTools suite [16], along with custom-written code written in C++. A user may optionally provide an email address to receive notification of a completed CressInt analysis, or may simply leave the web browser open and wait for the job to complete.

3. Results

3.1. Overview of the CressInt system

CressInt is designed to be easily useable for non-computational experts, while also maintaining sufficient power to be suitable for advanced downstream computational analyses. The system accepts one of three different types of inputs (Fig. 1, top): (1) Genomic coordinates (in UC Santa Cruz BED3, BED4, BED5, or BED6 formats); (2) Gene lists (either common gene names or TAIR IDs); or (3) A set of phenotypes of interest, taken from a recent GWAS study [5]. The user can also choose to include or exclude functional genomics datasets based on data or tissue type (Fig. 1, middle). After error and format checking, CressInt converts the input into a set of labeled genomic coordinates (in BED file format) and intersects these coordinates with the selected datasets. Two sets of results are presented to users (Fig. 1, bottom): (1) The intersection results, which indicate all data sets in the system whose coordinates overlap with the input set; and (2) TF differential binding predictions, which identify genetic variants that might impact the binding of specific TFs.

The CressInt system currently includes several different types of datasets relevant to gene regulation (Table 1). TF DNA binding specificity models taken from the CisBP database [8] are used to identify the specific TFs that might differentially bind a given genetic variant. The models are based on a large collection of universal PBM experiments covering 575 *A. thaliana* TFs [8]. Briefly, universal PBMs are double-stranded microarrays whose probes are designed such that all possible 10 base sequences occur exactly once, and hence all non-palindromic 32,896 8-base sequences occur 32 times in diverse flanking sequence contexts [11]. The resulting data, which track well with binding affinity [15], there-

fore offer a robust estimate of the binding of the assayed TF to every possible 8-base sequence. Although *in vitro*-derived TF binding specificities are in general reflective of *in vivo* specificities [17], we note that there can be exceptions (*e.g.*, in cases where a TF's binding is modified by a co-factor). Using this collection of TF binding models, CressInt has the capability to systematically scan the alleles of a given genetic variant to identify the particular TFs whose binding it might affect (see Section 2).

In addition to TF binding models, CressInt incorporates ChIP-seq datasets taken from a variety of studies assaying either the binding of specific TFs (14 datasets), or histone marks that are indicative of chromatin state (three datasets) (Table 1). All ChIP-seq datasets were subjected to a rigorous three step quality control procedure before being considered for inclusion in the system (see Section 2). CressInt also includes DNase-seq datasets taken from a recent large-scale study [7] and the Plant Regulome database (http://plantregulome.org/public/). DNase-seq is a next-generation sequencing assay that identifies DNase hypersensitive regions on a genome scale, and hence is capable of identifying regions of open chromatin in a certain tissue type [18]. Thus, DNase-seq data are useful for identifying areas of the genome that are likely to function as regulatory regions bound by TFs.

CressInt also includes the full set of 4,355,790 genetic variants identified in a recent study comparing the genomic sequences of 19 *A. thaliana* strains [6]. Among these variants, 317,570 are *cis* expression quantitative trait loci (eQTLs) taken from the same study (at a cutoff of $P < 0.001$). *Cis* eQTLs are variants that affect the expression of a nearby gene as a function of genotype. Thus, eQTLs are useful for identifying functional variants that are likely to affect the binding of TFs. CressInt also includes a set of 1004 genetic variants that are associated with one of 107 traits and phenotypes analyzed in a recent genome-wide association study (GWAS) [5] (see Section 2). Such variants provide important clues for understanding genome function and biological diversity, due to their ability to modulate a phenotype in a genotype-dependent manner.

In the following sections, we demonstrate the power of CressInt by presenting case studies of how a user might use the system. First, we show how it can be used to identify TFs that bind to the promoter of a gene of interest. Next, we demonstrate how a user can input their own ChIP-seq data in order to identify genetic variants within the ChIP peaks that might impact the binding of the ChIP'ed TF. Finally, we show how CressInt can be used to find TF binding sites that might be impacted by genetic variants associated with a particular phenotype.

Input

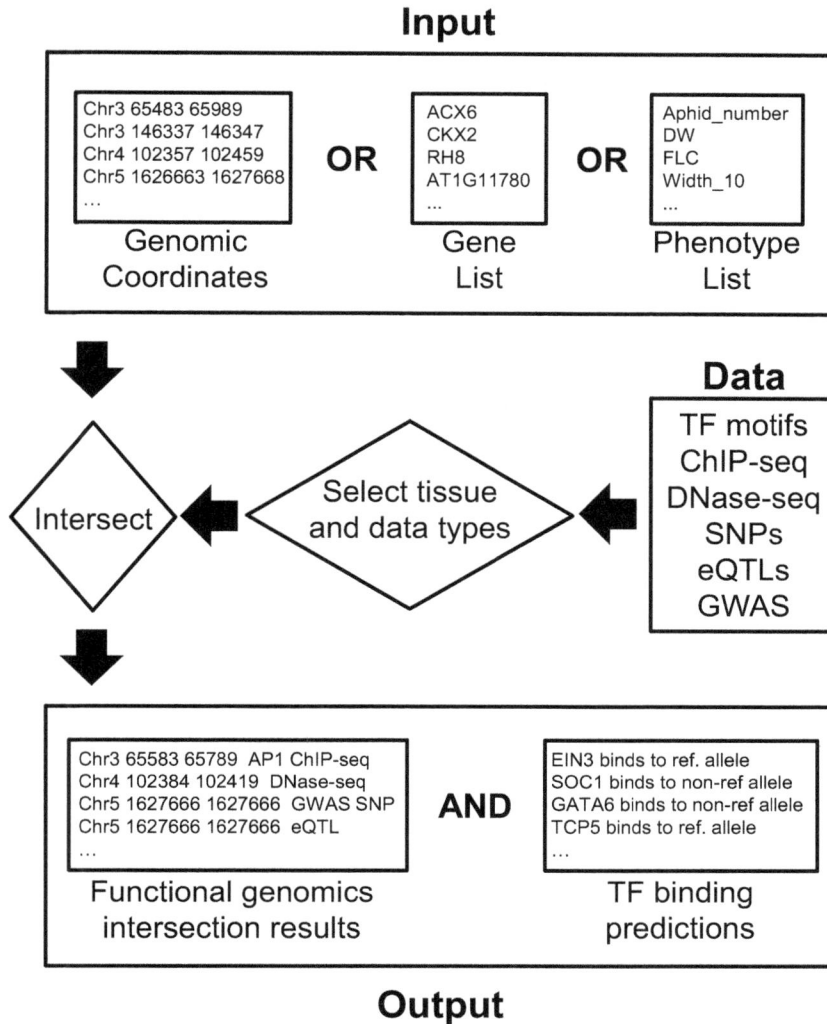

Fig. 1. Overview of the CressInt workflow.
As input, users can supply genomic coordinates, gene names, gene IDs, or phenotypes of interest (top). CressInt contains a wide range of genome-indexed data sources, which can be selected based on the data type or source tissue (middle—see also Table 1). Upon submitting a job, the user input is "intersected" with the selected data types, and results are displayed providing overlapping genomic coordinates (bottom left) and TF binding sites that overlap genetic variants (bottom right).

3.2. Case study 1: identifying TFs binding to a promoter of interest

A fundamental, powerful feature of CressInt is its ability to query genomic regions of interest to generate hypotheses. For example, consider the case where a user is interested in identifying potential regulators of the AGL20/SOC1 gene, which encodes a MADS box family TF that controls the flowering process. To identify possible candidates, the user simply enters "AGL20" and defines the desired promoter search space (for this example, we use the region starting at the AGL20 TSS and extending 1000 bases upstream). Upon completion of the job, the user is provided with data indicating that five different MADS box family TFs all bind the AGL20 promoter, based on ChIP-seq experiments performed in flower and inflorescence tissues taken from three different studies (Fig. 2A). Strikingly, all five TFs also play established roles in flower development. Further, MADS box TFs form homo- and hetero-dimers upon binding DNA [19], suggesting that these TFs might cooperatively regulate AGL20. Further supporting the CressInt-generated hypothesis, MADS box TFs recognize the CArG-box upon binding DNA [20], and there are five putative CArG boxes in the AGL20 promoter region (Fig. 2A). A pair is located directly within the peak summits of all five TFs, with a third and fourth located just upstream, also near peak summits

for all five TFs. In summary, through this simple query, we have identified specific binding sites for TFs likely to regulate a gene of interest.

3.3. Case study 2: identifying genetic variants likely to affect the binding of a ChIP'ed TF

To demonstrate how CressInt can generate specific hypotheses from a user-provided genome-wide dataset, we submitted PIF1/PIL5ChIP-seq peak regions in seedlings to CressInt as input, and asked the system to identify all likely PIF1 binding sites within these peaks that overlap naturally occurring genetic variants. In total, CressInt identified 53 variants that have a strong predicted binding site (E-score > 0.45) in the Col-0 (reference) strain, and a weak site (E-score < 0.30) in at least one other strain. For example, one variant, located in the promoter region of the RGA gene (Fig. 2B), has a reference allele that is predicted to be very strongly bound (E-score = 0.499), with weak binding predicted for the alternative allele (E-score = 0.172). Fig. 2C depicts the reference and non-reference allele sequences of this variant, along with flanking genomic bases. The reference allele perfectly matches the ideal PIF1 binding site (top sequence), while the non-reference allele

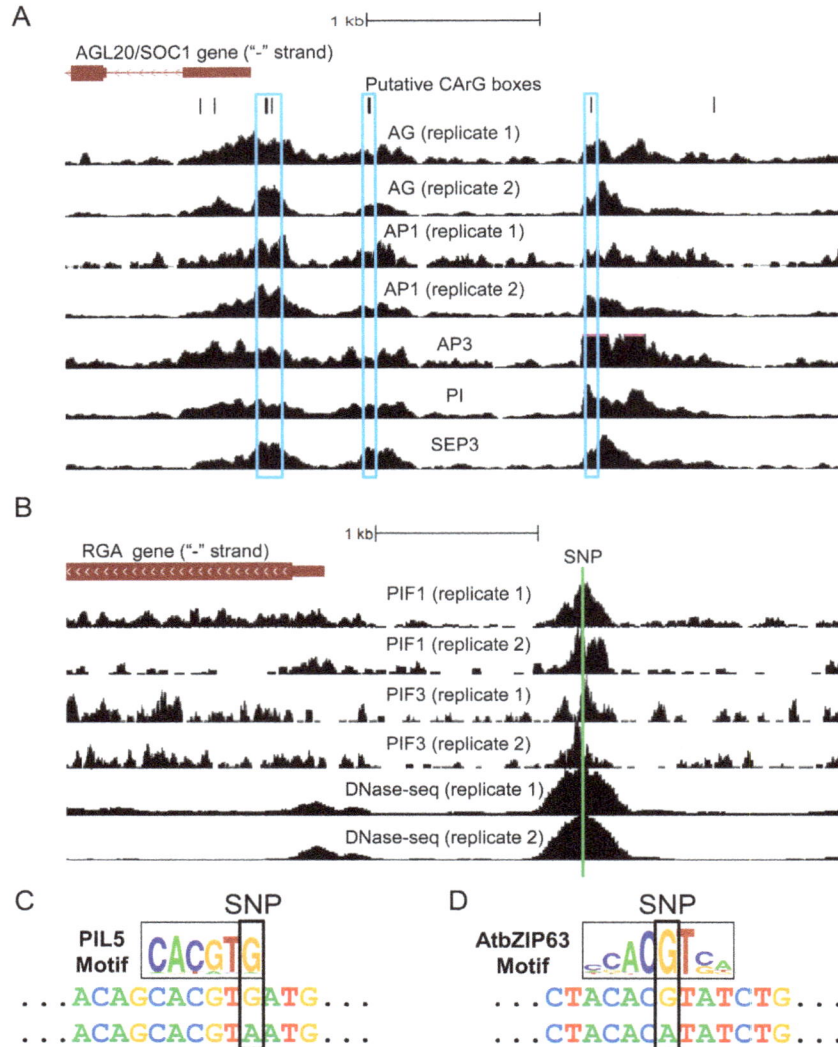

Fig. 2. Sample applications of CressInt.
A. Discovery of TFs binding a promoter of interest, as discussed in "case study 1". UC Santa Cruz genome browser [9] screenshot depicting locations of (top to bottom): AGL20/SOC1 gene, putative CArG boxes (recognized by MADS box family TFs), and ChIP-seq data for seven experiments describing the genome-wide binding of five MADS-family TFs in flower and inflorescence tissues. Blue boxes, which contain CArG boxes and ChIP-seq peaks for multiple TFs, indicate likely binding sites. **B**. Identification of genetic variants likely to impact the binding of a ChIP'ed TF, as discussed in "case study 2". See (A) for explanation. Green vertical line indicates location of the SNP discussed in "case study 2". **C**. Data supporting the differential binding of the PIL5 TF to the "case study 2" SNP. Sequence logo [23] at top indicates the preferred base at each position of the PIL5 DNA recognition sequence – taller nucleotides indicate preference for the corresponding base at the corresponding position. DNA sequences below indicate the two alleles of the SNP, along with flanking genomic bases. Note that the reference allele (top) is a strong match to the PIL5 DNA binding motif, but the non-reference sequence (bottom) is not. **D**. Data supporting the differential binding of the AtbZIP63 TF to the "case study 3" SNP. See (C) for explanation.

"breaks" this site (bottom sequence). Both PIF1 and RGA are TFs involved in negative regulation of seed germination through participation in the gibberellic acid-mediated signaling pathway. Further, PIF1 directly increases the expression of RGA by binding to the same site identified by CressInt, with binding being abolished upon mutation of this site [21]. This example demonstrates that CressInt can be used to identify naturally occurring genetic variants that might impact the functional binding of a particular TF. Specifically, it shows how a genome-wide ChIP-seq dataset can be used to formulate the specific hypothesis that in the Mt-0 strain, which harbors the alternative allele of this variant, the direct regulation of RGA by PIF1 is likely attenuated, due to decreased PIF1 binding at this site. Intriguingly, this locus also overlaps ChIP-seq peaks for PIF3/POC1, a bHLH family TF that also recognizes G-box motifs (Fig. 2**B**). Like PIF1, PIF3 is a member of the phytochrome interacting factor family of TFs, and the two proteins form heterodimers upon binding the G-box [22]. Thus, by intersecting the PIF1 ChIP-seq

dataset with other datasets available in CressInt, we have arrived at the testable hypothesis that PIF1 and PIF3 might cooperatively bind the RGA promoter, and that this interaction might be impacted by a naturally occurring genetic variant.

3.4. Case study 3: identifying TF binding sites likely to be affected by genetic variants associated with a phenotype of interest

As a final illustration, consider a user that is interested in the molecular mechanisms underlying the genetic determinants of flowering time. This user would start by selecting relevant phenotypes (*e.g.* "FT_field", which counts the number of days between germination date and appearance of the first flower). CressInt then finds all genetic variants associated with the selected phenotypes, and identifies potential TFs whose binding they might affect. One such example is illustrated in Fig. 2**D**, for a variant located 130 bases upstream of the ATCAF1B gene, which is expressed in flowers and

Fig. 3. The *CressInt* web server front page.
Screenshot of the CressInt user interface. Here, users can select the "mode" they would like to run (top), enter input data (middle top), select from available datasets (middle bottom), and provide information on the job being submitted (bottom).

plays a putative role in mRNA deadenylation. Although no clues for the function of this variant can be gleaned from the available functional genomics data (since it does not overlap any datasets), the expression of ATCAF1B is known to be affected by the genotype of the variant (*i.e.*, it was identified as an eQTL in Gan et al. [6]), suggesting that its functional impact on flowering time is likely due to TF binding events specific to one of its alleles. Based on

CressInt's output, the top TF candidate is AtbZIP63, which is predicted to strongly bind the "G" allele (E-score = 0.463), and not bind the "A" allele (E-score = 0.138) (Fig. **2D**). Importantly, AtbZIP63 is also expressed in flowers. Further, a different genetic variant also associated with flowering time is located proximal to the AtbZIP63 gene locus. Collectively, these results implicate a potential role for AtbZIP63 in flowering time determination, and specifically suggest that a flowering time-associated variant located in the promoter of ATCAF1B acts by causing differential binding of this TF. Further, they demonstrate how CressInt can be used to generate testable hypotheses for mechanisms underlying a particular phenotype, even without *a priori* knowledge of specific genes or genomic regions of interest.

3.5. The CressInt web interface

CressInt is available at http://cressint.cchmc.org. The CressInt web server has been tested on several web browsers, including Google Chrome, Firefox, and Internet Explorer 10 and 11. In addition to the main page for creating a new job, the web site has several additional useful features, including Help and FAQ pages, details on the incorporated datasets, an update log, links to a variety of other *Arabidopsis* web resources, and a contact page.

The CressInt web interface was designed to be easy to use, yet flexible (Fig. 3). At the top of the page, a user can select between three modes of operation (corresponding to the three case studies above): *Intersect*, *Find TF/SNPs*, and *Phenotypes to TF/SNPs*. *Intersect* identifies all datasets that overlap the user query. *Find TF/SNPs* identifies genetic variants located within the user query regions, and predicts the TFs whose binding they might affect. *Phenotypes to TF/SNPs* starts with a phenotype of interest, and identifies TFs whose binding might be affected by variants associated with that phenotype. As described above, users can select from multiple input format options, including genomic regions, gene names, and phenotypes. Users can paste or type entries into the online form, or upload text files. In the 'Parameters' section, users can select the data and tissue types to be included in the query (by default, all datasets are included) (Fig. 3, middle). Before submitting a job, a user has the option to provide their email address for automatic notification upon completion of their job. There is also an option to name the job for future reference. Upon submitting a job, a new page appears that automatically refreshes while the jobs run, and posts the final results when they are ready.

There are two basic output pages of CressInt (Fig. 1): intersection results indicate all data in the system whose coordinates overlap with the input set, while TF differential binding predictions identify genetic variants that might impact the binding of specific TFs. Both outputs are formatted to be easily human-readable, with multiple visualization options. All input files and parameter choices are saved and documented for reference. Data can be easily sorted or filtered, and are downloadable in tab-delimited text format for processing in another application such as microsoft excel, or for additional downstream computational analysis.

4. Discussion and conclusions

We present the CressInt web server, a user-friendly system for leveraging *A. thaliana* functional genomics datasets to formulate testable hypotheses about gene regulation and genome function. Through three case studies, we offer examples of how the CressInt system can be used to explore *Arabidopsis* biology. The power of CressInt lies in its combination of intuitive design and its inclusion of a wide range of diverse genome-scale datasets. The flexibility of the CressInt system design enables easy inclusion of additional datasets as they become available, and we encourage members of the plant community to use the provided links to alert us of additional useful datasets. In the future, we plan to extend the CressInt system to other plant and non-plant model organisms. We expect that CressInt will be a useful addition to the *Arabidopsis* genomic toolkit, and anticipate that it will enable numerous insights into the function of plant genomes.

Acknowledgements

We thank Siddharth Dixit and Tuhin Mukherjee for computational support, Phillip Drewe for providing us with eQTL data, and Jeremy Riddell and other members of the Weirauch lab for helpful suggestions. We thank Leah Kottyan, Nathan Salomonis, and Stephen Waggoner for critical feedback on the manuscript. This work was supported in part by NIH grant R21HG008186 to MTW.

References

[1] Arabidopsis Genome Initiative, Analysis of the genome sequence of the flowering plant *Arabidopsis thaliana*, Nature 408 (6814) (2000) 796–815.
[2] M. de Lucas, N.J. Provart, S.M. Brady, Bioinformatic tools in *Arabidopsis* research, Methods Mol. Biol. 1062 (2014) 97–136.
[3] G.W. Bassel, et al., Systems analysis of plant functional: transcriptional, physical interaction, and metabolic networks, Plant Cell 24 (10) (2012) 3859–3875.
[4] S.M. Brady, N.J. Provart, Web-queryable large-scale data sets for hypothesis generation in plant biology, Plant Cell 21 (4) (2009) 1034–1051.
[5] S. Atwell, et al., Genome-wide association study of 107 phenotypes in *Arabidopsis thaliana* inbred lines, Nature 465 (7298) (2010) 627–631.
[6] X. Gan, et al., Multiple reference genomes and transcriptomes for *Arabidopsis thaliana*, Nature 477 (7365) (2011) 419–423.
[7] A.M. Sullivan, et al., Mapping and dynamics of regulatory DNA and transcription factor networks in *A. thaliana*, Cell Rep. 8 (September 25 (6)) (2014) 2015–2030.
[8] M.T. Weirauch, et al., Determination and inference of eukaryotic transcription factor sequence specificity, Cell 158 (6) (2014) 1431–1443.
[9] W.J. Kent, et al., The human genome browser at UCSC, Genome Res. 12 (6) (2002) 996–1006.
[10] P.J. Kersey, et al., Ensembl Genomes 2013: scaling up access to genome-wide data, Nucleic Acids Res. 42 (Database issue) (2014) D546–52.
[11] M.F. Berger, et al., Compact: universal DNA microarrays to comprehensively determine transcription-factor binding site specificities, Nat. Biotechnol. 24 (11) (2006) 1429–1435.
[12] T. Barrett, et al., NCBI GEO. Archive for functional genomics data sets—10 years on, Nucleic Acids Res. 39 (Database issue) (2011) D1005–D1010.
[13] S. Heinz, et al., Simple combinations of lineage-determining transcription factors prime *cis*-regulatory elements required for macrophage and B cell identities, Mol. Cell 38 (4) (2010) 576–589.
[14] L.C. Kottyan, et al., The IRF5-TNPO3 association with systemic lupus erythematosus has two components that other autoimmune disorders variably share, Hum. Mol. Genet. 24 (2) (2015) 582–596.
[15] M.F. Berger, et al., Variation in homeodomain DNA binding revealed by high-resolution analysis of sequence preferences, Cell 133 (7) (2008) 1266–1276.
[16] A.R. Quinlan, I.M. Hall, BEDTools: a flexible suite of utilities for comparing genomic features, Bioinformatics 26 (6) (2010) 841–842.
[17] J.M. Franco-Zorrilla, et al., DNA-binding specificities of plant transcription factors and their potential to define target genes, Proc. Natl. Acad. Sci. U. S. A. 111 (6) (2014) 2367–2372.
[18] A.P. Boyle, et al., High-resolution mapping and characterization of open chromatin across the genome, Cell 132 (2) (2008) 311–322.
[19] K. Kaufmann, R. Melzer, G. Theissen, MIKC-type MADS-domain proteins: structural modularity, protein interactions and network evolution in land plants, Gene 347 (2) (2005) 183–198.
[20] K. Huang, et al., Solution structure of the MEF2A-DNA complex: structural basis for the modulation of DNA bending and specificity by MADS-box transcription factors, EMBO J. 19 (11) (2000) 2615–2628.
[21] E. Oh, et al., PIL5: a phytochrome-interacting bHLH protein, regulates gibberellin responsiveness by binding directly to the GAI and RGA promoters in *Arabidopsis* seeds, Plant Cell 19 (4) (2007) 1192–1208.
[22] Q. Bu, et al., Dimerization and blue light regulation of PIF1 interacting bHLH proteins in *Arabidopsis*, Plant Mol. Biol. 77 (4-5) (2011) 501–511.
[23] G.E. Crooks, et al., WebLogo: a sequence logo generator, Genome Res. 14 (6) (2004) 1188–1190.
[24] C. Chica, et al., Profiling spatial enrichment of chromatin marks suggests an additional epigenomic dimension in gene regulation, Front. Life Sci. 7 (1–2) (2013) 80–87.
[25] E.-V. Willing, et al., Genome expansion of *Arabis alpina* linked with retrotransposition and reduced symmetric DNA methylation, Nat. Plants 1 (2015) 14023.
[26] D.S. ÓMaoiléidigh, et al., Control of reproductive floral organ identity specification in *Arabidopsis* by the C function regulator AGAMOUS, Plant Cell 25 (7) (2013) 2482–2503.
[27] A. Pajoro, et al., Dynamics of chromatin accessibility and gene regulation by MADS-domain transcription factors in flower development, Genome Biol. 15 (3) (2014) R41.
[28] S.E. Wuest, et al., Molecular basis for the specification of floral organs by APETALA3 and PISTILLATA, Proc. Natl. Acad. Sci. U. S. A. 109 (33) (2012) 13452–13457.

[29] E. Oh, et al., Cell elongation is regulated through a central circuit of interacting transcription factors in the Arabidopsis hypocotyl, Elife (May 27) (2014) 3.

[30] J. Heyman, et al., ERF115 controls root quiescent center cell division and stem cell replenishment, Science 342 (6160) (2013) 860–863.

[31] M. Fan, et al., The bHLH transcription factor HBI1 mediates the trade-off between growth and pathogen-associated molecular pattern-triggered immunity in *Arabidopsis*, Plant Cell 26 (2) (2014) 828–841.

[32] M.K. Zhiponova, et al., Helix-loop-helix/basic helix-loop-helix transcription factor network represses cell elongation in *Arabidopsis* through an apparent incoherent feed-forward loop, Proc. Natl. Acad. Sci. U. S. A. 111 (7) (2014) 2824–2829.

[33] E. Moyroud, et al., Prediction of regulatory interactions from genome sequences using a biophysical model for the *Arabidopsis* LEAFY transcription factor, Plant Cell 23 (4) (2011) 1293–1306.

[34] A. Pfeiffer, et al., Combinatorial complexity in a transcriptionally centered signaling hub in *Arabidopsis*, Mol. Plant 7 (11) (2014) 1598–1618.

[35] Y. Zhang, et al., A quartet of PIF bHLH factors provides a transcriptionally centered signaling hub that regulates seedling morphogenesis through differential expression-patterning of shared target genes in Arabidopsis, PLoS Genet. 9 (1) (2013) e003244.

[36] R. Brandt, et al., Genome-wide binding-site analysis of REVOLUTA reveals a link between leaf patterning and light-mediated growth responses, Plant J. 72 (1) (2012) 31–42.

[37] W. Huang, et al., Mapping the core of the *Arabidopsis* circadian clock defines the network structure of the oscillator, Science 336 (6077) (2012) 75–79.

9

Growth enhancement and drought tolerance of hybrid poplar upon inoculation with endophyte consortia☆

Zareen Khan[a], Hyungmin Rho[a], Andrea Firrincieli[b], Shang Han Hung[a], Virginia Luna[c], Oscar Masciarelli[c], Soo-Hyung Kim[a], Sharon L Doty[a,*]

[a] School of Environmental and Forest Sciences, University of Washington, Seattle, WA 98195-2100, USA
[b] Department for Innovation in Biological, Agro-food and Forest systems, University of Tuscia, Viterbo 01100, Italy
[c] Laboratorio de Fisiología Vegetal, Departamento de Ciencias Naturales, Fac. de Cs. Exactas, Universidad Nacional de Río Cuarto, Ruta 36 Km 601, 5800 Río Cuarto, Argentina

ARTICLE INFO

Keywords:
Endophytes
Hybrid poplar
Drought tolerance
Biomass
Reactive oxygen species (ROS)
Photosynthesis
Stomatal conductance
Phytohormones
Volatile organic compounds (VOCs)
Drought tolerance genes

ABSTRACT

With increasing effects of global climate change, there is a strong interest in developing biofuels from trees such as poplar (*Populus* sp.) that have high C sequestration rates and relatively low chemical inputs. Using plant-microbe symbiosis to maximize plant growth and increase host stress tolerance may play an important role in improving the economic viability and environmental sustainability of poplar as a feedstock. Based on our previous research, a total of ten endophyte strains were selected as a consortium to investigate the effects of inoculation on commercial hardwood cuttings of *Populus deltoides* × *P. nigra* clone OP-367. After one and a half months of growth under non-stress conditions followed by one month under water stress, there was substantial growth promotion with improved leaf physiology of poplar plants in response to the endophyte inoculation. Furthermore, inoculated plants demonstrated reduced damage by reactive oxygen species (ROS) indicating a possible mechanism for symbiosis-mediated drought tolerance. Production of important phytohormones by these endophytes and identification of microbial genes involved in conferring drought tolerance suggests their potential roles in the modulation of the plant host stress response.

1. Introduction

Hybrid poplars are increasingly being considered as the premier woody perennial candidate for bioenergy feedstock production because of their ability to produce a significant amount of biomass in a short period of time and their high cellulose and low lignin content [1–4]. Hybrid poplar tree farms are established where there is sufficient water as increased productivity is associated with adequate growing-season precipitation [5,6]. As a consequence, productivity closely depends on water availability and could seriously limit yields at plantation sites where water availability is insufficient. Climate change models suggest that more frequent drought events of greater severity and length can be expected in the coming decades. Consequently, commercial genotypes that have high water use efficiency or increased drought tolerance, in addition to the traits such as high productivity, resistance to pests and insects, and improved wood quality are being used in poplar selection. However, this may not be simple to achieve because some of these beneficial traits may need to be compromised for others [7].

It has been demonstrated that in areas with abiotic stress factors, plants are more dependent on microorganisms that are able to enhance their ability to combat stress [8–12]. Among these plant-associated microbes, the role of endophytes (bacteria or fungi living inside plants) in stimulating plant growth and nutrition, in addition to increasing stress tolerance of their host plants is gaining more attention [13–16]. These microbial symbionts may confer benefits to their host plants via multiple mechanisms including biological nitrogen fixation [17,18] enhancing the bioavailability of phosphorous (P), iron (Fe) and other mineral nutrients [19], production of phytohormones including indole acetic acid (IAA), abscisic acid (ABA), gibberellic acid (GA), brassinosteroids (BR), jasmonates (JA), salicylic acid (SA) [20–23], generation of antioxidants [24–27] for increased plant productivity and tolerance to biotic or abiotic stresses. Another key factor may be microbial production of

☆ This article is part of a special issue entitled "Plants and global climate change: a need for sustainable agriculture", published in the journal Current Plant Biology 6, 2016.
* Corresponding author.
   E-mail address: sldoty@uw.edu (S.L. Doty).

1-aminocyclopropane-1-carboxylate (ACC) deaminase to decrease plant stress [10,28]. Although the interaction between endophytes and their host plants is not fully understood, several studies have demonstrated the positive effects of inoculation of endophytes to increase plant productivity and enhance drought tolerance as a result of multiple mechanisms [10,12,26,27,29]. Recently, the availability of genome sequences of important plant growth promoting endophytes is providing new insight into the biosynthetic pathways involved in plant-endophyte symbiosis, leading to optimization of this technology to increase plant establishment and biomass production.

The aim of the present study was to test the ability of an endophyte consortium to confer drought tolerance and to investigate the underlying mechanisms of endophyte-induced drought tolerance of a commercially-important hybrid poplar clone by monitoring physiological parameters, assaying for ROS activity and analyzing phytohormone production by endophytes in axenic medium. Finally, genome annotations of *Rhodotorula graminis* WP1, *Burkholderia vietnamensis* WPB, *Rhizobium tropici* PTD1, *Rahnella* sp. WP5, *Acinetobacter calcoaceticus* WP19 and *Enterobacter asburiae* PDN3 allowed identification of genes known to be involved in improving plant growth under drought stress.

2. Materials and methods

2.1. Endophyte strains and inoculum preparation

9 bacteria and 1 yeast strain previously isolated from poplar and willow trees growing in stressful environments [30–32] were selected based on their plant growth promoting abilities under nitrogen-limitation and drought stress on a variety of plants and grasses [33–37]. These are as follows: WP1 (*Rhodotorula graminis*), WPB (*Burkholderia vietnamiensis*), PTD1 (*Rhizobium tropici*), WP19 (*Acinetobacter calcoaceticus*), WP5 (*Rahnella* sp.), WP9 (*Burkholderia* sp.), PDN3 (*Enterobacter asburiae*), WW5 (*Sphingomonas yanoikuyae*), WW6 (*Pseudomonas* sp.), and WW7 (*Curtobacterium* sp.). For inoculum preparation, each isolate was grown in 25 ml MG/L [38] and incubated at 30 °C under shaking conditions for 24 h. To prepare the inoculum, cells were harvested by centrifugation at 8000 rpm at 4 °C for 10 min, resuspended in half strength Hoagland's solution [39] and the cell density of each strain was adjusted to produce an inoculum with a final optical density (OD_{600}) of 0.1.

2.2. Plant materials, growth conditions, drought stress

Woody stem cuttings of *Populus deltoides* x *P. nigra* clone OP367 were obtained from the Boardman Research Site near Boardman, Oregon (GreenWood Resources Inc,). Two groups of cuttings (approx. 15 cm) were washed and soaked overnight in sterile water. The next day, twenty cuttings were transplanted into Sunshine Mix #4 soil (Steubers Inc. USA) and grown in the greenhouse under the following conditions: average temperature of 22.3 °C, average relative humidity of 61.42% and the average photosynthetic photon flux density (PPFD) of 290.9 μmol m^2/s and 14/10-h light/dark photoperiod with supplementary high-pressure sodium light bulbs. After two weeks, 100 ml of the inoculum was poured at the base of the stem to one group (n = 10) of randomly selected plants. The control plants (n = 10) were mock-inoculated with 100 ml of half strength Hoagland's solution. After one and a half months of colonization, all the plants were subjected to drought by withholding water for one month after which they were harvested and separated into roots and stems. The samples were oven-dried at 70 °C, ground and weighed. For total nitrogen analysis, the oven dried root samples were ground by a plant grinder, passed through a 20 mesh screen

and analyzed on a PE 2400 series II CHN analyzer (University of Washington, SEFS Chemical Analysis Center).

2.3. In-vitro production of phytohormones by the endophytic isolates

The endophyte strains were grown in M9 minimal medium (with tryptophan added for the IAA analysis) to exponential growth phase (10E + 9) and centrifuged separately at 8000 rpm at 4 °C for 15 min. Supernatants were acidified at pH 2.5 with acetic acid solution (1%v/v), and 50 ng of deuterated $^2H_6$-ABA, $^2H_4$-SA, $^2H_2$-GA$_3$, $^2H_6$-JA, and 2H5-AIA (OlChemlm Ltd, Olomouc, Czech Republic) were added as internal standards. Each sample in triplicate was partitioned four times with the same volume of acetic acid-saturated ethyl acetate (1%, v/v). After the last partition, acidic ethyl acetate was evaporated to dryness at 36 °C in a Speed-Vac concentrator. Dried samples were dissolved in 1500 μl methanol, filtered and resuspended in 50 μl methanol (100%), and placed in vials. Analysis was done by Liquid Chromatography with Electrospray Ionization (LC) (Waters Corp., New York, NY, USA). The instrumental parameters are described elsewhere [40].

2.4. Effects of endophytic colonization on Fv/Fm, chlorophyll and stomatal conductance

The following plant physiological parameters were recorded from fully expanded second or third youngest leaves of both irrigated and drought-stressed poplar cuttings at midday (between 12-1pm) every 4–5 days before and after the drought stress treatment.

2.4.1. Photochemical efficiency of PSII (F_v/F_m)
Maximal photochemical efficiency is inversely proportional to damage to photosystem II (PSII) and this parameter was used to assess photosynthetic stress experienced by the poplar plants grown under drought stress. This was performed by using a portable fluorometer OS-30P+ (Opti-Sciences, Inc., Hudson, NH, USA). The samples were dark-adapted for 30 min before taking minimal fluorescence, F_o, followed by illuminating a saturating light flash to gain maximal fluorescence, F_m. Variable fluorescence, $F_v = (F_m - F_o)$, was calculated by a built-in program to estimate maximal photochemical efficiency of PSII (F_v/F_m) [41].

2.4.2. Indirect measurement of chlorophyll content using SPAD
Leaf chlorophyll content in vivo was measured using a SPAD 502 (Konica Minolta Sensing Americas, Inc., Ramsey, NJ, USA) hand-held chlorophyll meter. The instrument measures 'greenness of leaves' which is tightly correlated with the in vitro chlorophyll content of samples [42].

2.4.3. Measurement of stomatal conductance (g_s)
A steady state leaf porometer SC-1 (Decagon Devices, Inc., Pullman, WA, USA) was used to measure stomatal conductance (g_s) of poplar leaves at the midday. This time point was chosen based on preliminary results that indicated that the inoculation effects on g_s were most remarkable from 12 p.m. to 3 p.m. (data not shown).

2.5. Reactive oxygen species (ROS) assay

Using a cork borer, 3 leaf disks (2 mm) were obtained from each of 3 plants from the inoculated or control group and incubated in a solution of 1 μM of the herbicide paraquat (N,N'-Dimethyl-4,4'-bipyridinium dichloride) and incubated at 22 °C under fluorescent lights [43,44]. After 48 h exposure to paraquat, leaf disks were photographed to document chlorophyll oxidation visualized by tissue discoloration. All assays were performed three times.

2.6. Genomic analysis

The poplar bacterial endophyte strains, PTD1, WPB, WP5, WP9, WP19 and PDN3 were grown on NL-CCM agar medium [45]. Five ml MG/L broth or TYC broth [25] were inoculated with the strains and genomic DNA was subsequently isolated and purified using protocols provided by the Joint Genome Institute (JGI). All bacterial endophyte strains were sequenced by the Joint Genome Institute (JGI) and were annotated using JGI's microbial annotation pipeline [46]. Annotations and comparative analysis of all genomes are available through the Integrated Microbial Genome system [47]. The WP1 yeast genomic DNA was isolated, sequenced, and analyzed as described [48]. The JGI annotated genomes were analyzed for genes that have been reported to have beneficial effects during plant growth under drought stress conditions and for tolerance against oxidative stress. In this context, the presence of the following genes: *acdS*, *pqqABCDEF*, *budABC* genes involved in the biosynthesis of ethylene, pyrrolo-quinolone quinine (PQQ) and 2R,3R-butanediol, respectively, were assessed [49,50]. Moreover, genes involved in the trehalose biosynthesis were also considered due to their importance in tolerance against desiccation and drought-like stress [51–53]. Potential annotation gaps were detected as described [54]. For the analysis of putative genes encoding for a functional ACC deaminase, the amino acid sequences computed from *acdS* genes and D-cysteine desulfhydrase encoding genes were aligned in ClustalW2 software using the *acdS* gene from *Pseudomonas putida* strain UW4 as a reference (AY823987.1). For the detection of *pqqA* in *Acinetobacter calcoaceticus* WP19, the intergenic region of 396 bp upstream to *pqqB* was compared to all the genomes present in IMG database using BLASTX 2.2.26+ [55]. The annotated genome of *Rahnella aquatilis* HX2 (NCBI/RefSeq: CP003403; CP003404; CP003405; CP003406) and *Rahnella* sp. WP5 were compared to each other to identify genes without homologues in WP5. These genes of interest were compared against the WP5 genome by using TBlastn.

2.7. Statistical analysis

The effects of the drought stress and inoculation were evaluated using the paired t-test procedure for the time series data (SPAD, F_v/F_m, and g_s). Initial values measured before the drought treatment (0 days after drought stress, 0 DDS) were used to compare the drought effects on the variables at each time point. Control vs. Inoculated comparison at each time point was conducted using the same paired t-test procedure. The biomass allocation and N content data were collected only after the drought stress at harvest, so the simple two-tailed t-test procedure was used to test the inoculation effects on these variables. All the statistical analyses were applied to ten replicated samples per treatment. RStudio v.0.98.945[56] was used for conducting the statistical analyses.

3. Results

3.1. Biomass, N content, wilting response

At harvest, the inoculated plants had a significant 28% (P < 0.001) increase in total biomass resulting from a 42% stimulation in root biomass (P < 0.001), 43% (P < 0.001) higher total plant nitrogen, and a 21% (P < 0.001) increase in shoot biomass (Fig. 1). The large increase in the root biomass relative to shoot biomass resulted in a significant 23% (P < 0.001) increase in root to shoot ratio. At 20 days of drought stress (DDS), the leaves of the control plants started browning whereas the leaves of the inoculated plants remained

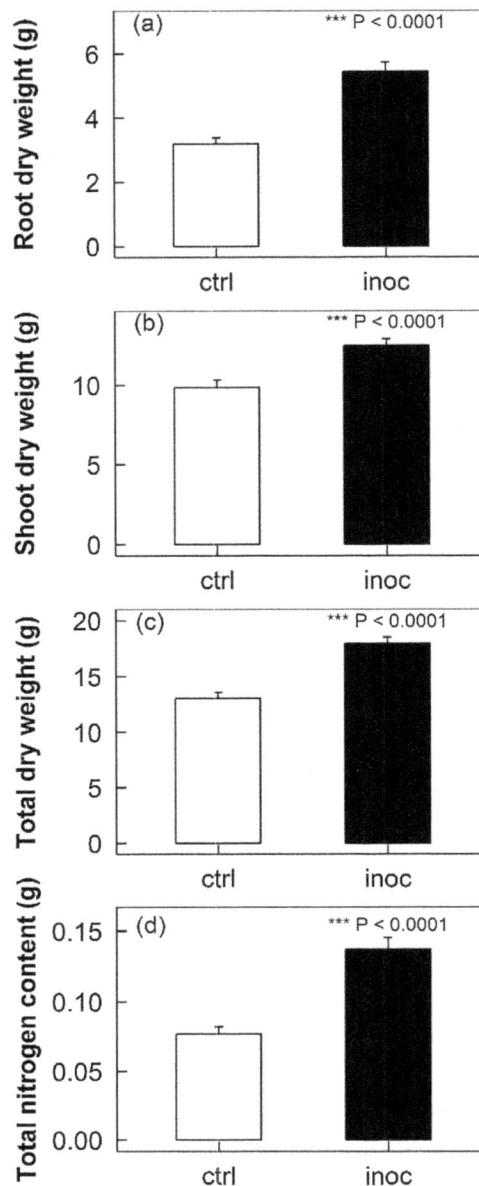

Fig. 1. Biomass and nitrogen content of the poplar plants in response to the drought stress treatment at harvest (31 days after the drought treatment). (a) Root dry weight; (b) shoot dry weight; (c) total dry weight; (d) total nitrogen. Open and closed bars represent the means of each response from the mock-inoculated control (ctrl) and the inoculated (inoc) samples, respectively. The bars represent means from ten replicated samples of each group along with error bars standing for standard errors of the means. The P-values for the t-statistics are presented where the differences are significant.

green. Complete wilting of all the mock-inoculated control plants was observed at 1 month of drought stress whereas most of the endophyte-treated plants had retained turgor (Fig. 2).

3.2. Analysis of culture filtrates for quantification of phytohormones

As seen in Table 1, all the selected endophyte strains produced IAA, GA₃, SA, ABA, JA and Brs in different concentrations. However, each microorganism varied in the type and quantity of the compound produced in our experimental conditions. The yeast endophyte-strain WP1 was the major producer of IAA and GA₃ while willow endophyte WW7 produced the most ABA. Isolate

Fig. 2. Photo of four representative plants showing the effects of one month of drought stress in inoculated poplars (on the left) and mock-inoculated poplars (on the right) under greenhouse condition.

Table 1

Phytohormone production (SA, salicylic acid; ABA, abscisic acid; IAA, indole-3-acetic acid; JA, jasmonic acid; GA₃, gibberellins-3-acid; Brs, epibrassinolides) in exponential growth cultures of endophytes.

Endophyte strains and 16S rRNA match	SA	JA	IAA	ABA	GA₃	Brs
WP1 (*Rhodotorula graminis*)	8.224 ± 1.12	0.613 ± 0.05	61.308 ± 1.34	0.435 ± 0.02	2.694 ± 0.25	4.306 ± 0.17
WPB (*Burkholderia vietnamiensis*)	1.729 ± 0.13	1.799 ± 0.23	3.293 ± 0.26	0.416 ± 0.02	0.729 ± 0.02	nd
PTD1 (*Rhizobium tropici*)	2.853 ± 0.18	2.469 ± 0.21	55.847 ± 2.02	0.436 ± 0.03	0.972 ± 0.03	7.699 ± 0.65
WP19 (*Acinetobacter calcoaceticus*)	9.76 ± 0.2	0.165 ± 0.02	17.727 ± 1.84	0.66 ± 0.03	2.275 ± 0.21	0.426 ± 0.04
WP5 (*Rahnella* sp.)	4.627 ± 0.37	0.171 ± 0.02	2.429 ± 0.26	0.405 ± 0.02	1.189 ± 0.25	4.563 ± 0.42
WP9 (*Burkholderia* sp.)	0.904 ± 0.009	0.462 ± 0.04	0.569 ± 0.02	0.405 ± 0.02	1.674 ± 0.16	2.26 ± 0.2
WW5 (*Sphingomonas yanoikuyae*)	48.519 ± 2.86	0.468 ± 0.04	5.627 ± 0.42	0.411 ± 0.02	0.562 ± 0.04	3.226 ± 0.32
WW6 (*Pseudomonas* sp.)	2.459 ± 0.46	0.194 ± 0.02	1.666 ± 0.2	0.404 ± 0.02	0.964 ± 0.03	8.849 ± 0.63
WW7 (*Curtobacterium* sp.)	1.378 ± 0.26	0.171 ± 0.02	9.141 ± 0.24	0.831 ± 0.03	1.045 ± 0.2	0.677 ± 0.05

Phytohormones are expressed in $\mu g/ml^{-1}$.

PTD1 produced significant amounts of IAA and JA and isolate WW5 produced high levels of SA.

3.3. Changes in plant physiology- greenness, chlorophyll fluorescence and stomatal conductance

Before withholding water from the poplar cuttings, indirect chlorophyll content measurements (SPAD) between the mock-inoculated control and inoculated poplar plants were comparable. After 19 DDS the difference of SPAD values of the control and inoculated plants began to increase, becoming largest at 31 DDS ($P = 0.059$) (Fig. 3a and Table 2). The SPAD value of the inoculated cuttings was 66.0% higher than that of the control at 31 DDS. Up to 15 DDS there was no effect of the inoculation on potential quantum yield of photosystem II (PSII) (F_v/F_m) of the poplar leaves. At 19 DDS, however, the inoculated plants maintained the same level of F_v/F_m value compared to an abrupt decrease of F_v/F_m of the control plants. This resulted in 32.9% higher F_v/F_m in the inoculated poplars compared to the mock-inoculated control plants ($P = 0.071$) (Fig. 3b and Table 2). Stomatal conductance (g_s) in the midday of both the control and inoculated groups showed a common response to water deficit; it steeply decreased 11 days after the imposition of the drought stress. However, it is noteworthy that the inoculated samples always had lower g_s throughout the drought period which was highlighted at 19 DDS up to 32.6% lower than the control (as the lowest $P = 0.110$) (Fig. 3c and Table 2). This substantial decrease of g_s in the inoculated leaves coincided with the higher F_v/F_m values at 19 DDS.

3.4. Decreased ROS activity in response to inoculation

Chlorophyll bleaching of photosynthetic tissue by the herbicide paraquat is indicative of oxidation damage due to production of superoxide ions. When leaf disks were exposed to paraquat, after 48 h the tissues of endophyte inoculated plants remained green, indicating an absence of ROS generation whereas the mock-inoculated tissues no longer remained green, indicative of chlorophyll bleaching (Fig. 4).

3.5. Putative endophytic genes involved in drought tolerance

Genomic analysis revealed the presence of microbial genes characterized for having beneficial effects against host plant drought stress and improving tolerance to oxidative stresses (Table 3). The genome annotation of PTD1, WP1, WP19 and PDN3, revealed the presence of putative genes involved in the biosynthesis of (*R,R*)-butane-2,3-diol and acetoin. Putative genes predicted to encode ACC deaminase were detected in WP19, PDN3 and WP5. However, the amino acid sequences lacked residues Glu295 and Leu322 known to be important for ACC-deaminase activity, thus it likely has only D-cysteine desulfhydrase activity as reported in other systems [57]. Only the gene in WPB predicted to encode ACC-deaminase showed a perfect match in amino acid residues (Supplementary Fig. S1). Genes required for PQQ biosynthesis (*pqqABCDEF*) that have been related to biological functions such as phosphate solubilization, antimicrobial activity and tolerance to oxidative stress were found in WP5 and WP19. In *Rahnella* sp. WP5, the *pqqA* gene that encodes for a small peptide of 25 amino acids was detected in the intergenic region upstream to the *pqqBCDE operon* (scaffold32: 29265-29660) by using the "Phylogenetic Pro-

Table 2
Paired t-test results of the means ± 1 S.E. ($n = 10$) of three plant physiological response variables (chlorophyll content, Fv/Fm, and g_s) over the drought stress period (DDS, days after drought stress). P-values of t-statistics of the drought effects are reported in a horizontal direction. (0 DDS vs. ~). P-values of the inoculation effects are reported in a vertical direction at each time point on the third rows (Control vs. Inoculated). See Section 2.7 for details. NA indicates that the measurements were not available due to seriously wilted leaf surfaces from the drought responses. Numbers in bold mean significant differences at $P < 0.05$ level.

Treatment	0 DDS	7 DDS	11 DDS	15 DDS	19 DDS	23 DDS	27 DDS	31 DDS
	Chlorophyll content (SPAD)							
Control	28.53 ± 1.434	27.67 ± 1.057 (0.354)	26.93 ± 1.143 (0.062)	25.34 ± 0.881 (**0.028**)	21.60 ± 2.774 (**<0.001**)	12.44 ± 2.914 (**<0.001**)	6.64 ± 1.523 (**<0.001**)	6.29 ± 1.079 (**<0.001**)
Inoculated	30.59 ± 1.647	30.18 ± 1.623 (0.535)	28.04 ± 1.460 (**0.029**)	27.40 ± 1.340 (**0.002**)	20.42 ± 3.924 (**0.023**)	14.83 ± 3.762 (**0.004**)	11.33 ± 3.101 (**<0.001**)	10.44 ± 1.665 (**<0.001**)
$P > |t|$	(0.295)	(0.240)	(0.571)	(0.2763)	(0.833)	(0.4031)	(0.211)	(0.059)
	Fv/Fm (unitless)							
Control	0.754 ± 0.012	0.770 ± 0.006 (0.293)	0.774 ± 0.007 (0.154)	0.778 ± 0.007 (0.065)	0.596 ± 0.100 (0.165)	0.433 ± 0.177 (0.141)	NA	NA
Inoculated	0.754 ± 0.013	0.766 ± 0.013 (0.304)	0.770 ± 0.006 (0.256)	0.787 ± 0.004 (**0.013**)	0.792 ± 0.008 (0.061)	0.605 ± 0.147 (0.346)	NA	NA
$P > |t|$	(0.931)	(0.773)	(0.732)	(0.310)	(0.071)	(0.495)	NA	NA
	g_s (mmol H$_2$O m^{-2} s^{-1})							
Control	153.4 ± 21.97	229.3 ± 23.21 (**<0.001**)	251.1 ± 27.75 (**0.002**)	179.9 ± 22.92 (0.413)	110.8 ± 15.31 (0.253)	123.4 ± 25.34 (0.434)	67.6 ± 18.16 (**0.024**)	NA
Inoculated	129.0 ± 26.24	212.3 ± 19.79 (**0.006**)	233.9 ± 24.39 (**0.001**)	193.8 ± 25.42 (**0.030**)	75.7 ± 11.26 (0.112)	96.7 ± 23.54 (0.465)	74.9 ± 22.61 (0.198)	NA
$P > |t|$	(0.204)	(0.571)	(0.657)	(0.751)	(0.110)	(0.423)	(0.821)	NA

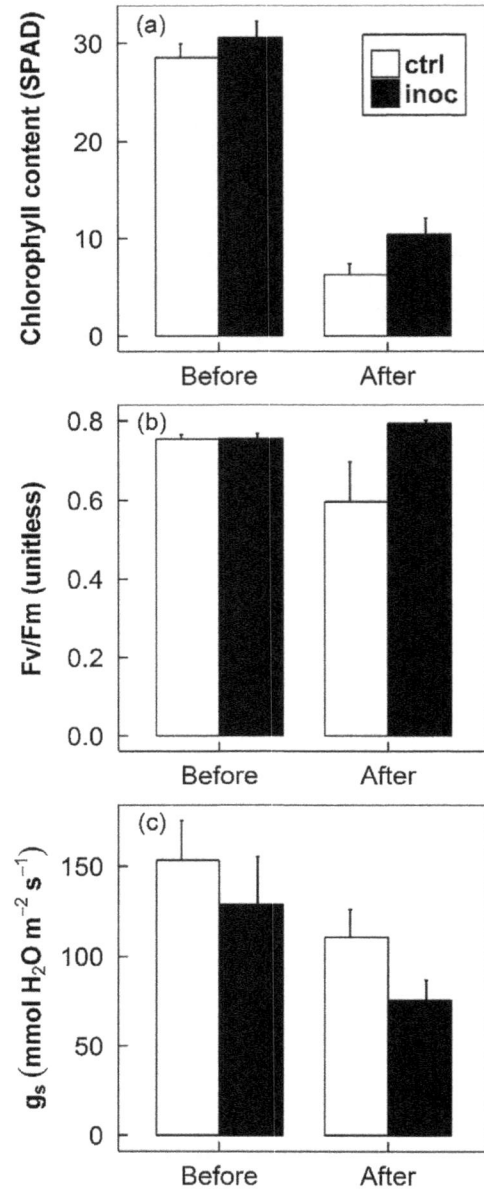

Fig. 3. Photosynthesis parameters of poplar leaves before and after the drought stress treatment. (a) Chlorophyll content; (b) potential quantum yield of PSII (Fv/Fm); (c) stomatal conductance (gs). The After values for chlorophyll content was chosen at 31 days after drought stress, while as those for Fv/Fm and gs were chosen at 19 days after drought stress because the rates of the physiological responses to drought differ by the process. The bars represent the means of each response from ten replicated mock-inoculated control (open bars) and inoculated groups (closed bars). Error bars stand for ±standard errors.

filer for Single Genes" tool. In a similar way *pqqA* was detected in the intergenic region of 396 bp (scaffold32: 29265-29660), upstream to *pqqB* in *A. calcoaceticus* WP19 (see Materials and methods). Finally, putative genes involved in the biosynthesis of trehalose via OtsA-OtsB from UDP-glucose and glucose 6-phosphate, were detected in WP1, PTD1, WP5, WP19 and PDN3 [58]. Two putative genes encoding for a malto-oligosyltrehalose trehalohydrolase (TreZ) and a malto-oligosyltrehalose synthase (TreY), respectively, were detected in WP5, PTD1 and PDN3. These genes are specific to the biosynthetic pathway TreY-TreZ which synthetizes trehalose from malto-oligosaccharides or alpha-1,4-glucans [59].

Table 3
List of putative drought tolerance genes in strains WP19, PDN3, WP5, PTD1, WPB and WP1.

Strain	Biological process	[2]Locus Tag	IMG Product Name
WP19	(R,R)-butane-2,3-diol Synthesis	EX32DRAFT_00261	acetolactate synthase, small subunit
		EX32DRAFT_00262	acetolactate synthase, large subunit
		EX32DRAFT_03428	threonine dehydrogenase and related Zn-dependent dehydrogenases
	Trehalose synthesis via OtsA-OtsB	EX32DRAFT_01928	trehalose-phosphatase (OtsB)
		EX32DRAFT_01927	trehalose-6-phosphate synthase (OtsA)
		EX32DRAFT_03479	coenzyme PQQ biosynthesis enzyme PqqE
		EX32DRAFT_03480	coenzyme PQQ biosynthesis protein PqqD
	PQQ synthesis	EX32DRAFT_03481	coenzyme PQQ biosynthesis protein C
		EX32DRAFT_03482	coenzyme PQQ biosynthesis protein B
		[1]EX32DRAFT_03482.1	coenzyme PQQ peptide PqqA
PDN3	Trehalose syntheis via OtsA-OtsB	EX28DRAFT_0655	trehalose-phosphatase (OtsB)
		EX28DRAFT_0656	alpha,alpha-trehalose-phosphate synthase (OtsA)
	Trehalose synthesis via TreY-TreZ	EX28DRAFT_1155	malto-oligosyltrehalose trehalohydrolase (TreZ)
		EX28DRAFT_1156	malto-oligosyltrehalose synthase (TreY)
		EX28DRAFT_2401	acetoin reductases (BudC)
		EX28DRAFT_2402	acetolactate synthase, large subunit (BudB)
		EX28DRAFT_2403	alpha-acetolactate decarboxylase (BudA)
	(R,R)-butane-2,3-diol and (R)-acetoin Synthesis	EX28DRAFT_3673	acetolactate synthase, large subunit
		EX28DRAFT_3674	acetolactate synthase, small subunit
		EX28DRAFT_3953	acetolactate synthase, small subunit
		EX28DRAFT_3954	acetolactate synthase, large subunit
		EX28DRAFT_4440	acetolactate synthase, large subunit
		EX28DRAFT_4441	acetolactate synthase, small subunit
WP5	PQQ synthesis	EX31DRAFT_01699	coenzyme PQQ biosynthesis probable peptidase PqqF
		EX31DRAFT_01700	coenzyme PQQ biosynthesis enzyme PqqE
		EX31DRAFT_01701	coenzyme PQQ biosynthesis protein PqqD
		EX31DRAFT_01702	coenzyme PQQ biosynthesis protein C
		EX31DRAFT_01703	coenzyme PQQ biosynthesis protein B
		[1]EX31DRAFT_01703.1	coenzyme PQQ biosynthesis protein A
	Trehalose synthesis via TreY-TreZ	EX31DRAFT_01717	malto-oligosyl trehalose synthase (TreY)
		EX31DRAFT_01718	malto-oligosyl trehalose hydrolase (TreZ)
		EX31DRAFT_02312	acetolactate synthase, small subunit
		EX31DRAFT_02313	acetolactate synthase, large subunit
	(R,R)-butane-2,3-diol and (R)-acetoin Synthesis	EX31DRAFT_02434	alpha-acetolactate decarboxylase
		EX31DRAFT_04636	acetolactate synthase, large subunit
		EX31DRAFT_04637	acetolactate synthase, small subunit
PTD1	Trehalose syntheis via OtsA-OtsB	EX06DRAFT_01128	trehalose-phosphatase (OtsB)
		EX06DRAFT_01129	alpha,alpha-trehalose-phosphate synthase (OtsA)
	Trehalose syntheis via TreY-TreZ	EX06DRAFT_01921	malto-oligosyltrehalose synthase (TreY)
		EX06DRAFT_01922	malto-oligosyltrehalose trehalohydrolase (TreZ)
	(R,R)-butane-2,3-diol and (R)-acetoin Synthesis	EX06DRAFT_05002	acetolactate synthase, large subunit
		EX06DRAFT_05003	acetolactate synthase, small subunit
WPB	Trehalose syntheis via OtsA-OtsB	Ga0008009_10698	Putative trehalose-phosphatase (OtsB)
		Ga0008009_10699	alpha,alpha-trehalose-phosphate synthase (OtsA)
	Trehalose syntheis via TreY-TreZ	Ga0008009_10872	malto-oligosyltrehalose synthase (TreY)
		Ga0008009_10874	malto-oligosyltrehalose trehalohydrolase (TreZ)
	(R,R)-butane-2,3-diol and R-acetoin Synthesis	Ga0008009_118101	acetolactate synthase, large subunit
		Ga0008009_118102	acetolactate synthase, small subunit
		Ga0008009_10299	Threonine dehydrogenase or related Zn-dependent dehydrogenase
	1-Aminocyclopropane-1-carboxylic acid degradation	Ga0008009_11518	1-aminocyclopropane-1-carboxylate deaminase
WP1	Trehalose synthesis via OtsA-OtsB	scaffold_19:208485-212329	Putative trehalose-phosphatase (OtsB)
		scaffold_9:870873-873744	alpha,alpha-trehalose-phosphate synthase (OtsA)
	(R,R)-butane-2,3-diol and (R)-acetoin Synthesis	scaffold_15:276834-278070	Putative acetolactate synthase, large subunit
		scaffold_5:1501147-1504056	Putative acetolactate synthase, small subunit
		scaffold_14:338156-339969	(R,R)-butanediol dehydrogenase

4. Discussion and conclusions

These endophytes either as single strain or multi-strain inoculum have previously exhibited mutualistic behavior when added to other plant species including grasses, corn, rice, Douglas-fir and a variety of crop plants [33–37,60]. Combined application of endophytes can result in larger effects than those possible with individual inoculations [34,35,61]. Rogers et al. [62] reported growth enhancements of hybrid poplar clone OP-367 using a single endophyte *Enterobacter* sp. strain 638 isolated from poplar growing at a phytoremediation field site. In our study, inoculation with a consortium of endophytes resulted in 28% higher biomass com-

pared to mock-inoculated controls. Some of these endophytes were isolated from wild poplar growing on rocks and gravel in their native riparian habitat. It can be expected that these trees have selectively recruited the most beneficial endophytes for their survival in that challenging environment, therefore these strains may harbor adaptive traits and have a superior potential to enhance host plant growth under stressful conditions. The inoculated plants also had a doubling of root biomass and this may reduce water requirements and increase survivability during the costly establishment phase of short rotation energy crops, thereby improving the economic viability of poplar as a feedstock for biofuel applications [63–66]. The auxin, IAA, is a well-known plant phytohormone that

| Leaf disks from inoculated plants (retained greenness) | Leaf disks from mock-inoculated plants (lost greenness) |

Fig. 4. Effect of inoculation on paraquat induced oxidative stress (ROS) under laboratory conditions. After 48 h of exposure to paraquat, leaves from plants that were inoculated with endophytes remained green (left) while leaves from the mock-inoculated plants were photobleached (right). Leaf disks were sampled from leaf tissues of similar size, developmental age, and location for comparison.

is involved in multiple plant growth processes and stress responses [67,68]. Since the endophytes produced significant amounts of IAA in culture, it is possible that the enhanced root growth and drought stress resistance may be via auxin related mechanisms. Other stress responsive hormones such as GA₃, JA, SA and Brs [69–74] were also produced by these endophytes which may have led to the morphological changes in the host plant.

It has been well established that drought stress causes oxidative damage by producing reactive oxygen species such as $O_2^{\bullet-}$, H_2O_2, and $^\bullet OH$ [75–77] causing oxidative damage to DNA, lipids, and proteins. In this study we assessed tissue tolerance to ROS by exposing photosynthetic tissue to the herbicide paraquat (mimics endogenous ROS production) which is oxidized by molecular oxygen resulting in the generation of superoxide ions and subsequent photobleaching/discoloration of chloroplasts [44]. When exposed to stress, mock-inoculated plant tissues lost their greenness indicating ROS activity while the inoculated tissues remained green. It is likely that endophytes may be helping plants to cope with drought stress by either efficiently scavenging ROS or preventing ROS production under drought stress [78,79]. Production of pyrroloquinoline is correlated to ROS activity [80]. Interestingly, genes involved in PQQ synthesis were identified in WP5 and WP19 suggesting a possible direct involvement by reducing the oxidative stress in cells. Biochemical analysis of ROS activity during water stress and mutant analysis will aid in confirming this hypothesis.

Besides an improved survival of the host plant, the survival of a microorganism under water-limited conditions may also represent an important trait for a stable interaction with the host plant. A well-known osmolyte used by microorganisms and plants

during dessication stress is trehalose, a disaccharide that protects biomolecules during osmotic stress [81]. The possible role of trehalose in beneficial plant-microbe interactions was explored in two studies involving engineered bacteria overexpressing trehalose biosynthesis genes. Bean plants inoculated with *Rhizobium etli* overexpressing *otsA* had increased drought tolerance, grain yields, and biomass [82]. In a similar study, the beneficial effect of trehalose against drought stress was assessed in maize plants inoculated with a genetically modified *Azospirilium brasilense* which over-accumulated trehalose through the overexpression of a chimeric trehalose biosynthetic gene as well as an exogenous copy of *otsA* from *Rhizobium etli* [83]. While only 40% of uninoculated maize survived the drought stress, 85% of those inoculated with the trehalose overexpressing strain survived. In our study the presence of trehalose biosynthesis genes in endophytic strains WP5, WP1, WPB, WP19 and PDN3 provide an opportunity to test its involvement in conferring drought stress tolerance.

Another suggested mechanism for microbially-conferred plant host drought tolerance is production of ACC deaminase that reduces host stress ethylene [10]. While ethylene is normally produced by plants at low levels, when exposed to biotic and abiotic stresses, production is significantly elevated, leading to reduced plant growth. Plants inoculated with ACC deaminase-producing bacterial strains have improved stress tolerance. For example, tomato and pepper plants exhibited increased drought tolerance when inoculated with the ACC deaminase producing strain ARV8 [84]. Genomic sequences of plant-associated microorganisms are commonly misannotated as having ACC deaminase, requiring a close inspection of the sequence encoding the active site of the enzyme [85]. Strain WPB does contain the putative ACC deaminase sequence, and likely does have this activity, although this must be confirmed biochemically. However, strains WP5, WP19, and PDN3, while annotated as having this enzyme, have substitutions that would likely make the resulting enzyme incapable of breaking the cyclopropane ring of ACC but would have D-cysteine desulfhydrase activity [86]. These two different enzyme activities, interchangeable with two amino acid residue alterations, may both have roles in protection of the plant host. D-cysteine desulhydrase converts D-cysteine into pyruvate, hydrogen sulfide and ammonia, and may be involved in sulfur induced pathogen resistance [87].

The genomes of PDN3, WP19, PTD1, WP5, WPB and WP1 carried genes involved in synthesis of the volatile organic compounds (VOCs), acetoin and 2,3-butanediol (Table 3). Production of these VOCs by plant growth promoting bacteria was reported to increase systemic resistance and drought tolerance [88]. *Enterobacter* sp. 638, an endophyte isolated from poplar stems increased drought tolerance of hybrid poplar, and annotation of the genome revealed the presence of genes for acetoin and 2,3-butanediol synthesis [88]. The involvement of 2,3-butanediol in inducing stomatal closure and drought tolerance was shown by Cho, et al. [49]. Under drought stress, root colonization of *Arabidopsis* plants by *Pseudomonas chlororaphis* strain O6 increased plant survival and reduced stomatal aperture. Exogenous application of this volatile compound to the plants gave similar results suggesting that the bacterial volatile may be a key determinant in increased drought tolerance.

The genomic analysis reported here is the first step in beginning to decipher the microbial mechanisms for improving host plant drought tolerance. Identification of genes that may be involved in the process provided the necessary insight that is leading to functional analysis using a directed mutational approach.

We also evaluated the response of poplars to water deficit based on physiological traits such as chlorophyll content, photochemical efficiency of PSII, and stomatal conductance which have been used as indicators of plant stress [89–91]. Previous studies demonstrated an increase of F_v/F_m due to beneficial microsymbionts under drought stress conditions [26,27,92]. Likewise, endophyte inoculation improved F_v/F_m in the present study. It is possible that the positive effects of endophyte inoculation may be linked to photosynthesis through production/stimulation of specific phytohormone(s) associated with leaf development [93]. Other physiological aspects that might also be influencing the system are water-use efficiency related to photosynthesis, xylem hydraulic conductivity, and partitioning of photoassimilates. The delay of the decreasing photosynthetic activities by the inoculated endophytes corresponded with the delay of degradation of chlorophyll determined by SPAD in the present study. Chlorophyll concentration is positively connected to photosynthetic activities of PSII, especially under water stressed conditions [94]. It was reported by Lawler et al. [95] that the drought-adapted ecotypes and cultivars of *Andropogon gerardii* had higher SPAD values in the drier habitats. Under severe drought stress conditions, the inoculated endophyte consortia may help the host plants preserve chlorophyll molecules in the leaves by providing fixed nitrogen and phytohormones which eventually increase the host plants' adaptation to the limited water environment.

Stomatal regulation is one of the fastest physiological mechanisms in plants responding to water deficit, limiting CO_2 diffusion and leading to decrease of the CO_2 assimilation rate of the plants [96]. Our results showed typical stomatal responses of plants after the drought stress imposed; at 15 DDS g_s of both control and inoculated groups started to rapidly decrease to 49.9% (Fig. 3c). Interestingly, however, even before the drought, g_s appeared to be decreased by the inoculation of the endophytes although the differences were not statistically significant. This response was highlighted at 19 DDS with the lowest $P = 0.110$. The decrease of g_s by the inoculation coincided with the delay of reduced F_v/F_m values at 19 DDS as described above. These combined physiological responses indicate that the endophytes may trigger the host plants to close the stomata, thereby losing less water and yet maintaining the photosynthetic efficiency of PSII. ABA is a key signal for stomatal closure of plants induced by drought or salt stresses as discussed by Fricke et al. [97], Zeng et al. [98] and Park et al. [99]. Especially since *Populus* is an isohydric species, stomatal control is strongly dependent on water status of the plants [97]. Therefore, the role of ABA in stomatal control under drought stress becomes more crucial in the present study. The amount of endophyte-produced ABA in vitro in our hormonal profile assay is enough to signal the stomatal reactions when compared to the amount of endogenously produced ABA *in planta* under severe drought reported by Fricke et al. [97]. An intriguing hypothesis for endophyte-mediated drought tolerance related to stomatal closure is that the colonized host plants may be able to utilize CO_2 respired by the endophytes, enabling continued photosynthesis with closed stomata while avoiding the water losses normally incurred through open stomata. It was shown by Bloemen et al. [100] that root-respired CO_2 could be incorporated into photosynthetic CO_2 assimilation up to 2.7% in *P. deltoides*. Since we saw extensive intercellular colonization by the endophytes, it is possible that CO_2 released by endophytic microbial respiration in the intercellular spaces might be another source of the CO_2 assimilation, and this combined with phytohormone modulation and reduced ROS, could eventually increase the host plants' adaptation to the limited water environment.

In conclusion, the results presented here demonstrate that inoculation of a commercial hybrid poplar with a consortia of beneficial endophytes can significantly enhance plant growth and tolerance of water deficit stress under greenhouse conditions. Field trials are underway to test if these findings will translate into the production environment. Since poplar production systems must be optimized to produce stable high yields despite the increased stresses imposed by climate change, a better understanding of plant-microbe interactions could be a key to adapting plants to a water limited environment. The availability of genomic sequences will greatly promote the progress of the research into the fundamental mechanisms of symbiosis and may yield ways to further increase biomass in an environmentally sustainable way.

5. Nucleotide sequence accession numbers

The sequence data (16S/28S rDNA) of the selected strains have been deposited in NCBI GenBank under accession numbers: EU563924 (WP1), JN634853(PDN3), EU563924(WPB), KT962907 (PTD1), KU523563(WP19), KU497675(WP5), KU523562(WP9), KT984987(WW5), KU557506(WW6), KU523564(WW7).

Acknowledgements

This work was funded by USDA AFRI grant #2010-05080. We thank Rehmatullah Arif and Sherry Zheng for their lab and greenhouse assistance, GreenWood Resources, Inc. (Portland, OR) for providing the plant material, and Brian Stanton for critical reading of this manuscript. We also wish to thank Andrew W. Sher and Pierre M. Joubert for performing PCR and sequence analysis of 16S rRNA of the endophyte strains. The work conducted by the U.S. Department of Energy Joint Genome Institute, a DOE office of Science User Facility, is supported by the Office of Science of the U.S. Department of Energy under Contract No. DE-AC02-05CH11231.

References

[1] G.A. Tuskan, Short-rotation woody crop supply systems in the United States: what do we know and what do we need to know? Biomass Bioenergy 14 (1998) 307–315.

[2] G. Deckmyn, I. Laureysens, J. Garcia, B. Muys, R. Ceulemans, Poplar growth and yield in short rotation coppice: model simulations using the process model SECRETS, Biomass Bioenergy 26 (2004) 221–227.

[3] A. Karp, I. Shield, Bioenergy from plants and the sustainable yield challenge, New Phytol. 179 (2008) 15–32.

[4] W. Nasso DiN. Guidi, G. Ragaglini, C. Tozzini, Biomass production and energy balance of a 12-year old short-rotation coppice poplar stand under different cutting cycles, GCB Bionergy 2 (2010) 89–97.

[5] R. Ceulemans, I.S.V. Impens, Genetic variation in aspects of leaf growth of Populus clones, using the leaf plastochron index, Can. J. Forest Res. 18 (1988) 1069–1077.

[6] L. Zsuffa, E. Giordano, L.D. Pryor, Trends in poplar culture:some global and regional perspectives, in: R.F. Stettler, H.D. Bradshaw, P.E. Heilman Jr., T.M. Hinckley (Eds.), Biology of Populus and Its Implications for Management and Conservation. Part II, Ch. 19, NRC Research Press, National Research Council of Canada, Ottawa, 1996, pp. 515–539.

[7] M. Larcheveque, M. Maurel, A. Desrochers, G.R. Larocque, How does drought tolerance compare between two improved hybrids of balsam poplar and an unimproved native species? Tree Physiol. 31 (2011) 240–249, http://dx.doi.org/10.1093/treephys/tpr011.

[8] B.R. Glick, D.M. Penrose, J. Li, A model for the lowering of plant ethylene concentrations by plant growth promoting bacteria, J. Theor. Biol. 190 (1998) 63–68.

[9] B.R. Glick, C.L. Patten, G. Holguin, D.M. Penrose, Biochemical and Genetic Mechanism Used by Plant Growth Promoting Bacteria, Imperial College Press, London, 1999.

[10] B.R. Glick, Plant growth-promoting bacteria: mechanisms and applications, Scientifica (2012), 963401, http://dx.doi.org/10.6064/2012/963401, 15 p.

[11] A.M. Pirttila, C. Frank, Endophytes of Forest Trees Biology and Applications, 80, Springer, Netherlands, 2011, ISBN: 978-94-007-1599-8.

[12] S. Timmusk, I.A. Abd El-Daim, L. Capolovici, Triin Tanilas, A. Kannaste, L. Behers, E. Nevo, G. Seisenbaeva, E. Stenstrom, U. Ninemets, Drought tolerance of wheat improved by rhizosphere bacteria from harsh environments: enhanced biomass production and reduced emission of stress volatiles, PLoS One 9 (5) (2014) e96086, http://dx.doi.org/10.1371/journal.pone.0096086.

[13] Y.C. Kim, B. Glick, Y. Bashan, C.M. Ryu, Enhancement of plant drought tolerance by microbes, in: R. Aroca (Ed.), Plant Responses to Drought Stress, Springer Verlag, Berlin, 2013.

[14] J.R. Gaiero, C.A. McCall, K. Thompson, N.J. Day, A.S. Best, K.E. Dunfield, Inside the root microbiome: bacterial root endophytes and plant growth promotion, Am. J. Bot. 100 (2013) 1738–1750, http://dx.doi.org/10.3732/ajb.1200572

[15] P.R. Hardoim, L.S. van Overbeek, G. Berg, A.M. Pirttila, S. Compant, A. Campisano, M. Doring, A. Sessitsch, The hidden world within plants:ecological and evolutionary considerations for defining funtioning of microbial endophytes, Microbiol.

[16] Z.A. Wani, N. Ashraf, T. Mohiuddin, S. Riyaz-Ul-Hassan, Plant Endophyte symbiosis, an ecological perspective, Appl. Microbiol. Biotechnol. 99 (2015) 2955–2965.

[17] V.C. Pankievicz, F.P. do Amaral, K.F.D.N. Santos, B. Agtuca, Y. Xu, M.J. Schueller, A.C.M. Arisi, M.B.R. Steffens, E.M. deSouza, F.O. Pedrosa, G. Stacey, R.A. Ferrieri, Robust biological nitrogen fixation in a model grass-bacterial association, Plant J. 81 (2015) 907–919.

[18] S.L. Doty, A.W. Sher, N.D. Fleck, M. Khorasani, Z. Khan, A.W.K. Ko, S.H. Kim, T.H. DeLuca, Variable nitrogen fixation in wild *Populus*, PLoS One 11 (5) (2016) e0155979, http://dx.doi.org/10.1371/journal.pone.0155979.

[19] D. Bulgarelli, K. Schlaeppi, S. Spaepen, E.V.L. van Themaat, P. Schulze-Lefert, Structure and functions of the bacterial microbiota of plants, Ann. Rev. Plant Biol. 64 (2013) 807–838, http://dx.doi.org/10.1146/annurev-arplant-050312-120106.

[20] N. Sharma, S.R. Abrams, Uptake, movement, activity, and persistence of an abscisic acid analog (80 acetylene ABA methyl ester) in marigold and tomato, J. Plant Growth Regul. 24 (2005) 28–35.

[21] M.G. Javid, A. Sorooshzadeh, F. Moradi, S.A.M.M. Sanavy, I. Allahdadi, The role of phytohormones in alleviating salt stress in crop plants, Aust. J. Crop Sci. 5 (2011) 726–734.

[22] D. Straub, H. Yang, Y. Liu, T. Tsap, U. Ludewig, Root ethylene signalling is involved in Miscanthus sinensis growth promotion by the bacterial endophyte Herbaspirillum frisingense GSF30T, J. Expt. Bot. 64 (2013) 4603–4615, http://dx.doi.org/10.1093/jxb/ert276.

[23] S. Fahad, S. Hussain, A. Banu, S. Saud, S. Hassan, D. Shaan, F.A. Khan, F. Khan, Y. Chen, C. Wu, M.A. Tabassum, M.X. Chun, M. Afzal, A. Jan, M.T. Jan, Potential role of phytohormones and plant growth-promoting rhizobacteria in abiotic stresses; consequences for changing environment, Environ. Sci. Pollut. Res. 22 (2015) 4907–4921.

[24] J.F. White Jr., M.S. Torres, Is plant-endophyte-mediated defensive mutualism the result of oxidative stress protection? Physiol. Plant. 138 (2010) 440–446.

[25] B. Mitter, A. Petrik, N.W. Shin, P.S. Chain, Lotte. Hauberg, B. Reinhold-Hurek, J. Nowak, A. Sessitsch, Comparative genome analysis of Burkholderia phytofirmans PsJN reveals a wide spectrum of endophytic lifestyles based on interaction strategies with host plants, Front. Plant Sci. 4 (2013) 120.

[26] M. Naveed, M.B. Hussain, Z.A. Zahir, B. Mitter, A. Sessitsch, Drought stress amelioration in wheat through inoculation with Burkholderia phytofirmans strain PsJN, Plant Growth Regul. 73 (2013) 121–131, http://dx.doi.org/10.1007/s10725-013-9874-8.

[27] M. Naveed, B. Mitter, T.G. Reichenauer, K. Wieczorek, A. Sessitsch, Increased drought stress resilience of maize through endophytic colonization by Burkholderia phytofirmans PsJN and Enterobacter sp. FD17, Environ. Expt. Bot. 97 (2014) 30–39, http://dx.doi.org/10.1016/j.envexpbot.2013.09.014.

[28] M.W. Yaish, I. Antony, B.R. Glick, Isolation and characterization of endophytic plant growth promoting bacteria from date palm tree (*Phoenix dactylifera* L.) and their potential role in salinity tolerance, Antonie Van Leeuwenhoek 107 (2015) 1519–1532.

[29] S. Vardharajula, A. Zulfikar, M. Grover, G. Reddy, B. Venkateswarlu, Drought-tolerant plant growth promoting Bacillus spp.: effect on growth, osmolytes, and antioxidant status of maize under drought stress, J. Plant Interact. 6 (1) (2011) 2011.

[30] S.L. Doty, M.R. Dosher, G.L. Singleton, A.L. Moore, B. Van Aken, R. Stettler, S. Strand, M. Gordon, Identification of an endophytic Rhizobium in stems of *Populus*, Symbiosis 39 (2005) 27–35.

[31] S.L. Doty, B. Oakley, G. Xin, J.W. Kang, G. Singleton, Z. Khan, A. Vajzovic, J.T. Staley, Diazotrophic endophytes of native black cottonwood and willow, Symbiosis 47 (2009) 23–33.

[32] J.W. Kang, Z. Khan, S.L. Doty, Biodegradation of trichloroethylene by an endophyte of hybrid poplar, Appl. Environ. Microbiol. 78 (2012) 3504–3507.

[33] G. Xin, G. Zhang, J.W. Kang, J.S. Staley, S.L. Doty, A diazotrophic, indole-3-acetic acid-producing endophyte from wild cottonwood, Biol. Fert. Soils 45 (2009) 669–674.

[34] J.L. Knoth, S.H. Kim, G.J. Ettl, S.L. Doty, Effects of cross host species inoculation of nitrogen-fixing endophytes on growth and leaf physiology of maize, GCB Bioenergy 5 (2013) 408–418, http://dx.doi.org/10.1111/gcbb.12006.

[35] J.L. Knoth, S.H. Kim, G.J. Ettl, S.L. Doty, Biological nitrogen fixation and biomass accumulation within poplar clones as a result of inoculations with diazotrophic endophyte consortia, New Phytol. (2014) 599–609, http://dx.doi.org/10.1111/nph.12536.

[36] Z. Khan, G. Guelich, H. Phan, R. Redman, S.L. Doty, Bacterial and yeast endophytes from poplar and willow promote growth in crop plants and grasses, ISRN Agron. (2012) 1–11, http://dx.doi.org/10.5402/2012/890280.

[37] Z. Khan, S.L. Kandel, D. Ramos, G.J. Ettl, S.H. Kim, S.L. Doty, Increased biomass of nursery grown Douglas-fir seedlings upon inoculation with diazotrophic endophytic consortia, Forests 6 (2015) 3582–3593.

[38] G.A. Cangelosi, E.A. Best, G. Martinetti, E.W. Nester, Genetic analysis of *Agrobacterium*, Methods Enzymol. 204 (1991) 384–397.

[39] D.R. Hoagland, D.I. Arnon, The water culture method for growing plants without soil Circular, 347, The College of Agriculture, University of California, Berkeley, California Agricultural Experiment Station, 1950.

[40] O. Masciarelli, L. Urbani, H. Reinoso, V. Luna, Alternative mechanism for the evaluation of indole-3-acetic acid (IAA) production by Azospirillum brasilense strains and its effects on the germination and growth of maize seedlings, J. Microbiol. 51 (2013) 590–597, http://dx.doi.org/10.1007/s12275-013-3136-3

[41] N.R. Baker, Chlorophyll fluorescence: a probe of photosynthesis in vivo, Ann. Rev. Plant Biol. 59 (2008) 89–113, http://dx.doi.org/10.1146/annurev.arplant.59.032607.092759.

[42] Z.G. Cerovic, G. Masdoumier, N. Ben Ghozlen, G. Latouche, A new optical leaf-clip meter for simultaneous non-destructive assessment of leaf chlorophyll and epidermal flavonoids, Physiol. Plant. 146 (2012) 251–260, http://dx.doi.org/10.1111/j.1399-3054.2012.01639.x.

[43] K.C. Vaughn, S.O. Duke, *In situ* localization of the sites of paraquat action, Plant Cell Environ. 6 (1983) 13–20.

[44] R.S. Redman, Y.O. Kim, C.J.D. Woodward, C. Greer, L. Espino, S.L. Doty, R.J. Rodriguez, Increased fitness of rice plants to abiotic stress via habitat adapted symbiosis: a strategy for mitigating impacts of climate change, PLoS One 6 (2011) 14823, http://dx.doi.org/10.1371/journal.pone.0014823.

[45] R.J. Rennie, A single medium for the isolation of acetylene-reducing (dinitrogen fixing) bacteria from soils, Can. J. Microbiol. 27 (1981) 8–14.

[46] M. Huntemann, N.N. Ivanova, K. Mavromatis, H.J. Tripp, D. Paez-Espino, K. Palaniappan, E. Szeto, M. Pillay, I.M.A. Chen, A. Pati, T. Nielsen, V.M. Markowitz, N.C. Krypides, The standard operating procedure of the DOE-JGI microbial genome annotation pipeline (MGAP v.4), Stand. Genom. Sci. 10 (2015) 86, http://dx.doi.org/10.1186/s40793-015-0077-y.

[47] V.M. Markowitz, I.M. Chen, K. Palaniappan, IMG: the Integrated Microbial Genomes database and comparative analysis system, Nucleic Acids Res. 40 (Database issue) (2012) D115–22.

[48] A. Firrincieli, R. Otillar, A. Salamov, J. Schmutz, Z. Khan, R. Redman, N.D. Fleck, E. Lindquist, I.V. Grigoriev, S.L. Doty, Data report: genome sequence of the plant growth promoting endophytic yeast *Rhodotorula graminis* WP1, Front. Micribiol. 6 (2015) 978.

[49] S.M. Cho, S.H. Han, A.J. Anderson, J.Y. Park, Y.H. Lee, B.H. Cho, K.Y. Yang, C.M. Ryu, Y.C. Kim, 2R,3R-butanediol, a bacterial volatile produced by *Pseudomonas chlororaphis* O6 is involved in induction of systemic tolerance to drought in *Arabidopsis thaliana*, Mol. Plant. Microbe. Interact. 8 (2008) 1067–1075.

[50] O. Choi, J. Kim, J.G. Kim, Y. Jeong, J.S. Moon, C.S. Park, I. Hwang, Pyrroloquinoline quinone is a plant growth promotion factor produced by *Pseudomonas fluorescens* B16, Plant Physiol. 146 (2008) 657–668.

[51] A.K. Garg, J.K. Kim, T.G. Owens, A.P. Ranwala, Y.D. Choi, L.X. Kochian, R.J. Wu, Trehalose accumulation in rice plants confers high tolerance levels to different abiotic stresses, Proc. Natl. Acad. Sci. 99 (2002) 15898–15903.

[52] R. Suárez, A. Wong, M. Ramírez, A. Barraza, C. Orozco Mdel, M.A. Cevalios, M. Lara, G. Hernandez, G. Iturriaga, Improvement of drought tolerance and grain yield in common bean by overexpressing trehalose-6-phosphate synthase in rhizobia, Mol. Plant Microbe Interact. 21 (2008) 958–966.

[53] H.J. McIntyre, H. Davies, T.A. Hore, S.H. Miller, J.P. Dufour, C.W. Ronson, Trehalose biosynthesis in Rhizobium leguminosarum bv trifolii and its role in desiccation tolerance, Appl. Environ. Microbiol. 73 (2007) 3984–3992.

[54] U.M. Markowitz, E. Szeto, K. Palaniappan, Y. Grechkin, K. Chu, I.M. Chen, et al., The integrated microbial genomes (IMG) system in 2007: data content and analysis tool extensions, Nucleic Acids Res. 36 (2008) D528–D533.

[55] S.F. Altschul, T.L. Madden, A.A. Schäffer, J. Zhang, g.Z. Zhan, W. Miller, D.J. Lipman, Gapped BLAST and PSI-BLAST: a new generation of protein database search programs, Nucleic Acids Res. 25 (1997) 3389–3402.

[56] R. Core Team, R: A Language and Environment for Statistical Computing, R Foundation for Statistical Computing, Vienna, Austria, 2014.

[57] B. Todorovic, B.R. Glick, The interconversion of ACC deaminase and D-cysteine desulfhydrase by directed mutagenesis, Planta 229 (2008) 193–205.

[58] M. Sugawara, E.J. Cytryn, M.J. Sadowsky, Functional role of Bradyrhizobium japonicum trehalose biosynthesis and metabolism genes during physiological stress and nodulation, Appl. Environ. Microbiol. 76 (2010) 1071–1081.

[59] R. Ruhal, R. Kataria, B. Choudhury, Trends in bacterial trehalose metabolism and significant nodes of metabolic pathway in the direction of trehalose accumulation, Microb. Biotechnol. 6 (2013) 493–502.

[60] S.L. Kandel, N. Herschberger, S.H. Kim, S.L. Doty, Diazotrophic endophytes of poplar and willow for growth promotion of rice plants in nitrogen-limited conditions, Crop. Sci. 55 (2015) 1–8.

[61] E.B. Silveira, A.M.A. Gomes, R.L.R. Mariano, E.B. Silva Neto, Bacterization of seeds and development of cucumber seedlings, Hortic. Bras. 22 (2004) 217–221.

[62] A. Rogers, K. Mc Donald, F. Muehlbauer, A. Hoffman, K. Koenig, L. Newman, S. Taghavi, D. vander Lelie, Inoculation of hybrid poplar with the endophytic bacterium Enterobacter sp. 638 increases biomass but does not impact leaf level physiology, GCB Bioenergy 4 (3) (2012) 364–370.

[63] D. Mazzoleni, S. Dickmann, Differential physiological and morphological responses of two hybrid populus clones to water stress, Tree Physiol. 4 (1988) 61–70.

[64] L. Ibrahim, M.F. Proe, A.D. Cameron, Main effects of nitrogen supply and drought stress upon whole-plant carbon allocation in poplar, Can. J. Forest. Res. 27 (1997) 1413–1419, http://dx.doi.org/10.1139/cjfr-27-9-1413.

[65] C.A. Souch, W. Stephens, Growth, productivity and water use in three hybrid poplar clones, Tree Physiol. 18 (1998) 829–835, http://dx.doi.org/10.1093/treephys/18.12.829.

[66] A.W. Bauen, A.J. Dunnett, G.M. Richter, A.G. Dailey, M. Avlott, E. Casella, G. Taylor, Modelling supply and demand of bioenergy from short rotation coppice and Miscanthus in the UK, Bioresour. Technol. 101 (2010) 8132–8143.

[67] H. Shi, L. Chen, T. Ye, X. Liu, K. Ding, Z. Chan, Modulation of auxin content in Arabidopsis confers improved drought stress resistance, Plant Physiol. Biochem. 82 (2014) 209–217, http://dx.doi.org/10.1016/j.plaphy.2014.06.008.

[68] J. Krasensky, C. Jonak, Drought, salt, and temperature stress-induced metabolic rearrangements and regulatory networks, J. Expt. Bot. 63 (2012) 1593–1608, http://dx.doi.org/10.1093/jxb/err460.

[69] R.K. Sairam, Effect of homobrassinolide application on plant metabolism and grain yield under irrigated and moisture stress conditions of two wheat varieties, Plant Growth Regul. 14 (1994) 173–181.

[70] G. Forchetti, O. Masciarelli, S. Alemano, D.A.G. Alvarez, Endophytic bacteria in sunflower (Helianthus annus L.): isolation, characterization, and production of jasmonates and abscisic acid in culture medium, Appl. Microbiol. Biotechnol. 76 (2007) 1145–1152.

[71] C.C. Dai, B.Y. Yu, X. Li, Screening of endophytic fungi that promote the growth of Euphorbia pekinensis, Afr. J. Biotechnol. 7 (2008) 3505–3510.

[72] A. Maggio, G. Barbieri, G.D.P.S. Raimondi, Contrasting effects of GA3 treatments on tomato plants exposed to increasing salinity, J. Plant Growth Regul. 29 (2010) 63–72.

[73] K.R. Li, H.H. Wang, G. Han, Q.J. Wang, J. Fan, Effects of brassinolide on the survival, growth and drought resistance of Robinia pseudoacacia seedlings under water-stress, New Forests 35 (3) (2007) 255–266, http://dx.doi.org/10.1007/s11056-007-9075-2.

[74] U. Kutschera, Z.Y. Wang, Growth-limiting proteins in maize coleoptiles and the auxin-brassinosteroid hypothesis of mesocotyl elongation, Protoplasma (2015), http://dx.doi.org/10.1007/s00709-015-0787-4.

[75] K. Apel, H. Hirt, Reactive oxygen species: metabolism, oxidative stress, and signal transduction, Annu. Rev. Plant Biol. 55 (2004) 373–399, http://dx.doi.org/10.1146/annurev.arplant.55.031903.141701.

[76] G. Miller, V. Shulaev, R. Mittler, Reactive oxygen signaling and abiotic stress, Physiol. Plant. 133 (2008) 481–489, http://dx.doi.org/10.1111/j.1399-3054.2008.01090.x.

[77] P. Sharma, A.B. Jha, R.S. Dubey, M. Pessarakli, Reactive oxygen species, oxidative damage and antioxidative defense mechanism in plants under stressful conditions, J. Bot. 2012 (2012) 1–26.

[78] C.M. Creus, R.J. Sueldo, C. Barassi, Water relations and yield in Azospirillum-inoculated wheat exposed to drought in the field, Can. J. Bot. 82 (2004) 273–281.

[79] J. Kohler, J.A. Hernández, F. Caravaca, A. Roldán, Plant-growth-promoting rhizobacteria and arbuscular mycorrhizal fungi modify alleviation biochemical mechanisms in water-stressed plants, Funct. Plant. Biol. 35 (2008) 141–151.

[80] H.S. Misra, N.P. Khairnar, A. Barik, K. Indira Priyadarshini, H. Mohan, S.K. Apte, Pyrroloquinoline-quinone: a reactive oxygen species scavenger in bacteria, FEBS Lett. 578 (2004) 26–30.

[81] G. Iturriaga, R. Suarez, B. Nova-Franco, Trehalose metabolism: from osmoprotection to signaling, Int. J. Mol. Sci. 10 (9) (2009).

[82] R. Suarez, A. Wong, M. Ramirez, A. Barraza, M.C. Orozco, M. Cevallos, et al., Improvement of drought tolerance and grain yield in common bean by overexpressing trehalose-6-phosphate synthase in Rhizobia, MPMI 21 (7) (2008) 958–966.

[83] J.1 Rodríguez-Salazar, R. Suárez, J. Caballero-Mellado, G. Iturriaga, Trehalose accumulation in Azospirillum brasilense improves drought tolerance and biomass in maize plants, FEMS Microbiol. Lett. 296 (2009) 52–59.

[84] S. Mayak, T. Tirosh, B.R. Glick, Plant growth promoting bacteria confer resistance in tomato plants to salt stress, Plant Physiol. Biochem. 42 (6) (2004) 565–572.

[85] B. Todorovic, B.R. Glick, The interconversion of ACC deaminase and D-cysteine desulfhydrase by directed mutagenesis, Planta 229 (2008) 193–205.

[86] K. Papenbroc, A. Riemenschneider, A. Kamp, H.N. Schulz-Vogt, A. Schmidt, Characterization of cysteine-degrading and H2S-releasing enzymes of higher plants-from field to the test tube and back, Plant Biol. 9 (5) (2007) 582–588.

[87] L. Ping, W. Boland, Signals from underground:bacterial volatiles promote growth in Arabidopsis, Trends Plant Sci. 9 (6) (2004) 263–366.

[88] S. Taghavi, C. Garafola, S. Monchy, L. Newman, A. Hoffman, N. Weyens, T. Barac, J. Vangronsveld, D. vander Lelie, Genome survey and characterization of endophytic bacteria exhibiting a beneficial effect on growth and development of poplar trees, Appl. Environ. Microbiol. 75 (2009) 748–757, http://dx.doi.org/10.1128/AEM.;1; 02239-08.

[89] A.J. Golding, G.N. Johnson, Down-regulation of linear and activation of cyclic electron transport during drought, Planta 218 (2003) 107–114, http://dx.doi.org/10.1007/s00425-003-1077-5.

[90] H. Bae, R.C. Sicher, M.S. Kim, S.H. Kim, M.D. Strem, R.C. Melrick, B.A. Bailey, The beneficial endophyte Trichoderma hamatum isolate DIS 219b promotes growth and delays the onset of the drought response in Theobroma cacao, J. Exp. Bot. 60 (2009) 3279–3295, http://dx.doi.org/10.1093/jxb/erp165.

[91] K. Bürling, Z.G. Cerovic, G. Cornic, J.M. Dacruet, G. Noga, M. Hunsche, Fluorescence-based sensing of drought-induced stress in the vegetative phase of four contrasting wheat genotypes, Environ. Exp. Bot. 89 (2013) 51–59, http://dx.doi.org/10.1016/j.envexpbot.2013.01.003.

[92] M. Hubbard, J.J. Germida, Fungal endophytes enhance wheat heat and drought tolerance in terms of grain yield and second-generation seed viability, J. Appl. Microbiol. 116 (2014) 109–122.

[93] Y. Shi, K. Lou, C. Li, Growth and photosynthetic efficiency promotion of sugar beet (Beta vulgaris L.) by endophytic bacteria, Photosynthesis Res. 105 (2010) 5–13, http://dx.doi.org/10.1007/s11120-010-9547-7.

[94] L. Zulini, M. RubinRubinigg, R.B.M. Zorer, Effects of drought stress on chlorophyll fluorescence and photosynthetic pigments in grapevine leaves (Vitis vinifera cv White Riesling), Acta Hortic. (2007) 289–294.

[95] D.W. Lawlor, W. Tezara, Causes of decreased photosynthetic rate and metabolic capacity in water-deficient leaf cells: a critical evaluation of mechanisms and integration of processes, Ann. Bot. 103 (2009) 561–579, http://dx.doi.org/10.1093/aob/mcn244.

[96] F. Tardieu, T. Simonneau, Variability among species of stomatal control under fluctuating soil water status and evaporative demand: modelling isohydric and anisohydric behaviours, J. Exp. Bot. 49 (1998) 419–432.

[97] W. Fricke, G. Akhiyarova, D. Veselov, G. Kudoyarova, Rapid and tissue-specific changes in ABA and in growth rate in response to salinity in barley leaves, J. Expt. Bot. 55 (2004) 1115–1123, http://dx.doi.org/10.1093/jxb/erh117.

[98] D.E. Zeng, P. Hou, F.L.Y. Xiao, Overexpression of Arabidopsis XERICO gene confers enhanced drought and salt stress tolerance in rice (Oryza Sativa L.), J. Plant Biochem. Biotechnol. 24 (2013) 56–64.

[99] S.Y. Park, F.C. Peterson, A. Mosquna, Agrochemical control of plant water use using engineered abscisic acid receptors, Nature (2015), http://dx.doi.org/10.1038/nature14123.

[100] J. Bloemen, M.A. McGuire, D.P. Aubrey, R.O. Teskey, Transport of root-respired CO2 via the transpiration stream affects aboveground carbon assimilation and CO2 efflux in trees, New Phytol. 197 (2013) 555–565, Tables.

10

SNP-Seek II: A resource for allele mining and analysis of big genomic data in *Oryza sativa*☆

Locedie Mansueto[a], Roven Rommel Fuentes[a], Dmytro Chebotarov[a], Frances Nikki Borja[a], Jeffrey Detras[a], Juan Miguel Abriol-Santos[a], Kevin Palis[a,b], Alexandre Poliakov[c,d], Inna Dubchak[c,d], Victor Solovyev[e], Ruaraidh Sackville Hamilton[a], Kenneth L. McNally[a], Nickolai Alexandrov[a], Ramil Mauleon[a,*]

[a] *International Rice Research Institute, College, Los Baños, Laguna, 4031, Philippines*
[b] *Boyce Thompson Institute, Ithaca, NY 14853, USA*
[c] *Lawrence Berkeley National Laboratory, Berkeley, CA 94720, USA*
[d] *DOE Joint Genome Institute, Walnut Creek, CA 94598, USA*
[e] *Softberry, Inc., Mount Kisco, NY 10549, USA*

ARTICLE INFO

Keywords:
Allele mining
Oryza
SNP
Indel
Genotype database
Genetic diversity

ABSTRACT

The 3000 Rice Genomes Project generated a large dataset of genomic variation to the world's most important crop, *Oryza sativa* L. Using the Burrows-Wheeler Aligner (BWA) and the Genome Analysis Toolkit (GATK) variant calling on this dataset, we identified ~40 M single-nucleotide polymorphisms (SNPs). Five reference genomes of rice representing the major variety groups were used: Nipponbare (temperate *japonica*), IR 64 (*indica*), 93–11 (*indica*), DJ 123 (*aus*), and Kasalath (*aus*).

The results are accessible through the Rice SNP-Seek Database (http://snp-seek.irri.org) and through web services of the application programming interface (API). We incorporated legacy phenotypic and passport data for the sequenced varieties originating from the International Rice Genebank Collection Information System (IRGCIS) and gene models from several rice annotation projects. The massive genotypic data in SNP-Seek are stored using hierarchical data format 5 (HDF5) files for quick retrieval. Germplasm, phenotypic, and genomic data are stored in a relational database management system (RDBMS) using the Chado schema, allowing the use of controlled vocabularies from biological ontologies as query constraints in SNP-Seek.

In this paper, we discuss the datasets stored in SNP-Seek, architecture of the database and web application, interoperability methodologies in place, and discuss a few use cases demonstrating the utility of SNP-Seek for diversity analysis and molecular breeding.

1. Introduction

One of the biggest challenges facing rice farming is increasing worldwide production by at least 25% to meet the demands imposed by the projected increase in global population by 2030, side by side with the constraints brought by reduction of arable land, less available water, and more severe environmental stresses due to climate change [1]. Genetic gains from current breeding methods are insufficient to achieve the target yield increase, but solutions such as molecular breeding technologies and use of allelic diversity for important rice traits could potentially increase genetic gains of ongoing rice breeding programs. Molecular breeding technologies have been utilized to improve disease resistance, drought tolerance, and agronomically important traits [2–4]. The rice gene bank collections serve as a potential source of allelic diversity for important genes. In 2014, the 3k RG Project [5] completed sequencing of 3024 rice genomes from the International Rice Genebank Collection at the International Rice Research Institute and The China National Crop Gene Bank (CNCGB), generating over 17 terabytes of raw sequence data. Bioinformatic analyses for discovery

Abbreviations: 3k RG, 3000 Rice Genomes; API, Application Programming Interface; DAO, Data Access Object; HDF5, Hierarchical Data Format 5 (file format); HDRA, High Density Rice Array; indel, insertion or deletion in genomic region; IRGCIS, International Rice Genebank Collection Information System (http://www.irgcis.irri.org:81/grc/irgcishome.html); IRRI, International Rice Research Institute; RDBMS, Relational Database Management System; SNP, Single Nucleotide Polymorphism.

☆ This article is part of a special issue entitled "Genomic resources and databases", published in the journal Current Plant Biology 7–8, 2016.

* Corresponding author at: International Rice Research Institute, College, Los Baños, Laguna, 4031, Philippines.
E-mail address: r.mauleon@irri.org (R. Mauleon).

of sequence variants using GATK Unified Genotyper (GATK-UG) [6] has identified over 40 million SNP variants and over 2.4 million short indels (< = 50 bases long). Information about the accessions sequenced, and the variants discovered have been made available through Rice SNP-Seek database [7]. The database aims to provide easy access, through a user-friendly web interface, to the SNPs and indels from the 3k RG. Data and tools built into SNP-Seek allow for exploratory discoveries of genomic variant – trait associations and examine allelic diversity at genome regions of interest (e.g. known genes, QTLs). One of the features that sets SNP-Seek apart from other publicly accessible databases such as dbSNP at NCBI [8], Gramene [9], RiceVarMap [10], IC4R [11], and RMBreeding [12], is the interactive real-time visualization of millions of SNPs in thousands of rice varieties. This makes SNP-Seek a unique tool for allele mining [6]. We are committed to provide continuous development support to SNP-Seek to incorporate new analyses results, datasets, viewers and query interfaces for multiple reference genomes and assemblies, as well as include features that will be useful to the broader rice research community.

2. Materials and methods

2.1. Data and code and availability

The SNP discovery pipeline scripts are available at https://github.com/IRRI-Bioinformatics/snp-discovery-pipeline with information on running the pipeline at https://github.com/IRRI-Bioinformatics/snp-discovery-pipeline/wiki/How-to-run-the-pipeline. The scripts are configured to work on an high-performance computing (HPC) cluster with a SLURM Workload Manager [13].

The SNP-Seek web application is implemented in Java Spring and ZK frameworks. The codes and instructions for development or installation are at our repository https://bitbucket.org/irridev/iric_portal. The application requires access to the SNP-Seek PostgreSQL or Oracle database server. Raw and analyzed files, including genotypic data are available for bulk download at the Download page.

2.2. Variety (Passport) and phenotype data

Information for each variety including country of origin, genetic stock account number, and variety group is documented in the IRIC information site (http://iric.irri.org/resources/3000-genomes-project). Further description of the selected germplasm is provided by the 3k RG Project [5]. The phenotypic and passport data are from IRGCIS [14] at IRRI.

2.3. Variant data generation

2.3.1. Alignment and variant calling
The SNP discovery pipeline was used to generate SNPs for 3024 rice accessions. The rice accessions were aligned with five reference genomes: Nipponbare [15], 93-11 [4], Kasalath [16], DJ 123 and IR 64 [17], details of which are in Table 1. The reference genomes were indexed with BWA version 0.7.10[18] and SAMtools version 1.0[19]. The sequence dictionary was created using Picard Tools version 1.119 [20]. Alignment was done using BWA-MEM with the parameters '-M' for Picard compatibility and '-t 8' for 8 threads. Each SAM alignment file was compressed to a sorted BAM file using Picard Tools then processed for marking duplicates, fixing mate-pair information and adding or replacing read groups. The processed BAM file was realigned for local indels using the GATK Realigner Target Creator and Indel Realigner. After realignment, the BAM files for each read-pair were merged for each accession using SAMtools. SNP calling was then performed using the GATK-UG version 3.2-2 [21] with parameters 'glm BOTH', '-mbq 20', '-genotyping_mode DISCOVERY' and '-out_mode EMIT_ALL_SITES'. One variety, PUTTIGE:IRGC 5258801 (IRIS 313-8921), with low sequence coverage had 98% missing calls so it was excluded from further analyses.

2.3.2. Whole-genome DNA alignment
We used the VISTA pipeline infrastructure [22,23] for the construction of genome-wide pairwise DNA alignments between the five reference genomes. To align genomes, we used an efficient combination of global and local alignment methods. First, we obtained a map of large blocks of conserved synteny between genome-pairs by applying Shuffle-LAGAN glocal chaining algorithm [24] to local alignments produced by translated BLAT [25]. After that we used Supermap, the fully symmetric whole-genome extension to the Shuffle-LAGAN. Then, for each syntenic block, we applied Shuffle-LAGAN a second time to obtain a more fine-grained map of small-scale rearrangements such as inversions.

The constructed genome-wide pairwise alignments can be downloaded from http://pipeline.lbl.gov/downloads.shtml .The alignments are accessible for browsing and performing various types of analysis through the VISTA browser at http://pipeline.lbl.gov/or through the SNP-Seek menu.

In each of the 10 resulting alignments, we calculated overall coverage, coverage of each reference genome in the alignments, coverage of different annotated sequences [26], the fraction of unique sequence for each genome, and mapping rates of the 3023 genomes to the reference genomes (Table 1). The reference genomes demonstrate high levels of similarity among them, with the total genome coverage among alignments at 70–92% levels (Supplementary Table 1). For each reference genome, there are unique regions (from 12.3 Mbp to 79.6 Mbp) that may harbor genes found only in these variety-group-specific genome segments. Consequently, allele variants cannot be discovered in accessions that are aligned to a reference genome that belong to a different variety group. This is shown by the reduced read mapping rate of accessions aligned to a reference genome that is not of the same variety group, and the highest read mapping occurring when accessions and reference genome belong to the same variety group (Table 1). These unique regions were used in the discovery of SNPs unique to a particular reference genome.

2.3.3. SNP and indel universe creation
The 3023 emit-all-sites Variant Call Format (VCF) files generated by GATK-UG [21] were analyzed to make the union of variants (SNP and indel "universe"). Due to the size of the data and the independent calling of SNPs and indels, we developed a joint genotyping tool to merge the variants instead of constructing a large VCF file that would be difficult to query. A variant position is reported and added to the list if at least one sample supports it and if the phred-scaled quality score is at least 30. The program ran in parallel by grouping the samples and assigning each group to a processor. The resulting position lists were merged before retrieving the respective calls from each sample before finally exporting the corresponding SNP and indel universe matrices to Hierarchical Data Format v5 (HDF5) [27] files. Two versions of HDF5 files were created for each dataset for efficient retrieval: one version has varieties set in rows and variant position in columns, optimized for queries made on a given set of varieties, returning variants in a long genome region. The other is a transposed version (variant positions in rows, varieties in columns) used when querying for a given set of variant positions/loci across all varieties.

We incrementally discovered new variants from unique genome regions of the other reference genomes from the pairwise genome

Table 1

Summary statistics of five published rice genomes used in SNP discovery and sequencing read mapping rates of 3k RG accessions (binned by 3 major variety groups) to the genomes.

Variety	Genome Publication	Assembly length (bps)	Number of contigs/scaffolds/ chromosomes	Number of annotated genes	Length of unique genome region from pairwise genome alignment (bps)	% reads mapping rate of 3k RG binned to 3 major variety groups		
						Aus	Indica	Temperate Japonica
IR 64 (Indica)	[17]	345,209,449	2919 scaffolds	37,758 (MAKER genes)	14,171,198	95.6	96.4	94.6
93-11 (Indica)	[4]	423,026,874	12 chromosomes 12,718 scaffolds	40,464 (GLEAN genes mRNA)	22,642,740	96.8	97.7	96.1
DJ 123 (Aus)	[17]	345,981,746	2819 scaffolds	37,812 (MAKER genes)	12,327,785	96.9	96.4	95.4
Kasalath (Aus)	[16]	401,141,708	12 chromosomes, 1 unmapped contig (14,822 contigs concatenated by 1000 Ns)	20,869 RAP-aligned genes; 53,662 genes predicted by tophat-cufflinks-cuffmerge	79,626,875	96.7	96.5	95.2
Nippon-bare (Japonica)	[15]	373,245,519	12 chromosomes + chloro-plast + mitochondria + unmapped contig	37,869 representative, 8118 predicted genes (IRGSP1.0/RAP); 16,979 TE & 39,102 non TE genes/loci (RGAP 7)	32,056,098	96.1	96.5	97.2

alignment results (Supplementary Table 1). Variants were progressively reduced following the schema in Supplementary Fig. 1.

2.3.4. Generation of SNP subsets

In addition to the full dataset of 32 million Nipponbare-based SNP calls, we provide the following PLINK-format subsets: All Biallelic SNPs (29 M), Base (18 M), Filtered (4.8 M), and Core (404k). Each subset resulted from the successive application of filtering criteria fit for different analyses. The details of these are described in the SNP-Seek Download page (http://snp-seek.irri.org/_download. zul). In brief, starting from biallelic SNP set, three rounds of filtering were applied since despite following best practices in SNP calling, there can still be false positive SNPs ("pseudoSNPs") due to many factors such as alignment ambiguity and undetected paralogs. A common signature of such SNPs is an elevated proportion of heterozygous calls. The Base SNP set was obtained from Full Biallelic SNP set by removing SNPs that have excessive heterozygosity for a given allele frequency and level of inbreeding. These SNPs have a higher chance of being false positives, although some of them may be true SNPs having high heterozygosity due to selection.

The Filtered SNP set is the subset of Base SNP set comprising the SNPs that have minor allele frequency > 1% and less than 20% of missing genotypes per SNP. The Filtered SNP set is the default dataset in SNP-Seek. It is recommended for most analyses where rare variation is not prioritized. Researchers interested in rare SNPs should use either the Base SNP set or the complete set (if sensitivity is prioritized over specificity).

Additionally, a smaller Core SNP set was created by LD-pruning the filtered set with $r2 > 0.8$. This dataset still shares the diversity present in Filtered SNP set with much lower number of SNPs. It is useful for getting a first-look genome-wide overview of features of 3k RG in a less data intensive setting.

2.4. Genomic data integration

2.4.1. Gene models

We merged gene models from four annotations for Nipponbare – Rice Genome Annotation Project 7 (RGAP7) [15], Rice Annotation Project Database (RAP-DB) representative, RAP-DB predicted [28], and FGenesh++ [29] as one set of loci and loaded these into SNP-Seek, simplifying query and analysis of genome and variant features. Gene loci from different annotations were merged as one common locus (assigned with a new unifying locus ID, described in Supplementary Materials II) if (1) there is at least 50% overlap from the first and last coding regions (CDS) and (2) the gene loci are in the same strand. Gene function annotation from RGAP7 is used for the merged locus, unless it is determined as 'hypothetical', 'unknown',

or 'expressed'. For these cases, the functional annotation from RAP-DB is used. The merging created 72,520 gene loci. The contribution of each source is shown in the Venn diagram in Fig. 1. The values in parenthesis are the number of loci from the source, while the number directly below it is the number of merged loci they generated. This number is lower because some loci from the same source are merged into one after the gene models from all sources are compared for overlaps. The numbers inside the Venn diagram refer to the number of merged loci at every possible intersection.

We also loaded the gene models of the other four reference genomes as generated by their respective projects. However, the loci designation used the name auto-generated by their corresponding annotation pipelines. So, we introduce a uniform gene model naming convention applicable to any sequenced variety of rice as described in Supplementary Materials II.

2.4.2. Marker and gene annotation data

To facilitate the annotation of SNP markers, we also incorporated data from additional analyses we made. This data includes the promoter regions from the FGenesh++ gene prediction pipeline that generated the gene models mentioned above. We also identified the effects of SNP variant on the RGAP7 gene models using the SNPEff software[30]. Data from public rice genomic databases relevant to gene-trait association discovery were also incorporated including,

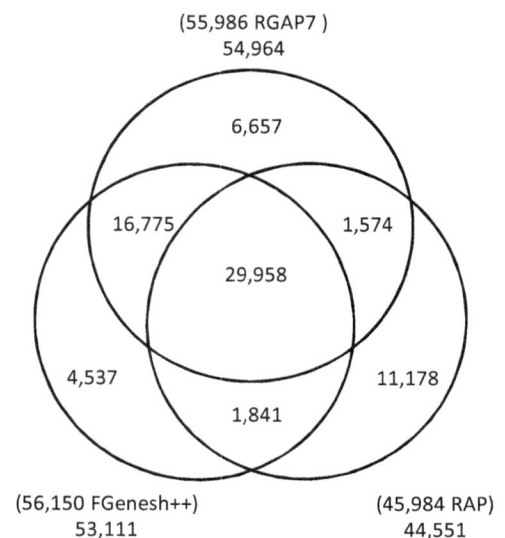

Fig. 1. Number of merged loci from the three sources of rice gene annotations.

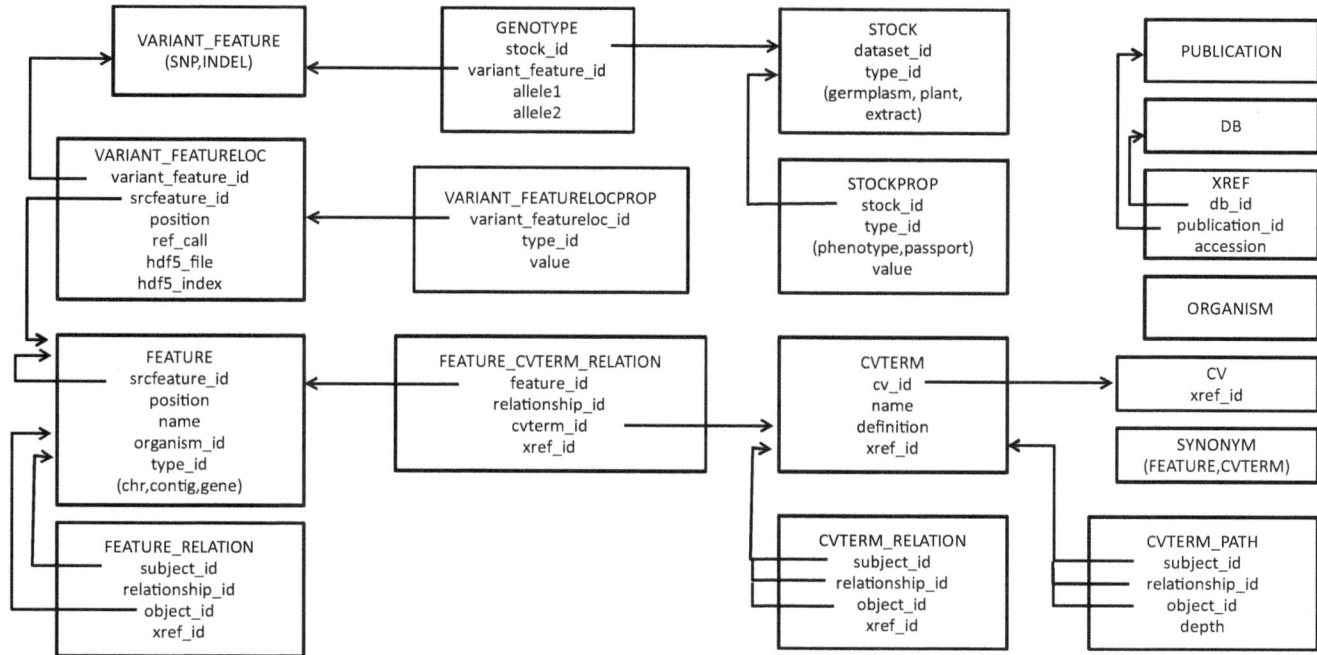

Fig. 2. Overview of SNP-Seek schema based on Chado.
Some relationships are not shown for clarity including: all relationship_id, type_id, dataset_id are referenced to CVTERM.cvterm_id; xref_id to XREF.xref_id; organism_id to ORGANISM.organism_id. Tables can be an abstract to more specific class of entities: VARIANT_FEATURE can be SNP_FEATURE or INDEL_FEATURE; SYNONYM can be CVTERM_SYNONYM or FEATURE_SYNONYM. The OGANISM table holds the different reference varieties. CVTERM_PATH is the pre-computed transitive closure relationships

but not limited to: gene ontology terms from RGAP7, Oryzabase [31], and OGRO [32]; Plant Trait (TO) and Plant (PO) ontology terms from Oryzabase; trait genes and QTL from Q-TARO [33]; rice promoter regions from PlantProm DB [34]; interacting genes from RiceNetv2 [35], and rice protein interactions from PRIN[36].

2.4.3. Biological ontologies

We designed SNP-Seek to be biologist-friendly especially in querying the diverse sets of biological data available. A simple but powerful approach is to use the biological concepts defined by the several ontology projects. This has simplified the querying of molecular markers using controlled biological terms and their transitive closure. Using the annotation data described in the Section 2.4.2 [*Marker and gene annotation data*], terms and relationships from Gene [37], Plant [38], Plant Trait [39], Sequence [40] Ontologies were loaded and referenced to gene models. We also mapped our legacy phenotypic data to Plant Trait and Crop Ontology [41] terms so that varieties can be queried using controlled trait terms.

2.4.4. Public rice genotypic data

We foresee hosting more publicly available data from other rice genotyping projects. A test case is the loading of the genotypic and variety data from the High Density Rice Array (HDRA) project [42]. When legacy passport and phenotypic data for the varieties are found in the IRGCIS, they are also loaded into SNP-Seek.

2.5. Information system design

2.5.1. Database schema

SNP-Seek uses the Chado [43] schema for storing most of the data. Chado is designed to be flexible to handle constantly evolving biological entities and relationships. Through the use of ontologies, controlled terms and hierarchy can be used to define biological entities and relationships in the relational database. A simplified overview to the schema is in Fig. 2. The major modules used are: STOCK for germplasm, phenotype and passport, CV for controlled terms and relations, FEATURE for genomic features, GENOTYPE for genotypic data and modules for referencing like PUBLICATION, DB, and SYNONYM. The X_RELATION tables handle relationships between two FEATURES, two CVTERMS or a FEATURE and a CVTERM, where the relationships are also controlled and evolving terms defined in CVTERM. The data loaded are listed in Table 2.

We made several modifications to the original Chado schema for the main purpose of performance. Ideally, SNPs and indels can be included in the FEATURE table but we separated them into SNP_FEATURE and INDEL_FEATURE. Also, SNPs or indels from different datasets can be merged into a single table but we created separate tables for each dataset. The major modification is our use of HDF5 files to store some of the genotypic dataset. The task of using the right RDBMS table or HDF5 file to query is assigned to the web application, particularly to the Data Access Object (DAO) layer in our software design.

An additional layer of abstraction between the DAO and the relational database is the use of views. Instead of querying the relational tables, the DAOs use views defined in the RDBMS so that changes in the schema only require a change in the views and not the DAO code, simplifying use of the SNP-Seek tables in either PostregSQL or Oracle RDBMS.

2.5.2. Web-application architecture

The web application architecture in Fig. 3 follows the Model-View-Controller (MVC) design to separate the viewers from the business logic, and the logic from the data storage. This is accomplished by defining interfaces in between layers and implemented in Java Spring Framework (https://spring.io). We used the ZK Framework (https://www.zkoss.org) to implement the user interfaces. The Model follows the Data Service and Data Access Layers pattern[44]. The middleware can be grouped into four modules based on the query features of SNP-Seek. The Genotype module handles SNPs and indels. The Variety module handles variety, phenotype, and passport data. The Genomics module handles genes,

Table 2
Data loaded to the Chado tables and their utility in the three (Genotype, Variety, Gene) SNP-Seek search features. Utility code: R-result, C-constraint, A-annotation/attribute, I-internal.

DB Table	Data	Source	Pub.	Genotype	Variety	Gene
STOCK	3024 varieties	3k RGP	[5]	C	R	
	1529 varieties	HDRA project	[42]			
STOCKPROP	Passport data	IRGCIS	[14]		C,A	
	Phenotype data	IRGCIS	[14]	A	C,A	
SNP_FEATURE, SNP_FEATURELOC	42,553,103 positions	this project		C		C
	700k positions	HDRA project	[42]	C		C
INDEL_FEATURE, INDEL_FEATURELOC	3,218,491 indel positions	this project		C		C
SNP_FEATUREPROP	SNP effects on RGAP7gene models using SNPEff	this project		A		
SNP_GENOTYPE	HDRA alleles	HDRA project	[42]	R		
INDEL_GENOTYPE	3k indel alleles	this project		R		
FEATURE, FEATURELOC	chromosomes for Nipponbare	IRGSP v1	[15]			
	RGAP7 gene models	MSU RGAP7	[15]			
	RAP gene models	RAP-DB	[28]			
	FGenesh++ gene models	this project	[29]			
	Merged RGAP7, RAP, FGenesh++ gene models	this project		C		R
	chromosomes, contigs and gene models for 93-11		[4]			
	Chromosomes and gene models for Kasalath		[16]			
	Contigs and gene models for DJ 123		[17]			
	Contigs and gene models for IR 64		[17]			
	QTL	http://qtaro.abr.affrc.go.jp (July 2015)	[33]			
	GO: gene ontology		[37]	A		C
	SO: sequence ontology	http://obofoundry.org (updated July 2015)	[40]	I		I
	PO: plant ontology		[38]	A		C
CV, CVTERM, CVTERM_RELATION, CVTERM_PATH, CVTERM_SYNONYM	TO: plant trait ontology		[39]	A	C	C
	RO: relationship ontology			I	I	I
	CO: crops ontology (rice)	http://cropontology.org	[41]	A	C	C
	Q-TARO traits	http://qtaro.abr.affrc.go.jp (July 2015)	[33]	A	C	C
	SNP-Seek controlled terms			I	I	I
FEATURE_CVTERM_RELATION	Trait genes	http://shigen.nig.ac.jp/rice/oryzabase, http://qtaro.abr.affrc.go.jp (July 2015)	[31,32]	A		C
	Plant anatomy, development genes	http://shigen.nig.ac.jp/rice/oryzabase, (July 2015)	[31]	A		C
	Gene ontology genes	RGAP7, Oryzabase		A		C
FEATURE_RELATION	Ricenet interations	RiceNetv2	[35]	A		C
	PRIN interactions	PRIN	[36]	A		C
	Merged RGAP7, RAP, FGenesh++ gene models mapping	this project		A		C
FEATURE_SYNONYM	gene symbols, accessions	http://shigen.nig.ac.jp/rice/oryzabase, Uniprot	[31]			C

marker annotations, and other sequence features. The Workspace module handles user lists and bulk downloads processing. The Genotype, Variety, and Genomics modules use ontology services for ontology-related queries.

3. Results and discussion

3.1. SNPs and indels

The number of SNPs and indels generated from the 3023 accessions on all five reference genomes are summarized in Table 3. Bulk of the primary SNP discovery was made on the gold standard reference genome for rice, Nipponbare, since there is high degree of similarity of the other four genomes relative to Nipponbare (77–87%, Supplementary Table 1). Reference genome-specific SNPs are discovered mainly in unique genome region (13–20%, Supplementary Table 1), hence a much lower number was observed (Table 3). The quality of genome assembly also affects the discovery rate of SNPs. Kasalath and 93-11 utilized longer read technology [16,4], allowing sequencing across repetitive genome regions. On the other hand, IR 64 and DJ 123 are from short-reads technology [17] which is problematic for highly repetitive genomes such as rice.

The distribution of SNP counts across the 3k RG accessions partitioned into the six major subpopulations is shown in Fig. 4. As expected, SNP density is lowest in accessions belonging to the same group (temperate japonica, subtropical) or closely related to Nipponbare (which is temperate japonica), and it increases as the accessions become more genetically distant from the reference genome (indica, aus).

Majority of the insertions discovered are from 1 to 5 bases long, followed by those of 6–17 bases. Insertions of 18–25 bases are rare (Fig. 4). For deletions, majority was also of size 1–5 bases, followed by sizes of 6–28 bases, and the rarest ones were those between sizes 29–36 bases (Fig. 4). We did not see any significant difference in the discovery rate of indels in these size ranges between the six *Oryza* subgroups (data not shown). The overall discovery rate may be biological in nature, but detection of longer indels may be incomplete, likely caused by short sequencing read length (83 bases), depth of coverage, and the GATK-UG detection method, which is best for detecting small indels only ($<= 50$ bases).

3.2. The SNP-Seek web interface

The rice variant and related data can be accessed by the rice research community through the Rice SNP-Seek website (http://snp-seek.irri.org). We describe here the major features of the site.

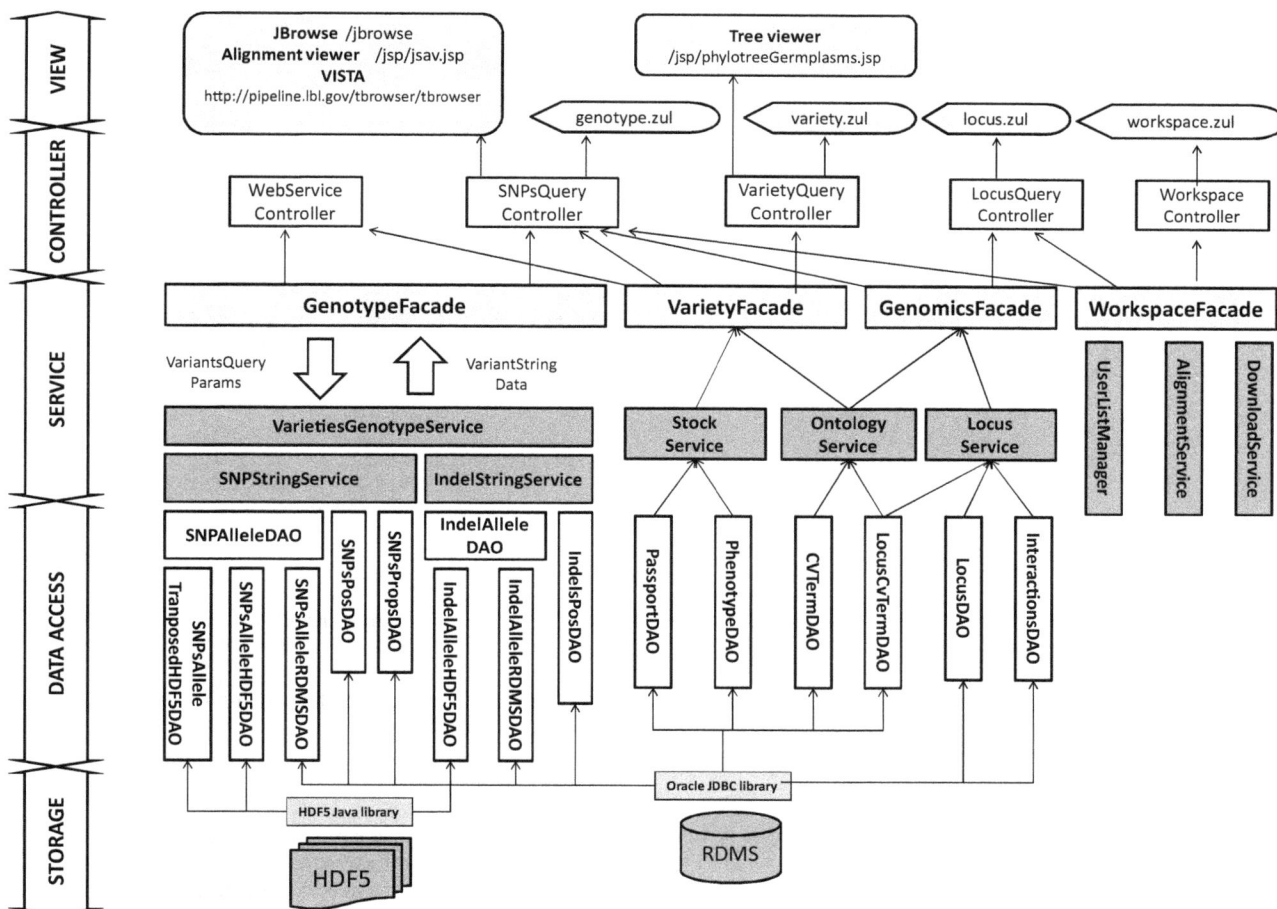

Fig. 3. Web-application software architecture.
The web-application follows the Model-View-Controller design. The VIEW layer are jsp or ZK templates with UI components data and events handled by the CONTROLLER Java classes. The MODEL is based on Data Service and Data Access Objects layers from core J2EE patterns [44]. The storage uses HDF5 for the large 3kRGP genotype matrix, and RDBMS for the rest.

Table 3
Count of small indels and SNPs discovered from Nipponbare and unique regions of four additional genomes.

Genome assembly	Variety group	Count of indels	Count of SNPs
Nipponbare (gold standard)	Temperate japonica	Biallelic: 2,354,934 Multiallelic: 344,545 Total: 2,699,479	Biallelic: 29,635,225 Multiallelic: 2,428,992 Total: 32,064,217
93-11	Indica	308,834	5,540,366
IR 64	Indica	24,041	163,879
Kasalath	Aus	173,892	4,744,760
DJ 123	Aus	12,245	39,881
Total		3,218,491	42,553,103

3.2.1. Data browsers

Several visualization tools are accessible from the Browse menu to display genotypic data. The JBrowse genome browser [45] displays genomic features for each of the five reference genomes used. The gene models and annotation data described in Section 2.4 [*Genomic data integration*] and the BAM and VCF files from the variant calling pipeline described in Section 2.3.1 [*Alignment and variant calling*] are added as tracks to the Nipponbare genome browser. We also generated the phylogenetic tree for all 3024 varieties and display it using the jsPhyloSVG [46] library. Another way to visualize the evolutionary relationships between the varieties is through the display of multidimensional scaling (MDS) plots. The results of the pairwise genome alignments between the five reference genomes described in Section 2.3.2 [*Whole-genome DNA alignment*] can be viewed in VISTA browser [22].

3.2.2. Search features

The major queries available in SNP-Seek are genotype, varieties, and gene loci. Genotype queries return the allele matrix for a given region or list of SNP positions. The varieties can be set to all or constraint by subpopulation or a list. It is also possible to rank the varieties by genotype similarity with a list of positions and alleles, or with any of the varieties in a particular region, or SNP positions. More options are available as shown by the Genotype Query interface in Fig. 5.

The Variety Query interface in Fig. 6 shows the constraints to query varieties including name, genebank accession, and country of origin or subpopulation. It is also possible to query a set of varieties to satisfy phenotypic or passport value. When several constraints are set, the results satisfy all these values. The Gene Locus Query interface returns gene loci from the selected gene model name,

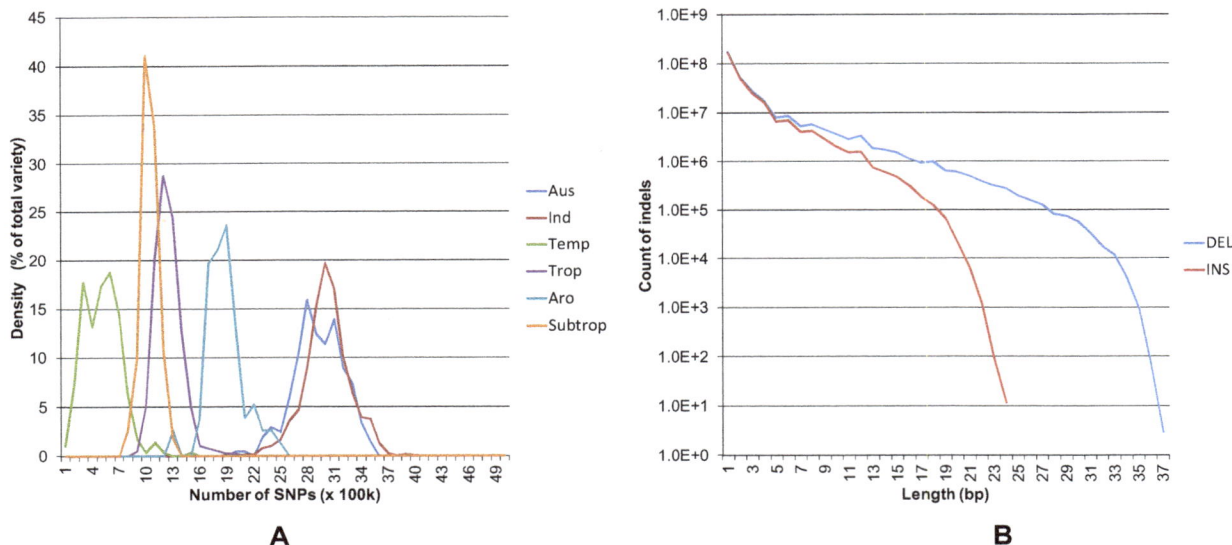

Fig. 4. Distribution of SNP and indels in the 3k RG accessions.
Panel A shows the distribution of SNPs in the 3k RG accessions partitioned to the six major variety groups. Panel B plots the count of indels against indel length (in basepairs) for all 3K RG accessions

function, ontology term, chromosome region, or sequence as constraint. When the constraint is a list of SNP positions, this feature annotates the positions with the available data described in Section 2.4.2 [*Marker and gene annotation data*].

The utility of data in any of the queries described is summarized in Table 2. The use of the data is marked as follows: Result (denoted as R) if it is the returned entity; Constraint (C) when it is used as query constraint; Annotation/Attribute (A) if it is added as an attribute or annotation to the returned entities; and Internal (I) if it is used internally in the database.

Another useful SNP-Seek feature is the possibility to create a list of varieties, SNP positions, or gene loci from query results. The list can also be defined manually by the user through the My List page. Once defined, the list can be used as constraint in succeeding queries. The list can be downloaded for reuse in future sessions.

3.2.3. Data download
All raw and analyses files generated for each variety and the genotype matrices are available through the Download page. There is also a feature to generate the alternate sequence from the VCF

file for any variety and specific regions using GATK's FastaAlternateReferenceMaker option.

3.3. SNP-Seek use cases

We present some use cases of SNP-Seek for rice research. The instructional details, with screenshots, are in Supplementary Material IV. SNP-Seek has also been highly utilized by scientists at IRRI for rice allele mining exercises [6].

3.3.1. Use case 1: given a region or gene of interest, get diversity of varieties
One of the important traits in domesticated rice is seed shattering. This is encoded by the *sh-4* gene (LOC_Os04g57530). Examining the variation of the region for this gene across the 3k RG may elucidate more information about this trait.

Utilizing the Genotype search function, accessions that showed variation in the region of interest were identified. In the Genotype search page, the subpopulation was chosen and the gene locus was supplied. After supplying this information, the start and end position of the gene will be automatically filled in. The table matrix showing the accession with variation in the region will be returned.

Fig. 5. Genotype query interface.

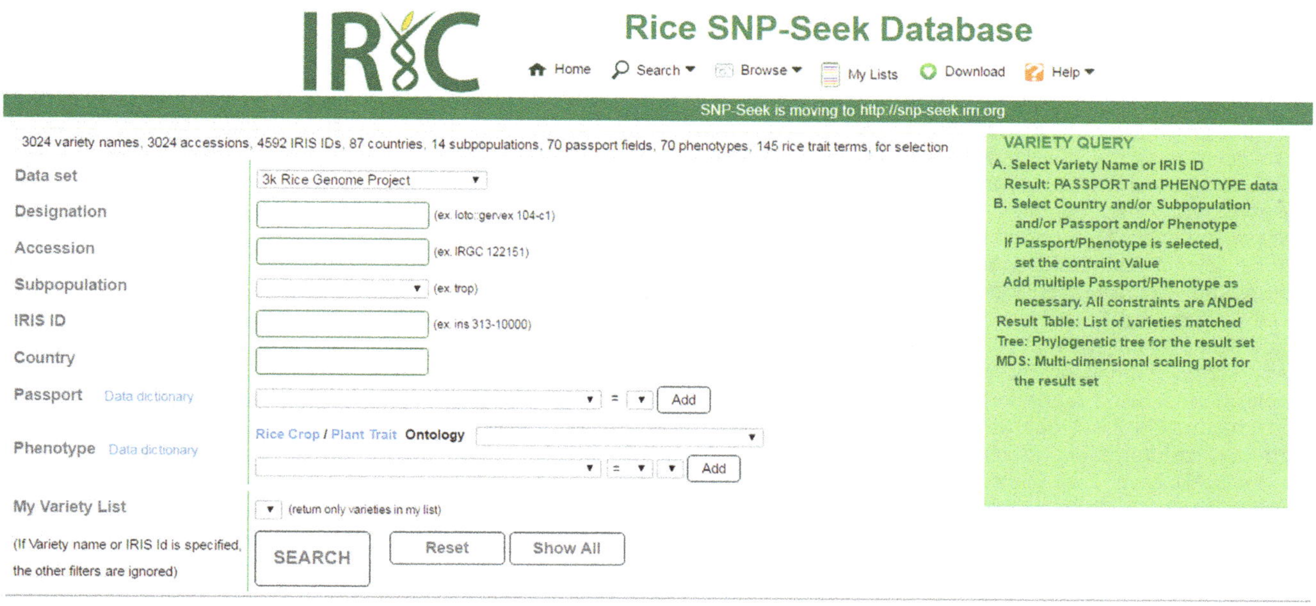

Fig. 6. Variety query interface.

There are 78 varieties that show variation in this region. The table matrix is downloaded in comma-separated value format to further examine the region.

3.3.2. Use case 2: find varieties from the 3k RG panel that is similar to a particular variety

Grain yield is an important trait in rice breeding. One of the components in determining grain yield is grain weight. In this case study, we will find similar accessions to an accession with the heaviest 100-grain weight.

The Genotype search function was used to search across all varieties in the region of interest. Additional information was added to specifically identify varieties with specific phenotypic data. The table matrix for 3024 accessions will be returned. The accession of interest is identified by sorting the table matrix according to the phenotypic data. This accession is used as a reference for the rest of the 3k RG panel. A new table matrix is returned and can be downloaded for further examination.

3.3.3. Use case 3: SNP discovery from a region not found in the nipponbare reference genome

There are genes for important traits in rice that can only be found in genome regions of the donor variety but not in the Nipponbare genome. An example is *Pstol1*, a gene encoding a protein kinase responsible for the phosphorus-uptake efficiency phenotype [47]. It is located within a major rice QTL, Pup1, which is associated with tolerance to P deficiency in soils [48,49]. Pup1 QTL was first identified to be 90 kb region in chromosome 12 in the P-deficiency tolerant variety Kasalath, and it is absent (deleted) in P-deficiency intolerant variety Nipponbare. Variant discovery in this region should be done in the Kasalath reference. Using the Genotype search function, over a thousand SNP positions were identified in the 3k RG accessions. This was done by selecting Kasalath as reference genome and inputting the published region of *Pstol1*. Using Nipponbare as reference resulted in no SNPs being discovered. Marker assays could now be designed around these SNP positions for molecular breeding purposes.

3.4. Web service APIs

It is possible to query data from SNP-Seek programmatically using the RESTful web service APIs we defined. One example is our in-house Galaxy instance (http://galaxy.irri.org) which queries genotypic data. We also implemented the Breeding API (http://docs.brapi.apiary.io) to adhere to the standards in phenotype/genotype databases. This allows other data providers and consumers to interact with SNP-Seek programmatically in the future. An intuitive and functional documentation page is available in SNP-Seek's Help page.

4. Conclusion

We described the development of the Rice SNP-Seek database and data portal for the 3k RG and HDRA Projects. The massive genotypic data required the adoption of modern storage and retrieval technologies. With the integration of diverse biological data, from legacy genebank information system to results from recent high-throughput genomic and computational analyses projects, the need for flexible database and middleware software design is ineviTable SNP-Seek is designed to be adaptive and responsive to the data from the two aforementioned projects as well as that from future sequencing and high-density genotyping projects. Trait data from germplasm panels with curated genotype data can also be accommodated, and legacy passport and phenotypic data for germplasm from the IRGCIS can also be accessed by SNP-Seek. We also demonstrated the utility of ontologies to simplify biological database design and for users to perform effective queries.

Based on user-community feedback and access statistics, SNP-Seek is being heavily utilized as a tool for allele discovery across the diverse 3k RG panel. It is also used for the discovery of SNPs and indels for QTL mapping experiments, and finding candidate molecular markers associated to important agronomic traits of interest for molecular breeding use. SNP-Seek promises to be an indispensable resource and tool for rice genomics and allele discovery. In the future, SNP-Seek will (1) integrate more publicly available rice genotypic and phenotypic data and (2) develop and integrate analysis and visualization tools for genome-wide association study and genomic selection studies.

Acknowledgements

We would like to thank Rolando Santos, Jr. for the administration of the database, Rogelio Alvarez, Denis Diaz and the IRRI ITS team for the operation of the web application servers. Variant calling was supported by the Philippine Genome Center Core Facility for Bioinformatics and DNAnexus, Inc. Computing and data/results hosting were graciously provided by the Extreme Science and Engineering Discovery Environment (XSEDE, which is supported by National Science Foundation grant number ACI-1053575), the Computing and Archiving Research Environment project of the Advanced Science and Technology Institute, Department of Science and Technology of the Philippines, and Amazon Public Data. The work (software development, operation of the database and web application) is being continuously funded by IRRI, the CGIAR Research Program CRP 3.3 (Global Rice Science Partnership) and the International Rice Informatics Consortium (IRIC). Special thanks are in order to CIRAD, Bayer Crop Sciences and Syngenta for providing IRIC funding, and to the Taiwan government for the grant to IRRI.

References

[1] P.A. Seck, A. Diagne, S. Mohanty, M.C.S. Wopereis, Crops that feed the world 7: Rice, Food Secur. 4 (2012) 7–24, http://dx.doi.org/10.1007/s12571-012-0168-1.

[2] S. Fahad, L. Nie, F.A. Khan, Y. Chen, S. Hussain, C. Wu, D. Xiong, W. Jing, S. Saud, F.A. Khan, Y. Li, W. Wu, F. Khan, S. Hassan, A. Manan, A. Jan, J. Huang, Disease resistance in rice and the role of molecular breeding in protecting rice crops against diseases, Biotechnol. Lett. 36 (2014) 1407–1420, http://dx.doi.org/10.1007/s10529-014-1510-9.

[3] H. Hu, L. Xiong, Genetic engineering and breeding of drought-resistant crops, Annu. Rev. Plant Biol. 65 (2014) 715–741, http://dx.doi.org/10.1146/annurev-arplant-050213-040000.

[4] Z.Y. Gao, S.C. Zhao, W.M. He, L.B. Guo, Y.L. Peng, J.J. Wang, X.S. Guo, X.M. Zhang, Y.C. Rao, C. Zhang, G.J. Dong, F.Y. Zheng, C.X. Lu, J. Hu, Q. Zhou, H.J. Liu, H.Y. Wu, J. Xu, P.X. Ni, D.L. Zeng, D.H. Liu, P. Tian, L.H. Gong, C. Ye, G.H. Zhang, J. Wang, F.K. Tian, D.W. Xue, Y. Liao, L. Zhu, M.S. Chen, J.Y. Li, S.H. Cheng, G.Y. Zhang, J. Wang, Q. Qian, Dissecting yield-associated loci in super hybrid rice by resequencing recombinant inbred lines and improving parental genome sequences, Proc. Natl. Acad. Sci. U. S. A. 110 (2013) 14492–14497, http://dx.doi.org/10.1073/pnas.1306579110.

[5] 3k RGP, The 3,000 rice genomes project, (2014) 1–6. http://dx.doi.org/10.1186/2047-217X-3-7.

[6] H. Leung, C. Raghavan, B. Zhou, R. Oliva, I.R. Choi, V. Lacorte, M.L. Jubay, C.V. Cruz, G. Gregorio, R.K. Singh, V.J. Ulat, F.N. Borja, R. Mauleon, N.N. Alexandrov, K.L. McNally, R. Sackville Hamilton, Allele mining and enhanced genetic recombination for rice breeding, Rice 8 (2015) 34, http://dx.doi.org/10.1186/s12284-015-0069-y.

[7] N. Alexandrov, S. Tai, W. Wang, L. Mansueto, K. Palis, R.R. Fuentes, V.J. Ulat, D. Chebotarov, G. Zhang, Z. Li, R. Mauleon, R.S. Hamilton, K.L. McNally, SNP-Seek database of SNPs derived from 3000 rice genomes, Nucleic Acids Res. 43 (2015) D1023–D1027, http://dx.doi.org/10.1093/nar/gku1039.

[8] S.T. Sherry, M.H. Ward, M. Kholodov, J. Baker, L. Phan, E.M. Smigielski, K. Sirotkin, dbSNP: the NCBI database of genetic variation, Nucleic Acids Res. 29 (2001) 308–311, http://dx.doi.org/10.1093/nar/29.1.308.

[9] M.K. Tello-Ruiz, J. Stein, S. Wei, J. Preece, A. Olson, S. Naithani, V. Amarasinghe, P. Dharmawardhana, Y. Jiao, J. Mulvaney, S. Kumari, K. Chougule, J. Elser, B. Wang, J. Thomason, D.M. Bolser, A. Kerhornou, B. Walts, N.A. Fonseca, L. Huerta, M. Keays, Y.A. Tang, H. Parkinson, A. Fabregat, S. McKay, J. Weiser, P. D'Eustachio, L. Stein, R. Petryszak, P.J. Kersey, P. Jaiswal, D. Ware, Gramene 2016: Comparative plant genomics and pathway resources, Nucleic Acids Res. 44 (2016) D1133–D1140, http://dx.doi.org/10.1093/nar/gkv1179.

[10] H. Zhao, W. Yao, Y. Ouyang, W. Yang, G. Wang, X. Lian, Y. Xing, L. Chen, W. Xie, RiceVarMap: a comprehensive database of rice genomic variations, Nucleic Acids Res. 43 (2015) D1018–D1022, http://dx.doi.org/10.1093/nar/gku894.

[11] The IC4R project consortium, information commons for rice (IC4R), Nucleic Acids Res. 44 (2015) gkv1141, http://dx.doi.org/10.1093/nar/gkv1141.

[12] T. Zheng, H. Yu, H. Zhang, Z. Wu, W. Wang, S. Tai, L. Chi, J. Ruan, C. Wei, J. Shi, Y. Gao, B. Fu, Y. Zhou, X. Zhao, F. Zhang, K.L. McNally, Z. Li, G. Zhang, J. Li, D. Zhang, J. Xu, Z. Li, Rice functional genomics and breeding database (RFGB)-3K-rice SNP and InDel sub-database, Chinese Sci. Bull. 60 (2015) 367, http://dx.doi.org/10.1360/N972014-01231 (Chinese Version).

[13] M. Jette, M. Grondona, SLURM: simple linux utility for resource management, Clust Conf. Expo CWCE (2003) 44–60, http://dx.doi.org/10.1007/10968987.

[14] M.T. Jackson, Conservation of rice genetic resources: the role of the International Rice Genebank at IRRI, Plant Mol. Biol. 35 (1997) 61–67 http://www.ncbi.nlm.nih.gov/pubmed/9291960.

[15] Y. Kawahara, M. delaBastide, J.P. Hamilton, H. Kanamori, W.R. McCombie, S. Ouyang, D.C. Schwartz, T. Tanaka, J. Wu, S. Zhou, K.L. Childs, R.M. Davidson, H. Lin, L. Quesada-Ocampo, H. Vaillancourt, H. Sakai, S.S. Lee, J. Kim, H. Numa, T. Itoh, C.R. Buell, T. Matsumoto, Improvement of the *Oryza sativa* Nipponbare reference genome using next generation sequence and optical map data, Rice (N. Y) 6 (4) (2013) 397–405, http://dx.doi.org/10.1186/1939-8433-6-4.

[16] S. Hiroaki, K. Hiroyuki, A.K. Yuko, S.H. Mari, E. Kaworu, O. Youko, K. Kanako, F. Hiroko, K. Satoshi, M. Yoshiyuki, H. Masao, I. Takeshi, M. Takashi, K. Yuichi, W. Kyo, Y. Masahiro, W. Jianzhong, Construction of pseudomolecule sequences of the aus rice cultivar Kasalath for comparative genomics of Asian cultivated rice, DNA Res. 21 (2014) 397–405, http://dx.doi.org/10.1093/dnares/dsu006.

[17] M.C. Schatz, L.G. Maron, J.C. Stein, A. Hernandez Wences, J. Gurtowski, E. Biggers, H. Lee, M. Kramer, E. Antoniou, E. Ghiban, M.H. Wright, J. Chia, D. Ware, S.R. McCouch, W.R. McCombie, Whole genome de novo assemblies of three divergent strains of rice, *Oryza sativa*, document novel gene space of aus and indica, Genome Biol. 15 (2014) 506, http://dx.doi.org/10.1186/PREACCEPT-2784872521277375.

[18] H. Li, R. Durbin, Fast and accurate short read alignment with Burrows-Wheeler transform, Bioinformatics 25 (2009) 1754–1760, http://dx.doi.org/10.1093/bioinformatics/btp324.

[19] H. Li, B. Handsaker, A. Wysoker, T. Fennell, J. Ruan, N. Homer, G. Marth, G. Abecasis, R. Durbin, The sequence alignment/map format and SAMtools, Bioinformatics 25 (2009) 2078–2079, http://dx.doi.org/10.1093/bioinformatics/btp352.

[20] Broad Ins,titute, Picard tools, (2016). http://broadinstitute.github.io/picard/.

[21] A. McKenna, M. Hanna, E. Banks, A. Sivachenko, K. Cibulskis, A. Kernytsky, K. Garimella, D. Altshuler, S. Gabriel, M. Daly, M.A. DePristo, The genome analysis toolkit: a MapReduce framework for analyzing next-generation DNA sequencing data, Genome Res. 20 (2010) 1297–1303, http://dx.doi.org/10.1101/gr.107524.110.

[22] K.A. Frazer, L. Pachter, A. Poliakov, E.M. Rubin, I. Dubchak, VISTA: Computational tools for comparative genomics, Nucleic Acids Res. 32 (2004) 273–279, http://dx.doi.org/10.1093/nar/gkh458.

[23] I. Dubchak, A. Poliakov, A. Kislyuk, M. Brudno, Multiple whole-genome alignments without a reference organism, Genome Res. 19 (2009) 682–689, http://dx.doi.org/10.1101/gr.081778.108.

[24] M. Brudno, S. Malde, A. Poliakov, C.B. Do, O. Couronne, I. Dubchak, S. Batzoglou, Glocal alignment: finding rearrangements during alignment, Bioinformatics (2003), http://dx.doi.org/10.1093/bioinformatics/btg1005.

[25] W.J. Kent, BLAT – The BLAST-like alignment tool, *Genome Res.* **12**, (2002) 656–664. http://dx.doi.org/10.1101/gr.229202. Article published online before March 2002.

[26] S. Schwartz, W.J. Kent, A. Smit, Z. Zhang, R. Baertsch, R.C. Hardison, D. Haussler, W. Miller, Human-mouse alignments with BLASTZ, Genome Res. 13 (2003) 103–107, http://dx.doi.org/10.1101/gr.809403.

[27] M. Folk, G. Heber, Q. Koziol, An overview of the HDF5 technology suite and its applications, Proc. EDBT (2011) 36–47, http://dx.doi.org/10.1145/1966895.1966900.

[28] H. Sakai, S.S. Lee, T. Tanaka, H. Numa, J. Kim, Y. Kawahara, H. Wakimoto, C.C. Yang, M. Iwamoto, T. Abe, Y. Yamada, A. Muto, H. Inokuchi, T. Ikemura, T. Matsumoto, T. Sasaki, T. Itoh, Rice annotation project database (RAP-DB): An integrative and interactive database for rice genomics, Plant Cell Physiol. 54 (2013), http://dx.doi.org/10.1093/pcp/pcs183.

[29] V. Solovyev, P. Kosarev, I. Seledsov, D. Vorobyev, Automatic annotation of eukaryotic genes, pseudogenes and promoters, Genome Biol. 7 (Suppl. 1) (2006) S10.1–S10.12, http://dx.doi.org/10.1186/gb-2006-7-s1-s10.

[30] P. Cingolani, A. Platts, L.L. Wang, M. Coon, T. Nguyen, L. Wang, S.J. Land, X. Lu, D.M. Ruden, A program for annotating and predicting the effects of single nucleotide polymorphisms, SnpEff: SNPs in the genome of *Drosophila melanogaster* strain w 1118; iso-2; iso-3, Fly (Austin) 6 (2012) 80–92, http://dx.doi.org/10.4161/fly.19695.

[31] N. Kurata, Y. Yamazaki, Oryzabase. An integrated biological and genome information database for rice, Plant Physiol. 140 (2006) 12–17, http://dx.doi.org/10.1104/pp.105.063008.

[32] E. Yamamoto, J. Yonemaru, T. Yamamoto, M. Yano, OGRO: The Overview of functionally characterized Genes in Rice online database, Rice 5 (2012) 26, http://dx.doi.org/10.1186/1939-8433-5-26.

[33] J. ichi Yonemaru, T. Yamamoto, S. Fukuoka, Y. Uga, K. Hori, M. Yano, Q-TARO: QTL annotation rice online database, Rice 3 (2010) 194–203, http://dx.doi.org/10.1007/s12284-010-9041-z.

[34] I.A. Shahmuradov, A.J. Gammerman, J.M. Hancock, P.M. Bramley, V.V. Solovyev, PlantProm: a database of plant promoter sequences, Nucleic Acids Res. 31 (2003) 114–117, http://dx.doi.org/10.1093/nar/gkg041.

[35] T. Lee, T. Oh, S. Yang, J. Shin, S. Hwang, C.Y. eong Kim, H. Kim, H. Shim, J.E. unShim, P.C. Ronald, I. Lee, RiceNet v2: an improved network prioritization server for rice genes, Nucleic Acids Res 43 (2015) W12–W127, http://dx.doi.org/10.1093/nar/gkv253.

[36] H. Gu, P. Zhu, Y. Jiao, Y. Meng, M. Chen, PRIN: A predicted rice interactome network, BMC Bioinf. 12 (2011), http://dx.doi.org/10.1186/1471-2105-12-161, 13 str.

[37] The Gene Ontology Consortium, Gene Ontology Consortium: going forward, *Nucleic Acids Res.* **43**, 2015, D1049–D1056. http://dx.doi.org/ doi: 10.1093/nar/gku1179.

[38] P. Jaiswal, S. Avraham, K. Ilic, E.A. Kellogg, S. McCouch, A. Pujar, L. Reiser, S.Y. Rhee, M.M. Sachs, M. Schaeffer, L. Stein, P. Stevens, L. Vincent, D. Ware, F.

[39] E. Arnaud, L. Cooper, R. Shrestha, Towards a reference plant trait ontology for modeling knowledge of plant traits and phenotypes, Proc. Int. Conf. Knowl. Eng. Ontol. Dev. (2012) 220–225, http://dx.doi.org/10.5220/0004138302200225.

[40] C.J. Mungall, C. Batchelor, K. Eilbeck, Evolution of the Sequence Ontology terms and relationships, J. Biomed. Inform. 44 (2011) 87–93, http://dx.doi.org/10.1016/j.jbi.2010.03.002.

[41] R. Shrestha, E. Arnaud, R. Mauleon, M. Senger, G.F. Davenport, D. Hancock, N. Morrison, R. Bruskiewich, G. McLaren, Multifunctional crop trait ontology for breeders' data: field book, annotation, data discovery and semantic enrichment of the literature, AoB Plants (2010), http://dx.doi.org/10.1093/aobpla/plq008.

[42] S.R. McCouch, M.H. Wright, C.-W. Tung, L.G. Maron, K.L. McNally, M. Fitzgerald, N. Singh, G. DeClerck, F. Agosto-Perez, P. Korniliev, A.J. Greenberg, M.B. Elizabeth Naredo, S.Q. Mae Mercado, S.E. Harrington, Y. Shi, D.A. Branchini, P.R. Kuser-Falcão, H. Leung, K. Ebana, M. Yano, G. Eizenga, A. McClung, J. Mezey, Open access resources for genome-wide association mapping in rice, Nat. Commun. 7 (2016) 10532, http://dx.doi.org/10.1038/ncomms10532.

[43] C.J. Mungall, D.B. Emmert, W.M. Gelbart, A. de Grey, S. Letovsky, S.E. Lewis, G.M. Rubin, S.Q. Shu, C. Wiel, P. Zhang, P. Zhou, A Chado case study: an ontology-based modular schema for representing genome-associated biological information, Bioinformatics 23 (2007) 337–346, http://dx.doi.org/10.1093/bioinformatics/btm189

[44] D. Alur, J. Crupi, D. Malks, Core J2EE patterns, Design (2003) 650, http://dx.doi.org/10.1017/CBO9781107415324.004.

[45] M.E. Skinner, A.V. Uzilov, L.D. Stein, C.J. Mungall, I.H. Holmes, JBrowse: a next-generation genome browser, Genome Res. 19 (2009) 1630–1638, http://dx.doi.org/10.1101/gr.094607.109.

[46] S.A. Smits, C.C. Ouverney, jsPhyloSVG: A javascript library for visualizing interactive and vector-based phylogenetic trees on the web, PLoS One 5 (2010), http://dx.doi.org/10.1371/journal.pone.0012267.

[47] M. Wissuwa, J. Wegner, N. Ae, M. Yano, Substitution mapping of Pup1: A major QTL increasing phosphorus uptake of rice from a phosphorus-deficient soil, Theor. Appl. Genet. 105 (2002) 890–897, http://dx.doi.org/10.1007/s00122-002-1051-9.

[48] M. Wissuwa, M. Yano, N. Ae, Mapping of QTLs for phosphorus-deficiency tolerance in rice (Oryza sativa L.), Theor. Appl. Genet. 97 (1998) 777–783, http://dx.doi.org/10.1007/s001220050955.

[49] J.H. Chin, R. Gamuyao, C. Dalid, M. Bustamam, J. Prasetiyono, S. Moeljopawiro, M. Wissuwa, S. Heuer, Developing rice with high yield under phosphorus deficiency: Pup1 sequence to application, Plant Physiol. 156 (2011) 1202–1216, http://dx.doi.org/10.1104/pp.111.175471.

11

Making the right connections: Network biology and plant immune system dynamics ☆

Maggie E. McCormack, Jessica A. Lopez, Tabitha H. Crocker, M. Shahid Mukhtar*

Department of Biology, University of Alabama at Birmingham, Birmingham, AL, USA

ARTICLE INFO

Keywords:
Proteome
Protein–protein interaction
Bioinformatics
Network
Disease–disease
Host–pathogen

ABSTRACT

Network analysis has been a recent focus in biological sciences due to its ability to synthesize global visualizations of cellular processes and predict functions based on inferences from network properties. A protein–protein interaction network, or interactome, captures the emergent cellular states from gene regulation and environmental conditions. Given that proteins are involved in extensive local and systemic molecular interactions such as signaling and metabolism, understanding protein functions and interactions are essential for a systems view of biology. However, in plant sciences these network-based approaches to data integration have been few and far between due to limited data, especially protein–protein interaction data. In this review, we cover network construction from experimental data, network analysis based on topological properties, and finally we discuss advances in networks in plants and other organisms in a comparative approach. We focus on applications of network biology to discover the dynamics of host–pathogen interactions as these have potential agricultural uses in improving disease resistance in commercial crops.

1. Introduction

Cells are the fundamental units of life. Cells contain a complex internal architecture with highly organized compartments capable of responding to diverse signals. In order to maintain homeostasis, cells must be both stable and flexible. This robustness is achieved through a wide range of proteins that function in variety of capacities to receive environmental cues and respond to them in an appropriate manner. Proteins generally have functional roles that contribute to a certain process within the cell. In this way, proteins can be viewed as members of functional communities that may have differential roles given certain cellular conditions. These roles usually involve interactions with other proteins or substrates that directly or indirectly activate cellular responses to environmental stimuli. The conditional probabilities of these interactions influence the direction of cellular behavior at any given time. Protein interactions have different functional roles during steady state and fluctuating environmental conditions. Additionally, protein interactions can be transient or long-term, adding another dimension to the dynamic behavior of proteins. The biological information contained within the relationships between both individual proteins and the communities of proteins can be visualized as a network of possible interactions between each protein within a cell [1,2]. Network biology seeks to map and understand these networks as systems-level views of cell behavior. In the context of protein interactions, network biology provides the tools to answer questions such as the specific spatiotemporal arrangement of proteins in the cellular landscape, the effects of diverse environmental conditions on the flow of biological information between proteins, and the phenotypic results of perturbations of protein communities.

The widespread use of high-throughput methods in recent decades has facilitated the accumulation of large datasets. The need for analytical approaches that can handle this informational load has driven the rise of a class of biological disciplines called "omics" that attempt to integrate large-scale data such as genomes, proteomes, transcriptomes, and others, into descriptive models that interpret molecular functions at a systems level [3]. One of the main goals of -omics is mapping associative relationships between genotypes and phenotypes in order to understand the cellular hierarchy of interactions between genes, transcriptional products, and proteins. Obtaining a sufficient understanding of these systems will enable the development of effective genetic engineering techniques such as personalized medicine and customized cultivars for agronomically important crops [4,5]. However, a growing

☆ This article is part of a special issue entitled "Protein networks – a driving force for discovery in plant science".
 * Corresponding author at: University of Alabama at Birmingham, Department of Biology, 1300 University Boulevard, Birmingham, AL 35294, USA.
   E-mail address: smukhtar@uab.edu (M.S. Mukhtar).

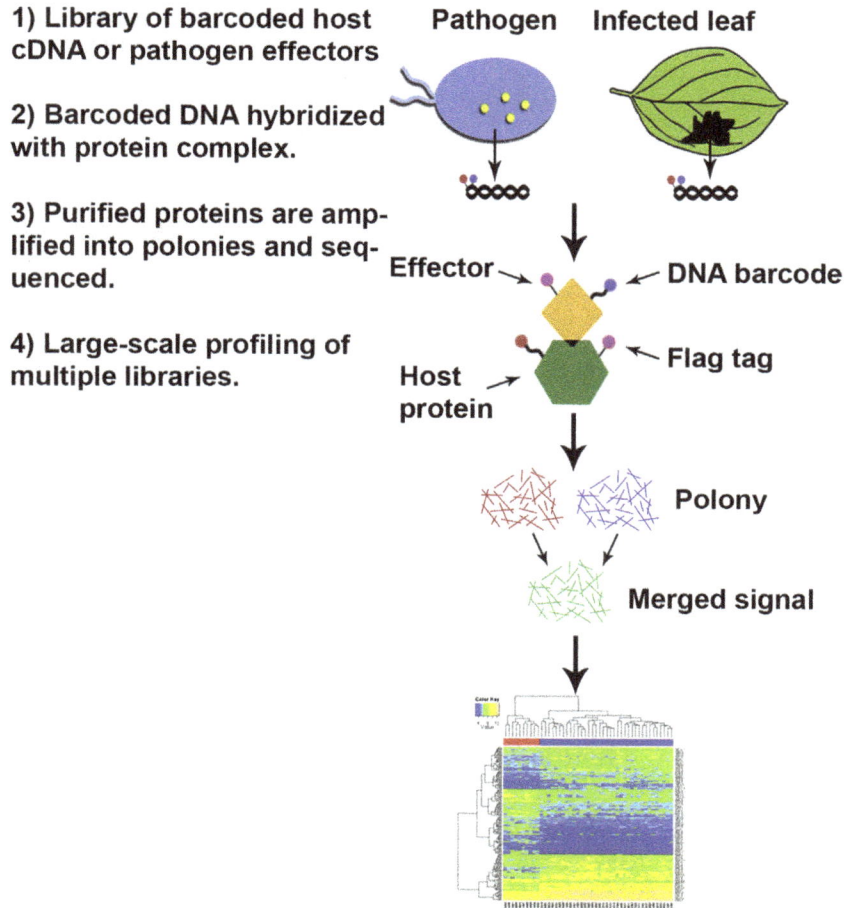

1) **Library of barcoded host cDNA or pathogen effectors**

2) **Barcoded DNA hybridized with protein complex.**

3) **Purified proteins are amplified into polonies and sequenced.**

4) **Large-scale profiling of multiple libraries.**

Pathogen   Infected leaf

Effector → ← **DNA barcode**

Host protein → ← **Flag tag**

Polony

Merged signal

Fig. 1. A step-by-step visualization of SMI-seq applied to elucidating pathogen-host interactions.

concern is that while the aggregation of experimental data into public databases is increasingly popular, most of this data remains under-utilized [6]. The disparity between used and unused data in analysis and integration reflects a limitation in the ability to infer meaningful information from large, noisy datasets. To alleviate this problem, bioinformatic approaches have attained increasing importance and relevance in biological sciences because of the ability to streamline statistical analyses. One output of computational methods is a network or map of relationships between cellular components that reveals novel correlations between datasets. These network-guided analyses offer integrative approaches to biological sciences that are beginning to challenge the previously accepted model of one gene-one protein for describing cellular phenotypes. Indeed, much of the data produced by genomic analysis over recent decades has pointed towards the increasing complexity of functional molecular systems [7–11]. Bioinformatics has been at the forefront of revealing the underlying principles and mechanisms of these relationships.

Protein networks, or interactomes, are less characterized compared to genomic or gene regulatory networks due to limited availability of cost-effective and statistically powerful methods for characterizing protein functions and interactions. This is especially problematic in plant systems because robust methods for observing protein functions and interactions are additionally limited by sensitivity of assays to detect transient interactions or weak binding affinities [12]. However, considering the varied roles of protein dynamics in cellular processes like metabolism, signaling, and gene regulation, determining protein functions and interactions is essential in understanding the pathway from genotype to phenotype.

Interactomes have a variety of functional purposes in plants such as elucidating signaling pathways, stress responses, and differential expression levels during developmental stages. Additionally, a main goal of describing interactomes is understanding the perturbations and maintenance of cellular homeostasis and the effects of environmental factors, especially in the context of pathophysiologies. Although much work has been done in defining interactomes in other organisms such as yeast [13] and humans [14,15], the current state of protein network analysis in plants is underwhelming despite recent accomplishments in some areas [16–18]. The *Arabidopsis thaliana* (hereafter, *Arabidopsis*) and human genome projects were completed in 2000 and 2001, respectively, however network models in humans are much more developed compared to *Arabidopsis*. The success of network analyses in other organisms in predicting functional relationships indicates that these methods need to be utilized more in plant sciences. These networks have the potential to contribute to practical applications, such as improved crop performance, especially when applied to plant development and disease resistance, the latter of which is the focus of this article.

A particularly lucrative role for proteomic networks in plants is the inference of host–pathogen interactions and the elucidation of immune system dynamics. The classical model of plant-pathogen co-evolutionary immune interactions is a zig–zag model [19]. In this model, conserved bacterially-derived microbe-associated molecular patterns (MAMPs) trigger MAMP-triggered immunity (MTI) upon detection by plant pattern recognition receptors [20]. MTI is a broad-spectrum immune response that is effective against most pathogens. However, since hosts and pathogens are in an evolutionary arms race, specialized pathogens have developed several strategies to subvert or suppress the MTI responses. Effector proteins secreted by pathogens target components at various layers of MTI to disrupt the immune response [21]. To circumvent this, plants equipped with the sensors to recognize effectors can initiate effector-triggered immunity (ETI). ETI is manifested by robust defense responses that generally leads to the hypersensitive response, a form of localized cell death. ETI is a similar but amplified version of MTI. While several MTI and ETI players have been discovered in recent decades, the global understanding of how pathogen effectors cross communicates with MTI and ETI is lacking. Additionally, there is a limited understanding of the full range of plant-microbe interactions and the cellular components involved in the determination of cell fate decisions based on these interactions. Computational approaches to studying plant immune networks are scarce due to both insufficient datasets and limited technical ability. However, as there is a strong functional role of protein interactions between plant receptors and bacterial effectors in directing the immune response, the predictive capability of network integration and analysis may help to fill in the gaps and generate hypotheses for future research. A useful benchmark for the potential utility of these bioinformatic tools is seen in the efforts towards developing and understanding the human disease network [8,11]. These networks have revealed novel disease–disease relationships and possible genetic variants responsible for pathological phenotypes.

In this review, we demonstrate techniques for the integration of available datasets and synthesis of protein–protein interaction networks. We discuss the components of these networks as well as analyses through statistical metrics and topological properties. Further, we cover a few examples of networks that have predicted previously unknown protein–protein interactions and have since been experimentally validated in independent studies. We conclude with speculations towards the future utility of network biology with an emphasis on the elucidation of dynamic host–pathogen interactions in plant immunity and possible agricultural applications of a completed plant disease network.

2. Construction of biological networks

Experimental identification of protein–protein interactions relies primarily on the construction of cDNA libraries and mapping of heterologous expression. Many methods for determining protein functions and interactions are based on systems utilizing yeast. To generate large-scale networks in *Arabidopsis*, two of the most widely used yeast-based techniques are GAL4-based yeast two-hybrid (Y2H) and split-ubiquitin (sUbq). Although both systems rely upon the activation of a reporter gene, in the Y2H system the transcription factor is reconstituted in the nucleus whereas in sUbq the transcription factor is cleaved and translocated to the nucleus. GAL4-based Y2H screens were used to generate the Arabidopsis Interactome-1 (AI-1) and the Plant-Pathogen Interaction Network-1 (PPIN-1) [22,23]. AI-1 used a Y2H-based screening platform coupled with extensive validation to identify 5664 interactions between 2661 proteins. AI-1 was compared to a literature curated network that contained 4252 interactions between 2160 proteins and was shown to exhibit similar characteristics. Similarly, PPIN-1 identified 1358 interactions among 926 proteins, indicating the ability of Y2H assays to screen large-scale protein interactions.

Y2H assays are widely used due to their versatility. Many modified versions of Y2H have been reported to help identify specific classes of protein interactions [24]. The sUbq system is based on Y2H principles but is designed specifically for understanding interactions of transmembrane proteins. This method was used to generate an interactome of membrane-bound proteins in *Arabidopsis* [25,26]. This network, the membrane interactome database

version 1 (MIND1), screened a test space of 6.4 million potential interactions and found 12,102 reliable interactions among 1523 proteins. Of these interactions, 99% were previously unknown [26]. Despite the numerous applications of Y2H systems, one major drawback is the inability of Y2H screens to identify transient interactions, which have been shown to be important in many signaling cascades. Thus, there is a growing need for more sensitive screening methods.

One such method is the coupling of affinity purification with mass spectrometry (AP-MS) to isolate protein complexes. Purification can be done in one-step or two-step methods depending on the number of tags used. A popular two-step method is tandem affinity purification (TAP) that has been shown to have a high degree of accuracy and low false positive rates due to the second purification step. A benefit of these methods is the ability to test interactions *in vivo* [27,28]. AP-MS and TAP have been employed in the construction of a cell cycle network in plants [29]. TAP methods coupled with Y2H screens were used to characterize a rice kinase interactome [30]. The rice kinome is larger than both *Arabidopsis* and humans and remains largely uncharacterized [31]. The kinase interactome identified 378 interactions among 274 proteins and suggested new functional roles for E3 ubiquitin ligases in pathogen defense recognition and kinase interaction. However, one limitation in the purification assays is the disruption of weak interactions during the isolation steps as well as degradation of interactions during long processing times [32]. An additional concern is the absence of overlap between the interactions identified using TAP and Y2H screens [30]. In the rice kinome, TAP identified more interactors when the interactions are stable, but Y2H identified more transient interactions. Previous studies have also reported this discrepancy [33], indicating potential limitations in the use of a single method. Future studies should implement multiple screening methods to identify more protein–protein interactions in addition to comparing the accuracy of individual screens.

Protein microarrays are often used in conjunction with affinity purification methods. Microarrays are heterologous systems that allow for the parallel profiling of protein interactions. Purified protein complexes are loaded onto chemically treated glass slides and assayed for the detection of interactions and other features [27]. Microarray experiments were used to determine phosphorylation targets of mitogen-activated protein kinases (MAPKs) described in an *Arabidopsis* protein phosphorylation network [34]. MAPKs are important signaling proteins conserved among eukaryotes. A variety of MAPK signaling complexes are involved in innate immunity in plants, including detection of MAMPs, hormone signaling, and abiotic stress response. Despite their essential role in several cellular processes, MAPK signaling pathways are not well characterized. Popescu et al. generate a network of MAPK phosphorylation targets to elucidate the differential roles of these proteins. MAPKs were shown to respond universally to a common core of effectors, however, differential gene expression is induced through specific combinations of MAPK phosphorylation events initiated by specific effectors. In this way, MAPKs can respond to diverse substrates by changing the downstream targets of phosphorylation. A human MAPK interactome has also been generated [35]. MAPKs in humans have been implicated in cancer and autoimmune disorders. This network sought to identify interactions outside the canonical model of MAPKKK-MAPKK-MAPK pathway that may contribute to the functionality of the MAPK signaling cascade and reveal potential drug targets. Analysis of this network revealed a potential scaffold protein that may integrate multiple signals into distinct signaling complexes. Although this study used Y2H screens to identify MAPK interactions, the potential interactions were subjected to multiple sampling and cross-species comparison, which may be used in future studies to determine the quality of protein interactions identified during large-scale screens.

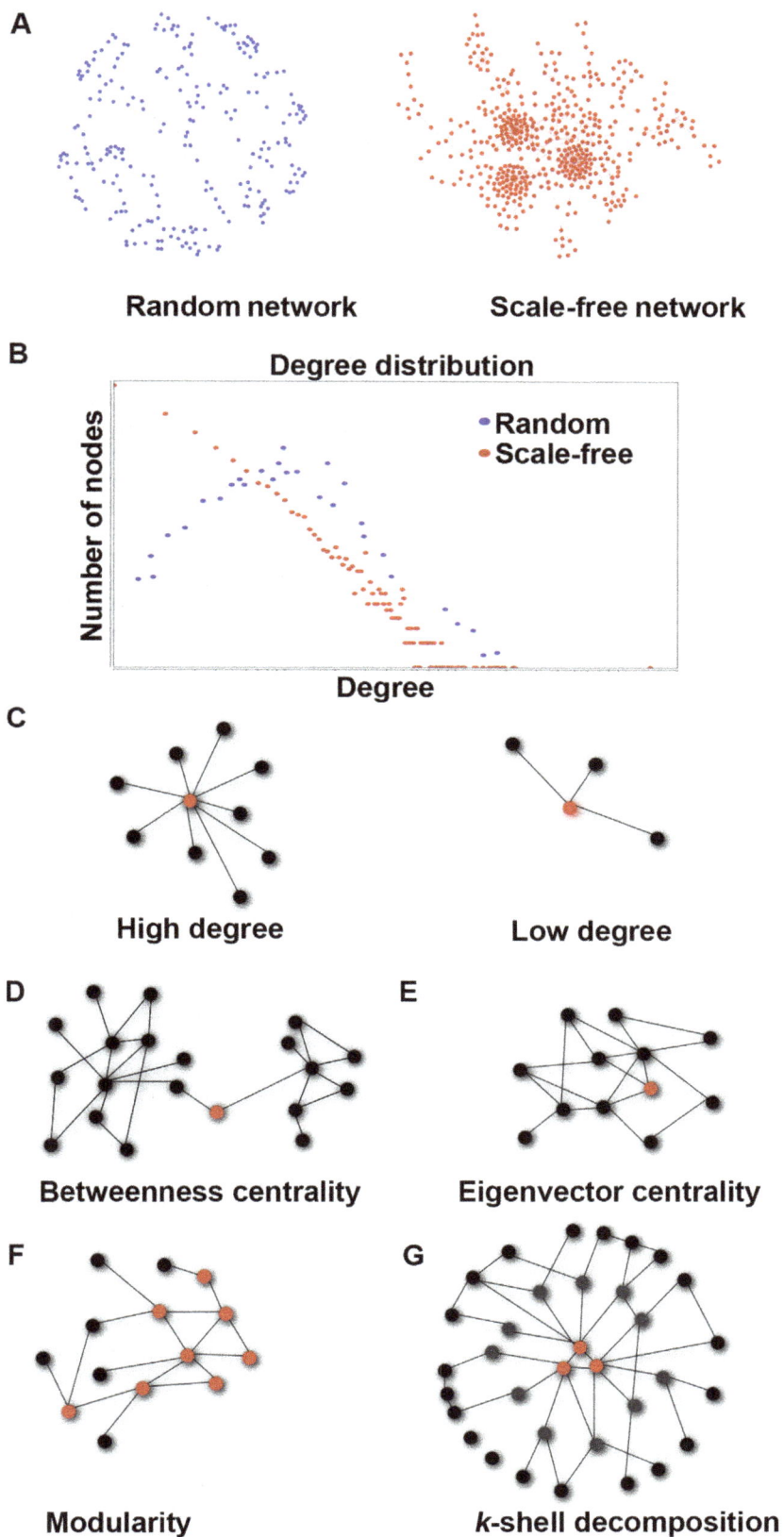

Fig. 2. Overview of topological features of biological networks. (A) Comparison of the topology of a scale-free network and a random network. (B) High degree and low degree. (C) Betweenness centrality. (D) Eigenvector centrality. (E) Modularity. (F) k-shell decomposition.

In response to the growing need for massive screening of protein functions and interactions, next-generation sequencing (NGS) methods have been adapted for proteomic analyses. One such method is

Stitch-seq, which is based on the Y2H system and derives its name from PCR-based stitching of two cDNA fragments of interacting proteins. Stitch-seq was used to generate Human Interactome with NGS (HI-NGS) that mapped 1166 interactions between proteins coded by 1147 genes, a 42% increase in interaction detection over previous human interactomes [36]. This method improves cost-effectiveness and throughput of protein interaction screening compared to traditional Y2H systems and Sanger sequencing methods. Inspired by Stitch-seq, a method employing a sUbq system coupled with NGS called quantitative interaction screening (QIS-Seq) was developed and utilized in *Arabidopsis* [37]. QIS-seq was used to screen for host targets of *Pseudomonas syringae* effector family HopZ. In their study, QIS-seq identified MLO2 as a virulence target of *P. syringae* effector HopZ2. This interaction was confirmed *in vivo* using fluorescent microscopy and revealed novel functions of MLO2 in *P. syringae* resistance. MLO2 had previously been characterized in the resistance of powdery mildew but had not been implicated in immune responses to bacterial pathogens.

Recently, a heterologous method for large-scale identification of protein interactions has been described called single-molecule interaction sequencing (SMI-seq) [38]. In this method, cDNA libraries are barcoded with unique tags and hybridized with proteins of interest. Protein-cDNA hybrids can be generated at the whole library level using ribosome display [39] wherein polymerase is stalled on the cDNA-mRNA hybrid strand to prevent the release of the protein from its assigned DNA barcode. These complexes are protein-ribosome-mRNA-cDNA (PMRC) complexes. Additionally, proteins can be individually barcoded using fusion proteins modified with HaloTags that conjugate with cDNA strands containing complementary HaloTag ligands. The protein-DNA hybrids are purified using a two-step process and immobilized on polyacrylamide gel where the DNA barcodes are amplified and quantified based on a previously described method called polony sequencing [40,41]. SMI-seq is notable for its versatile applications and adaptability to several types of protein interactions. SMI-seq can be used to profile library versus library interactions, transient interactions, and the binding affinities of interactions, indicating its sensitivity as a screening assay. A schematic of SMI-seq applied to plant-pathogen protein interaction profiling is presented in Fig. 1.

NGS provides flexible platforms that are adaptable to several types of assays, for example membrane bound receptors and signaling cascades. As NGS methods are increasingly more cost-effective, these technologies will be implemented in large-scale studies of protein interactions. However, no one method can describe all the protein interactions within a cell. As mentioned previously, many methods identify interactions that are not detected by other techniques [30,33]. An on-going criticism of protein–protein interaction datasets is the difficulty in determining which interactions are real and the limited coverage of individual assays. The accuracy of data generated from screening assays can be assessed by comparing new data to a reference dataset of confirmed interactions. Using multiple methods may reveal overlapping data that increases the confidence in those interactions. Multiple screens also widens the coverage of diverse interaction types (*e.g.*, transient, stable, abundant) and spatial localization of interactions. Moreover, there is a concern that interactions detected through screening assays may not be functionally relevant or represented *in vivo*. This criticism is due to both the lack of *in vivo* assays as well as the potential for identifying interactions that normally would not occur because of different spatial or temporal representations of proteins. Therefore, there is a strong need to develop *in vivo* assays to characterize real interactions. Benchmarking will continue to be important in characterizing the validity and utility of new technologies. The next steps will be integrating data from multiple techniques to gain a comprehensive understanding of protein functions within the cell.

3. Topological properties of biological networks

Generally, a network is a map that visualizes connections between its components. Protein–protein interaction networks describe functional relationships between proteins. In the context of a network visualization, a protein is represented by a node and the interactions between proteins are described by lines, or edges, connecting the nodes. A grouping of highly connected nodes is referred to as a hub and generally indicates that the nodes within a hub are involved in a similar functional process. As the field of network biology has developed over recent years, two important universal characteristics of biological networks that have been described are their scale-free distribution and small world property. The scale-free distribution describes how the connections are represented in the network. A network is said to be scale-free if the distributions of its edges follows a power law, where most of the interactions are between a small subset of the proteins in the network (Fig. 2A) [42]. The scale-free property has been reported in all biological networks, including examples from humans, yeast, and plants [43]. This concept denotes the tendency of objects to cluster or form hubs within a network. Building on this idea, the small world property adds that proteins in distal parts of a network have relatively short path lengths due to the combinatorial effects of nearby proteins being connected to hubs and other highly connected proteins [44,45]. In this view, nodes have a multitude of indirect connections through nearby interactions with nodes contained in highly connected in hubs. In a biological context, proteins that interact in signaling cascades have indirect interactions with the related proteins that also participate in that cascade. These features form a high level of robustness within the network that makes the overall system resistant to individual node failure but sensitive to hub-related node failure [46]. Random networks do not exhibit these properties. Instead they have an even distribution of edges that creates distant average relationships between its nodes. A visual comparison is presented in Fig. 2B. Although the scale-free distribution and small world property are universal features of biological networks, several other topological properties contain information about how proteins and interactions behave in a network.

In its simplest form, a network can be viewed as a graph that has certain topological properties. In a protein network, the topology is a visualization of various aspects of the relationships between proteins. When building such a network, the first question is: which proteins have the highest number of interactions? These proteins generally have central roles in cellular processes. The number of connections extending from and incoming to a protein is termed its degree (Fig. 2C). Important biological implications are derived from a protein's degree. Proteins with a high degree are typically evolutionary conserved, essential components of overall cellular function [47,48]. Malfunctions in these proteins result in disease phenotypes and may be targets for pathogen virulence. Therefore these proteins are of experimental interest. Relationships between degree and connectivity have been analyzed in the context of power grids and grid failure [49]. In this study, redundant connections between substations provided greater overall stability within the network. However, removal of 2% of substations with a high number of connections resulted in over 60% loss of global connectivity of the network. These results show how malfunctions in a small set of nodes can propagate rapidly across the network to perturb its overall topology. Understanding how deleterious effects on target proteins influences distal interactions and essential pathways will be an important aspect of network biology. Interestingly, the node degree had differential correlation with essentiality in a yeast interactome comparing separate datasets using Y2H and AP-MS [50]. Yu et al. attributed this difference to bias in the data, concluding that overall there was no correlation between degree and essential function. Further characterizing the importance of topological features such as degree in both network and cellular structure and function will be an important issue in future studies. Moreover, if these properties are vulnerable to experimental biases, developing rigorous standards for confidence scoring will be an essential feature of network biology in order to improve the validity of future networks.

When looking at the effects of degree distribution as a global property of network topology, some important features emerge that describe the network's overall connectivity, stability, and functional organization. The centrality of a protein in a protein–protein interaction (PPI) network relates its position in relationship to other proteins in the network. Two widely used measures of centrality are betweenness centrality and eigenvector centrality [51]. Betweenness centrality quantifies a protein's frequency in mediating connections to other proteins (Fig. 2D). Biologically, these proteins serve as control steps in the flow of information or signals. Disruption of these proteins may have severe downstream effects on other pathways, especially if that protein is an important link to hubs. Eigenvector centrality, on the other hand, scores each connection between proteins to determine which proteins have more influence within the network based on the types of connections and not strictly the number of connections (Fig. 2E). Connections to structurally important nodes or hub-related nodes provide stronger weight to the interaction compared to those with nonessential nodes at the periphery of the network. Centrality measures are often applied in the modeling of disease spreading [52–54]. One study focused on the spread of disease among social groups of chimpanzees to determine which individuals are more likely to maintain disease populations and spread disease to new communities [55]. Although many chimpanzees form tight-knit familial communities, the interactions between juveniles and nursing mothers of separate family groups resulted in the rapid and widespread propagation of disease, indicating that connectivity is a feature of interactions between hubs and not necessarily within hubs. In a separate but similar study, centrality was used to identify targets for vaccination to reduce disease spreading among chimpanzee groups [56]. Rushmore et al. additionally found that pathogens are more or less infectious depending on the season. Targeted vaccinations based on chimpanzees that were more central in the social network reduced disease spread by up to 35%. Furthermore, pathogens can affect up to 30% of the social network if introduced to central nodes, whereas pathogens infecting peripheral nodes were less likely to spread. Centrality can be used to understand not only epidemic spread of disease in a clinical setting but also in an agricultural setting. Understanding how infected individuals spread disease as well as how virulence is spread among pathogens is critical to our ability to prevent and lessen effects of disease. Moreover, given that infection severity has a temporal dependency in plants [57], it will be interesting to see if network structure and topology resembles those described by Rushmore et al. and if similar targeted approaches can be adapted to limiting the spread of crop disease.

Similar to measures of centrality is the application of k-shell (or k-core) degeneracy to network structure. This method uses a stepwise process to systematically eliminate layers, or shells, of a network to identify nodes that are central in network connectivity (Fig. 2F). This approach assigns k-values to nodes according to their influence within the network [58]. Higher k-values constitute more influential spreaders. The k-shell is a description of how information spreads within a network and identifies components that are the most influential in the spread of information. The k-values can be weighted to favor relationships that are essential for the inherent structure of the network, similar to the eigenvector centrality measure [59]. Notably, nodes with high k-values are typically at the core of the network and do not necessarily need to be part of a hub, rather they only need to be essential to the accessibility of the hubs to other network communities. These core nodes can be assessed to determine the patterns in the spread of biological information and are useful metrics for organismal and population level studies of the spread of disease. Recently, a comparison of various measures of centrality, including betweenness centrality, eigenvector centrality, k-shell decomposition, and others, was reported [60]. In this study, eight different metrics were applied to a variety of biological and non-biological networks to assess the patterns in the flow of information in disease and social contexts. The metrics identified different core components depending on the conditions and structure of the network. The k-shell method was the most accurate in identifying the structure of epidemic spreading, as reported earlier [61]. Nevertheless, the context of the network itself to cellular function as well as the conditions and parameters applied to the network must be carefully considered when analyzing the topological properties as the methods may differentially select which nodes are most important. A recent application of this method was used to compare pathogen targets in PPIN-1 and a human-pathogen network [62]. Surprisingly, Shakarian and Wickiser found that the effector targets identified in the Arabidopsis and human PPI networks did not occupy the core of the network after k-shell decomposition, rather they occupied shells immediately outside the core. The authors propose this may be an infection strategy to maximize the survivability of both pathogens and hosts in order to avoid triggering a lethal response in host cells and thus destroying nutrient sources and mechanisms for reproduction. Interestingly, though, pathogen targets occupied similar shells across both plant and human networks, indicating infection strategies may be conserved not only between pathogen species but also for diverse host types. Further clarification is needed to understand the correlations between k-shell structure and pathogen infection. Additional studies in other host species, as well as studies exploring the structural differences in necrotrophic pathogens that may benefit from triggering lethal disruption of host cells, may help to resolve some of these issues.

Centrality measures and k-shell values attempt to elucidate the structural organization of a network and determine the roles of network components in contributing to its robustness and stability. As stated earlier, these metrics are derived from the tendency of biological networks to form sub-clusters that inherently skews the importance of some nodes over others. In addition to structural analysis of the network topology, there are some properties that are used to describe the functional roles of the network sub-clusters. Although hubs form important clusters of connections within a network that contribute to overall stability, they also represent functional communities or modules of proteins with similar cellular roles. Modularity is a metric that compares the probabilities of interactions compared to expected values if the network were random (Fig. 2G). In a PPI network, proteins that have related interactions at a frequency higher than expected form communities that tend to share functional roles within a cell [63]. Modularity has been reported in several types of networks, including yeast, AI-1, and the human interactome (HI-II-14) [14,22,64]. Over 20 protein communities were expressed in AI-1, including hormone signaling pathways, water transport pathways, ubiquitination pathways, and others [22]. Analysis of these modules reveals that biologically distinct pathways may be linked in previously unknown ways and provide a tool for hypothesis generation. An important aspect of network biology is the ability to view the systems-level organization of pathways to understand how different pathways may function synergistically or antagonistically to respond to stimuli such as abiotic or biotic stress. Network topology of AI-1 indicates that transcriptional regulators are organized in modules to integrate and respond to diverse signals. In HI-II-14, modules were identified and correlated with cancer-related mutations to understand how disturbances in various pathways may contribute to tumorgenesis [14]. Importantly, whereas centrality measures describe the directional flow of biological information, modularity describes the rate of information exchange. A useful example is a signaling cascade within a cell. Information flows rapidly through a signaling pathway but may have delayed effects on distant cellular processes that are not directly involved in the reception and response to a signal.

One perspective on the organization of high-degree nodes into hubs is the concept of party and date hubs [64]. Party hubs are groups of proteins that have constant interactions with their neighbors and are represented by high Pearson correlation coefficients (PCCs). In contrast, date hubs have low PCCs and have diverse interactions activated under specific spatiotemporal conditions. This bimodal distribution of hubs seeks to characterize the overall organization of protein networks through general behavioral features of protein modules. Although the party hubs have a more constant representation within the network, the date hubs are more central to overall network structure because they interact with a variety of pathways and direct the flow of biological information based on cellular conditions. However, recently there has been some contention over the modeling of protein interactions as dichotomous modules [65]. Much of the dispute over the presence of two distinct module types has resulted from inconsistent experimental design and data. A recent study resolved these inconsistencies using models from both yeast and human protein networks. Although these authors did confirm the presence of party and date hubs, they advise that this may be an oversimplified view of real protein function as the properties of date hubs in particular were more complex than previously described.

No one network feature can present a complete model of the degree of biological relevance of a given network or its components. The integration of a variety of measurements is essential in understanding the correlations between the structure and function of interactions. Additionally, there is an increasing emphasis on representing dynamic processes that reflect the natural progression of cellular processes rather than looking at static representations of conditional states. Network biology has emerged as an important tool for modeling the adaptability of functional groups within a cell to respond to potentially adverse environmental conditions. A key application of dynamic modeling is the elucidation of cellular responses over the course of pathogen infections, from the point of invasion until cell recovery or initiation of programmed cell death. A framework for such a study is presented in Fig. 3. The metrics discussed in this section will provide quantitative measurements of the flexibility of protein communities and interactions in the temporal progression of diverse cell responses.

4. Protein–protein interactions in plant immune system dynamics

Plant diseases are responsible for billions of dollars of crop yield losses worldwide every year. Moreover, commercial use of pesticides is costly and not always effective. Thus, there is a high demand for efficient methods for preventing crop diseases. Manipulating plant innate immunity presents the potential to engineer crops that have increased disease resistance, attenuating both crop yields and costly disease treatments. An important aspect of plant immunity is the interaction between host proteins and pathogen effectors. These interactions have downstream effects

in gene expression that can alter cellular phenotypes locally and systemically. However, a challenge in designing biotechnologies to respond to pathogen challenge is characterizing how plants interact with diverse pathogen species. Network biology provides the tools to model these dynamic interactions and differential responses over the course of an infection. Such a network could potentially reveal vulnerable components of plant cells that contribute to disease susceptibility. Moreover, understanding how plant-pathogen relationships evolve over time enables us to predict functional changes that may alter virulent states in pathogens or disease resistance in plants. Although plant immunity occurs at several genetic, regulatory, metabolic, and spatial layers within the cell, protein interactions are an extensive component of modulating plant-pathogen interactions and direct many of the immune responses in plants. Therefore, PPI networks provide a useful interface for understanding the scope of these interactions.

PPI networks focusing on aspects of plant immune responses have only recently been explored. An evolutionary perspective on the dynamic host–pathogen interactions in *Arabidopsis* was mapped using experimental data [23]. This network, the plant-pathogen immune network version 1 (PPIN-1), focused on the targets of pathogen effectors from the bacteria *P. syringae* and the oomycete *Hyaloperonospora arabidopsidis*, two evolutionarily distant pathogens. Network analysis of PPIN-1 found that effectors consistently target proteins in the host that aggregate into hubs, indicating their functional centrality in immune responses and the sharing of effector targets among diverse pathogen species. A human-pathogen network was recently reported that compared the interactions between human proteins and proteins from three bacterial pathogens: *Bacillus anthracis*, *Francisella tularensis*, and *Yersinia pestis* [66]. Similar to previous studies in humans [67] and PPIN-1 in plants, pathogens target host proteins that are hub-related or are important connection points for other proteins and groups of proteins. These results indicate the importance of comparative studies to reveal more detailed information about how host–pathogen interactions may perturb host pathways to contribute to susceptibility and resistance. Moreover, these studies reveal that networks exhibit universal properties among diverse kingdoms of life. Similar studies conducted in plants may reveal downstream effects of effector-mediated disruption of host pathways. PPIN-1 also describes how plants evolve in response to pathogens. Selective pressures favor guarding high profile effector targets rather than altering the effector target itself since these are typically valuable proteins with essential cellular functions. One example of this is the representation of nucleotide binding leucine-rich repeat (NB-LRR) proteins that play a role in disease resistance [68]. In PPIN-1, NB-LRRs interact directly with pathogen effectors less than 0.07% of the time, however, over 40% of NB-LRR interactors are effector targets, including several hub-related proteins. NB-LRRs act as sensors to detect perturbations in the functional capacity of effector targets and induce signaling cascades leading to defense responses.

Expanding the scope of PPIN-1, a later study repeated the network analysis with an additional pathogen, the ascomycete *Golovinmyces orontii* [69]. This led to the identification of 46 effector candidates in *G. orontii* with 122 interactions and 60 interactors in *Arabidopsis*. A subset of this test space was used to generate a *G. orontii* effector-host protein interaction network, which was integrated with PPIN-1, AI-1, and interaction evidence from the literature to create PPIN-2. This subset includes 41 *G. orontii* effector candidates with 92 interactions and 45 interactors. PPIN-2 contains 178 *Arabidopsis* proteins, 123 effector proteins, and 421 host-effector interactions. Complementing the evolutionary convergence described in PPIN-1, the effectors from *G. orontii* targeted similar host proteins compared to *P. syringae* and *H. arabidopsidis*. Additionally, effectors from individual pathogens also converge

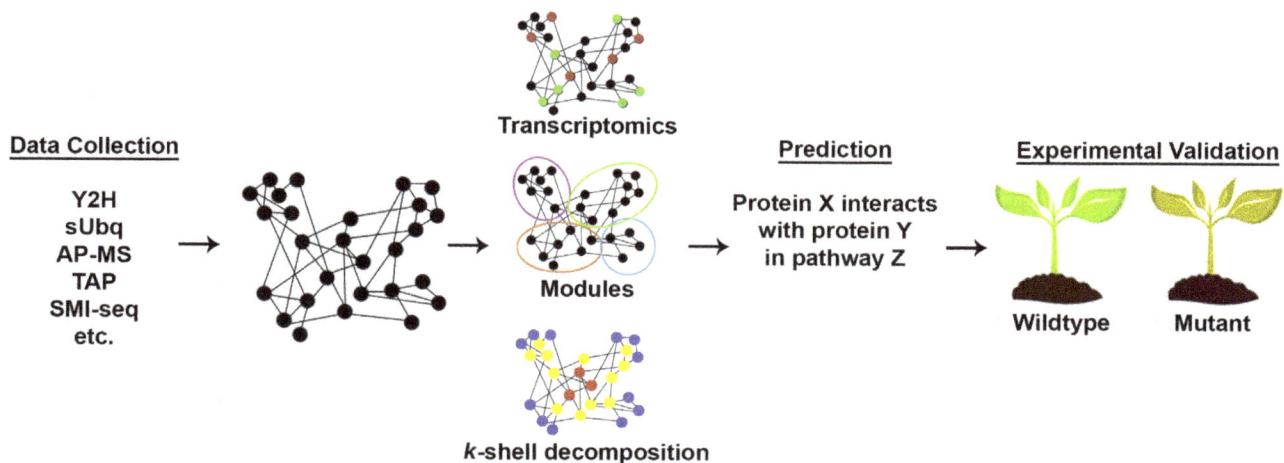

Fig. 3. Pipeline describing network generation from experimental data and subsequent validation.

onto repeatedly targeted host proteins. These results indicate that evolutionary pressures influencing the dynamics of host–pathogen interactions are conserved among pathogen types and that the targets of virulent effectors represent important functional components within host cells. The most common effector targets in PPIN-2 are a suite of TCP transcription factors important in development, hormone signaling, and the circadian clock. TCP14 has the highest number of interactions with effectors in PPIN-2, with 23 from *G. orontii*, 25 from *H. arabidopsidis*, and 4 from *P. syringae*. Of 11 interactions tested between TCP14 and pathogen effectors, 67% were shown to colocalize *in planta*. Thus, PPIN-2 expands the scope of PPIN-1 by showing that there are positive selective pressures on effector targets that are shared by distant pathogen types.

Although not directly part of PPIN-1 or PPIN-2, viral PPI networks have also been explored. Potyvirus is a large family of viruses accounting for 30% of plant viral infections. Potyviruses have 11 well-defined proteins that have dynamic functions during the viral life stages. This small protein content allows for effective modeling of virus-host interactions to explore the progression of viral infections. Bosque et al. utilized a step-wise method for visualizing the range of effects in host cells after viral invasion [70]. Viral proteins can perturb nearly the entire host protein network, albeit by different strategies. Targeting hub-related proteins spread viral infection faster as the combinatorial effects of disrupting highly-connected proteins compound at a quicker rate compared to targeting peripheral proteins. Notably, several viral proteins had no reported host interactions, emphasizing the possibility for other functions (*e.g.*, interactions only between viral proteins) or interactions with DNA, RNAs, or other macromolecules in host cells. In humans, several host-virus protein interaction networks have been described [67,71,72]. Like plant-virus networks, human viruses tend to target hub-related proteins and viral classes have different strategies for infection [73,74]. These diverse infection strategies may reveal distinct evolutionary paths among viral classes. Mapping these interactions will help elucidate how host machinery can be manipulated to propagate viral infections. Furthermore, understanding patterns in both viral evolution and different viral infection strategies will be important for developing biotechnology to confer broad spectrum resistance to these pathogens.

In contrast to the relatively recent development of PPI and disease networks in plants, models in humans are well-described. Extensive mappings of the relationships between different disease states and the genetic causers of disease have been an increasing focus of human network biology [8,11,75]. An early network, the human disease network (HDN), connects disorders caused by mutations within shared genes [8]. Conversely, the disease-

gene network (DGN) links disease-related genes when they share common disorders. These complementary views can be integrated into a comprehensive diseasome describing the connections between disease related phenotypes and genotypes. Notably, the HDN and DGN revealed that individual disease states arise from mutations in a few genes whereas a single disrupted gene can be linked to many different disorders. This view is presented in contrast to earlier methods of studying diseases individually and describing phenotypes as unique, separate conditions relative to disease states. An integrative approach, however, shows that diseases can be linked to one another based on the shared mutations in functional groups and gene communities. These studies have been expanded to analyze the relationships between diseases and symptoms (the human symptoms disease network, HSDN) [76]. Network analysis of the HSDN shows that diseases that share symptoms share disease genes and protein–protein interactions. This model has apparent applications in clinical settings to understand complex polyphenotypic diseases. A recent study generalized this idea to characterize relationships between diseases based on modularity within the network [75]. Relatedness of distinct disease modules was represented by the mean shortest distance between protein interactions linking disease pairs. In this model, compounding symptomatic characteristics are explained by the perturbations of influential elements that are centrally located in the network. Thus, disruptions in one disease hub can affect the functionality of neighboring hubs resulting in similar phenotypes because disease states are linked by protein communities.

Disease networks have several clinical applications. A better understanding of the connectedness of diverse diseases and the characterization of phenotypes will lead to more specific drug targets and markers for more effective diagnoses and monitoring of treatment effectiveness [11]. Furthermore, mapping the relationships between disease genes and disease states may offer insights into the differential responses to drugs, as unique mutations may affect distal pathways in novel ways. The human disease networks described here represent models that could be translated into plant sciences to understand the relationships between genotypes and disease-related phenotypes. For example, is there a relationship between symptoms of plant disease and perturbations in cellular activity common among different disease types? Additionally, how are protein interaction networks involved in the relationships between genotypes and phenotypes? As networks in many organisms become more developed and increasingly validated, a potential next step will be integrating networks from multiple layers of cell activity into "meta-networks" exploring the complex interactions between genes, transcriptional regulatory elements, proteins, and metabolism.

Although complete interactomes in many model species are still a few decades away from being completed, high quality networks can be produced from existing data that provide useful information about global interactions between functional groups and models for hypothesis generation. As seen in the human disease networks discussed in this section, these models can dramatically alter current ideas about complex interactions within cells that give rise to multiple phenotypes. As high-throughput methods are continually improved in terms of efficiency, sensitivity, and cost effectiveness, the availability of data grows at a near exponential rate. Thus, there is a strong need for flexible models that can integrate large-scale datasets in meaningful ways. Network biology has proven useful in mapping functional groups and describing global effects of perturbations within individual modules. As experimental datasets grow, more complete networks can be generated to clarify the complex organization of cell functions. Considering that many effector targets have been shown to be conserved among pathogen types, these networks may reveal putative targets to enhance broad-spectrum disease resistance for important crop species. Since early steps contributing to either susceptibility or resistance involve the interactions between host and pathogen proteins, these networks will be essential for understanding how these initial interactions affect the eventual outcome of host–pathogen interactions.

5. Predictive modeling and network evolution

Network modeling is commonly used to predict functional properties of poorly characterized nodes. In static networks, prediction is typically achieved through "guilt-by-association" where nodes are assumed to have similar functional roles as their neighbors. However, this assumption may lead to mis-predictions when nodes interact with functionally diverse parts of the network, so called "moonlighting proteins" [77,78]. Another method for function prediction is using conserved interacting orthologs, or interlogs. In plants, one of the earliest protein networks was built using Y2H data [16]. This network used data from humans, yeast, flies, and worms to identify orthologs in *Arabidopsis* that were likely to share functional roles. This study implemented PCC to analyze the linear interactions detected in the network between 3617 proteins analyzed. A total of 1159 high-confidence interactions were identified based on references to experimental evidence, and several thousand more medium and low-confidence interactions were additionally detected. These interactions were further analyzed to determine the cellular localization and contributions of co-expressed genes to the spatiotemporal interactions of linked proteins. A similar orthological approach was repeated in *Arabidopsis* a few years later [79].

Network biology has recently shifted towards developing dynamic rather than static models. An important aspect of understanding global cellular functions is characterizing the differential roles of proteins given specific spatiotemporal conditions. Although static models are certainly useful, they do not represent the fluctuating environment within cells. In the context of protein interactions, static models visualize the presence of an interaction but do not characterize functional changes in the interaction over time. Thus, static models fail to describe which interactions may be transient or dependent on specific cellular conditions. Moreover, active and inactive states are important features of many proteins that are difficult to represent in a network. Some interactions, such as cell cycle dependent interactions, may be underrepresented because of the low frequency of the interactions although they are essential for cell viability. From a disease perspective, identifying the changes in functionality of protein communities is critical in understanding systems level cell responses to pathogen invasion. Proteins that have multifunctional roles may be poorly characterized in static networks where the differential functions are not represented as dependent on certain cell states and are instead visualized as a lump sum of interactions. For example, molecular switches involved in cell fate decisions during pathogen infection participate in diverse pathways and understanding how these pathways are directed over the course of an infection are important in distinguishing disease susceptibility and resistance. Thus, dynamic models represent the next step in network biology to gain a more comprehensive and sensitive view of changes in protein behavior and interaction in response to diverse conditions.

Although many attempts have been made to describe methods for building dynamic networks [80–82], a recurrent limitation in these methodologies is modeling networks at individual, separate time points instead of as a continuous network. A recent study proposed a novel method for merging time series data called Time Smooth Overlapping Complex Detection (TS-OCD), which is based on principles of nonnegative matrix factorization (NMF) [83]. NMF is a method for decomposing a matrix or dataset to extrapolate functional modules based on the restriction to positive values that allows for the additive combination of feature vectors. NMF has been used extensively for graph clustering [84]. Ou-Yang et al. measure the efficiency of TS-OCD against six previously described algorithms for modeling dynamic networks. Notably, TS-OCD uses stable protein interactions as a benchmark for representing transient interactions according to their probability densities. TS-OCD performed with greater precision and accuracy compared to the other methods tested and reflected realistic cellular states by modeling the transition between distinct protein interactions. Expanding network generation methods to include more sensitive representations of protein interactions will be the focus of future research.

Related to dynamic modeling of time course data is the analysis of network evolution, which offers insights into the evolution of protein communities and overall network stability over time at a population level. Network evolution studies have several applications. Comparative analyses of protein networks from distantly related species can identify conserved functional groups and the divergence and subsequent specialization of pathways. Orthological approaches to network construction have been used to discover conserved functional groups among *Arabidopsis*, yeast, and human and serve as predictions for functional roles when similar protein communities are better described in other organisms [79]. Interestingly, network evolution can be used to study dynamic properties of networks themselves. AI-1 and PPIN-1 and -2, although mapping evolutionary strategies between organisms, also explore the effects of selective pressures on network rewiring [22,23]. Analysis of AI-1 shows that network rewiring follows a power-law decay, wherein rewiring of protein–protein interactions occurs rapidly after divergence but slows as the interactions attain functionality. Moreover, the edges evolve faster than the nodes, indicating the tight selective pressures on protein sequences compared to the functional roles. Notably, despite the comparatively large size of the predicted complete interactome in *Arabidopsis* (~300,000 interactions), the network topology and behavior of protein interactions are conserved among different organisms. This shows potential utility for comparative studies between topological features of networks originating from different species.

In humans, the human PPIN was decomposed according to the phylogenetic age groups of proteins [85]. Older proteins occupied more central positions in the network, whereas evolutionary young proteins existed at the periphery. This may be explained by the differential selective pressures on young and old proteins. Young proteins may not have acquired essential functional roles yet due to weaker pressures. Moreover, proteins in similar age groups interacted more with each other than with proteins in different age groups. One important conclusion from these results is that older, more central proteins tend to not develop new interactions over time. Again, this may be due to extreme selective pressures on essential proteins. However, this avoidance of perturbation of aged proteins implies an overall network structure that may be conserved over time. Since these aged proteins perform essential cellular functions and are integral for the overall connectivity of the network, duplications of these proteins are highly unfavorable as they may disrupt network organization and thus cell function. Simulated models of network growth support this idea in both humans and yeast [86,87]. Understanding how networks evolve is important in characterizing cellular organization and evidence suggests that organizational schemes may be conserved among species [85]. In the context of plant-pathogen interactions, such evolutionary network studies may additionally reveal how host–pathogen relationships develop over time. As explained earlier, comparative studies have thus far shown that diverse pathogen species target similar host proteins. A phylogenetic study conducted on plant networks may elucidate how plants respond to pathogen evolution by rewiring network structure to shift peripheral proteins towards more central locations to increase overall network and cellular stability.

6. Conclusions

Recent advances in molecular and cellular techniques for analyzing protein function and interactions have resulted in large public domain datasets. The utilization of these datasets, although representatively low, is facilitated by bioinformatic approaches, primarily network construction, that are increasingly important in biological sciences. A hallmark of molecular biology is approaching the study of individual pathways from a reductionist point-of-view, however, a pitfall of reductionism is the inability to understand large-scale interactions of individual components within cellular and environmental contexts. Network biology attempts to alleviate these limitations by instead taking an integrative view of genes, regulatory elements, proteins, and metabolites using powerful statistical methods. So far analyses from various types of networks have been instrumental in understanding gaps in regulatory pathways, reordering conceptual approaches to gene organization, and elucidating the development of pathophysiologies. In plants, network analysis has been limited compared to other organisms, however recent attempts at network generation has revealed novel functional groups and interactions between diverse pathways. One area that could benefit from network analysis is the characterization of protein interactions involved in plant immune responses. PPIN-1 and -2 described relationships among pathogen effectors and host proteins among evolutionary distinct pathogen species, revealing conserved targeting of host proteins. Coupled with analysis of the general protein–protein interaction map AI-1, these networks reveal general patterns of network topology represented in several model organisms.

The dynamic modeling of cellular processes will continue to be an important area of molecular and systems biology, not only for retroactively visualizing existing experimental data, but also for predicting properties of poorly understood proteins. A promising yet challenging aspect of network generation will be the ability to integrate data from different experimental techniques. Improvements in these areas may point towards a more comprehensive view of cellular states and how disruptions within a small group of proteins affect global cellular functions.

Although protein network synthesis in plants is limited, the work done in other model organisms indicates the potential contribution of computational methods to understanding plant biology.

Comparative approaches may serve as benchmarks for the stability of network-based analysis in the future. Increasing coverage of protein networks in aspects of plant biology such as signal transduction, defense responses, developmental stages, and cell cycle dependent processes has several practical applications in crop science [88]. A complete interactome may inform agricultural biotechnology to improve disease and pest resistance, stress response, crop yields and other desirable traits.

References

[1] A.-L. Barabási, Z.N. Oltvai, Network biology: understanding the cell's functional organization, Nat. Rev. Genet. 5 (2) (2004) 101–113.
[2] M. Pellegrini, D. Haynor, J.M. Johnson, Protein interaction networks, Expert Rev. Proteomics 1 (2) (2004) 239–249.
[3] R.P. Horgan, L.C. Kenny, 'Omic' technologies: genomics, transcriptomics, proteomics and metabolomics, Obstet. Gynaecol. 13 (3) (2011) 189–195.
[4] K. Offit, Personalized medicine: new genomics, old lessons, Hum. Genet. 130 (1) (2011) 3–14.
[5] C. Kole, et al., Application of genomics-assisted breeding for generation of climate resilient crops: progress and prospects, Front. Plant Sci. 6 (2015) 563.
[6] B. Palsson, K. Zengler, The challenges of integrating multi-omic data sets, Nat. Chem. Biol. 6 (11) (2010) 787–789.
[7] A.J. Butte, I.S. Kohane, Creation and implications of a phenome-genome network, Nat. Biotechnol 24 (1) (2006) 55–62.
[8] K.I. Goh, et al., The human disease network, Proc. Natl. Acad. Sci. U. S. A. 104 (21) (2007) 8685–8690.
[9] J. Loscalzo, I. Kohane, A.L. Barabasi, Human disease classification in the postgenomic era: a complex systems approach to human pathobiology, Mol. Syst. Biol. 3 (2007) 124.
[10] P. Sebastiani, et al., Genome-wide association studies and the genetic dissection of complex traits, Am. J. Hematol. 84 (8) (2009) 504–515.
[11] A.-L. Barabási, N. Gulbahce, J. Loscalzo, Network medicine: a network-based approach to human disease, Nat. Rev. Genet. 12 (1) (2011) 56–68.
[12] P.L. Kastritis, A.M.J.J. Bonvin, On the binding affinity of macromolecular interactions: daring to ask why proteins interact, J. R. Soc. Interface 10 (79) (2012) 20120835.
[13] T.R. Hughes, C.G. de Boer, Mapping yeast transcriptional networks, Genetics 195 (1) (2013) 9–36.
[14] T. Rolland, et al., A proteome-scale map of the human interactome network, Cell 159 (5) (2014) 1212–1226.
[15] N. de Souza, Systems biology: an expanded human interactome, Nat. Methods 12 (2) (2015) 107.
[16] J. Geisler-Lee, et al., A predicted interactome for Arabidopsis, Plant Physiol. 145 (2) (2007) 317–329.
[17] Y. Zhang, P. Gao, J.S. Yuan, Plant protein–protein interaction network and interactome, Curr. Genomics 11 (1) (2010) 40–46.
[18] J. Yang, et al., Inferring the Brassica rapa interactome using protein–protein interaction data from Arabidopsis thaliana, Front. Plant Sci. 3 (2013).
[19] J.D. Jones, J.L. Dangl, The plant immune system, Nature 444 (7117) (2006) 323–329.
[20] M.A. Newman, et al., MAMP (microbe-associated molecular pattern) triggered immunity in plants, Front. Plant Sci. 4 (2013) 139.
[21] M. Guo, et al., The majority of the type III effector inventory of Pseudomonas syringae pv. tomato DC3000 can suppress plant immunity, Mol. Plant Microbe. Interact. 22 (9) (2009) 1069–1080.
[22] M. Dreze, et al., Evidence for network evolution in an Arabidopsis interactome map, Science 333 (6042) (2011) 601–607.
[23] M.S. Mukhtar, et al., Independently evolved virulence effectors converge onto hubs in a plant immune system network, Science 333 (6042) (2011) 596–601.
[24] A. Bruckner, et al., Yeast two-hybrid, a powerful tool for systems biology, Int. J. Mol. Sci. 10 (6) (2009) 2763–2788.
[25] A.M. Jones, et al., Border control—a membrane-linked interactome of Arabidopsis, Science 344 (6185) (2014) 711–716.
[26] S. Lalonde, et al., A membrane protein/signaling protein interaction network for Arabidopsis version AMPv2, Front Physiol. 1 (2010) 24.
[27] P. Braun, et al., Plant protein interactomes, Annu. Rev. Plant Biol. 64 (2013) 161–187.
[28] M. Gstaiger, R. Aebersold, Applying mass spectrometry-based proteomics to genetics, genomics and network biology, Nat. Rev. Genet. 10 (9) (2009) 617–627.
[29] J. Boruc, et al., Systematic localization of the Arabidopsis core cell cycle proteins reveals novel cell division complexes, Plant Physiol. 152 (2) (2010) 553–565.
[30] X. Ding, et al., A rice kinase-protein interaction map, Plant Physiol. 149 (3) (2009) 1478–1492.
[31] S.H. Shiu, A.B. Bleecker, Receptor-like kinases from Arabidopsis form a monophyletic gene family related to animal receptor kinases, Proc. Natl. Acad. Sci. U. S. A. 98 (19) (2001) 10763–10768.
[32] S.C. Popescu, M. Snyder, S.P. Dinesh-Kumar, Arabidopsis protein microarrays for the high-throughput identification of protein–protein interactions, Plant Signal.Behav. 2 (5) (2007) 416–420.

[33] J.S. Rohila, et al., Protein–protein interactions of tandem affinity purification-tagged protein kinases in rice, Plant J. 46 (1) (2006) 1–13.

[34] S.C. Popescu, et al., MAPK target networks in *Arabidopsis* thaliana revealed using functional protein microarrays, Genes Dev. 23 (1) (2009) 80–92.

[35] S. Bandyopadhyay, et al., A human MAP kinase interactome, Nat. Methods 7 (10) (2010) 801–805.

[36] H. Yu, et al., Next-generation sequencing to generate interactome datasets, Nat. Methods 8 (6) (2011) 478–480.

[37] J.D. Lewis, et al., Quantitative Interactor Screening with next-generation Sequencing (QIS-Seq) identifies *Arabidopsis thaliana* MLO2 as a target of the *Pseudomonas syringae* type III effector HopZ2, BMC Genomics 13 (1) (2012) 8.

[38] L. Gu, et al., Multiplex single-molecule interaction profiling of DNA-barcoded proteins, Nature 515 (7528) (2014) 554–557.

[39] C. Zahnd, P. Amstutz, A. Plückthun, Ribosome display: selecting and evolving proteins in vitro that specifically bind to a target, Nat. Methods 4 (3) (2007) 269–279.

[40] G.J. Porreca, J. Shendure, G.M. Church, Polony DNA. Sequencing, in Current Protocols in Molecular Biology, Wiley-Blackwell, 2001.

[41] J. Shendure, Accurate multiplex polony sequencing of an evolved bacterial genome, Science 309 (5741) (2005) 1728–1732.

[42] A.L Barabasi, Scale-free networks: a decade and beyond, Science 325 (5939) (2009) 412–413.

[43] M. Vidal, E. Michael Cusick, A.-L. Barabási, Interactome networks and human disease, Cell 144 (6) (2011) 986–998.

[44] V. Latora, M. Marchiori, Efficient behavior of small-world networks, Phys. Rev. Lett. 87 (19) (2001).

[45] Z. Zhang, J. Zhang, A big world inside small-world networks, PLoS ONE 4 (5) (2009) e5686.

[46] S. Mizutaka, K. Yakubo, Robustness of scale-free networks to cascading failures induced by fluctuating loads, Phys. Rev. E 92 (1) (2015).

[47] N. Rangarajan, P. Kulkarni, S. Hannenhalli, Evolutionarily conserved network properties of intrinsically disordered proteins, PLoS One 10 (5) (2015) e0126729.

[48] H. Jeong, et al., Lethality and centrality in protein networks, Nature 411 (6833) (2001) 41–42.

[49] R. Albert, I. Albert, G.L. Nakarado, Structural vulnerability of the North American power grid, Phys. Rev. E 69 (2) (2004).

[50] H. Yu, et al., High-quality binary protein interaction map of the yeast interactome network, Science 322 (5898) (2008) 104–110.

[51] M. Ghasemi, et al., Centrality measures in biological networks, CBIO 9 (4) (2014) 426–441.

[52] S. Rautureau, B. Dufour, B. Durand, Structural vulnerability of the French swine industry trade network to the spread of infectious diseases, Animal 6 (07) (2012) 1152–1162.

[53] J.M. Gomez, C.L. Nunn, M. Verdu, Centrality in primate-parasite networks reveals the potential for the transmission of emerging infectious diseases to humans, Proc. Natl. Acad. Sci. 110 (19) (2013) 7738–7741.

[54] R.M. Christley, Infection in social networks: using network analysis to identify high-risk individuals, Am. J. Epidemiol. 162 (10) (2005) 1024–1031.

[55] J. Rushmore, et al., Social network analysis of wild chimpanzees provides insights for predicting infectious disease risk, J. Animal Ecol. 82 (5) (2013) 976–986.

[56] J. Rushmore, et al., Network-based vaccination improves prospects for disease control in wild chimpanzees, J. R. Soc. Interface 11 (97) (2014) 20140349.

[57] V. Bhardwaj, et al., Defence responses of *Arabidopsis thaliana* to infection by *Pseudomonas syringae* are regulated by the circadian clock, PLoS One 6 (10) (2011) e26968.

[58] D. Miorandi, F. De Pellegrini, K-shell Decomposition for Dynamic Complex Networks in Modeling and Optimization in Mobile, Ad Hoc and Wireless Networks (WiOpt), 2010 Proceedings of the 8th International Symposium on. IEEE (2010).

[59] A. Garas, F. Schweitzer, S. Havlin, A k-shell decomposition method for weighted networks, New J. Phys. 14 (8) (2012) 083030.

[60] G.F. de Arruda, et al., Role of centrality for the identification of influential spreaders in complex networks, Phys. Rev. E 90 (3) (2014).

[61] M. Kitsak, et al., Identification of influential spreaders in complex networks, Nat. Phys. 6 (11) (2010) 888–893.

[62] P. Shakarian, J.K. Wickiser, Similar pathogen targets in *Arabidopsis thaliana* and homo sapiens protein networks, PLoS One 7 (9) (2012) e45154.

[63] F. Luo, et al., Modular organization of protein interaction networks, Bioinformatics 23 (2) (2006) 207–214.

[64] J.-D.J. Han, et al., Evidence for dynamically organized modularity in the yeast protein–protein interaction network, Nature 430 (6995) (2004) 88–93.

[65] X. Chang, et al., Dynamic modular architecture of protein–protein interaction networks beyond the dichotomy of 'date' and 'party' hubs, Sci. Rep. 3 (2013).

[66] M.D. Dyer, et al., The human-bacterial pathogen protein interaction networks of *Bacillus anthracis, Francisella tularensis,* and *Yersinia pestis,* PLoS One 5 (8) (2010) e12089.

[67] M.D. Dyer, T.M. Murali, B.W. Sobral, The landscape of human proteins interacting with viruses and other pathogens, PLoS Pathog. 4 (2) (2008) e32.

[68] H.-A. Lee, S.-I. Yeom, Plant NB-LRR proteins: tightly regulated sensors in a complex manner, Brief. Funct. Genomic. 14 (4) (2015) 233–242.

[69] R. Weßling, et al., Convergent targeting of a common host protein-network by pathogen effectors from three kingdoms of life, Cell Host Microbe 16 (3) (2014) 364–375.

[70] G. Bosque, et al., Topology analysis and visualization of Potyvirus protein–protein interaction network, BMC Syst. Biol. 8 (1) (2014).

[71] E.A. White, et al., Systematic identification of interactions between host cell proteins and E7 oncoproteins from diverse human papilloma viruses, Proc. Natl. Acad. Sci. U. S. A. 109 (5) (2012) E260–E267.

[72] S. Jager, et al., Global landscape of HIV-human protein complexes, Nature 481 (7381) (2012) 365–370.

[73] A. Pichlmair, et al., Viral immune modulators perturb the human molecular network by common and unique strategies, Nature 487 (7408) (2012) 486–490.

[74] H.R. Rachita, H.A. Nagarajaram, Viral proteins that bridge unconnected proteins and components in the human PPI network, Mol. BioSyst. 10 (9) (2014) 2448.

[75] J. Menche, et al., Uncovering disease–disease relationships through the incomplete interactome, Science 347 (6224) (2015) 1257601.

[76] X. Zhou, et al., Human symptoms–disease network, Nat. Commun. 5 (2014).

[77] I. Khan, K. shita, D. Kihara, Computational characterization of moonlighting proteins: table 1, Biochem. Soc. Trans. 42 (6) (2014) 1780–1785.

[78] H.N. Chua, W.K. Sung, L. Wong, Exploiting indirect neighbours and topological weight to predict protein function from protein–protein interactions, Bioinformatics 22 (13) (2006) 1623–1630.

[79] S. De Bodt, et al., Predicting protein–protein interactions in *Arabidopsis thaliana* through integration of orthology, gene ontology and co-expression, BMC Genomics 10 (1) (2009) 288.

[80] J. Wang, et al., Construction and application of dynamic protein interaction network based on time course gene expression data, Proteomics 13 (2) (2013) 301–312.

[81] Y. Kim, et al., Inference of dynamic networks using time-course data, Brief. Bioinform. 15 (2) (2013) 212–228.

[82] J. Das, J. Mohammed, H. Yu, Genome-scale analysis of interaction dynamics reveals organization of biological networks, Bioinformatics 28 (14) (2012) 1873–1878.

[83] L. Ou-Yang, et al., Detecting temporal protein complexes from dynamic protein–protein interaction networks, BMC Bioinformatics 15 (1) (2014) 335.

[84] C. Ding, X. He, H.D. Simon, On the Equivalence of Nonnegative Matrix Factorization and Spectral Clustering, in: In Proceedings of the 2005 SIAM International Conference on Data Mining, Society for Industrial & Applied Mathematics (SIAM), 2005, pp. 606–610.

[85] C.-Y. Chen, et al., Dissecting the human protein–protein interaction network *via* phylogenetic decomposition, Sci. Rep. 4 (2014) 7153.

[86] R.D. George, et al., Trans genomic capture and sequencing of primate exomes reveals new targets of positive selection, Genome Res. 21 (10) (2011) 1686–1694.

[87] S. Park, B. Lehner, Epigenetic epistatic interactions constrain the evolution of gene expression, Mol. Syst. Biol. 9 (1) (2013) 645.

[88] C. Rampitsch, N.V. Bykova, Proteomics and plant disease: advances in combating a major threat to the global food supply, Proteomics 12 (4–5) (2012) 673–690.

12

A molecular tug-of-war: Global plant proteome changes during viral infection[☆]

Mariko M. Alexander[a,b], Michelle Cilia[a,b,c,*]

[a] *Plant Pathology and Plant-Microbe Biology Section, School of Integrative Plant Science, Cornell University, Ithaca, NY, USA*
[b] *Boyce Thompson Institute for Plant Research, Ithaca, NY, USA*
[c] *Robert W. Holley Center for Agriculture and Health, USDA-ARS, Ithaca, NY, USA*

ARTICLE INFO

Keywords:
Plant virology
Plant pathology
Plant immunity
Host-virus interactions
Host-pathogen interactions
Plant proteomics

ABSTRACT

Plant pathogenic viruses cause a number of economically important diseases in food, fuel, and fiber crops worldwide. As obligate parasites with highly reduced genomes, viruses rely heavily on their hosts for replication, assembly, intra- and intercellular movement, and attraction of vectors for dispersal. Therefore, viruses must influence or directly utilize many host proteins and processes. While many general effects of virus infection have long been known (e.g., reduction in photosynthesis, alterations in carbon metabolism and partitioning, increased expression of pathogenesis-related proteins), the precise underlying mechanisms and functions in the viral life cycle are largely a mystery. Proteomic studies, including studies of differential protein regulation during infection as well as studies of host–viral protein–protein interactions, can help shed light on the complex and varied molecular interactions between viruses and plant hosts. In this review, we summarize current literature in plant-virus proteomics and speculate on why viruses have been selected to manipulate these diverse biochemical pathways in their plant hosts.

Abbreviations: 2D DIGE, two dimensional difference in gel electrophoresis; BNYVV, *Beet necrotic yellow vein virus*; BYV, *Beet yellows virus*; CABYV, *Cucurbit aphid-borne yellows virus*; CaLCuV, *Cabbage leaf curl virus*; CaMV, *Cauliflower mosaic virus*; CLCuV, *Cotton leaf curl virus*; CMV, *Cucumber mosaic virus*; CSDaV, *Citrus sudden death-associated virus*; CYDV-RPV, *Cereal yellow dwarf virus*, RPV strain; CymMV, *Cymbidium mosaic virus*; ER, endoplasmic reticulum; GAPDH, glyceraldehyde-3-phosphate dehydrogenase; GFLV, *Grapevine fanleaf virus*; GLP, germin-like protein; GLRaV-1, *Grapevine leafroll-associated virus 1*; GO, gene ontology; GST, glutathione-*S*-transferase; GVA, *Grapevine virus A*; HR, hypersensitive response; HSP, heat shock protein; LC/MS-MS, liquid chromatography coupled to mass spectrometry; MNSV-1, *Melon necrotic spot virus*; MP, movement protein; MYMIV, *Mungbean yellow mosaic India virus*; ORMV, *Oilseed rape mosaic virus*; ORSV, *Odontoglossum ringspot virus*; PLRV, *Potato leafroll virus*; PMeV, *Papaya meleira virus*; PMMoV, *Pepper mild mottle virus*; PMTV, *Potato mop-top virus*; PPV, *Plum pox virus*; PR protein, pathogenesis-responsive protein; PSV, *Peanut stunt virus*; PVX, *Potato virus X*; RBSDV, *Rice black-streaked dwarf virus*; RNP, ribonucleoprotein complex; ROS, reactive oxygen species; RSPaV, *Rupestris stem pitting-associated virus*; RuBisCO, ribulose-1,5-bisphosphatase carboxylase/oxygenase; RYMV, *Rice yellow mottle virus*; SCMV, *Sugarcane mosaic virus*; SCPMV, *Southern cowpea mosaic virus*; SMV, *Soybean mosaic virus*; SOD, superoxide dismutase; SqLCV, *Squash leaf curl virus*; SqMV, *Squash mosaic virus*; TEV, *Tobacco etch virus*; TMV, *Tobacco mosaic virus*; ToC-MoV, *Tomato chlorotic mottle virus*; ToMV, *Tomato mosaic virus*; TuMV, *Turnip mosaic virus*; TVCV, *Turnip vein clearing virus*; ZYMV, *Zucchini yellow mosaic virus*.

[☆] This article is part of a special issue entitled "Protein networks – a driving force for discovery in plant science".

[*] Corresponding author at: USDA ARS, Robert W. Holley Center for Agriculture and Health, 538 Tower Road, Ithaca, NY 14853, USA.
E-mail addresses: michelle.cilia@ars.usda.gov, mlc68@cornell.edu (M. Cilia).

1. Introduction

Plant diseases caused by viruses incur enormous costs to growers each year, both directly, in the form of yield and quality loss, and indirectly, in the forms of time and funds spent on scouting and disease management. Compared to even the smallest known bacterial genome, the genomes of plant viruses are tiny, sometimes encoding fewer than ten proteins. Therefore, they are masterful at co-opting host cell components to complete their life cycle. Many aspects of the life cycles of plant pathogenic viruses remain a mystery.

Due to the barrier of the cell wall, plant pathogenic viruses require outside assistance to infect a new host. Mechanically transmissible viruses are carried on tools, equipment and herbivores to infect a new host through contact with wounds. Other viruses require a vector for transmission. The most prolific vectors are sap-feeding insects, such as aphids, whiteflies, and leafhoppers, although some viruses are transmitted by beetles, nematodes, mites, or plasmodiophorids. Insect-transmitted plant viruses can be broadly categorized by the length of time they remain associated with their vector. Stylet- and foregut-borne viruses associate transiently with the cuticle lining the stylet or foregut, and may be transmissible for only hours or days after acquisition, respectively. In contrast, circulative viruses are acquired into the insect hemolymph, where they circulate until they reach salivary tissues. Once acquired, circulative viruses remain associated with their vector for the remainder of the insect's life. Unlike stylet- and foregut-borne viruses, an extended feeding period is required for both the acquisition of circulative viruses from infected plant hosts and the inoculation of healthy hosts. Evidence shows that some plant pathogenic viruses manipulate their host and/or vector to promote vector behavior conducive to their transmission [1–3].

After entering a plant cell, the virus must uncoat and transit to its replication site, which may be the nucleus (for viruses with DNA genomes) or cytoplasmic membranes (for viruses with RNA genomes). With assistance from host proteins, viral proteins and new viral genomes are produced. Progeny virions and ribonucleoprotein complexes (RNPs; complexes of viral nucleic acid and proteins, which are different from transmissible virions) are assembled and translocated to plasmodesmata. For viruses with single-stranded RNA genomes, formation of replication sites near plasmodesmata is facilitated by interactions between viral movement proteins (MPs) and plant synaptotagmin-family proteins, which create contact sites between the ER and plasma membranes [4]. Viral MPs promote callose degradation in plasmodesmata to facilitate passage of virions or RNPs into a neighboring cell [5], where the process starts again. Viruses use the phloem to travel to distal regions of the plant to achieve a systemic infection. The majority of circulative viruses infect only the phloem tissue during a natural infection. Phloem tropism may facilitate plant-to-plant transmission by phloem-feeding insect vectors [6]. Viruses must also evade host defenses and ensure an environment conducive to their replication. Often, infection results in the production of symptoms in plants, including chlorosis, necrosis, tissue proliferation, phyllody, leaf curling, and other physiological changes, although the selection pressures and underlying molecular mechanisms for these symptoms remain largely uncharacterized.

Host responses to viral infection can be broadly categorized in two ways: compatible versus incompatible, or susceptible versus resistant. A compatible response results in successful virus infection, replication, and spread to other cells. An incompatible response occurs when the virus is recognized by the host, resulting in the hypersensitive response (HR; localized programmed cell death), preventing virus spread [7–10]. Susceptibility and resistance, in contrast, are defined in terms of the ability of the virus to cause disease in a given host. A susceptible reaction to a virus results in disease—replication of the virus and production of symptoms by the host. A resistant reaction does not result in the production of symptoms, but may still permit viral replication if the host exhibits tolerance to the virus. In some cases, a host may be said to be partially resistant if the virus is able to cause a reduced level of disease as compared to susceptible hosts of the virus. This review considers proteomic studies from the full spectrum of host responses: tolerant, partially resistant, and resistant.

Most publications in plant-virus proteomics use 2-dimensional electrophoresis or 2D difference in gel electrophoresis (2D DIGE) to look for proteins or protein isoforms which are differentially regulated during virus infection, although studies have also been published that use shotgun proteomics, where the entire proteome is digested with trypsin and analyzed by liquid chromatography coupled to tandem mass spectrometry (LC–MS/MS). New advances include characterization of virus–plant protein interactions using co-immunoprecipitation coupled to LC–MS/MS. Structural proteomics using chemical cross-linking has also been used to identify regions in the viral capsid that regulate host–virus interactions [11]. In this review, we survey these proteomic data to discuss impacts on plant health during virus infection and speculate on how selection has favored viruses to tap into these host pathways. For a review of common techniques in plant proteomics, their limitations, and a summary of some previous literature in plant–virus proteomic studies, see Ref. [12].

2. Manipulation of intracellular trafficking

Plant viruses associate with a variety of subcellular structures for replication and movement, including the endomembrane system and the cytoskeleton. It is sometimes difficult to separate associations important for inter- and intra-cellular movement of plant viruses from associations important for replication, as noted by several recent reviews on the subject [13,14]. It is possible that these two important aspects of the viral lifecycle are inextricably linked in plant infections.

2.1. Endomembrane systems

RNA viruses, which make up the majority of plant pathogenic viruses, replicate in the cytoplasm in concert with ER, vacuole, chloroplast, peroxisome, or other membranes, which may be recruited or remodeled to form inclusion bodies or complex structures [15–17]. Endomembrane systems are also important for transport of some viruses and viral proteins.

Plant viral MPs enable plant viruses to move from cell to cell through specialized, ER-lined intercellular channels called plasmodesmata. Understanding how plant viral MPs function has been a major focus of the plant virology field for the past two decades. A synaptotagmin-family protein (*At*SYTA) was found by yeast two-hybrid to interact with the movement proteins of *Cabbage leaf curl virus* (CaLCuV; *Geminiviridae*: *Begomovirus*), *Squash leaf curl virus* (SqLCV; *Geminiviridae*: *Begomovirus*), and *Tobacco mosaic virus* (TMV; *Virgaviridae*: *Tobamovirus*), and to be important for cell-to-cell movement of CaLCuV and TMV MPs [18]. The native functions of *At*SYTA are regulation of endocytosis and formation of ER-plasma membrane contact sites which support ER structure. Interestingly, a Rab GTPase (also involved in membrane trafficking) was found in a separate study to be upregulated during TMV infection [19]. Further studies with *Turnip vein clearing virus* (TVCV; *Virgaviridae*: *Tobamovirus*) led to a paradigm-shifting model for MP function linking viral replication, intercellular movement, and endomembrane transport: TVCV MP hijacks *At*SYTA to remodel membrane contact sites near plasmodesmata, where the virus forms replication complexes and moves from cell-to-cell [4].

A recent publication by DeBlasio et al. [20] identified a number of proteins involved in clathrin-mediated endocytosis as co-immunoprecipitating with the aphid-transmitted *Potato leafroll virus* (PLRV; *Luteoviridae*: *Polerovirus*), and PLRV also directly interacts with golgin and a dymeclin-like protein (DeBlasio et al., in revision). PLRV has been previously observed by transmission electron microscopy in cytoplasmic vesicles, which fuse with the nucleus, mitochondria, vacuoles, and sites in the ER near plasmodesmata [21,22]. Although the function of these vesicles is unknown, clathrin-mediated endocytosis is also thought to be used by PLRV to traffic across tissue barriers in aphids [23] and may use these pathways in their plant hosts as well. This possibility is supported by the fact that the same viral capsid protein, a translational readthrough product from the coat protein open reading frame called the readthrough protein (RTP), is required for movement in both plant hosts and aphid vectors.

Aside from the aforementioned, endomembrane and related proteins tend to be identified only rarely in proteomic studies. This may be due to experimental bias—membrane proteins are often poorly soluble and difficult to extract with conventional protocols, and may be low in abundance to begin with—or simply because viruses are able to hijack these pathways without altering the levels or post-translational modifications of the relevant proteins. Such proteins would not be easily identified in quantitative proteomics studies looking at differential expression.

2.2. Cytoskeleton: microtubules

The cell cytoskeleton is a dynamic network of microtubules and microfilaments. Use of the cytoskeleton by viruses has been established for a number of animal viruses; both directly, by interaction of viral proteins with microtubules or microfilaments as they polymerize, or indirectly, by interaction with motor proteins that traffic various cargo along the cytoskeleton [24,25]. The best-studied example of a similar association in a plant virus exists for TMV (for review: see Ref. [26]). TMV forms replication complexes of genomic RNA, MP, and replication-associated proteins at ER-plasma membrane contact sites as discussed in Section 2.1 [4,18]. Multiple lines of evidence show an association between the TMV MP and microtubules [27–29], and a region of the TMV MP shows sequence similarity to tubulin [30]. However, the function of this association is uncertain. Although some evidence suggests that the MP-microtubule association is important for intracellular movement of the replication complex [30–32], pharmacological disruption of microtubules does not inhibit TMV movement [33,34]. To wit, an MP mutant which does not bind microtubules still localizes strongly to plasmodesmata [33], suggesting that MP function at PD is not dependent on the microtubule network. It has been suggested that microtubules actually function to promote degradation of MP by the proteasome [31,33,35,36]; however, further studies are necessary to confirm this hypothesis. In addition to the TMV MP, some evidence suggests that the MPs of *Tomato mosaic virus Ob* (ToMV; *Virgaviridae: Tobamovirus*) and *Potato mop-top virus* (PMTV; *Virgaviridae: Pomovirus*) also interact with microtubules [35,37], and *Grapevine fanleaf virus* (GFLV; *Secoviridae: Nepovirus*) requires intact microtubules for cell–cell movement in some hosts [38]. Microtubule interactions are also important for transmission of *Cauliflower mosaic virus* (CaMV; *Caulimoviridae: Caulimovirus*), which forms inclusion bodies key for aphid acquisition in a microtubule-dependent manner [39,40].

Several proteomic studies have found a link between other virus species and microtubules: PLRV was recently reported to co-immunoprecipitate with tubulin [20], and β-tubulin was shown to be upregulated in papaya leaves during infection with *Papaya meleira virus* (PMeV; unclassified), as well as in grape berries during mixed infection with *Grapevine leafroll-associated virus 1* (GLRaV-1; *Closteroviridae: Ampelovirus*), *Grapevine virus A* (GVA; *Betaflexiviridae: Vitivirus*), and *Rupestris stem pitting-associated virus* (RSPaV; *Betaflexiviridae: Foveavirus*) [41,42]. Although the role of microtubules in infection with these viruses is yet unknown, their identification is unsurprising given their importance for other diverse virus species.

2.3. Cytoskeleton: microfilaments

Ample evidence also exists for involvement of the other component of the cytoskeleton—actin microfilaments—in plant virus movement. The replication complexes of TMV and *Turnip mosaic virus* (TuMV; *Potyviridae: Potyvirus*) traffic along microfilaments [43,44], TMV MP binds to microfilaments *in vitro* [29], and some evidence suggests that intact microfilaments are required for cell–cell movement of TMV and *Potato virus X* (PVX; *Potyviridae: Potyvirus*) [34,45]. Several other diverse viruses, including CaMV, PMTV, and *Tobacco etch virus* (TEV; *Potyviridae: Potyvirus*) form granules or other small structures which traffic along microfilaments [46–48]. Interestingly, impairing the ability of the MPs of both TMV and CaMV to sever microfilaments also prevents these proteins from affecting plasmodesmata pore size [49], suggesting that some viral MPs may utilize the cytoskeleton for manipulation of plasmodesmata. Although the mechanism by which this may occur is unknown, we can hypothesize that microfilaments may be important for MP targeting of plasmodesmata, or that interfering with the ability of MPs to sever microfilaments also impairs the ability of these proteins to recruit callose-degrading enzymes. Motor proteins that traffic along microfilaments have also been shown to be important for movement of some plant viruses. Silencing of certain myosins inhibits the intercellular movement of TMV, and movement-associated tubule formation in GFLV [45,50].

The frequency with which actin and related motor proteins have been identified in plant virus proteomic studies underscores their importance in viral movement. Levels of actin are increased in both resistant and susceptible sugar beets six weeks after germination in soils inoculated with *Beet necrotic yellow vein virus* (BNYVV; *Benyviridae: Benyvirus*) [51]. Infection of papaya with PMeV induces an increase in one isoform of actin, but a decrease in another, as well as a decrease in actin polymerizing factor [41]. Grapevines co-infected with GLRaV-1, GVA, and RSPaV also show a decrease in fimbrin, a microfilament cross-linking protein, but an increase in alpha actin [42], suggesting that cross-linked and free microfilaments may play opposing roles in infection with these viruses. Specific functions of fimbrim have been understudied in plants, but include formation of intestinal microvilli in vertebrates and cytokinesis in yeast, among other functions [52–54]. PLRV was found to co-immunoprecipitate with multiple actin and myosin homologs in *N. benthamiana* [20], and a direct interaction with the PLRV CP/RTP has been shown (DeBlasio et al., in revision). Although the decrease in some actin isoforms and/or related proteins seen in the infected papaya and grapevines may seem counterintuitive if microfilaments are used for virus transport, these plants were in relatively late stages of infection, unlike many other proteomic studies reviewed here. The decrease in actin and related proteins in these plants may have been related to decreased cell health rather than a targeted effect of the viruses. It is also possible, given the increase in tubulin in these plants, that viral trafficking by these species or during late infection uses primarily microtubules, rather than microfilaments.

3. Manipulation of photosynthesis and primary metabolism

3.1. Photosynthesis and carbon fixation

Chlorosis and net reduction in photosynthesis are among the most commonly observed symptoms of virus infection in plants. It is unknown whether viruses directly manipulate the photosynthetic machinery to promote a successful infection or whether the impact on photosynthesis during infection is an indirect effect of the virus. As several studies have also found photosynthetic proteins in complex with viral particles, it is possible that virus proteins themselves may regulate photosynthesis, either directly or as part of a complex of interacting proteins. It is also possible that down-regulation of photosynthesis is partially an effect of damage done to chloroplasts, directly or indirectly, by the virus, as infection with a number of diverse viruses has been observed to alter chloroplast structure, size, or number [55–58].

Photosynthesis is also tightly linked to the production of reactive oxygen species (ROS), as chloroplasts are the primary site of ROS production in plants (for review: see Refs. [59,60]). During photosynthesis, oxygen produced can be reduced by electrons passing through the electron transport chain, forming superoxide. Under normal conditions, superoxide and other ROS byproducts of photosynthesis are detoxified by ROS-scavenging enzymes; however, under conditions of biotic or abiotic stress, or when photosynthesis is perturbed, ROS may be allowed to build up. These accumulated ROS are important for defense and stress responses (see Section 5.1), but may also damage organelles and cellular components. In cases where HR is triggered, ROS are a key component of cell death signaling pathways.

The interplay between viruses and the photosynthetic machinery, including ROS signaling, during infection is complex. During infection with some viruses, the decrease in net photosynthetic capacity is due not to a decrease in proteins involved in light capture, but rather an increase in amount or activity of proteins involved in carbon fixation [61,62], which may contribute to chlorosis by buildup of assimilates [63]. In other cases, however, carbon fixation acts as the rate-limiting step that inhibits photosynthesis during virus infection [62,64]. There is also evidence to support the involvement of relative levels of the different proteins in the oxygen evolving complex as important in determining photosynthetic rates during infection [65], which is supported by the frequency with which these proteins have been found in proteomic studies [20,41,51,66–75] (DeBlasio et al., under revision). Alterations in ferredoxin levels, also found in several proteomic studies reviewed here [20,66,69,72,76,77], have been shown to be associated with symptom development in TMV-infected plants [78].

Substantiating the ample literature linking viral infection to photosynthesis, photosynthetic proteins make up a major category of virus-interacting or differentially regulated proteins in most plant-virus proteomic studies. This is likely due both to photosynthesis being commonly exploited (or altered) by plant pathogenic viruses and to the relatively high abundance of photosynthetic proteins in green tissues. A number of photosynthetic proteins have been identified as co-purifying with Rice yellow mottle virus (RYMV; Unclassified: Sobemovirus), including: phosphoenolpyruvate carboxylase, a RuBisCO binding protein, the RuBisCO large subunit, PsbQ, PsbP, and subunits of ATP synthase [66]. Photosynthetic enzymes, including a putative transketolase, components of ATP synthase, and ferredoxin NADP(H) oxidoreductase were also found to co-purify with the RPV strain of Cereal yellow dwarf virus (CYDV-RPV; Luteoviridae: Luteovirus) [76]. PLRV was recently shown to co-immunoprecipitate with a number of proteins involved in photosynthesis and gluconeogenesis, including the oxygen-evolving enhancer proteins PsbP and PsbQ, proteins from both photosystem I and II, subunits of ATP synthase, multiple chlorophyll-binding proteins, and transketolase [20], and a direct interaction with the CP/RTP could be demonstrated for PsbQ (DeBlasio et al., under revision). A number of photosynthetic proteins were found exclusively or were significantly enriched in co-immunopurifications of wild-type PLRV compared to a mutant form of PLRV lacking the readthrough domain of the RTP (the minor structural protein), showing that the readthrough domain mediates protein interactions with the photosynthetic machinery. These interactions may lead to the development of chlorosis during infection or suppression of host immune responses [79]. Finally, ORSV co-immunoprecipitates with RuBisCO and related proteins, three subunits of photosystem I, and several proteins involved in photorespiration [80].

CMV infection has been shown to downregulate a subunit of photosystem II, PsbO, the large subunit of RuBisCO, RuBisCO activase, and carbonic anhydrase, as well as four proteins involved in photorespiration, and a plastidic aldolase [67]. Unexpectedly, grape berries co-infected with GLRaV-1, GVA, and RSPaV show an increase in a subunit of ATP synthase, although it is unknown if this trend holds for leaves and other photosynthetically-active tissue [42]. Similarly Larson and colleagues [51] reported a relative increase in levels of several members of the oxygen-evolving complex in susceptible sugar beet roots, as well as an increase in the RuBisCO large subunit in both resistant and susceptible roots, during BNYVV infection. Mixed regulation of proteins related to photosynthesis and carbon fixation was found in maize leaves infected with Rice black-streaked dwarf virus (RBSDV; Reoviridae: Fijivirus) [81,82]. Infection of Nicotiana benthamiana with Pepper mild mottle virus

(PMMoV; Virgaviridae: Tobamovirus) caused a decrease in PsbP, but not PsbO [68]. Changes specifically in chloroplastic protein levels during infection of N. benthamiana with PMMoV were assessed by Pineda et al. [69], who identified 16 down-regulated polypeptides, including cytochrome F, ATP synthase, RuBisCO, and phosphoglycerate kinase. Infection of papaya with PMeV decreases levels of the small chain of RuBisCO, RuBisCO activase, a member of the oxygen-evolving complex, and beta hydroxyacyl ACP dehydratase [41]. Analysis of proteins responsive to Peanut stunt virus (PSV; Bromoviridae: Cucumovirus) infection found multiple proteins involved in photosynthesis and gluconeogenesis to be differentially regulated by the virus, its satellite RNA, or both [70], and the Calvin cycle enzyme ribose-5-phosphate isomerase was found to be downregulated during infection of tomato with TMV [71]. Transketolase, a Calvin cycle enzyme, was found to be upregulated during SCMV infection of susceptible maize [77], and all but one isoform was also upregulated during infection of maize with RBSDV [81]. Differential regulation of ferredoxins, ATP synthase, Psb proteins, RuBisCO, and other photosynthetic proteins was also shown during SCMV infection of maize [72]. Several photosynthetic and carbon fixation proteins are differentially regulated during Cymbidium mosaic virus (CymMV; Alphaflexiviridae: Potexvirus) and Ondontoglossum ringspot virus (ORSV; Virgaviridae: Tobamovirus) infection of Phalaenopsis amabilis orchids, both in single and double infections [73]. Mixed regulation of proteins important for photosynthesis, carbon fixation, and chlorophyll biosysnthesis was shown during Zucchini yellow mosaic virus (ZYMV; Potyviridae: Potyvirus) infection of partially resistant zucchini [74]. Proteins important for carbon fixation, including ATP synthase and RuBisCO, were found to be upregulated in a resistant cultivar of soybean during infection with Soybean mosaic virus (SMV; Potyviridae: Potyvirus), although one homolog of RuBisCO appeared downregulated [83]. A Psb protein, subunits of ATP synthase, and a Calvin cycle enzyme are differentially regulated during transient expression of the AC2 protein from Tomato chlorotic mottle virus (ToCMoV; Geminiviridae: Begomovirus) in N. benthamiana [75]. Photosynthetic and carbon fixation proteins are also differentially regulated during Mungbean yellow mosaic India virus (MYMIV; Geminiviridae: Begomovirus) infection of Vigna mungo [84].

3.2. Carbon partitioning and metabolism

Proteomic studies show that the alteration of primary metabolism in virus-infected plants is widespread and complex, a finding substantiated by enzyme activity studies in CMV-infected Cucurbita pepo L. two decades ago [85,86]. Multiple enzymes involved in carbon metabolism were found to co-immunoprecipitate with PLRV [20], and to co-purify with CYDV-RPV and RYMV [66,76]. One or more enzymes important for carbon metabolism were found to be upregulated during infection with TMV and ORSV [71,73], but downregulated during infection with CMV, PMMoV, PMeV, and CymMV [41,67,69,73]. Differential regulation of carbon metabolic enzymes was observed during infection with RBSDV, RYMV, SCMV, SMV, PSV, ZYMV, MYMIV, and during triple infection of grapes with GLRaV-1, GVA, and RSPaV [42,70,72,74,77,81–83,87,88]. All differentially regulated carbon metabolic enzymes were decreased during infection with RBSDV, except for glyceraldehyde-3-phosphate (GAPDH), which was increased [81]. Phosphoglycerate kinase was downregulated in skin of grape berries co-infected with GLRaV-1, GVA, and RSPaV, but upregulated in in infected fruit pulp [42]. These data highlight the importance of tissue choice when performing and interpreting proteomic studies, as virus infection may have different effects even in tissues in proximity to one another. Such differences may also be observed in resistant vs. susceptible hosts: GAPDH levels were increased during RYMV infection of resistant rice, but unchanged in susceptible rice; whereas aldolase levels were increased in susceptible rice but unchanged in resistant [88]. Most differentially regulated carbon metabolic enzymes in SCMV-infected maize were downregulated during infection of a susceptible cultivar, although GAPDH was increased in the resistant cultivar [77]. GAPDH levels were decreased during infection of a resistant soybean cultivar with SMV, while NADPH-specific isocitrate dehydrogenase levels were increased [83]. Changes in several carbon metabolic enzymes are induced by transient expression of the ToCMoV AC2 protein in *N. benthamiana* [75]. It is important to note here that many carbon metabolic enzymes are common to glycolysis and carbon fixation/gluconeogenesis; without further information it is not possible to say which process is being targeted by viral infection. Viral targets in these pathways may vary by virus species, host species and cultivar, infection time point, and plant age. These variables likely account for some of the proteome variation observed in carbon metabolic enzymes during infection. However, these proteome data paint a compelling picture that carbon metabolism is a key hub for viral manipulation during infection.

A number of plant-pathogenic viruses are known to also have an effect on carbon partitioning and allocation or phloem biology/physiology. Some plant pathogenic viruses cause damage to or blockage of phloem [89]. Many plant pathogenic viruses have been observed to cause an alteration in starch content of infected leaves [57,61,89–91] or roots [92]. Transgenic expression of the TMV MP has been shown to increase sugar and starch content in source leaves by preventing export to phloem, and decreases plant biomass allocated to roots [93,94]. Interestingly, the effect of the MP on biomass partitioning has been shown by mutational studies to be independent of the ability of the MP to affect plasmodesmata pore size [93,95]. The starch biosynthetic enzymes ADP/UDP-glucose pyrophosphorylase have been found by proteomic studies to be upregulated during infection of maize with RBSDV [81,82] and infection of tomato with TMV [71], and RYMV from both resistant and infected rice plants co-purifies with a putative 4-alpha glucanotransferase (an enzyme involved in starch and sucrose metabolism) [66].

Manipulation of carbon metabolism, partitioning, and allocation may occur due to direct manipulation of involved proteins by viruses, or indirect effects of virus infection. In the case of insect-vectored viruses, we can hypothesize that alterations in carbon content may increase attractiveness of infected plants to insect vectors, encourage them to feed for a longer or shorter amount of time (depending on the vectoring strategy), or improve vector fitness. Changes to carbon partitioning may also have effects on photosynthesis if photoassimilates are allowed to build up. Similar to insect-vectored human diseases, plant viruses are known to manipulate the behavior of their vectors to their advantage, both by affecting the biology of diseased hosts and by altering the behavior of viruliferous insects [1–3].

4. Manipulation of amino acid metabolism

Metabolic changes during virus infection are not limited to carbon metabolism. Multiple studies point to an increase in amino acid metabolism in virus-infected plants. During infection of squash with *Squash mosaic virus* (SqMV; *Secoviridae: Comovirus*), chloroplastic amino acid biosynthesis has been shown to be increased [56], and TRV infection was shown to cause an increase in leaf amino acid content [96]. Analysis of combined microarray data revealed that amine biosynthetic processes and processes related to aromatic amino acid metabolism are overrepresented in the upregulated category during infection of plants with compatible viruses [97].

Amino acid biosynthetic enzymes and proteins involved in protein transport were found as part of a major network of proteins, centering around the 14-3-3 protein GRF2, found co-immunoprecipitating with PLRV [20]. Multiple members of the glycine cleavage system, a group of four proteins which degrade excess glycine, co-immunoprecipitate or co-purify with PLRV, RYMV, and CYDV-RPV [20,66,76], and glycine dehydrogenase is upregulated during PMeV infection [41]. Aminotransferases also co-immunoprecipitate or co-purify with PLRV, RYMV, and ORSV [20,66,80], and a subunit of isopropyl malate isomerase (part of the leucine biosynthesis pathway) co-immunoprecipitates with ORSV coat protein [80]. Phosphoglycerate dehydrogenase, another amino acid biosynthetic enzyme, was found to be upregulated in a susceptible rice cultivar during infection with RYMV [88], as well as in a resistant maize cultivar during SCMV infection [77]; and methionine synthase and ornithine carbamoyltransferase levels are increased during papaya infection with PMeV [41]. Serine hydroxymethyltransferase was found to be upregulated during MNSV-1 infection, and fumarylacetoacetate hydrolase, an enzyme important for tyrosine degradation, was downregulated [98]. Three enzymes involved in amino acid biosynthesis are upregulated during RBSDV infection [82]. In contrast to data showing an increase in amino acid content during virus infection, amino acid biosynthetic enzymes are also sometimes found to be downregulated in proteomic studies: Aminotransferases are downregulated during infection of maize with RBSDV or SCMV (susceptible cultivar only) [77,81]. Glutamine synthase is downregulated during PMeV infection, and cysteine synthase is downregulated during both PMeV and RBSDV infection [41,81]. This seemingly contradictory finding may be unique to these particular virus-host combinations or infection stages. Additionally, the RBSDV and PMeV studies were both performed on infected plants from field trials. Although both studies used uninfected plants from the same field as controls, it is still possible that unseen biotic or abiotic stresses may have had different effects on infected versus uninfected plants.

Increased amino acid biosynthesis may serve simply to provide amino acids for synthesis of viral proteins during replication. Alternatively, there is significant emerging evidence for modulation of defense responses to a broad spectrum of plant pathogens by amino acid homeostasis (for review: see Ref. [99]). Although most related studies focus on resistance to bacterial, fungal, and oomycete pathogens, many of the demonstrated downstream effects of perturbing amino acid homeostasis could certainly function in defense against viruses. For viruses transmitted by insects, it is also possible that manipulation of amino acid metabolism is related to host manipulation to improve attractiveness, nutrition, or palatability for insect vectors.

5. Manipulation of stress-responsive proteins

5.1. Reactive oxygen species

ROS and ROS-scavenging enzymes are an important part of plant response to both biotic and abiotic stress. ROS can participate in defensive signaling, act as a local microbicide, or assist in strengthening of cell walls. The most common enzymes implicated in generation of ROS during pathogen defense are peroxidases [100]. However, the highly reactive nature of ROS means that they can also be harmful to host cell molecules, membranes, and proteins. Chlorotic symptoms of virus infection have been proposed to be due, in full or part, to damage done to chlorophyll and/or chloroplasts by ROS [101]. To control ROS levels, ROS-scavenging enzymes, such as superoxide dismutases (SOD), catalases, peroxidases, and thioredoxins, detoxify hydrogen peroxide and superoxide anions.

Direct or indirect interactions of virions with ROS-scavenging and related enzymes have been shown for multiple species. Brizard et al. [66] found that SOD and four peroxidases co-purify with viruses from both RYMV resistant and susceptible rice, whereas a peroxidase co-purifies only with virus from resistant plants, and a peroxiredoxin purifies only with virus from susceptible plants. PLRV co-immunoprecipitates with at least one member of each major class of ROS-scavenging enzymes [20], and directly interacts with a peroxidase (DeBlasio et al., under revision), and the related *Cucurbit aphid-borne yellows virus* (CABYV; *Luteoviridae*: *Polerovirus*) was shown to bind to a peroxidase by far western [102]. Infection with SMV, RYMV, RBSDV, or CMV, causes an increase in one or more ROS-scavenging enzymes [67,81,83,88], although a second study instead found mixed differential regulation of ROS-scavenging enzymes during RBSDV infection [82]. Levels of two catalase isozymes are altered during ZYMV infection [74]. A peroxidase was found to be present at greater levels in leaves infected with an HR-causing strain of PMMoV than leaves infected with a non-HR-causing strain, and was absent in uninfected plants [103]. A comparison of the proteome in resistant versus susceptible maize cultivars infected with SCMV found a SOD to be downregulated in the susceptible cultivar, but a peroxiredoxin and a peroxidase to be upregulated in the resistant cultivar [77]. A similar study in BNYVV-infected beets found a SOD to be upregulated in the susceptible cultivar, and a peroxidase to be upregulated in both resistant and susceptible cultivars, as compared to the uninfected controls [51]; and, two peroxiredoxins were found to be upregulated during MYMIV infection of a resistant *V. mungo* cultivar, but not a susceptible [87]. A comparison of infected, but asymptomatic tomato fruits to uninfected tomato fruits showed mixed changes in regulation of four peroxidases [71]. During ZYMV infection of a resistant zucchini cultivar, peroxiredoxin levels are increased, but thioredoxin and superoxide dismutase levels are decreased [77]. During PMeV infection of papaya, catalase levels are decreased in leaves, but levels of a peroxidase and a peroxiredoxin are increased [41]. MNSV-1 infection causes an alteration in levels of two isoforms of phospholipid hydroperoxide glutathione peroxidase in phloem sap [98].

5.2. Chaperones and related proteins

Heat shock proteins (HSPs), are a class of chaperone proteins which aid in proper folding of other proteins, either after they are synthesized, or during stress conditions which promote protein misfolding [104]. Chaperones are broadly important for cell function under normal conditions as well as conditions of biotic and abiotic stress, and many are conserved across eukaryotes and prokaryotes. HSPs are classified into five major families: the Hsp70, or DnaK, family; the Hsp60, or chaperonin/GroEL family; the Hsp100, or Clp, family; the Hsp90 family; and the small Hsp (sHsp) family (for review: see Ref. [85]). In addition to their role in protein folding, chaperonins are also important for intercellular trafficking of transcription factors in plants [104]. Although not always classified as chaperones, some other proteins also perform functions in protein folding, including protein disulfide isomerases, calreticulins, calnexins, and lectins [105]. In recent years, the importance of HSPs and their co-chaperones in plant innate immunity has come to light (for review: see Refs. [106–108]). However, HSPs serve additional roles in plant-virus interactions. Host cell HSPs have been shown to be important factors in virus movement, folding of viral proteins, assembly of RNA replication complexes, and other functions in viral infection (for review: see Ref. [109]). Additionally, *Beet yellows virus* (BYV; *Closteroviridae*: *Closterovirus*) encodes a 65 kDa protein which is homologous to Hsp70, which seems to function in assembly, intracellular movement, and interactions with the cytoskeleton [110–116].

Chaperones and related proteins are frequently identified in plant-virus proteomic studies. Four Hsp70 homologs were found to co-purify with RYMV, and four were also found to co-immunoprecipitate with PLRV [20,66]. Levels of an Hsp70 homolog are increased during infection of a susceptible rice cultivar with RYMV [88], decreased during infection of papaya with PMeV [41], and are detectable in CMV-infected (but not control) melon phloem sap [117]. An Hsp90 homolog was found to co-purify with RYMV from both resistant and susceptible rice [66]. Direct interaction was shown between PLRV CP/RTP and a luminal binding protein HSP (DeBlasio et al., under revision), and two Hsp90s were found to co-immunoprecipitate with PLRV [20]. sHsps also co-immunoprecipitate with PLRV [20], and two sHsps are upregulated in resistant rice during RYMV infection [88]. Calreticulin was found to co-purify with RYMV from resistant rice [66] and to co-immunoprecipitate with PLRV [20], and is upregulated during infection with SMV [83] and infection with PMeV [41]. Three HSPs and a heat shock factor have been shown to interact with PVX stem loop 1 RNAs, indicating that PVX may have one or more HSP-responsive elements in its promoter(s) [118]. Other chaperonins were found to co-purify with RYMV [66] or co-immunoprecipitate with PLRV [20]. A TCP-1/cpn60 family chaperonin and two other chaperonins were found to be upregulated during SCMV infection of maize [77]. A 20 kDa chaperonin was upregulated during TMV infection of a partially tolerant cultivar [71] and was upregulated during transient expression of the AC2 protein from ToCMoV [75], and a chaperonin 60 is differentially regulated during MYMIV infection of resistant and susceptible cultivars of *V. mungo* [87].

5.3. Stress response and pathogen defense

Viruses, like other pathogens, trigger a number of inducible basal defense responses when recognized by plants, including the upregulation of a number of common defensive proteins. These proteins have broad functions, including beta-1,3-glucanases, chitinases, peroxidases (discussed above), defensins, and a number of proteins with poorly-understood functions. Some defensive proteins have been classified as pathogenesis-response (PR) proteins, which are typically small, protease-resistant proteins that are induced during pathogen attack (for review: see Refs. [119,120]). Defensive proteins, including PR proteins, have been shown to contribute to resistance against many diverse plant pathogens, including viruses.

Beta-1,3-glucanases/PR-2 proteins hydrolyze callose, and are hypothesized to function in pathogen defense primarily by regulating the size of plasmodesmal openings [121]. A beta-1,3-glucanase was found to co-immunoprecipitate with PLRV from *N. benthamiana* [20]. In a survey of PR proteins during compatible and incompatible interactions of PMMoV with hot pepper, two beta-1,3-glucanases were shown to be upregulated during both compatible and incompatible infections, while a third was only detectable during the incompatible reaction [103]. Beta-1,3-glucanase was also upregulated during CMV infection of both susceptible and transgenic resistant tomato [67], in asymptomatic tomato fruits during TMV infection [71], and during RBSDV infection of rice [82].

Chitinases, enzymes which break down chitin, are also classified as defensive proteins. The PR-3, 4, 8, and 11 classes all contain proteins with chitinase activity. Although chitinases function in defense against fungi and insects, proteomic studies revealed they are also differentially regulated during viral infection. In the aforementioned study by Elvira et al. [103], four chitinases were upregulated during both compatible and incompatible PMMoV infection, of which two were upregulated to a greater degree in the incompatible reaction Infection of papaya with PMeV upregulated one chitinase but downregulated another [41]. A chitinase was found to downregulated in the bark of *Citrus sudden death-associated virus* (CSDaV; *Tymoviridae*: *Marafivirus*) infected citrus of a susceptible cultivar, but not a tolerant cultivar [122], and a chitinase was also downregulated in asymptomatic TMV-infected tomato fruits [71]. Chitinases are upregulated during RBSDV infection of rice [82] and a chitinase is upregulated during transient expression of the ToCMoV AC2 protein [75]. While differential regulation of chitinases during viral infection could be due to triggering of non-specific defense responses, a class III chitinase co-purified with RYMV from a susceptible rice cultivar, and a different putative chitinase co-purified with RYMV when a resistant cultivar was used instead [66]. Simple induction of basal defense does not explain why different chitinases would be induced during resistant versus susceptible responses, indicating that a more nuanced explanation is needed.

The PR-5 class of proteins contains thaumatins, a class of proteins with antifungal properties [123] which are also associated with osmotic stress. Although no role for these proteins in viral infection has yet been identified, it is possible that an undiscovered function exists, or that there is overlap between antiviral and antifungal signaling or defense pathways in plants. Two thaumatins were found to co-immunoprecipitate with PLRV [20], and a thaumatin was found to be upregulated in the apoplast of *Plum pox virus* (PPV; *Potyviridae*: *Potyvirus*) infected peach cells [124]. A thaumatin was also found to be upregulated during infection of *Capsicum chinense* with an incompatible, but not a compatible, strain of PMMoV [103], and during RBSDV infection of rice [82,48].

Oxalate oxidase (PR-15) and germin-like proteins (GLPs; PR-16) have well-established roles in defense against a spectrum of plant pathogens. Both classes are part of the cupin superfamily of proteins, and bear homology to one another; however, oxalate oxidase is believed to be cereal-specific and catalyzes the degradation of oxalate to hydrogen peroxide, whereas GLPs are ubiquitous in plants and perform other functions, many of which are poorly understood [125]. Although no oxalate oxidases have been confirmed outside of cereal species, Rodrigues et al. [41] reports a putative oxalate oxidase in papaya which is downregulated during infection with PMeV, a finding which is supported by the observation of calcium oxalate crystals correlating with ROS production in latex [126]. A GLP was found to be upregulated in the roots of a resistant variety of sugar beet during BNYVV infection [51], and GLPs were downregulated during PMMoV, CMV, and SCMV infection [67,77,103]. A 24 k GLP was also found to co-immunoprecipitate with PLRV [20], showing that these proteins may function in complex with viruses.

Other PR proteins were found less frequently in proteomic studies, a trend which could be due to low abundance of these proteins rather than diminished importance in viral pathosystems. A PR-10 ribonuclease was found to be upregulated in a susceptible, but not a resistant, rice cultivar during RYMV infection [88]. Defensin/PR-12 was found co-immunoprecipitating with PLRV [20], and is upregulated in a resistant cultivar during SCMV infection [77]. The functions of proteins in the PR-1 and PR-17 families are yet unknown, but a PR-1 was found to be upregulated during infection of *C. chinense* with an incompatible strain of PMMoV, and a PR-17 protein was shown to be enriched during infection with both the compatible and incompatible strain [103]. The PR-6 and PR-7 classes encode proteinase inhibitors and endoproteinases, respectively, and will be covered in section 7, below.

Although not strictly defensive, 14-3-3-like proteins were also found in a significant number of proteomic studies. 14-3-3 proteins are ubiquitous in eukaryotes and are involved in signal transduction pathways related to environmental response, defense, response to light, brassinosteroid signaling, legume nodulation, and many others ([127,128] and for review: see Refs. [129,130]). Many 14-3-3 proteins regulate enzymes important for carbon and nitrogen

metabolism, making them prime targets for manipulation of host primary metabolism [131]. Six 14-3-3 proteins form protein complexes with PLRV from *N. benthamiana* [20]. A putative 14-3-3 protein co-purifies with RYMV in a susceptible cultivar [66], and a 14-3-3 protein interacts with PVX stem loop 1 RNAs [118]. In a study of differential regulation of nuclear proteins during TMV infection of *Capsicum annuum* L., a 14-3-3 protein was shown to be upregulated during infection [19], suggesting a role for 14-3-3 proteins in transcriptional responses that occur during viral infection.

Glutathione-*S*-transferases (GSTs) are stress-responsive proteins that perform a number of functions, including sequestration of toxins, mitigation of oxidative stress, and possibly hormone response [132]. A GST was found to co-immunoprecipitate with PLRV [20], and to co-purify with RYMV from resistant rice [66]. At least one GST was found to be upregulated during infection with TMV and PMeV [41,71], as well as during infection of resistant sugar beet roots with BNYVV [51] and during infection of susceptible maize seedlings with SCMV [77]. In older maize plants, different GST isoforms were downregulated in resistant and susceptible maize cultivars [72], and long-term RBSDV infection of rice induced an increase in a GST [82].

A number of other defense-related or stress-responsive proteins were also found to be targeted during virus infection, including dehydrin [88], unknown salt-stress induced proteins [66,88], the R-protein RPS2 [81], cinnamoyl CoA reductase CCR2 [73], and cystatin [51]. This is by no means an exhaustive list, as there are a staggering number of defense-related proteins in plants, but a sample to provide an idea of the diversity of defense responses triggered or manipulated during viral infection.

6. Manipulation of cell wall biogenesis and metabolism

Proteome studies show that the cell wall is a target of plant viruses during infection. Callase and callose synthase were found to be upregulated during transient expression of the ToCMoV AC2 protein [75]. Xyloglucan endo-transglycosylase, an enzyme involved in xyloglucan (a type of hemicellulose) metabolism, was found to be upregulated during infection of papaya with PMeV [41], and two putative xyloglucan endo-transglycosylases co-immunoprecipitate with PLRV [20]. Xyloglucan endo-transglycosylases are important for degradation of hemicellulose associated with loosening of the cell wall during growth, and are also believed to be important for fruit ripening and abscission (for review: see Ref. [133]). Despite these findings, enzymes involved in cell wall biogenesis, metabolism, and modification represent a minor category in plant-virus proteomic studies. This is likely due in large part to experimental bias, as apoplastic proteins are difficult to extract even using specialized techniques (for review: Ref. [134]; examples: see Refs. [135,136]), and are therefore likely to be undersampled. That this category of enzymes is also found infrequently in proteomic studies of bacterial pathogens of plants further substantiates this theory [137]. Additionally, the overwhelming majority of proteomic studies reviewed here are performed in leaf tissue. It is possible that cell wall modification is generally less important for viral pathogenesis in leaves than it is in fruits, roots, or other tissues. As per the details in Section 5.3, beta-1,3-glucanases, which hydrolyze callose, are commonly found to be differentially regulated during virus infection. Unlike most other cell-wall modifying proteins, beta-1,3-glucanases are often cytoplasmic, exempting them from the aforementioned difficulties [121]. Callase and callose synthase were also found to be upregulated during transient expression of the ToCMoV AC2 protein [75]

Lignin is actually the name for any of the many aromatic polymers which are important for cell wall rigidity and resistance against degradation by pathogens [138]. Several enzymes involved in lignin biosynthesis are highlighted in plant-virus proteomic studies: larreatricin hydroxylase was found to be downregulated, then upregulated at a later time point, during infection of beet roots with BNYVV [51]; cinnamyl alcohol dehydrogenase is upregulated during SCMV infection of a resistant maize cultivar [77], as well as during RBSDV infection of maize [81]; and caffeic acid 3-O-methyltransferase is also upregulated during RBSDV infection.

Pectin forms a gel-like polysaccharide matrix in cell walls, and has been shown to be important for plant growth and development, defense, cell–cell adhesion, wall porosity, and a variety of other functions. Its structure is highly complex, and its biosynthesis involves a multitude of enzymes, mainly transferases (for review: see Ref. [139]), making it difficult to pinpoint whether its biosynthesis is affected by virus infection. However, pectin methylesterases, which catalyze the demethylesterification of pectin, have been found in two proteomic studies: pectin methylesterase co-immunopurifies with PLRV [20]; and was found to be upregulated during GLRaV-1/GVA/RSPaV triple infection of grape berries [42]. Pectin methylesterases are involved in cell wall remodeling, and their expression has been shown to be correlated with a variety of biotic and abiotic stresses. Additionally, the interaction of pectin methylesterases with the TMV MP is required for cell–cell movement of TMV [140].

Several other proteins involved in cell wall metabolism or modification have been found more rarely in proteomic studies. A putative exoglucanase precursor (a cellulase) co-purifies with RYMV from both resistant and susceptible rice cultivars [66], and an expansin co-immunoprecipitates with PLRV [20].

An understanding of the impact of viral infection on the cell wall is important not only in agricultural systems, but in biofuel crops as well. In particular, biofuel crop breeders aim to reduce lignin content, increase biomass, and increase growth rate. However, evidence in switchgrass suggests that these changes may also result in an increased susceptibility to insect-vectored viruses [141]. Highly-selected modern switchgrass cultivars were more susceptible to Barley and Cereal yellow dwarf viruses than near-wild cultivars, both in greenhouse and field studies. This study highlights the need for future research in to consider cell wall proteome effects on disease susceptibility when breeding crops for biofuel, particularly as many of these species are perennial and could act as long-term reservoirs for viruses.

7. Manipulation of translation, protein processing, and protein degradation

The final category of interacting or differentially regulated proteins to be discussed in this review is that of peptide metabolism. One or more ribosomal proteins co-purify with RYMV [66] and CYDV-RPV [76], or co-immunoprecipitate or directly interact with PLRV [20] (DeBlasio et al., under revision). One or more ribosomal proteins are also differentially regulated during infection with SMV, SCMV, RYMV, PMeV, PMMoV, RBSDV, BNYVV, and MYMIV, and during transient expression of the ToCMoV AC2 protein [41,51,69,72,77,81,83,87,88]. Translation initiation and elongation factors, which are also co-opted by viruses for production of viral proteins, are differentially regulated during infection with SCMV, RYMV, PMeV, RBSDV, and MYMIV [41,77,81,82,87,88]. An elongation factor and a ribosomal protein were shown to interact with stem loop 1 RNAs from PVX [118]. One translation elongation factor co-purifies with RYMV from a resistant rice cultivar [66], and nearly 20 elongation or initiation factors co-immunoprecipitate with PLRV [20], at least one of which directly interacts with the CP/RTP (DeBlasio et al., under revision). Given the importance of protein synthesis for viral replication, and the strict dependence of viruses on host

machinery for this process, it is unsurprising that related proteins are so frequently differentially regulated.

The abundant differential regulation of proteases and related proteins and protease inhibitors during viral infection highlights the molecular tug-of-war between host and virus that occurs during infection. A ubiquitin-like protein was found to be upregulated during RYMV infection of a susceptible rice cultivar, while another was downregulated during infection of a resistant cultivar [88]. Ubiquitin fusion protein is upregulated in resistant beet roots during BNYVV infection [51], in tomato during TMV infection [19], and in rice during long-term RBSDV infection [82]. A ubiquitin fusion protein was also found to co-immunoprecipitate with PLRV, as did an E3 ubiquitin ligase [20], which interacts directly with the PLRV CP/RTP (DeBlasio et al., under revision). An E1 ubiquitin-activating enzyme was found to bind to bind in vitro to RYMV [66]. One or more subunits of the proteasome co-immunoprecipitate or co-purify with RYMV or PLRV, respectively [20,66], and are differentially regulated during infection with SMV, PMeV, RBSDV, TMV, GLRaV-1/GVA/RSPaV, ZYMV, and MNSV-1 [19,41,42,74,81,83,98]. Assorted proteases are differentially regulated during infection with SMV, PMeV, RBSDV, CMV, TMV, and in latex of PMeV-infected papaya [41,67,71,81–83,142]; co-purify with RYMV, CYDV-RPV, and ORMV [66,76,143]; and co-immunoprecipitate with PLRV [20]. Finally, putative protease inhibitors co-purify with RYMV [66], and are upregulated in leaves during PMeV infection [41]. A serine protease inhibitor is downregulated in latex sap of papaya during PMeV infection [142], but upregulated in phloem sap of melon during MNSV-1 infection [98]. A cystatin is upregulated in resistant sugar beet roots during BNYVV infection [51], and a trypsin inhibitor co-immunoprecipitates with ORSV CP [80].

8. Problems, pitfalls, and future directions

Differential regulation of proteins during viral infection is complex, and likely varies according to virus species and strain, host species and cultivar, infection stage, plant age, tissue, cellular compartment, and environmental conditions. In many cases, two homologs of the same protein will be regulated in different directions during infection with different viruses, in different hosts, or at different time points, and it is not unusual for two homologous proteins to be found to be regulated in opposite directions even in the same study at the same time point. This makes it extremely difficult to establish specific directional trends in proteins, protein classes, or pathways which are altered or exploited—for example, while we can certainly say that peroxidases are often differentially regulated during virus infection, it is much harder to make a generalization about the direction of their regulation, or even to identify the general set of conditions under which they are up or down regulated.

It is nonetheless clear that virus infection generally has large effects on core plant metabolism, including photosynthesis and carbon and amino acid metabolism. Some of these effects may be collateral damage as a result of general stress and defense responses during infection; however, as multiple virus species have been shown to interact directly or indirectly with metabolic proteins, it is likely that this regulation has been selected due to benefits obtained by the virus during infection or by the plant during the anti-viral defense response, to at least some extent. As viruses are entirely dependent on their hosts for replication, it is likely advantageous for the virus to utilize vital enzymes for its own life cycle whenever possible, as these host proteins cannot be easily deleted or mutated during the host-pathogen arms race. Additionally, the manipulation of core metabolism by insect-vectored viruses may occur as part of the host manipulation hypothesis [1,3], to enhance virus transmission to new hosts. Although plant anti-viral defense pathways only overlap with defense pathways against other pathogens to a limited extent, viral infection nonetheless has an influence on many proteins involved in biotic and abiotic stress responses. It is likely that a portion of these effects can be attributed to a basal defense response; however, some PR proteins, including HSPs and beta-1,3-glucanases, have well-studied roles in viral infection. Both proteomic and other studies highlight the importance of the cytoskeleton during viral infection, particularly for movement and formation of replication sites; yet, even in model systems there is still some debate about the precise role of each cytoskeletal component. Undoubtedly this is an area where significant advances will be made in the coming years. It is also likely that, as proteomics technologies continue to improve, further light will be shed on the effects of viral infection on proteins that are low in abundance or difficult to extract, which tend to be identified less frequently in current proteome approaches that rely on a relatively high threshold of abundance in protein extractions for detection.

The use of mass spectrometry for protein identification requires a well-annotated database for the host species of interest, which is not available for many plant species. In most cases this issue is solved by the use of a database from one or more related species; however, a database which more nearly approximates the actual possible proteins present in a sample will certainly improve the number and accuracy of protein identifications. One possible solution is to bypass genome sequencing and instead perform RNA-seq under the conditions of interest, as proteomic studies do not require any information about untranscribed regions of the genome. Recent work has shown that protein identification can be achieved using a transcriptome for searching, which is much faster and easier to obtain than a fully annotated genome [144]. An additional issue is the lack of annotation or known function for a not-insignificant proportion of proteins in any database, as "hypothetical proteins" and proteins with "unknown function" were identified as differentially regulated in a number of proteomic experiments reviewed here. A yeast two-hybrid study using the CaMV movement protein as bait (not included in this review) could not find significant structural homology for any of the three protein interaction partners discovered [145]. Structural information, including post-translational modifications, composition of oligomers, enzymatic active sites, and the three-dimensional structure of proteins, is also important to understand the proteome.

Another consideration for proteomics studies is choice of tissue. Although plant roots and fruit can be a major site of viral damage and/or replication, only three reviewed papers included one of these tissues [42,51,71]. Most experiments in this field have focused on leaf tissue, which, while informative, may be biased toward photosynthetic and related proteins due to their abundance relative to other proteins. In some cases, however, performing proteomics on other tissues may present unique challenges: some tissues are difficult to harvest in sufficient quantity, difficult to clean (i.e., of soil) or to grind, or are enriched in proteins or other compounds that complicate extraction or downstream sample preparation (protein digestion, sample clean-up, etc). These challenges are surmountable with careful planning and alteration of protocols [146,147]. As the field of plant virology advances, it will become increasingly important to move beyond the use of model systems and easy tissues to assess what occurs in the hosts and tissues that are most important for each pathosystem.

The overwhelming majority of publications in plant-virus proteomics use 2-dimensional electrophoresis or 2-dimensional fluorescence difference gel electrophoresis (2D DIGE) or, more rarely, mass spectrometry and spectral counting for protein quantification. Both of these approaches search for differences in the quantity of particular proteins or protein isoforms between treatments (i.e., infected vs. healthy). While these studies can be very informative, they do not necessarily account for proteins for which viral infection changes their subcellular localization, structure, post-translational modifications, or simply co-opts them for their own purposes. For example, viral remodeling of the host cytoskeleton likely plays an important role in intra- and intercellular trafficking of many viruses, but may be accomplished without altering levels of actin or tubulin. At the same time, these types of quantification-based analyses may be enriched for proteins far downstream in signaling pathways that are manipulated or perturbed by viruses. This may be part of the reason that some proteins, like beta-1,3-glucanases, are found to be upregulated in nearly all proteomic studies dealing with both viral and non-viral pathogens. To compliment these types of experiments, it will be imperative to elucidate which proteins interact with the viral proteins of interest, either directly or as part of a protein complex, and furthermore, to define the protein complexes that form with each viral protein so that the functions during infection can be elucidated in combination with traditional plant virology studies (for example: see Ref. [79]). This can be done either by co-immunoprecipitation of tagged or antibody-reactive viral components [20], far Western analysis [102], or using mass spectrometry-based technologies [11,76]. In the field of human and animal virus-host interactions, significant progress has been made through proteomic studies utilizing co-immunoprecipitation coupled to mass spectrometry, demonstrating the value of these approaches for studying these unique and highly recalcitrant systems [148–151].

Analysis of large data sets, like those often generated in proteomic experiments, remains a challenge in the "-omic" era. Some tools, such as gene ontology (GO) and STRING (http://string-db.org) analysis, are available to help identify the primary pathways, networks, or functions represented in a data set (for examples: see Refs. [20,77,80,152]), but teasing out candidate genes for validation and downstream analysis is a significant hurdle. Some groups, primarily in vertebrate biology, seek to solve this issue using systems biology: a computational modeling approach that aims to simulate the complex interconnected network of genes and proteins in a cell. Systems biology models can be used to predict effects of perturbing a particular gene/protein, predict disease outcomes for a given dataset [153], or identify novel or key genes in disease. Despite the potential applications, however, systems biology has not been appreciably applied to crop disease, likely due primarily to the difficulty in setting up these models, which require carefully curated databases containing multiple "-omics" data sets, as well as a significant knowledge of programming and mathematics. Systems biology in plant pathology, further application of proteomics to non-model hosts and tissues, and integrating information about the plant host, pathogen, and in some cases the vector, will open up new avenues for crop disease management.

Acknowledgment

The authors would like to thank the National Science Foundation grant 1354309 and NIH grant 5T15HD072999-04 for funding.

References

[1] S.M. Gray, M. Cilia, M. Ghanim, Circulative, nonpropagative virus transmission: an orchestra of virus-, insect-, and plant-derived instruments, Adv. Virus Res. 89 (2014) 141–199.
[2] L.L. Ingwell, S.D. Eigenbrode, N.A. Bosque-Pérez, Plant viruses alter insect behavior to enhance their spread, Sci. Rep. 2 (2012).
[3] J.C. Holmes, W.M. Bethel, Modification of intermediate host behaviour by parasites, in: E.U. Canning, C.A. Wright (Eds.), Behavioural Aspects of Parasite Transmission, Academic Press, 1972, pp. 123–149.
[4] A. Levy, Y. Zheng Judy, G. Lazarowitz Sondra, Synaptotagmin SYTA forms ER-plasma membrane junctions that are recruited to plasmodesmata for plant virus movement, Curr. Biol. 25 (2015) 1–8.

[5] A. Levy, M. Erlanger, M. Rosenthal, B.L. Epel, A plasmodesmata-associated β-1,3-glucanase in *Arabidopsis*, Plant J. 49 (2007) 669–682.

[6] K.A. Peter, F.E. Gildow, P. Palukaitis, S.M. Gray, The C terminus of the *Polerovirus* p5 readthrough domain limits virus infection to the phloem, J. Virol. 83 (2009) 5419–5429.

[7] J.L. Dangl, R.A. Dietrich, M.H. Richberg, Death don't have no mercy: cell death programs in plant-microbe interactions, Plant Cell 8 (1996) 1793–1807.

[8] J.T. Greenberg, N. Yao, The role and regulation of programmed cell death in plant-pathogen interactions, Cell Microbiol. 6 (2004) 201–211.

[9] R.I. Pennell, C. Lamb, Programmed cell death in plants, Plant Cell 9 (1997) 1157–1168.

[10] E. Lam, N. Kato, M. Lawton, Programmed cell death, mitochondria and the plant hypersensitive response, Nature 411 (2001) 848–853.

[11] J.D. Chavez, M. Cilia, C.R. Weisbrod, H.-J. Ju, J.K. Eng, Cross-linking measurements of the *Potato leafroll virus* reveal protein interaction topologies required for virion stability, aphid transmission, and virus-plant interactions, J. Proteom. Res. 11 (2012) 2968–2981.

[12] M. Di Carli, E. Benvenuto, M. Donini, Recent insights into plant-virus interactions through proteomic analysis, J. Proteom. Res. 11 (2012) 4765–4780.

[13] M. Heinlein, Plant virus replication and movement, Virology 479–480 (2015) 657–671.

[14] J. Tilsner, O. Linnik, M. Louveaux, I.M. Roberts, S.N. Chapman, K.J. Oparka, Replication and trafficking of a plant virus are coupled at the entrances of plasmodesmata, J. Cell Biol. 201 (2013) 981–995.

[15] R. Grangeon, J. Jiang, J.F. Laliberté, Host endomembrane recruitment for plant RNA virus replication, Current Opin. Virol. 2 (2012) 677–684.

[16] C. Patarroyo, J.-F. Laliberté, H. Zheng, Hijack it, change it: how do plant viruses utilize the host secretory pathway for efficient viral replication and spread? Front. Plant Sci. 3 (2012) 308.

[17] H. Sanfaçon, Replication of positive-strand RNA viruses in plants: contact points between plant and virus components, Can. J. Bot. 83 (2005) 1529–1549.

[18] J.D. Lewis, S.G. Lazarowitz, *Arabidopsis* synaptotagmin SYTA regulates endocytosis and virus movement protein cell-to-cell transport, Proc. Natl. Acad. Sci. 107 (2010) 2491–2496.

[19] B.-J. Lee, S.J. Kwon, S.-K. Kim, K.-J. Kim, C.-J. Park, Y.-J. Kim, O.K. Park, K.-H. Paek, Functional study of hot pepper 26S proteasome subunit RPN7 induced by *Tobacco mosaic virus* from nuclear proteome analysis, Biochem. Biophys. Res. Commun. 351 (2006) 405–411.

[20] S.L. DeBlasio, R. Johnson, J. Mahoney, A. Karasev, S.M. Gray, M.J. MacCoss, M. Cilia, Insights into the *Polerovirus*-plant interactome revealed by coimmunoprecipitation and mass spectrometry, Mol. Plant Microbe Interact. 28 (2015) 467–481.

[21] S. Shepardson, K. Esau, R. McCrum, Ultrastructure of potato leaf phloem infected with *Potato leafroll virus*, Virology 105 (1980) 379–392.

[22] W. Golinowski, K. Tomenius, P. Oxelfelt, Ultrastructural studies on potato phloem cells infected with *Potato leaf roll virus*—comparison of two potato varieties, Acta Agric. Scand. 37 (1987) 3–19.

[23] S.M. Gray, F.E. Gildow, *Luteovirus*-aphid interactions, Annu. Rev. Phytopathol. 41 (2003) 539–566.

[24] U.F. Greber, M. Way, A superhighway to virus infection, Cell 124 (2006) 741–754.

[25] B.M. Ward, The taking of the cytoskeleton one two three: how viruses utilize the cytoskeleton during egress, Virology 411 (2011) 244–250.

[26] C. Liu, R.S. Nelson, The cell biology of *Tobacco mosaic virus* replication and movement, Front. Plant Sci. 4 (2013) 12.

[27] M. Heinlein, B.L. Epel, H.S. Padgett, R.N. Beachy, Interaction of *Tobamovirus* movement proteins with the plant cytoskeleton, Science 270 (1995) 1983–1985.

[28] M. Heinlein, H.S. Padgett, J.S. Gens, B.G. Pickard, S.J. Casper, B.L. Epel, R.N. Beachy, Changing patterns of localization of the *Tobacco mosaic virus* movement protein and replicase to the endoplasmic reticulum and microtubules during infection, Plant Cell 10 (1998) 1107–1120.

[29] B.G. McLean, J. Zupan, P.C. Zambryski, *Tobacco mosaic virus* movement protein associates with the cytoskeleton in tobacco cells, Plant Cell 7 (1995) 2101–2114.

[30] V. Boyko, J. Ferralli, J. Ashby, P. Schellenbaum, M. Heinlein, Function of microtubules in intercellular transport of plant virus RNA, Nat. Cell Biol. 2 (2000) 826–832.

[31] P. Más, R.N. Beachy, Replication of *Tobacco mosaic virus* on endoplasmic reticulum and role of the cytoskeleton and virus movement protein in intracellular distribution of viral RNA, J. Cell Biol. 147 (1999) 945–958.

[32] V. Boyko, Q. Hu, M. Seemanpillai, J. Ashby, M. Heinlein, Validation of microtubule-associated *Tobacco mosaic virus* RNA movement and involvement of microtubule-aligned particle trafficking, Plant J. 51 (2007) 589–603.

[33] T. Gillespie, P. Boevink, S. Haupt, A.G. Roberts, R. Toth, T. Valentine, S. Chapman, K.J. Oparka, Functional analysis of a DNA-shuffled movement protein reveals that microtubules are dispensable for the cell-to-cell movement of *Tobacco mosaic virus*, Plant Cell 14 (2002) 1207–1222.

[34] S. Kawakami, Y. Watanabe, R.N. Beachy, *Tobacco mosaic virus* infection spreads cell to cell as intact replication complexes, Proc. Natl. Acad. Sci. U. S. A. 101 (2004) 6291–6296.

[35] H.S. Padgett, B.L. Epel, T.W. Kahn, M. Heinlein, Y. Watanabe, R.N. Beachy, Distribution of *Tobamovirus* movement protein in infected cells and

implications for cell-to-cell spread of infection, Plant J. 10 (1996) 1079–1088.

[36] C. Reichel, R.N. Beachy, *Tobacco mosaic virus* infection induces severe morphological changes of the endoplasmic reticulum, Proc. Natl. Acad. Sci. U. S. A. 95 (1998) 11169–11174.

[37] K.M. Wright, G.H. Cowan, N.I. Lukhovitskaya, J. Tilsner, A.G. Roberts, E.I. Savenkov, L. Torrance, The N-terminal domain of PMTV TGB1 movement protein is required for nucleolar localization, microtubule association, and long-distance movement, Mol. Plant Microbe Interact. 23 (2010) 1486–1497.

[38] C. Laporte, G. Vetter, A.-M. Loudes, D.G. Robinson, S. Hillmer, C. Stussi-Garaud, C. Ritzenthaler, Involvement of the secretory pathway and the cytoskeleton in intracellular targeting and tubule assembly of *Grapevine fanleaf virus* movement protein in tobacco BY-2 cells, Plant Cell 15 (2003) 2058–2075.

[39] S. Blanc, I. Schmidt, M. Vantard, H.B. Scholthof, G. Kuhl, P. Esperandieu, M. Cerutti, C. Louis, The aphid transmission factor of *Cauliflower mosaic virus* forms a stable complex with microtubules in both insect and plant cells, Proc. Natl. Acad. Sci. U. S. A. 93 (1996) 15158–15163.

[40] A. Martinière, D. Gargani, M. Uzest, N. Lautredou, S. Blanc, M. Drucker, A role for plant microtubules in the formation of transmission-specific inclusion bodies of *Cauliflower mosaic virus*, Plant J. 58 (2009) 135–146.

[41] S.P. Rodrigues, J.A. Ventura, C. Aguilar, E.S. Nakayasu, I.C. Almeida, P.M.B. Fernandes, R.B. Zingali, Proteomic analysis of papaya (*Carica papaya* L.) displaying typical sticky disease symptoms, Proteomics 11 (2011) 2592–2602.

[42] M. Giribaldi, M. Purrotti, D. Pacifico, D. Santini, F. Mannini, P. Caciagli, L. Rolle, L. Cavallarin, M.G. Giuffrida, C. Marzachì, A multidisciplinary study on the effects of phloem-limited viruses on the agronomical performance and berry quality of *Vitis vinifera* cv. Nebbiolo, J. Proteom. 75 (2011) 306–315.

[43] J.-Z. Liu, E.B. Blancaflor, R.S. Nelson, The *Tobacco mosaic virus* 126-kilodalton protein, a constituent of the virus replication complex, alone or within the complex aligns with and traffics along microfilaments, Plant Physiol. 138 (2005) 1853–1865.

[44] P.A. Harries, K. Palanichelvam, W. Yu, J.E. Schoelz, R.S. Nelson, The *Cauliflower mosaic virus* protein P6 forms motile inclusions that traffic along actin microfilaments and stabilize microtubules, Plant Physiol. 149 (2009) 1005–1016.

[45] P.A. Harries, J.-W. Park, N. Sasaki, K.D. Ballard, A.J. Maule, R.S. Nelson, Differing requirements for actin and myosin by plant viruses for sustained intercellular movement, Proc. Natl. Acad. Sci. U. S. A. 106 (2009) 17594–17599.

[46] S. Cotton, R. Grangeon, K. Thivierge, I. Mathieu, C. Ide, T. Wei, A. Wang, J.-F. Laliberté, *Turnip mosaic virus* RNA replication complex vesicles are mobile, align with microfilaments, and are each derived from a single viral genome, J. Virol. 83 (2009) 10460–10471.

[47] S. Haupt, T. Stroganova, E. Ryabov, S.H. Kim, G. Fraser, G. Duncan, M.A. Mayo, H. Barker, M.E. Taliansky, Nucleolar localization of *Potato leafroll virus* capsid proteins, J. Gen. Virol. 86 (2005) 2891–2896.

[48] H.-J. Ju, T.D. Samuels, Y.-S. Wang, E. Blancaflor, M. Payton, R. Mitra, K. Krishnamurthy, R.S. Nelson, J. Verchot-Lubicz, The *Potato virus X* TGBp2 movement protein associates with endoplasmic reticulum-derived vesicles during virus infection, Plant Physiol. 138 (2005) 1877–1895.

[49] S. Su, Z. Liu, C. Chen, Y. Zhang, X.-C. Wang, L. Zhu, L. Miao, X.-C. Wang, M. Yuan, *Cucumber mosaic virus* movement protein severs actin filaments to increase the plasmodesmal size exclusion limit in tobacco, Plant Cell 22 (2010) 1373–1387.

[50] K. Amari, A. Lerich, C. Schmitt-Keichinger, V.V. Dolja, C. Ritzenthaler, Tubule-guided cell-to-cell movement of a plant virus requires class XI myosin motors, PLoS Pathog. 7 (2011).

[51] R.L. Larson, W.M. Wintermantel, A. Hill, L. Fortis, A. Nunez, Proteome changes in sugar beet in response to *Beet necrotic yellow vein virus*, Physiol. Mol. Plant Pathol. 72 (2008) 62–72.

[52] M.S. Mooseker, Actin binding proteins of the brush border, Cell 35 (1983) 11–13.

[53] J.R. Glenney, P. Kaulfus, P. Matsudaira, K. Weber, F-actin binding and bundling properties of fimbrin, a major cytoskeletal protein of microvillus core filaments, J. Biol. Chem. 256 (1981) 9283–9288.

[54] J.-Q. Wu, J. Bähler, J.R. Pringle, Roles of a fimbrin and an α-actinin-like protein in fission yeast cell polarization and cytokinesis, Mol. Biol. Cell 12 (2001) 1061–1077.

[55] J.-G. Gao, A. Nassuth, Alteration of major cellular organelles in wheat leaf tissue infected with *Wheat streak mosaic Rymovirus* (*Potyviridae*), Phytopathology 83 (1993) 206–213.

[56] A.C. Magyarosy, B.B. Buchanan, P. Schürmann, Effect of a systemic virus infection on chloroplast function and structure, Virology 55 (1973) 426–438.

[57] L. Técsi, A. Maule, A. Smith, R. Leegood, Complex, localized changes in CO_2 assimilation and starch content associated with the susceptible interaction between *Cucumber mosaic virus* and a cucurbit host, Plant J. 5 (1994) 837–847.

[58] T. Mochizuki, S.T. Ohki, Single amino acid substitutions at residue 129 in the coat protein of *Cucumber mosaic virus* affect symptom expression and thylakoid structure, Arch. Virol. 156 (2011) 881–886.

[59] K. Asada, Production and scavenging of reactive oxygen species in chloroplasts and their functions, Plant Physiol. 141 (2006) 391–396.

[60] S.S. Gill, N. Tuteja, Reactive oxygen species and antioxidant machinery in abiotic stress tolerance in crop plants, Plant Physiol. Biochem. 48 (2010) 909–930.

[61] D. Shalitin, S. Wolf, Cucumber mosaic virus infection affects sugar transport in melon plants, Plant Physiol. 123 (2000) 597–604.

[62] A.T. Lehrer, E. Komor, Carbon dioxide assimilation by virus-free sugarcane plants and by plants which were infected by Sugarcane yellow leaf virus, Physiol. Mol. Plant Pathol. 73 (2009) 147–153.

[63] M.C. Gonçalves, J. Vega, J.G. Oliveira, M.M.A. Gomes, Sugarcane yellow leaf virus infection leads to alterations in photosynthetic efficiency and carbohydrate accumulation in sugarcane leaves, Fitopatologia Brasileira 30 (2005) 10–16.

[64] B. Sampol, J. Bota, D. Riera, H. Medrano, J. Flexas, Analysis of the virus-induced inhibition of photosynthesis in malmsey grapevines, New Phytol. 160 (2003) 403–412.

[65] J. Rahoutei, I. García-Luque, M. Barón, Inhibition of photosynthesis by viral infection: effect on PSII structure and function, Physiol. Plant. 110 (2000) 286–292.

[66] J.P. Brizard, C. Carapito, F. Delalande, A. Van Dorsselaer, C. Brugidou, Proteome analysis of plant-virus interactome: comprehensive data for virus multiplication inside their hosts, Mol. Cell. Proteom.: MCP 5 (2006) 2279–2297.

[67] M. Di Carli, M.E. Villani, L. Bianco, R. Lombardi, G. Perrotta, E. Benvenuto, M. Donini, Proteomic analysis of the plant-virus interaction in Cucumber mosaic virus (CMV) resistant transgenic tomato, J. Proteome. Res. 9 (2010) 5684–5697.

[68] M.L. Pérez-Bueno, J. Rahoutei, C. Sajnani, I. García-Luque, M. Barón, Proteomic analysis of the oxygen-evolving complex of photosystem II under biotec stress: Studies on Nicotiana benthamiana infected with Tobamoviruses, Proteomics 4 (2004) 418–425.

[69] M. Pineda, C. Sajnani, M. Barón, Changes induced by the Pepper mild mottle Tobamovirus on the chloroplast proteome of Nicotiana benthamiana, Photosynth. Res. 103 (2009) 31–45.

[70] A. Obrealska-Steplowska, P. Wieczorek, M. Budziszewska, A. Jeszke, J. Renaut, How can plant virus satellite RNAs alter the effects of plant virus infection? A study of the changes in the Nicotiana benthamiana proteome after infection by Peanut stunt virus in the presence or absence of its satellite RNA, Proteomics 13 (2013) 2162–2175.

[71] J. Casado-Vela, S. Sellés, R.B. Martínez, Proteomic analysis of Tobacco mosaic virus-infected tomato (Lycopersicon esculentum M.) fruits and detection of viral coat protein, Proteomics 6 (Suppl. 1) (2006) S196–S206.

[72] L. Wu, S. Wang, X. Chen, X. Wang, L. Wu, X. Zu, Y. Chen, Proteomic and phytohormone analysis of the response of maize (Zea mays L.) seedlings to Sugarcane mosaic virus, PLoS One 8 (2013).

[73] T. Lai, Y. Deng, P. Zhang, Z. Chen, F. Hu, Q. Zhang, Y. Hu, N. Shi, Proteomics-based analysis of Phalaenopsis amabilis in response toward Cymbidium mosaic virus and/or Odontoglossum ringspot virus infection, Am. J. Plant Sci. 4 (2013) 1853–1862.

[74] S. Nováková, G. Flores-Ramírez, M. Glasa, R. Danchenko, R. Fiala, L. Skultety, Partially resistant Cucurbita pepo showed late onset of the Zucchini yellow mosaic virus infection due to rapid activation of defense mechanisms as compared to susceptible cultivar, Front. Plant Sci. 6 (2015).

[75] L.S.T. Carmo, R.O. Resende, L.P. Silva, S.G. Ribeiro, A. Mehta, Identification of host proteins modulated by the virulence factor AC2 of Tomato chlorotic mottle virus in Nicotiana benthamiana, Proteomics 13 (2013) 1947–1960.

[76] M. Cilia, K.A. Peter, M.S. Bereman, K.J. Howe, T. Fish, D. Smith, F. Gildow, M.J. MacCoss, T.W. Thannhauser, S.M. Gray, Discovery and targeted LC–MS/MS of purified Polerovirus reveals differences in the virus–host interactome associated with altered aphid transmission, PLoS One 7 (2012) e48177.

[77] L. Wu, Z. Han, S. Wang, X. Wang, A. Sun, X. Zu, Y. Chen, Comparative proteomic analysis of the plant-virus interaction in resistant and susceptible ecotypes of maize infected with Sugarcane mosaic virus, J. Proteom. 89 (2013) 124–140.

[78] Y. Ma, T. Zhou, Y. Hong, Z. Fan, H. Li, Decreased level of ferredoxin I in Tobacco mosaic virus-infected tobacco is associated with development of the mosaic symptom, Physiol. Mol. Plant Pathol. 72 (2008) 39–45.

[79] S.L. DeBlasio, R. Johnson, M.M. Sweeney, A. Karasev, S.M. Gray, M.J. MacCoss, M. Cilia, Potato leafroll virus structural proteins manipulate overlapping, yet distinct protein interaction networks during infection, Proteomics 15 (2015) 2098–2112.

[80] P.-C. Lin, W.-C. Hu, S.-C. Lee, Y.-L. Chen, C.-Y. Lee, Y.-R. Chen, L.-Y.D. Liu, P.-Y. Chen, S.-S. Lin, Y.-C. Chang, Application of an integrated omics approach for identifying host proteins that interact with Odontoglossum ringspot virus capsid protein, Mol. Plant Microbe Interact. 28 (2015) 711–726.

[81] K. Li, C. Xu, J. Zhang, Proteome profile of maize (Zea mays L.) leaf tissue at the flowering stage after long-term adjustment to Rice black-streaked dwarf virus infection, Gene 485 (2011) 106–113.

[82] Q. Xu, H. Ni, Q. Chen, F. Sun, T. Zhou, Y. Lan, J. Zhou, Comparative proteomic analysis reveals the cross-talk between the responses induced by H2O2 and by long-term Rice black-streaked dwarf virus infection in rice, PLoS One 8 (2013) e81640.

[83] H. Yang, Y. Huang, H. Zhi, D. Yu, Proteomics-based analysis of novel genes involved in response toward Soybean mosaic virus infection, Mol. Biol. Rep. 38 (2011) 511–521.

[84] S. Kundu, D. Chakraborty, A. Pal, Proteomic analysis of salicylic acid induced resistance to Mungbean yellow mosaic India virus in Vigna mungo, J. Proteom. 74 (2011) 337–349.

[85] L.I. Tecsi, A.J. Maule, A.M. Smith, R.C. Leegood, Metabolic alteration in cotyledons of Cucurbita pepo infected by Cucumber mosaic virus, J. Exp. Bot. 45 (1994) 1541–1551.

[86] L.I. Tecsi, A.M. Smith, A.J. Maule, R.C. Leegood, A spatial analysis of physiological changes associated with infection of cotyledons of marrow plants with Cucumber mosaic virus, Plant Physiol. 111 (1996) 975–985.

[87] S. Kundu, D. Chakraborty, A. Kundu, A. Pal, Proteomics approach combined with biochemical attributes to elucidate compatible and incompatible plant-virus interactions between Vigna mungo and Mungbean yellow mosaic India virus, Proteome Sci. 11 (2013) 15.

[88] M. Ventelon-Debout, F. Delalande, J.P. Brizard, H. Diemer, A. Van Dorsselaer, C. Brugidou, Proteome analysis of cultivar-specific deregulations of Oryza sativa indica and O. sativa japonica cellular suspensions undergoing Rice yellow mottle virus infection, Proteomics 4 (2004) 216–225.

[89] M. Ashraf, Z.U. Zafar, T. McNeilly, C.J. Veltkamp, Some morpho–anatomical characteristics of cotton (Gossypium hirsutum L.) in relation to reistance to Cotton leaf curl virus (CLCuV), J. Appl. Bot. 73 (1999) 76–82.

[90] M. Ashraf, Z.U. Zafar, Patterns of free carbohydrates and starch accumulation in the leaves of cotton (Gossypium hirsutum L.) cultivars differing in resistance to Cotton leaf curl virus, Arch. Agron. Soil Sci. 45 (2000) 1–9.

[91] M.C. Arias, S. Lenardon, E. Taleisnik, Carbon metabolism alterations in sunflower plants infected with the Sunflower chlorotic mottle virus, J. Phytopathol. 151 (2003) 267–273.

[92] N. Ephraim, B. Yona, A. Evans, A. Sharon, A. Titus, Effect of Cassava brown streak disease (CBSD) on cassava (Manihot esculenta Crantz) root storage components, starch quantities and starch quality properties, Int. J. Plant Physiol. Biochem. 7 (2015) 12–22.

[93] A.A. Olesinski, W.J. Lucas, E. Galun, S. Wolp, Pleiotropic effects of Tobacco-mosaic-virus movement protein on carbon metabolism in transgenic tobacco plants, Planta (1995) 118–126.

[94] W.J. Lucas, A. Olesinski, R.J. Hull, J.S. Haudenshield, C.M. Deom, R.N. Beachy, S. Wolf, Influence of the Tobacco mosaic virus 30-kDa movement protein on carbon metabolism and photosynthate partitioning in transgenic tobacco plants, Planta 190 (1993) 88–96.

[95] S. Balachandran, R.J. Hull, Y. Vaadia, S. Wolf, W.J. Lucas, Alteration in carbon partitioning induced by the movement protein of Tobacco mosaic virus originates in the mesophyll and is independent of change in the plasmodesmal size exclusion limit, Plant Cell Environ. 18 (1995) 1301–1310.

[96] L. Fernandez-Calvino, S. Osorio, M.L. Hernandez, I.B. Hamada, F.J. del Toro, L. Donaire, A. Yu, R. Bustos, A.R. Fernie, J.M. Martinez-Rivas, et al., Virus-induced alterations in primary metabolism modulate susceptibility to Tobacco rattle virus in Arabidopsis, Plant Physiol. 166 (2014) 1821–1838.

[97] O.A. Postnikova, L.G. Nemchinov, Comparative analysis of microarray data in Arabidopsis transcriptome during compatible interactions with plant viruses, Virol. J. 9 (2012) 101.

[98] M. Serra-Soriano, J.A. Navarro, A. Genoves, V. Pallás, Comparative proteomic analysis of melon phloem exudates in response to viral infection, J. Proteom. 124 (2015) 11–24.

[99] J. Zeier, New insights into the regulation of plant immunity by amino acid metabolic pathways, Plant Cell Environ. 36 (2013) 2085–2103.

[100] M.R. Bauer, Role of reactive oxygen species and antioxidant enzymes in systemic virus infections of plants, J. Phytopathol. 148 (2000) 297–302.

[101] M. Rodriguez, E. Taleisnik, R. Lascano, Are Sunflower chlorotic mottle virus infection symptoms modulated by early increases in leaf sugar concentration? J. Plant Physiol. 167 (2010) 1137–1144.

[102] B. Bencharki, S. Boissinot, S. Revollon, V. Ziegler-Graff, M. Erdinger, L. Wiss, S. Dinant, D. Renard, M. Beuve, C. Lemaitre-Guillier, et al., Phloem protein partners of Cucurbit aphid borne yellows virus: possible involvement of phloem proteins in virus transmission by aphids, Mol. Plant Microbe Interact. 23 (2010) 799–810.

[103] M.I. Elvira, M.M. Galdeano, P. Gilardi, I. García-Luque, M.T. Serra, Proteomic analysis of pathogenesis-related proteins (PRs) induced by compatible and incompatible interactions of Pepper mild mottle virus (PMMoV) in Capsicum chinense L3 plants, J. Exp. Bot. 59 (2008) 1253–1265.

[104] X.M. Xu, J. Wang, Z. Xuan, A. Goldshmidt, P.G.M. Borrill, N. Hariharan, J.-Y. Kim, D. Jackson, Chaperonins facilitate KNOTTED1 cell-to-cell trafficking and stem cell function, Science 333 (2011) 1141–1144.

[105] D.B. Williams, Beyond lectins: the calnexin/calreticulin chaperone system of the endoplasmid reticulum, J. Cell Sci. 119 (2005) 615–623.

[106] J.A. O'Brien, A. Daudi, V.S. Butt, G.P. Bolwell, Reactive oxygen species and their role in plant defence and cell wall metabolism, Planta 236 (2012) 765–779.

[107] L. Chen, K. Shimamoto, Emerging roles of molecular chaperones in plant innate immunity, J. Gen. Plant Pathol. 77 (2011) 1–9.

[108] K. Shirasu, The HSP90-SGT1 chaperone complex for NLR immune sensors, Ann. Rev. Plant Biol. 60 (2009) 139–164.

[109] J. Verchot, Cellular chaperones and folding enzymes are vital contributors to membrane bound replication and movement complexes during plant RNA virus infection, Front. Plant Sci. 3 (2012) 275.

[110] A.A. Agranovsky, V.P. Boyko, A.V. Karasev, E.V. Koonin, V.V. Dolja, Putative 65 kDa protein of Beet yellows closterovirus is a homologue of HSP70 heat shock proteins, J. Mol. Biol. 217 (1991) 603–610.

[111] A.V. Karasev, A.S. Kashina, V.I. Gelfand, V.V. Dolja, HSP70-related 65 kDa protein of *Beet yellows closterovirus* is a microtubule-binding protein, FEBS Lett. 304 (1992) 12–14.

[112] A.J. Napuli, B.W. Falk, V.V. Dolja, Interaction between HSP70 homolog and filamentous virions of the *Beet yellows virus*, Virology 274 (2000) 232–239.

[113] A.I. Prokhnevsky, V.V. Peremyslov, A.J. Napuli, V.V. Dolja, Interaction between long-distance transport factor and Hsp70-related movement protein of *Beet yellows virus*, J. Virol. 76 (2002) 11003–11011.

[114] A.I. Prokhnevsky, V.V. Peremyslov, Actin cytoskeleton is involved in targeting of a viral Hsp70 homolog to the cell periphery, J. Virol. 79 (2005) 14421–14428.

[115] D. Avisar, A.I. Prokhnevsky, V.V. Dolja, Class VIII myosins are required for plasmodesmatal localization of a *Closterovirus* Hsp70 homolog, J. Virol. 82 (2008) 2836–2843.

[116] D.V. Alzhanova, A.I. Prokhnevsky, V.V. Peremyslov, V.V. Dolja, Virion tails of *Beet yellows virus*: coordinated assembly by three structural proteins, Virology 359 (2007) 220–226.

[117] D. Malter, S. Wolf, Melon phloem-sap proteome: developmental control and response to viral infection, Protoplasma 248 (2011) 217–224.

[118] S.Y. Cho, W.K. Cho, K.H. Kim, Identification of tobacco proteins associated with the stem-loop 1 RNAs of *Potato virus X*, Mol. Cells 33 (2012) 379–384.

[119] A. Edreva, Pathogenesis-related proteins: research progress in the last 15 years, Gen. Appl. Plant Physiol. 31 (2005) 105–124.

[120] L.C. van Loon, M. Rep, C.M.J. Pieterse, Significance of inducible defense-related proteins in infected plants, Annu. Rev. Phytopathol. 44 (2006) 135–162.

[121] A. Levy, D. Guenoune-Gelbart, B.L. Epel, Beta-1,3-glucanases: plasmodesmal gate keepers for intercellular communication, Plant Signal. Behav. 2 (2007) 404–407.

[122] M.D. Cantú, A.G. Mariano, M.S. Palma, E. Carrilho, N.A. Wulff, Proteomic analysis reveals suppression of bark chitinases and proteinase inhibitors in citrus plants affected by the citrus sudden death disease, Phytopathology 98 (2008) 1084–1092.

[123] A.J. Vigers, S. Wiedemann, W.K. Roberts, M. Legrand, C.P. Selitrennikoff, B. Fritig, Thaumatin-like pathogenesis-related proteins are antifungal, Plant Sci. 83 (1992) 155–161.

[124] P. Diaz-Vivancos, S. Rubio, V. Mesonero, P.M. Periago, A. Ros Barceló, P. Martínez-Gómez, J.A. Hernández, The apoplastic antioxidant system in *Prunus*: response to long-term *Plum pox virus* infection, J. Exp. Bot. 57 (2006) 3813–3824.

[125] J.M. Dunwell, J.G. Gibbings, T. Mahmood, S.M.S. Naqvi, Germin and germin-like proteins: evolution, structure, and function, Crit. Rev. Plant Sci. (2008) 37–41.

[126] S.P. Rodrigues, M. Da Cunha, J.A. Ventura, P.M.B. Fernandes, Effects of the *Papaya meleira virus* on papaya latex structure and composition, Plant Cell Rep. 28 (2009) 861–871.

[127] S.S. Gampala, T.-W. Kim, J.-X. He, W. Tang, Z. Deng, M.-Y. Bai, S. Guan, S. Lalonde, Y. Sun, J.M. Gendron, H. Chen, N. Shibagaki, R.J. Ferl, D. Ehrhardt, K. Chong, A.L. Burlingame, Z.-Y. Wang, An essential role for 14-3-3 proteins in brassinosteroid signal transduction in *Arabidopsis*, Dev. Cell 13 (2007) 177–189.

[128] M.H. Oh, X. Wang, S. Clouse, S. Huber, 14-3-3 proteins bind to the brassinosteroid receptor kinase, BRI1 and are positive regulators of brassinosteroid signaling, American Society of Plant Biologists Annual Meeting (2009).

[129] R. Lozano-Durán, S. Robatzek, 14-3-3 proteins in plant-pathogen interactions, Mol. Plant Microbe Interact. 28 (2015) 511–518.

[130] F.C. Denison, A.-L. Paul, A.K. Zupanska, R.J. Ferl, 14-3-3 proteins in plant physiology, Semin. Cell Dev. Biol. 22 (2011) 720–727.

[131] S.C. Huber, C. MacKintosh, W.M. Kaiser, Metabolic enzymes as targets for 14-3-3 proteins, Plant Mol. Biol. 50 (2002) 1053–1063.

[132] P.G. Sappl, A.J. Carroll, R. Clifton, R. Lister, J. Whelan, A. Harvey Millar, K.B. Singh, The *Arabidopsis* glutathione transferase gene family displays complex stress regulation and co-silencing multiple genes results in altered metabolic sensitivity to oxidative stress, Plant J. 58 (2009) 53–68.

[133] J. Braam, P. Campbell, Xyloglucan endotransglycosylases: diversity of genes, enzymes and potential wall-modifying functions, Trends Plant Sci. 4 (1999) 361–366.

[134] S.J. Lee, R.S. Saravanan, C.M. Damasceno, H. Yamane, B.D. Kim, J.K. Rose, Digging deeper into the plant cell wall proteome, Plant Physiol. Biochem. 42 (2004) 979–988.

[135] E. Ruiz-May, J.K. Rose, Progress toward the tomato fruit cell wall proteome, Front. Plant Sci. 4 (2013) 159.

[136] B. Printz, R. Dos Santos Morais, S. Wienkoop, K. Sergeant, S. Lutts, J.F. Hausman, J. Renaut, An improved protocol to study the plant cell wall proteome, Front. Plant Sci. 6 (2015) 237.

[137] A. Afroz, M. Zahur, N. Zeeshan, S. Komatsu, Plant-bacterium interactions analyzed by proteomics, Front. Plant Sci. 4 (2013) 21.

[138] R. Vanholme, B. Demedts, K. Morreel, J. Ralph, W. Boerjan, Lignin biosynthesis and structure, Plant Physiol. 153 (2010) 895–905.

[139] D. Mohnen, Pectin structure and biosynthesis, Curr. Opin. Plant Biol. 11 (2008) 266–277.

[140] M.H. Chen, J. Sheng, G. Hind, A.K. Handa, V. Citovsky, Interaction between the *Tobacco mosaic virus* movement protein and host cell pectin methylesterases is required for viral cell-to-cell movement, EMBO J. 19 (2000) 913–920.

[141] A.C. Schrotenboer, M.S. Allen, C.M. Malmstrom, Modification of native grasses for biofuel production may increase virus susceptibility, GCB Bioenergy 3 (2011) 360–374.

[142] S.P. Rodrigues, J.A. Ventura, C. Aguilar, E.S. Nakayasu, H. Choi, T.J.P. Sobreira, L.L. Nohara, L.S. Wermelinger, I.C. Almeida, R.B. Zingali, et al., Label-free quantitative proteomics reveals differentially regulated proteins in the latex of sticky diseased *Carica papaya* L. plants, J. Proteom. 75 (2012) 3191–3198.

[143] A. Niehl, Z.J. Zhang, M. Kuiper, S.C. Peck, M. Heinlein, Label-free quantitative proteomic analysis of systemic responses to local wounding and virus infection in *Arabidopsis thaliana*, J. Proteome. Res. 12 (2013) 2491–2503.

[144] G. Lopez-Casado, P.A. Covey, P.A. Bedinger, L.A. Mueller, T.W. Thannhauser, S. Zhang, Z. Fei, J.J. Giovannoni, J.K. Rose, Enabling proteomic studies with RNA-Seq: the proteome of tomato pollen as a test case, Proteomics 12 (2012) 761–774.

[145] Z. Huang, V.M. Andrianov, Y. Han, S.H. Howell, Identification of *Arabidopsis* proteins that interact with the *Cauliflower mosaic virus* (CaMV) movement protein, Plant Mol. Biol. 47 (2001) 663–675.

[146] T. Isaacson, C.M. Damasceno, R.S. Saravanan, Y. He, C. Catala, M. Saladie, J.K. Rose, Sample extraction techniques for enhanced proteomic analysis of plant tissues, Nat. Protoc. 1 (2006) 769–774.

[147] A. Mehta, B.S. Magalhães, D.S.L. Souza, E.A.R. Vasconcelos, L.P. Silva, M.F. Grossi-de-Sa, O.L. Franco, P.H.A. da Costa, T.L. Rocha, Rooteomics: the challenge of discovering plant defense-related proteins in roots, Curr. Protein Peptide Sci. 9 (2008) 108–116.

[148] I.M. Cristea, J.W. Carroll, M.P. Rout, C.M. Rice, B.T. Chait, M.R. MacDonald, Tracking and elucidating *Alphavirus*-host protein interactions, J. Biol. Chem. 281 (2006) 30269–30278.

[149] N.J. Moorman, R. Sharon-Friling, T. Shenk, I.M. Cristea, A targeted spatial-temporal proteomics approach implicates multiple cellular trafficking pathways in human *Cytomegalovirus* virion maturation, Mol. Cell. Proteom. 9 (2010) 851–860.

[150] D.L. Rowles, S.S. Terhune, I.M. Cristea, Discovery of host-viral protein complexes during infection, Methods Mol. Biol. 1064 (2013) 43–70.

[151] D.L. Rowles, Y.C. Tsai, T.M. Greco, A.E. Lin, M. Li, J. Yeh, I.M. Cristea, DNA methyltransferase DNMT3A associates with viral proteins and impacts HSV-1 infection, Proteomics 15 (2015) 1968–1982.

[152] G. Rodrigo, J. Carrera, V. Ruiz-Ferrer, F.J. del Toro, C. Llave, O. Voinnet, and S.F. Elena, Characterization of the *Arabidopsis thaliana* interactome targeted by viruses *Santa Fe Institute Working Paper* 11-10-049 (2011).

[153] J. Das, K.M. Gayvert, F. Bunea, M.H. Wegkamp, H. Yu, ENCAPP: elastic-net-based prognosis prediction and biomarker discovery for human cancers, BMC Genomics 16 (2015) 1–13.

13

A protein–protein interaction network linking the energy-sensor kinase SnRK1 to multiple signaling pathways in *Arabidopsis thaliana*☆

Madlen Nietzsche[a], Ramona Landgraf[a], Takayuki Tohge[b], Frederik Börnke[a,c,*]

[a] *Plant Metabolism Group, Leibniz-Institute of Vegetable and Ornamental Crops (IGZ), 14979 Großbeeren, Germany*
[b] *Max-Planck-Institute of Molecular Plant Physiology, Am Mühlenberg 1, D-14476 Potsdam-Golm, Germany*
[c] *Institute of Biochemistry and Biology, University of Potsdam, 14476 Potsdam, Germany*

ARTICLE INFO

Keywords:
Arabidopsis
SnRK1
Protein–protein interaction
Stress signaling

ABSTRACT

In plants, the sucrose non-fermenting (SNF1)-related protein kinase 1 (SnRK1) represents a central integrator of low energy signaling and acclimation towards many environmental stress responses. Although SnRK1 acts as a convergent point for many different environmental and metabolic signals to control growth and development, it is currently unknown how these many different signals could be translated into a cell-type or stimulus specific response since many components of SnRK1-regulated signaling pathways remain unidentified. Recently, we have demonstrated that proteins containing a domain of unknown function (DUF) 581 interact with the catalytic α subunits of SnRK1 (AKIN10/11) from *Arabidopsis thaliana* and could potentially act as mediators conferring tissue- and stimulus-type specific differences in SnRK1 regulation. To further extend the SnRK1 signaling network in plants, we systematically screened for novel DUF581 interaction partners using the yeast two-hybrid system. A deep and exhaustive screening identified 17 interacting partners for 10 of the DUF581 proteins tested. Many of these novel interaction partners are implicated in cellular processes previously associated with SnRK1 signaling. Furthermore, we mined publicly available interaction data to identify additional DUF581 interacting proteins. A protein–protein interaction network resulting from our studies suggests connections between SnRK1 signaling and other central signaling pathways involved in growth regulation and environmental responses. These include TOR and MAP-kinase signaling as well as hormonal pathways. The resulting protein–protein interaction network promises to be effective in generating hypotheses to study the precise mechanisms SnRK1 signaling on a functional level.

1. Introduction

Robustness is an inherent and essential property of all biological systems that enables to maintain phenotypic stability in the face of diverse perturbations arising from e.g., changing environmental conditions or genetic variation [1]. Due to their sessile lifestyle, plants have evolved a particular complex cellular signaling network that enables them to respond to a wide range of environmental signals in order to accommodate stress conditions leading to phenotypic robustness. Protein kinases are central to these signaling networks because through the phosphorylation of various substrates they relay extra and intracellular signals into a cellular response which could be e.g., transcriptional reprogramming, modulation of enzyme activity through phosphorylation and eventually mediating adjustment of the metabolic network to the novel conditions. In plants, the sucrose non-fermenting related kinase 1 (SnRK1) has emerged as a key regulator of cellular metabolism through activation of signaling cascades that are protective against various stresses. SnRK1 becomes activated by energy deprivation and abscisic (ABA) signals, and is inactivated by sugars that restore energy balance [2,3]. In addition, SnRK1 coordinates stress induced responses, including antiviral defense, and fundamental developmental processes, from germination and sprouting to reproduction and senescence [3]. Some of the known SnRK1 substrates are key metabolic enzymes such as sucrose phosphate synthase, nitrate reductase, and 3-hydroxy-3-methyl-glutaryl-CoA reductase [4]. Furthermore, activation of SnRK1 triggers extensive reprogramming of transcription, affecting over a thousand genes in *Arabidopsis*, that contributes to restoring homeostasis, promotes cell survival and long-term stress adaptation [2]. In general, SnRK1-mediated transcriptional reprogramming results in the

☆ This article is part of a special issue entitled "Protein networks – a driving force for discovery in plant science".
* Corresponding author at: Plant Metabolism Group, Leibniz-Institute of Vegetable and Ornamental Crops (IGZ), Theodor-Echtermeyer-Weg 1, 14979 Großbeeren, Germany.
E-mail address: boernke@igzev.de (F. Börnke).

down-regulation of energy consuming processes and the induction of catabolic pathways to provide alternative energy sources [5]. Although the broad effect on transcription and the specificity that is required to respond to a particular stress is likely to require modulation of a range of different downstream target proteins, only a few direct SnRK1 substrates have yet been identified [2,6–9]. In addition to the direct interactions between protein kinases and their substrates, sometimes the two proteins interact through the intermediacy of adaptors or scaffolds, which act as organizing platforms that recruit both the kinase and the substrate to the same complex [10]. We have previously demonstrated that proteins containing a domain of unknown function (DUF) 581 interact with the catalytic α subunits of SnRK1 (AKIN10/11) from *Arabidopsis thaliana* and could potentially act as mediators conferring tissue- and stimulus-type specific differences in SnRK1 regulation by recruitment of potential substrate proteins and the kinase into the same complex [11]. This hypothesis was based on the observation that expression of DUF581−containing proteins is highly responsive to hormones and various stress conditions and that one DUF581 protein tested had a common interaction partner with SnRK1 in the yeast two-hybrid system [11]. In the present study, we used the yeast two-hybrid system to assemble a comprehensive protein–protein interaction network comprising the two catalytic α subunits of *Arabidopsis* SnRK1, AKIN10 and AKIN11, and members of the DUF581-protein family with additional interaction partners that potentially could play a role in SnRK1 signaling in plants. The network architecture provides clues for connections between SnRK1 signaling and other central signaling pathways involved in growth regulation and environmental responses. These include TOR and MAP-kinase signaling as well as hormonal pathways and transcription factors. The resulting protein–protein interaction network promises to be effective in generating hypotheses to study the precise mechanisms of SnRK1 signaling in plants on a functional level.

2. Results and discussion

In order to construct a comprehensive SnRK1/DUF581-protein interaction network we conducted exhaustive yeast two-hybrid (Y2H) screenings of two different cDNA libraries from *Arabidopsis* using 16 of the 18 *Arabidopsis* DUF581-proteins as baits. Only DUF581-1 and DUF581-11 were excluded from the screen, the first because of auto-activation of the yeast reporter genes when fused to the GAL4 DNA binding domain and the latter because we were not able to amplify a corresponding cDNA fragment likely owing to its low expression level. To identify additional direct interaction partners of SnRK1 the two α subunits AKIN10 and AKIN11 were also used as baits. All potential DUF581-protein and AKIN10/11 interaction partners identified during the library screenings were retested in direct Y2H interaction assays including appropriate negative and positive controls. Furthermore, we mined publicly available protein–protein interaction data derived from high-throughput Y2H screens [12] to identify additional potential DUF581–protein interactors with a known function. These were then used in direct Y2H interaction assays to confirm their binding to a given DUF581-protein. Selected candidate interaction partners of individual DUF581-proteins identified during the library screens were also tested in direct assays against other DUF581-protein isoforms as well as against the SnRK1 subunits AKIN10 and AKIN11, respectively. Collectively, the interaction data reveal a highly interconnected network (41 nodes with 65 edges), with some proteins identified as interaction partners of two or more of the bait proteins (Fig. 1). For 6 of the DUF581-proteins (DUF581-1, -5, -6, -12, -13, -17) no novel interaction partners apart from SnRK1 could be identified. This could either be due to the absence of a potential

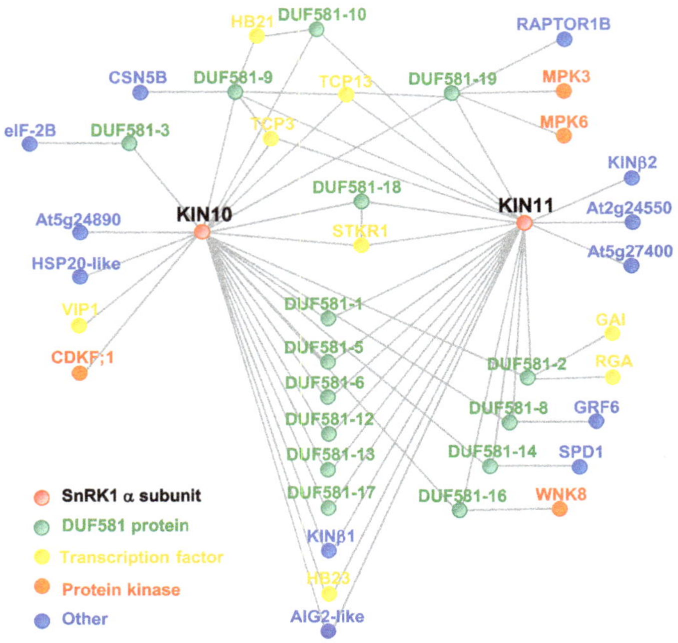

Fig. 1. *Arabidopsis* yeast two-hybrid protein–protein interaction network surrounding SnRK1 α subunits (KIN10/11) and DUF581-domain containing proteins. Nodes in the network represent proteins and are colored according to their functional class (see color key). The protein–protein interactions are indicated by lines ("edges"). The network was produced using Pajek software (http://vlado.fmf.uni-lj.si/pub/networks/pajek/). TAIR annotations of network nodes are listed in Supplementary Table S1.

interactor from the libraries used for the screening or because binding partners are not functional in the yeast assay (e.g., not localizing to the nucleus, integral membrane proteins). Three out of the 18 DUF581-proteins from *Arabidopsis* (DUF581-3, -8, -15) did not bind to AKIN10/11 when the SnRK1 subunits were used as bait proteins in direct yeast assays [11]. The screening identified AKIN10 as an interaction partner for BD-DUF581-3 and BD-DUF581-8 indicating that orientation of the BD/AD-fusion proteins in yeast can affect the outcome of the Y2H assay. For DUF581-15 no binding partners could be identified.

A comparison of the gene ontology annotations of the newly identified AKIN/DUF581-interacting proteins to the *Arabidopsis* genome indicated enrichment in these categories: nucleus, other cytoplasmic components, protein binding, kinase activity, transferase activity, and transcription factor activity (Supplementary Table S1). We have previously shown that several DUF581-proteins tested displayed a nucleo-cytoplasmic localization when transiently expressed as GFP fusion proteins in *Nicotiana benthamiana* and co-expression with SnRK1 shifted the fluorescence signal into sub-nuclear speckles [11]. Thus, the enrichment for interaction partners with a suspected nuclear localization can serve as a criterion for high-quality and biologically significant interactome data sets [13].

According to our hypothesis the DUF581 serves as a generic SnRK1 interaction domain while the non-conserved part of the DUF581-containing proteins mediates binding of additional partners specific for individual DUF581-protein isoforms. Thus, while all DUF581 proteins should interact with SnRK1, binding of additional proteins should display a certain degree of specificity. To test this, we analyzed the specificity of interaction between certain DUF581-proteins and some of the newly identified interactors. The experiment revealed that for instance TCP3 (TEOSINTE BRANCHED1, CYCLOIDEA, PROLIFERATING CELL FACTORS), which had been identified as an interaction partner of DUF581-9, did not

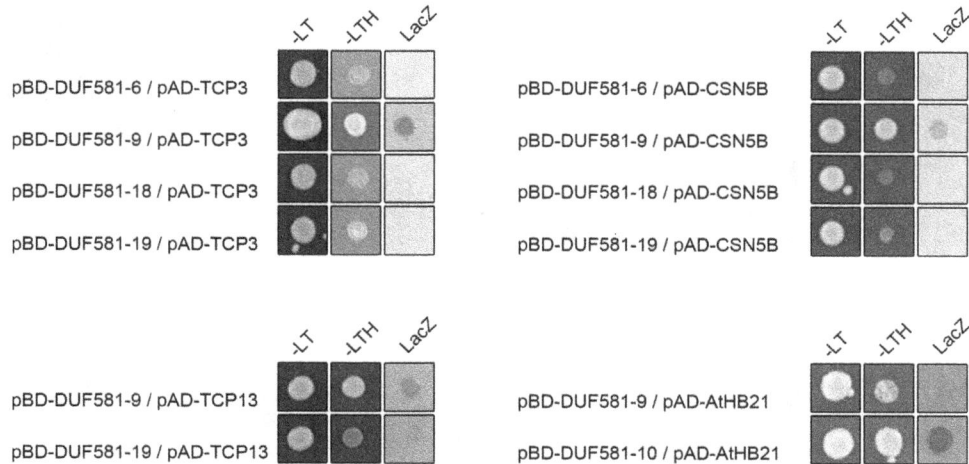

Fig. 2. Yeast two-hybrid assay for direct interaction between AKIN10/11 and DUF581-protein interaction partners. Cells were grown on selective media before a LacZ filter assay was performed. -LT, yeast growth on medium without Leu and Trp. -HLT, yeast growth on medium lacking His, Leu, and Trp, indicating expression of the HIS3 reporter gene. LacZ, activity of the lacZ reporter gene.

interact with DUF581-6, 18 or 19 in yeast, suggesting a high degree of specificity of the DUF581-9/TCP3 interaction (Fig. 2). Similarly, STOREKEEPER RELATED 1 (STKR1) a potential transcriptional regulator that was identified as a partner of DUF581-18, did not interact with DUF581-6, 9, and 19 in a direct Y2H assay [11]. Also CSN5B (COP9 signalosome 5B) specifically bound DUF581-9 but no other isoforms tested (Fig. 2). In contrast, TCP13 interacts with DUF581-9 and very weakly with DUF581-19 while homeobox-leucine zipper protein AtHB21 interacts strongly with DUF581-10 but weakly with DUF581-9 (Fig. 2). These, data suggest that the interaction of DUF581-proteins with partners other than SnRK1 displays a high degree of specificity of individual DUF581-protein isoforms, with only a few proteins interacting with more than one DUF581-protein.

2.1. In planta confirmation of selected protein–protein interactions using bimolecular fluorescence complementation

We applied the bimolecular fluorescence complementation (BiFC) approach to a subset of identified PPIs to assess whether the observed Y2H interactions occur in living plant tissues/cells. Transient co-expression of DUF581-2 fused to the C-terminal fragment of venus with GAI (gibberellic acid insensitive) or RGA (repressor of ga1-3), each fused to the N-terminal part of venus in leaves of N. benthamiana, yielded a fluorescence signal in discrete cellular regions corresponding to the nuclei (Fig. 3). In contrast, when GAI or RGA were co-expressed with DUF581-19-venus^{C155}, no signal was observed indicating that GAI and RGA specifically interact with DUF581-2 but not with DUF581-19 inside plant cells. Furthermore, the BiFC analyses confirmed an interaction of DUF581-19 with AKIN10 and AKIN11, respectively, within the plant cell nucleus (Fig. 3). Interestingly, a nucleo-cytoplasmic distribution of the fluorescence signal was observed when DUF581-19-venus^{C155} was combined either with MPK3-venus^{N173} or MPK6-venus^{N173} (Fig. 3). A combination of Venus^{N173}-FBPase with DUF581-19-venus^{C155} and VenusN^{173}-MPK3 or 6 with FBPase-Venus^{C155} induced no fluorescence and served as a negative control (Supplementary Fig. S1).

2.2. Some DUF581-protein binding partners also interact with SnRK1 directly

We have previously shown that STOREKEEPER RELATED 1 (STKR1), a potential transcriptional regulator that was identified in

a Y2H library screening using DUF581-18 as bait also interacts with SnRK1 in direct Y2H interaction assays [11]. This finding suggests that other DUF581-protein binding partners might also be direct interactors of the kinase subunits. In order to test this hypothesis randomly selected DUF581-protein interactors were tested for their ability to directly bind SnRK1 in yeast. As shown in Fig. 4A, the transcription factors TCP3 (TEOSINTE BRANCHED1, CYCLOIDEA, PROLIFERATING CELL FACTORS), TCP13, and AtHB21, which have initially been identified as interactors of different DUF581-proteins (Fig. 1), also interact with both SnRK1 subunits in direct assays. Interestingly, these proteins did not show up in unbiased library screenings using SnRK1 as bait, which might be due to the higher stringency of the library screening as compared to direct interaction assays. However, the protein kinases with no lysine 8 (WNK8) and MPK3, which have been identified as interaction partners of DUF581-16 and -19, respectively, do not interact with SnRK1 directly when tested in yeast (Fig. 4B).

2.3. The protein–protein interaction network links SnRK1 signaling to a range of previously unconnected cellular processes

The PPI network reveals many new leads to test a role of SnRK1 signaling in cellular function, some of which are known while others are new. A TAIR annotation of the proteins constituting the network is summarized in Supplementary Table S2. The potential significance of these novel interactions in relation to SnRK1 signaling is discussed below.

2.3.1. Protein kinases

DUF581-16 not only interacts with SnRK1 but also binds the protein kinase WNK8 (Fig. 1) and thus potentially enables cross-regulation between both kinases. WNK8 has been implicated in several aspects of plant development [14] and acts as a negative regulator of salt stress tolerance [15]. In addition, WNK8 is involved in glucose sensing via a G protein signaling pathway [16]. In response to glucose, the negative regulator of G protein signaling AtRGS1 becomes phosphorylated by WNK8 and subsequently undergoes endocytosis allowing initiation of down-stream glucose signaling. Thus, WNK8 acts as a positive regulator of glucose signaling which would be antagonistic to SnRK1 activation by glucose starvation. DUF581-16 expression is strongly induced by glucose and abiotic stress conditions [11] and hence could be involved in reciprocal regulation of both signaling pathways.

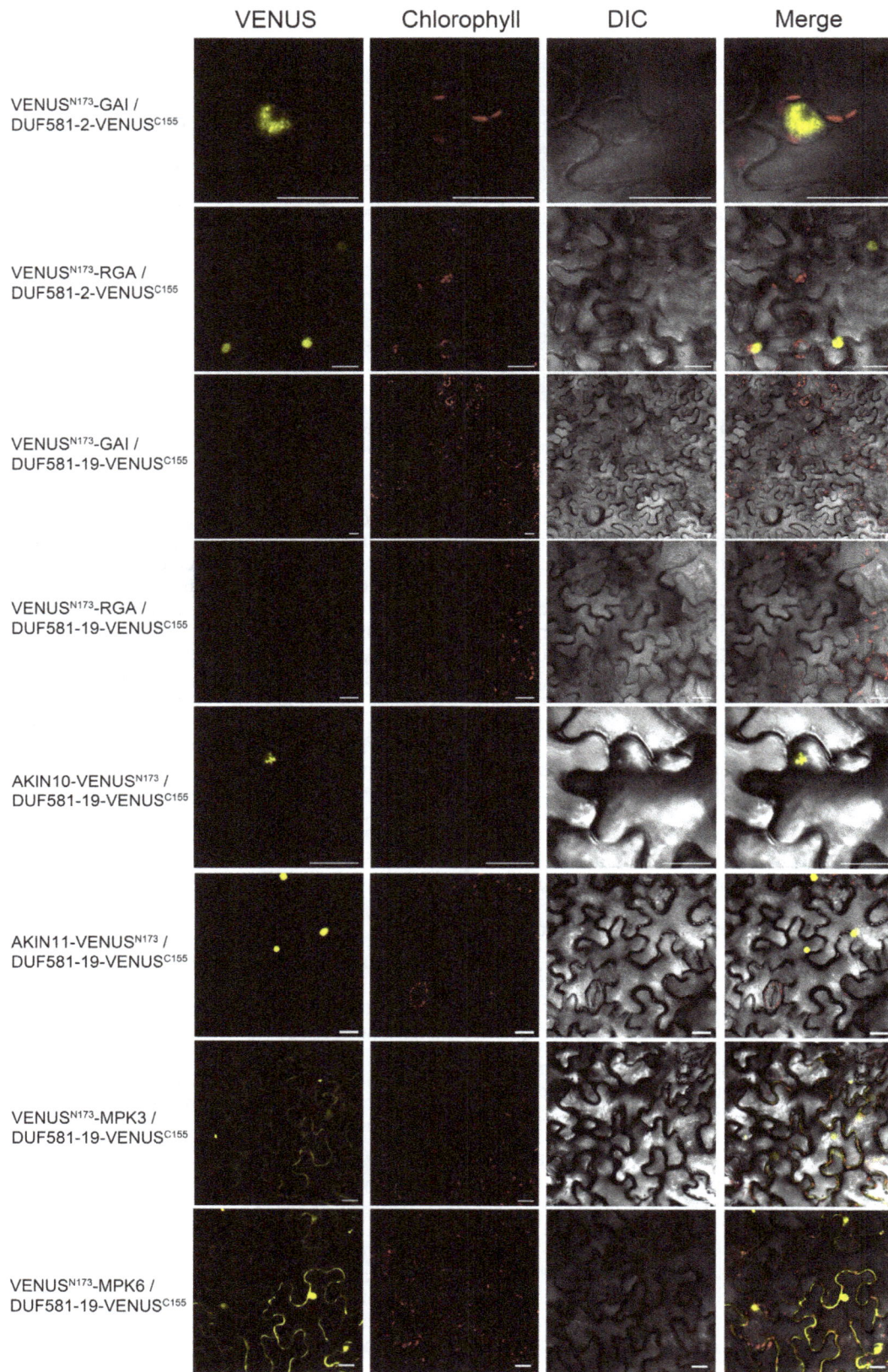

Fig. 3. *In planta* BiFC analysis of selected protein–protein interactions. Leaves of *N. benthamiana* were transiently transformed with mixtures of proteins pairs using *Agrobacterium*-infiltration. Bait and prey proteins were fused to the N- and C-terminal part of Venus, respectively. Fluorescence and localization were observed by confocal laser scanning microscopy. Bar represents 20 μm.

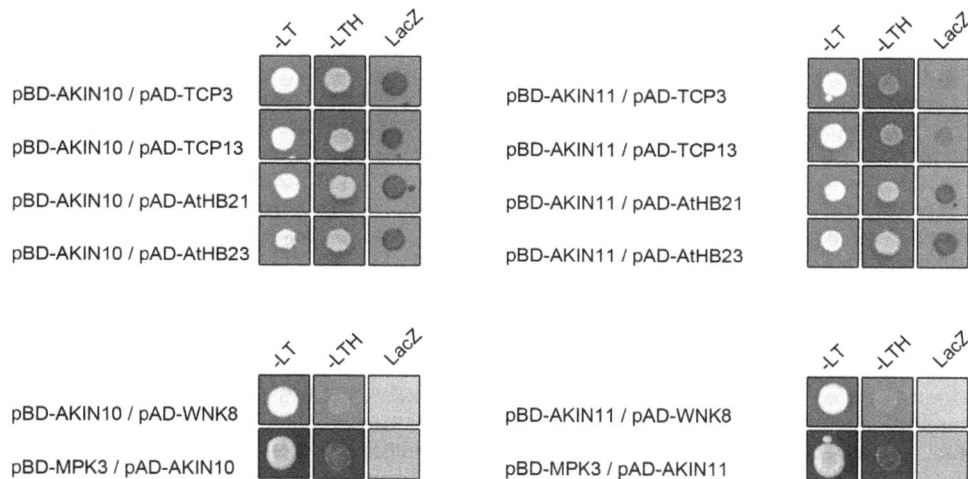

Fig. 4. Assay for direct interaction of DUF581-protein binding partners with AKIN10/11 in yeast. A, Transcription factors that have been identified as interaction partners of different DUF581-proteins also interact with AKIN10/11 in a yeast two-hybrid assay. B, The protein kinases WNK8 and MPK3 which have been shown to interact with DUF581-16 and 19, respectively, do not interact with AKIN10/11 in direct assays. Cells were grown on selective media before a LacZ filter assay was performed. -LT, yeast growth on medium without Leu and Trp. -HLT, yeast growth on medium lacking His, Leu, and Trp, indicating expression of the HIS3 reporter gene. LacZ, activity of the lacZ reporter gene.

A high-throughput yeast-two-hybrid screen that defined the *Arabidopsis* interactome identified the two mitogen-activated protein kinases, MPK3 and MPK6, as possible interactors for DUF581-19 [12]. We could confirm this interaction in direct yeast assays as well as *in planta* using BiFC (Figs. 1 and 3). MPK3 and MPK6 play critical roles in plant disease resistance by regulating multiple defense outputs and they become activated through phosphorylation by upstream kinases in response to bacterial and fungal pathogen-associated molecular patterns and ROS [17]. These induced defense responses represent a significant investment of resources for the production of secondary metabolites and antimicrobial proteins that would otherwise be utilized for growth and development. In order to prioritize resource allocation into defense responses upon pathogen attack growth needs to be halted. SnRK1 signaling acts as a negative regulator of growth and also releases alternative energy sources which could fuel into defense responses. Thus, MPK and SnRK1 signaling could be synergistically activated during defense for instance by *trans*-phosphorylation of SnRK1 by MPKs. An example for cross-regulation of MPK signaling with other kinase pathways has recently been described in rice where the calcium-dependent protein kinase 18 (CDPK18) was demonstrated to phosphorylate and activate MPK5, although the physiological consequences of this cross-activation remain unknown [18].

The CYCLIN-DEPENDENT KINASE F;1 (CDKF;1) was identified as an interaction partner of AKIN10 in a library screening using the SnRK1 subunit as bait. It was also found as an interactor of AKIN11 in the *Arabidopsis* interactome network [12]. Thus, it likely interacts with both SnRK1 isoforms. CDKF;1 is a plant specific CDK-activating kinase that has been shown to phosphorylate the T-loop of several CDKs, which themselves represent the main regulatory core of eukaryotic cell cycle. Indeed, a knockout mutant of CDKF;1 displayed severe defects in cell division, cell elongation and endoreduplication [19]. The homolog of SnRK1 in yeast, SNF1, has been shown to be involved in cell cycle regulation and over-expression of rye SnRK1 in yeast resulted in a dramatic reduction in yeast cell size, suggesting that the yeast cells were completing their cell cycles too early [20]. Thus, the interaction of SnRK1 with CDKF;1 could link energy signaling to cell cycle control.

2.3.2. Transcription factors

Activation of SnRK1 signaling is accompanied by a massive transcriptional reprogramming affecting the expression of approximately a thousand genes [2]. This implies the involvement of several transcriptional regulators in SnRK1 signaling. Previous studies have identified transcription factors from different families as potential components of the SnRK1 pathway. Direct interaction and phosphorylation of the B3-domain transcription factor FUSCA3 (FUS3), an essential regulator of seed maturation in *Arabidopsis*, by SnRK1 has been demonstrated [7]. In a cell-free system, AKIN10 positively regulates FUS3 stability, providing a possible mechanism of FUS3 regulation by SnRK1. Other TFs that have been shown to directly interact with both or either of the two SnRK1 α SUs in *Arabidopsis* include ATAF1 (*A. thaliana* activating factor 1), INDETERMINATE DOMAIN (IDD)-containing transcription factor IDD8, and PETAL LOSS (PTL) [21–23], although in these cases it remains unclear whether the TFs also serve a SnRK1 phosphorylation targets. Recently, ATAF1 was shown to trigger transcriptional responses largely overlap with expression patterns observed in plants starved for carbon or energy supply [24] lending additional support for the concept of ATAF1 acting as a component of SnRK1 signaling. Based on protoplast reporter assays members of the C/S1 group of basic leucine zipper (bZIP) TFs were proposed to be involved in low-energy signaling by SnRK1 [2]. A direct phosphorylation of one member of this TF family at different sites, namely bZIP63, by SnRK1 could recently be demonstrated [8]. Phosphorylation of these sites affects bZIP63's dimerization preferences with specific members of the bZIP family controlling its regulatory activity ultimately mediating changes in gene expression. Our own experiments identified the bZIP protein VIP1 as a direct interaction partner of AKIN10 (Fig. 1). VIP1 is involved in the process of *Agrobacterium tumefaciens* infection in *Arabidopsis* [25,26] and regulates the expression of *Agrobacterium*-responsive genes [27,28]. A Ser residue at position 79 (Ser-79) of VIP1 is phosphorylated by a MAPK, MPK3, which promotes the nuclear import of VIP1 [27]. Recently, VIP1 was also found to play a role during osmosensory signaling where it is shuttled to the nucleus independent of MAPK activation [29]. Given the direct interaction with SnRK1 it appears conceivable that VIP1 also phosphorylated by other kinases, including SnRK1, in response to changing turgor pressure.

Although we were not able to recapitulate the previously published SnRK1/TF interactions during Y2H screenings conducted in the present study, our protein–protein interaction network suggests the involvement of additional types of TFs in SnRK1 signaling that have not previously been implicated to act in this pathway. Two members from the plant specific TCP family of transcription factors namely TCP3 and TCP13 were identified as interaction partners of DUF581-9 and also directly interact with SnRK1 in yeast (Figs. 1 and 4A). Members of the TCP family have generally been implicated in the regulation of various aspects of plant development including growth, cell proliferation, and organ identity [30]. For instance, TCP3 has been shown to act redundantly with other members of the TCP family in regulating the expression of genes encoding cellular energy metabolism components, particularly those functioning in mitochondria [31]. However, regulation of TCP activity by phosphorylation has not been shown so far.

ATHB21 and ATHB23 are two members of a zinc finger-homeodomain (ZF-HD) TF subfamily which is represented by 14 genes in *Arabidopsis* [32]. The function of these TFs is generally unknown but they have been shown to homo- and heterodimerize which might contribute to greater selectivity in DNA binding while these proteins appear to lack an intrinsic activation domain [32]. Overexpression of one member, ATHB25, in *Arabidopsis* increases seed longevity likely by inducing the expression of a gibberellic acid biosynthetic enzyme and a resulting increase in GA levels [33]. ATHB21 was initially identified as an interaction partner of DUF581-10 but also interacts with AKIN10 and AKIN11 in direct Y2H assays (Fig. 4). Conversely, ATHB23 only interacted with both SnRK1 α SUs but not with DUF581-10 in yeast. This TF family has previously not been associated with SnRK1 signaling and it will be interesting to see whether other members will also interact with SnRK1 or DUF581 proteins and how this is functionally linked to SnRK1 signaling. Recently it has been shown that nuclear import and DNA binding activity of members of the ATHB TF family are negatively regulated through interaction with mini zinc finger (MIF) proteins [34]. It was proposed that the ATHB/MIF interaction is regulated by TF phosphorylation although this awaits further experimental confirmation.

Gibberellins (GAs) are a class of plant hormones modulating growth and development throughout the whole life cycle of the plant also in response to certain environmental stresses [35]. One of the major events during GA-mediated growth is the degradation of DELLA proteins, key negative regulators of GA signaling pathway that act immediately downstream of the GA receptor. The DELLA proteins are a subfamily of the plant-specific GRAS (for GAI, RGA and SCARECROW) family of regulatory proteins [36]. *Arabidopsis* contains 5 DELLA protein genes [RGA, GAI, RGA-Like1 (RGL1), RGL2 and RGL3]; each displays overlapping, but also some distinct functions in repressing GA responses. For example, GAI and RGA are important for stem elongation; RGL2 regulates seed germination; whereas RGA, RGL1, and RGL2 are involved in floral development. In the absence of GA DELLAs are localized in the nucleus where they interact with other transcription factors to inhibit the transcription of GA-responsive genes. Binding of GA to its soluble receptor, GID1, causes binding of GID1-GA to DELLAs and leads to their degradation via the ubiquitin-proteasome pathway and thus relieving their repressor activity on GA responsive genes. In addition to that, DELLA stability seems to be regulated by reversible protein phosphorylation and de-phosphorylation [37,38]. Recently, protein phosphatase TOPP4, a member of the protein phosphatase 1 (PP1) class was identified in *Arabidopsis* that de-phosphorylates DELLA proteins RGA and GAI, promoting the GA-induced destabilization of these two negative regulators [39]. The kinase responsible for DELLA phosphorylation has yet not been identified in *Arabidopsis* although in rice a casein kinase I, named early flowering 1 (EL1),

was identified and shown to stabilize the rice DELLA protein SLR1 by phosphorylation [37].

Inspection of the *Arabidopsis* interactome generated by high-throughput Y2H analysis [12] suggested that DUF581-2 interacts with GAI and thus might link the SnRK1 pathway to GA signaling. In order to investigate this in further detail, we tested the ability of DUF581-2 to interact with all five DELLA proteins in yeast. The analysis revealed that GAI and RGA bound DUF581-2, while RGL1, 2, and 3 did not show interaction (Supplementary Fig. S2). This finding was confirmed by an *in planta* BiFC assay, showing the interaction between DUF581-2 and the two DELLAs to occur within the nucleus (Fig. 3). Interestingly, a direct interaction test between SnRK1 and the DELLA proteins in yeast was negative (Supplementary Fig. S2), indicating that DUF581-2 could act as a mediator in SnRK1/DELLA interaction. In this hypothetical scenario, SnRK1 phosphorylation of GAI and RGA mediated by DUF581-2 would lead to DELLA protein stabilization and thus suppresses GA mediated growth responses involving GAI and RGA2 under conditions of activated SnRK1 signaling. This would fit to the proposition that SnRK1 acts as a negative regulator of growth.

2.3.3. Other regulatory proteins

SnRK1 is regarded as major energy-sensing kinase responding, once activated by various cellular stresses such as starvation or hypoxia, by enhancing catabolism and limiting anabolism, thus maintaining energy homeostasis [5]. Conversely, under energy rich conditions the TOR (target of rapamycin) kinase is globally activated and the TOR signaling pathway stimulates various energy-consuming cellular outputs, like mRNA translation or cell proliferation [40,41]. Thus, both pathways act antagonistically to each other to adjust growth and metabolism according to the energy level of the cell. TOR is a highly conserved nutrient-responsive regulator of cell growth found in all eukaryotes [40]. In mammals and yeast, TOR is found in two biochemically and functionally distinct signaling complexes [40], only one of which seems to be present in plants [42]. This heterotrimeric complex consists of the TOR kinase, the RAPTOR/KOG1 (Regulatory-associated protein of mTOR/Kontroller of growth 1) and LST8/GbetaL (Lethal with Sec13 8/Protein G beta subunit like). In mammals, it has been shown that the mammalian counterpart of SnRK1, the AMP-activated protein kinase (AMPK), inhibits cell cycle and growth in part through inhibition of the TOR pathway [43]. Upon activation of AMPK by nutrient loss, AMPK directly phosphorylates the TSC2 tumor suppress, which acts as an upstream regulator of mammalian TOR and its phosphorylation through AMPK eventually inhibits TOR activity. However, research has shown that the relationship between AMPK and inactivation of TOR is conserved across eukaryotes, including several that lack TSC2 orthologs such as *Caenorhabditis elegans* and *Saccharomyces cerevisiae*. This suggests that additional AMPK substrates may directly or indirectly modulate mTORC1 activity. It has recently been found that the critical mTOR binding partner RAPTOR is a direct substrate of AMPK in human cells, and that phosphorylation of RAPTOR by AMPK is required for suppression of TOR activity by energy stress [43]. Phosphorylation of raptor by AMPK occurs at a serine residue that is highly conserved across species, including plants. This opens the possibility that also in plants regulation of TOR signaling by SnRK1 could occur through the phosphorylation of RAPTOR. Although biochemical evidence for a direct interaction between RAPTOR and SnRK1 is yet lacking, high-throughput data suggested an interaction of RAPTOR1b with the SnRK1-interacting protein DUF581-19/MARD1 in yeast [12]. We could independently reproduce this interaction in a direct Y2H assay which tightly connects RAPTOR with AKIN10/11 via DUF581-19/MARD1 (Fig. 1). Hypothetically, DUF581-19/MARD1 could act as an adaptor protein that brings together several kinase signaling pathways allowing for their coordinated regulation.

14-3-3 proteins are a class of regulatory/effector proteins that modulate client protein function through phosphorylation dependent associations [44]. The *Arabidopsis* genome encodes 15 different 14-3-3 isoforms, 13 of which are transcriptionally expressed. Biochemical evidence suggests that several metabolic enzymes are altered in activity upon phosphorylation and subsequent 14-3-3 binding, including nitrate reductase and SPS (sucrose-phosphate synthase), which both have been proposed to be phosphorylated by SnRK1 [45]. On the contrary, 14-3-3 proteins appear to be phosphorylated themselves on multiple residues although the biochemical effects of this modification have remained elusive and only a few of the responsible kinases have been identified [46,47]. *Arabidopsis* 14-3-3 isoform λ was previously been shown to be phosphorylated by the somatic embryogenesis receptor-like kinase 1 (AtSERK1) [46] and was identified as an interacting protein of DUF581-8 in yeast (Fig. 1). DUF581-8 does not contain a predicted 14-3-3 binding site (data not shown) and thus it appears reasonable to assume that it could act as a scaffold protein to mediate the interaction between 14-3-3λ and SnRK1. This interaction could either serve to enable 14-3-3λ phosphorylation by the kinase or it could aid the binding of 14-3-3 target proteins that have previously been phosphorylated by SnRK1. This letter scenario is similar to what has recently been shown the *Nicotiana tabacum* Ca²⁺-dependent protein kinase NtCDPK1 which phosphorylates the transcription factor REPRESSION OF SHOOT GROWTH (RSG) to create a 14-3-3 binding site [48]. NtCDPK1 also binds a 14-3-3 protein that is transferred to RSG after phosphorylation and thus acts as a scaffolding kinase that is supposed to increase the specificity and efficiency of signaling by coupling catalysis with scaffolding on the same protein [49].

The COP9 signalosome (CSN) is a multiprotein complex that regulates the activity of CULLIN-RING E3 ubiquitin ligases (CRLs). CRLs ubiquitinate substrate proteins and thus target them for proteasomal degradation [50]. By controlling the activity of CRLs, the CSN integrates and fine-tunes a vast array of cellular processes including hormone signaling, the cell cycle, and regulation of growth, development, and defenses. The CRL CUL4 has been shown to be involved in proteasomal degradation of AKIN10 in a cell free assay and AKIN10 protein levels were increased in a *cul4* knock-out mutant [51]. Thus, the interaction of DUF581-9 with the CSN subunit CSN5b could be involved in controlling SnRK1 protein levels via the CSN. On the other hand, it has been shown that several protein kinases are associated with the CSN in mammalian cells which have been proposed to further regulate the ubiquitin-dependent degradation of various transcription factors [52]. Accordingly, the DUF581-9/CSN5b interaction could lead to the association of SnRK1 with the CSN to regulate SnRK1 target stability subsequent to phosphorylation.

We have previously shown that a SnRK1-GFP fusion protein shows a nuclear-cytoplasmic distribution in plant cells and co-expression with DUF581-containing proteins re-locates the SnRK1-GFP fluorescence exclusively to the nucleus [11]. This led to the hypothesis that DUF581-proteins are involved in the execution of SnRK1 signaling within the nucleus. In contrast, two proteins, named OsSKIN1 and OsSKIN2 (SnRK1 interacting negative regulators 1 and 1), have recently been identified as SnRK1 interaction partners in rice (*Oryza sativa*) and were shown to antagonize SnRK1 signaling at least partially by preventing SnRK1 nuclear localization under conditions of activated SnRK1 signaling [53]. A Y2H screening using *Arabidopsis* AKIN10 and 11 as bait has identified two previously uncharacterized proteins (At5g24890 for AKIN10 and At2g24550 for AKIN11) as SnRK1 interaction partners (Fig. 1) that show high similarity to the OsSKIN1 and 2 proteins. Future experiments will have to clarify whether these two *Arabidopsis* proteins play a similar role in SnRK1 regulation as their respective orthologs from rice.

2.3.4. Conclusions

We elaborated a protein–protein interaction network involving the two *Arabidopsis* SnRK1 subunits, AKIN10 and AKIN11, and their direct binding partners from the DUF581-protein family. The network complements and extends previous studies to identify components of SnRK1 signaling in plants using protein–protein interaction studies. Noticeably, using an unbiased Y2H library screening approach, we were not able to re-isolate interaction partners of *Arabidopsis* SnRK1 that have earlier been detected by different methods (Supplementary Table S3), except those representing non-catalytic SnRK1 subunits (AKINβ1 and AKINβ2; Fig. 1). In cases where these interaction partners have previously been isolated by a Y2H approach, this discrepancy is most likely due to technical differences between library preparations, vector systems and orientation of bait/prey pairs. However, we were able to confirm a range of SnRK1 and DUF581-protein interactions, respectively, which have previously been found during a partial pairwise plant interactome mapping effort (8000 by 8000 matrix; [12]). These establish possible links between SnRK1 signaling and MAPK and TOR signaling, respectively, as well as hormonal signaling by DELLA proteins. The observation that DUF581-proteins generally interact with the two SnRK1 subunits AKIN10 and 11 but display a high degree of specificity during secondary interaction reinforces the hypothesis that DUF581-containing proteins could serve as scaffolds to facilitate SnRK1—substrate interaction under specific cellular conditions [11]. Future experiments will have to clarify whether ternary complexes between SnRK1, DUF581-proteins and their interaction partners exist in plants and what the functional role of these complexes during SnRK1 signaling might be.

3. Materials and methods

3.1. Cloning

The construction of Y2H constructs to screen for DUF581-protein interaction partners has previously been described [11]. Plasmids containing the coding sequence of MPK3, MPK6, Raptor and the different DELLA proteins were constructed in an analogous manner using the primers listed in Supplementary Table S4. Constructs for bi-molecular complementation analysis are based on Gateway®-cloning compatible versions of pRB-C-VenusN¹⁷³ and pRB-C-VenusC¹⁵⁵.

3.2. Yeast two-hybrid

Yeast two-hybrid techniques were performed according to the yeast protocols handbook and the Matchmaker GAL4 Two- hybrid System 3 manual (both Clontech, Heidelberg, Germany) using the yeast reporter strains AH109 and Y187. The yeast strain Y187 carrying the bait construct was mated with AH109 cells pre-transformed with either a two-hybrid library from *Arabidopsis* inflorescence [54] (kindly provided by the *Arabidopsis* Biological Resource Center) or with a library derived from *Arabidopsis* source leaves (Vertis Biotechnology). Diploid cells were selected on medium lacking Leu, Trp, and His supplemented with 4 mM 3- aminotriazole. Cells growing on selective medium were further tested for activity of the *lacZ* reporter gene using filter lift assays. Library plasmids from *his3/lacZ* positive clones were isolated from yeast cells and transformed into *Escherichia coli* before sequencing of the cDNA inserts. All primary library isolates were reassessed with their respective baits and appropriate negative and positive controls in direct interaction assays. Direct interaction of two proteins was investigated by cotransformation of the respective plasmids in the yeast strain Y190, followed by selection of transformants on medium lacking Leu and Trp at 30 °C for 3 days and subsequent transfer to medium lacking Leu, Trp, and His for growth selection and *lacZ* activity testing of interacting clones.

3.3. Bimolecular fluorescence complementation (BiFC)

Constructs were transformed into *A. tumefaciens* C58C1 and transiently expressed by *Agrobacterium*-infiltration in *N. benthamiana*. The BiFC-induced YFP fluorescence was detected by CLSM (LSM510; Zeiss) after 48 hpi. The specimens were examined using the LD LCI Plan-Apochromat 253/0.8 water-immersion objective for detailed images with excitation using the argon laser (458- or 488-nm line for BiFC and chlorophyll autofluorescence). The emitted light passed the primary beam- splitting mirrors at 458/514 nm and was separated by a secondary beam splitter at 515 nm. Fluorescence was detected with filter sets as follows: on channel 3, 530–560 band pass; and on channel 1, for red autofluorescence of chlorophyll.

References

[1] J. Stelling, U. Sauer, Z. Szallasi, F.J. Doyle 3rd, J. Doyle, Robustness of cellular functions, Cell 118 (2004) 675–685.
[2] E. Baena-González, F. Rolland, J.M. Thevelein, J. Sheen, A central integrator of transcription networks in plant stress and energy signalling, Nature 448 (2007) 938–942.
[3] C. Polge, M. Thomas, SNF1/AMPK/SnRK1 kinases, global regulators at the heart of energy control? Trends Plant Sci. 12 (2007) 20–28.
[4] N.G. Halford, S.J. Hey, D. Jhurreea, S. Laurie, R.S. McKibbin, Y. Zhang, M.J. Paul, Dissection and manipulation of metabolic signalling pathways, Ann. appl. Biol. 142 (2003) 25–31.
[5] E. Baena-Gonzalez, J. Sheen, Convergent energy and stress signaling, Trends Plant Sci. 13 (2008) 474–482.
[6] J. Hanson, M. Hanssen, A. Wiese, M.M. Hendriks, S. Smeekens, The sucrose regulated transcription factor bZIP11 affects amino acid metabolism by regulating the expression of ASPARAGINE SYNTHETASE1 and PROLINE DEHYDROGENASE2, Plant J. 53 (2008) 935–949.
[7] A.Y. Tsai, S. Gazzarrini, AKIN10 and FUSCA3 interact to control lateral organ development and phase transitions in *Arabidopsis*, Plant J. 69 (2012) 809–821.
[8] A. Mair, et al., SnRK1-triggered switch of bZIP 63 dimerization mediates the low-energy response in plants, Elife (2015) 4.
[9] Y. Zhang, P.J. Andralojc, S.J. Hey, L.F. Primavesi, M. Specht, J. Koehler, M.A.J. Parry, N.G. Halford, *Arabidopsis* sucrose non-fermenting-1-related protein kinase-1 and calcium-dependent protein kinase phosphorylate conserved target sites in ABA response element binding proteins, Ann. Appl. Biol. 153 (2008) 401–409.
[10] R.P. Bhattacharyya, A. Remenyi, B.J. Yeh, W.A. Lim, Domains, motifs, and scaffolds: the role of modular interactions in the evolution and wiring of cell signaling circuits, Annu. Rev. Biochem. 75 (2006) 655–680.
[11] M. Nietzsche, I. Schiessl, F. Börnke, The complex becomes more complex: protein–protein interactions of SnRK1 with DUF 581 family proteins provide a framework for cell- and stimulus type-specific SnRK1 signaling in plants, Front. Plant Sci. (2014) 5.
[12] A.I.M. Consortium, Evidence for network evolution in an *Arabidopsis* interactome map, Science 333 (2011) 601–607.
[13] U. Stelzl, et al., A human protein–protein interaction network: a resource for annotating the proteome, Cell 122 (2005) 957–968.
[14] T. Tsuchiya, T. Eulgem, The *Arabidopsis* defense component EDM2 affects the floral transition in an FLC-dependent manner, Plant J. 62 (2010) 518–528.
[15] B. Zhang, K. Liu, Y. Zheng, Y. Wang, J. Wang, H. Liao, Disruption of AtWNK8 enhances tolerance of *Arabidopsis* to salt and osmotic stresses via modulating proline content and activities of catalase and peroxidase, Int. J. Mol. Sci. 14 (2013) 7032–7047.
[16] D. Urano, N. Phan, J.C. Jones, J. Yang, J. Huang, J. Grigston, J.P. Taylor, A.M. Jones, Endocytosis of the seven-transmembrane RGS1 protein activates G-protein-coupled signalling in *Arabidopsis*, Nat. Cell Biol. 14 (2012) 1079–1088.
[17] A. Pitzschke, A. Schikora, H. Hirt, MAPK cascade signalling in plant defence, Curr. Opin. Plant Biol. 12 (2009) 421–426.
[18] K. Xie, J. Chen, Q. Wang, Y. Yang, Direct phosphorylation and activation of a mitogen-activated protein kinase by a calcium-dependent protein kinase in rice, Plant Cell 26 (2014) 3077–3089.
[19] H. Takatsuka, R. Ohno, M. Umeda, The *Arabidopsis* cyclin-dependent kinase-activating kinase CDKF;1 is a major regulator of cell proliferation and cell expansion but is dispensable for CDKA activation, Plant J. 59 (2009) 475–487.
[20] J.R. Dickinson, D. Cole, N.G. Halford, A cell cycle role for a plant sucrose nonfermenting-1-related protein kinase (SnRK1) is indicated by expression in yeast, Plant Growth Regul. 28 (1999) 169–174.

[21] T. Kleinow, S. Himbert, B. Krenz, H. Jeske, C. Koncz, NAC domain transcription factor ATAF1 interacts with SNF1-related kinases and silencing of its subfamily causes severe developmental defects in *Arabidopsis*, Plant Sci. 177 (2009) 360–370.
[22] E.Y. Jeong, P.J. Seo, J.C. Woo, C.M. Park, AKIN10 delays flowering by inactivating IDD8 transcription factor through protein phosphorylation in *Arabidopsis*, BMC Plant Biol. 15 (2015) 110.
[23] M. O'Brien, R.N. Kaplan-Levy, T. Quon, P.G. Sappl, D.R. Smyth, PETAL LOSS, a trihelix transcription factor that represses growth in *Arabidopsis thaliana*, binds the energy-sensing SnRK1 kinase AKIN10, J. Exp. Bot. 66 (2015) 2475–2485.
[24] P. Garapati, R. Feil, J.E. Lunn, P. Van Dijck, S. Balazadeh, B. Mueller-Roeber, Transcription factor arabidopsis activating factor1 integrates carbon starvation responses with trehalose metabolism, Plant Physiol. 169 (2015) 379–390.
[25] T. Tzfira, M. Vaidya, V. Citovsky, VIP1, an *Arabidopsis* protein that interacts with Agrobacterium VirE2, is involved in VirE2 nuclear import and Agrobacterium infectivity, EMBO J. 20 (2001) 3596–3607.
[26] J. Li, A. Krichevsky, M. Vaidya, T. Tzfira, V. Citovsky, Uncoupling of the functions of the *Arabidopsis* VIP1 protein in transient and stable plant genetic transformation by *Agrobacterium*, Proc. Natl. Acad. Sci. U. S. A. 102 (2005) 5733–5738.
[27] A. Djamei, A. Pitzschke, H. Nakagami, I. Rajh, H. Hirt, Trojan horse strategy in *Agrobacterium* transformation: abusing MAPK defense signaling, Science 318 (2007) 453–456.
[28] A. Pitzschke, A. Djamei, M. Teige, H. Hirt, VIP1 response elements mediate mitogen-activated protein kinase 3-induced stress gene expression, Proc. Natl. Acad. Sci. U. S. A. 106 (2009) 18414–18419.
[29] D. Tsugama, S. Liu, T. Takano, A bZIP protein, VIP1, is a regulator of osmosensory signaling in *Arabidopsis*, Plant Physiol. 159 (2012) 144–155.
[30] P. Cubas, N. Lauter, J. Doebley, E. Coen, The TCP domain: a motif found in proteins regulating plant growth and development, Plant J. 18 (1999) 215–222.
[31] E. Giraud, et al., TCP transcription factors link the regulation of genes encoding mitochondrial proteins with the circadian clock in *Arabidopsis thaliana*, Plant Cell 22 (2010) 3921–3934.
[32] Q.K. Tan, V.F. Irish, The *Arabidopsis* zinc finger-homeodomain genes encode proteins with unique biochemical properties that are coordinately expressed during floral development, Plant Physiol. 140 (2006) 1095–1108.
[33] E. Bueso, et al., *ARABIDOPSIS THALIANA* HOMEOBOX25 uncovers a role for gibberellins in seed longevity, Plant Physiol. 164 (2014) 999–1010.
[34] S.Y. Hong, O.K. Kim, S.G. Kim, M.S. Yang, C.M. Park, Nuclear import and DNA binding of the ZHD5 transcription factor is modulated by a competitive peptide inhibitor in *Arabidopsis*, J. Biol. Chem. 286 (2011) 1659–1668.
[35] C.M. Fleet, T.P. Sun, A DELLAcate balance: the role of gibberellin in plant morphogenesis, Curr. Opin. Plant Biol. 8 (2005) 77–85.
[36] L.D. Pysh, J.W. Wysocka-Diller, C. Camilleri, D. Bouchez, P.N. Benfey, The GRAS gene family in *Arabidopsis*: sequence characterization and basic expression analysis of the SCARECROW-LIKE genes, Plant J. 18 (1999) 111–119.
[37] C. Dai, H.W. Xue, Rice early flowering1, a CKI, phosphorylates DELLA protein SLR1 to negatively regulate gibberellin signalling, EMBO J. 29 (2010) 1916–1927.
[38] F. Wang, et al., Biochemical insights on degradation of *Arabidopsis* DELLA proteins gained from a cell-free assay system, Plant Cell 21 (2009) 2378–2390.
[39] Q. Qin, W. Wang, X. Guo, J. Yue, Y. Huang, X. Xu, J. Li, S. Hou, *Arabidopsis* DELLA protein degradation is controlled by a type-one protein phosphatase, TOPP4, PLoS Genet. 10 (2014) e1004464.
[40] S. Wullschleger, R. Loewith, M.N. Hall, TOR Signaling in growth and metabolism, Cell 124 (2006) 471–484.
[41] Y. Xiong, J. Sheen, The role of target of rapamycin signaling networks in plant growth and metabolism, Plant Physiol. 164 (2014) 499–512.
[42] C. Robaglia, M. Thomas, C. Meyer, Sensing nutrient and energy status by SnRK1 and TOR kinases, Curr. Opin. Plant Biol. 15 (2012) 301–307.
[43] D.M. Gwinn, D.B. Shackelford, D.F. Egan, M.M. Mihaylova, A. Mery, D.S. Vasquez, B.E. Turk, R.J. Shaw, AMPK phosphorylation of raptor mediates a metabolic checkpoint, Mol. Cell 30 (2008) 214–226.
[44] C. Oecking, N. Jaspert, Plant 14-3-3 proteins catch up with their mammalian orthologs, Curr. Opin. Plant Biol. 12 (2009) 760–765.
[45] N.G. Halford, S.J. Hey, Snf1-related protein kinases (SnRKs) act within an intricate network that links metabolic and stress signalling in plants, Biochem. J. 419 (2009) 247–259.
[46] I.M. Rienties, J. Vink, J.W. Borst, E. Russinova, S.C. de Vries, The *Arabidopsis* SERK1 protein interacts with the AAA-ATPase AtCDC48, the 14-3-3 protein GF14lambda and the PP2C phosphatase KAPP, Planta 221 (2005) 394–405.
[47] K.N. Swatek, R.S. Wilson, N. Ahsan, R.L Tritz, J.J. Thelen, Multisite phosphorylation of 14-3-3 proteins by calcium-dependent protein kinases, Biochem. J. 459 (2014) 15–25.
[48] D. Igarashi, S. Ishida, J. Fukazawa, Y. Takahashi, 14-3-3 proteins regulate intracellular localization of the bZIP transcriptional activator RSG, Plant Cell 13 (2001) 2483–2497.
[49] T. Ito, M. Nakata, J. Fukazawa, S. Ishida, Y. Takahashi, Scaffold function of Ca²⁺-dependent protein kinase: tobacco Ca²⁺-DEPENDENT PROTEIN KINASE1 Transfers 14-3-3 to the substrate REPRESSION OF SHOOT GROWTH after phosphorylation, Plant Physiol. 165 (2014) 1737–1750.

[50] J.W. Stratmann, G. Gusmaroli, Many jobs for one good cop—the COP9 signalosome guards development and defense, Plant Sci. 185–186 (2012) 50–64.

[51] J.H. Lee, et al., Characterization of *Arabidopsis* and rice DWD proteins and their roles as substrate receptors for CUL4-RING E3 ubiquitin ligases, Plant Cell 20 (2008) 152–167.

[52] O. Harari-Steinberg, D.A. Chamovitz, The COP9 signalosome: mediating between kinase signaling and protein degradation, Curr. Protein Pept. Sci. 5 (2004) 185–189.

[53] C.R. Lin, et al., SnRK1A-interacting negative regulators modulate the nutrient starvation signaling sensor SnRK1 in source-sink communication in cereal seedlings under abiotic stress, Plant Cell 26 (2014) 808–827.

[54] H.Y. Fan, Y. Hu, M. Tudor, H. Ma, Specific interactions between the K domains of AG and AGLs, members of the MADS domain family of DNA binding proteins, Plant J. 12 (1997) 999–1010.

14

From the proteomic point of view: Integration of adaptive changes to iron deficiency in plants☆

Hans-Jörg Mai[a], Petra Bauer[a,b,*]

[a] Institute of Botany, Heinrich Heine University Düsseldorf, Universitätsstraße 1, Building 26.13, 02.36, 40225 Düsseldorf, Germany
[b] CEPLAS Cluster of Excellence on Plant Sciences, Heinrich Heine University Düsseldorf, Düsseldorf, Germany

ARTICLE INFO

Keywords:
Arabidopsis
Iron
Proteome
Post-transcriptional regulation

ABSTRACT

Knowledge about the proteomic adaptations to iron deficiency in plants may contribute to find possible new research targets in order to generate crop plants that are more tolerant to iron deficiency, to increase the iron content or to enhance the bioavailability of iron in food plants. We provide this update on adaptations to iron deficiency from the proteomic standpoint. We have mined the data and compared ten studies on iron deficiency-related proteomic changes in six different Strategy I plant species. We summarize these results and point out common iron deficiency-induced alterations of important biochemical pathways based on the data provided by these publications, deliver explanations on the possible benefits that arise from these adaptations in iron-deficient plants and present a concluding model of these adaptations. Furthermore, we demonstrate the close interdependence of proteins which were found regulated across multiple studies, and we pinpoint proteins with yet unknown function, which may play important roles in iron homeostasis.

1. Introduction

As a micronutrient, iron plays important roles in the biochemistry in animals and plants, and iron deficiency belongs to the most prevalent nutritional disorders in humans [1]. Despite the high soil iron content, iron deficiency also frequently occurs in plants because only a small portion of iron is bioavailable. In the high pH environment of calcareous soils Fe^{3+} is nearly insoluble which ultimately leads to iron deficiency in plant species that are not adapted [2]. Although essential for plants and other organisms, iron is potentially toxic [2–4] and excess iron may also cause yield loss in crop plants [5]. Therefore iron homeostasis in plants is tightly regulated.

As a result of iron deficiency, plants adapt and induce or repress a number of iron-related gene and protein functions. A primary response to iron deficiency is that plants stimulate iron acquisition via two distinct strategies. Strategy II used by grasses is based on the secretion of phytosiderophores that form complexes with Fe^{3+}. The complexes are transported into the root cells by Fe^{3+}-phytosiderophore transporters [6]. Iron acquisition in Strategy I plants, which are all other studied plants, comprises three basic steps [6–8]: solubilization of Fe^{3+} by rhizosphere acidification by a P-type H^+-ATPase [9,10], reduction of Fe^{3+} to Fe^{2+} by the ferric chelate reductase FRO2 (FERRIC REDUCTION OXIDASE 2) [11,12] and uptake of Fe^{2+} into the root cells by the divalent metal transporter IRT1 (IRON-REGULATED TRANSPORTER 1) [13–16]. In Arabidopsis, expression of these three genes is regulated in an iron-dependent manner by the bHLH transcription factor FIT [17–19]. *fit* knock-out mutants display a severe phenotype of iron starvation with reduced growth and chlorosis and they are lethal at the seedling stage [17,18]. The iron deficiency response is also accompanied by accumulation of organic acids such as citrate which serve as chelators for enhancing iron mobilization. Root-to-shoot organic acid transport is also discussed to provide a carbon source to keep basic mechanisms such as respiration going in iron-deficient leaves [20]. Phenylpropanoids and riboflavin are synthesized in iron-deficient roots and, depending on the species, may contribute to enhanced iron uptake and mobilization [21–29]. Since induced iron acquisition also facilitates uptake of other transition metals [30–32], protective mechanisms against heavy metal stress are switched on [33,34]. Furthermore, iron can be reversibly interchanged between the ferrous (Fe^{2+}) and ferric (Fe^{3+}) state which enables it to function as a cofactor of oxidoreductases or oxygenases most of which are heme or iron–sulfur proteins [35]. Besides iron–sulfur and heme proteins there is a third iron-containing class

---

☆ This article is part of a special issue entitled "Protein networks – a driving force for discovery in plant science".

* Corresponding author at: Institute of Botany, Heinrich Heine University Düsseldorf, Universitätsstraße 1, Building 26.13, 02.36, 40225 Düsseldorf, Germany.

E-mail address: petra.bauer@hhu.de (P. Bauer).

of proteins referred to as mono- and dinuclear non-heme iron proteins [36]. Aside from enzymes that can be found in all kingdoms, plant-specific iron–sulfur or heme proteins are involved in the synthesis of many compounds such as phenylpropanoids [37], oxylipins [38], gibberellins [39] or brassinosteroids [40,41] and many other primary and secondary metabolites. Iron is also needed by enzymes involved in nutrient assimilation such as nitrate and nitrite reductase or sulfite reductase that require heme, siroheme and [4Fe–4S] cluster cofactors [42–44]. Given the multitude of processes that directly or indirectly depend on iron, expression of a multitude of genes and proteins acting in different metabolic and regulatory pathways is adapted as plants encounter iron deficiency.

Large scale transcriptomic analysis based on DNA microarrays or RNA shotgun sequencing (RNA-Seq) are tools to gain overview of the plant's global responses to physiological stimuli. Gene expression responses could be divided into clusters of co-expressed genes [19], and indeed separate transcription factor cascades are known to regulate these clusters [45–54]. However, due to post-transcriptional regulation, protein stability and turnover and post-translational modifications, protein expression often does not correlate with transcript levels and this has also been shown for *Arabidopsis thaliana* with regard to iron homeostasis [22,55]. Indeed, AHA2 and FRO2 activity [10,11] as well as IRT1 subcellular location and turnover are post-translationally controlled [56–59]. Also the activity of transcription factors is controlled at the protein level [60]. Therefore investigation of the proteome is very crucial to uncover the components for plant iron adaptation.

For this review we used the data of 10 previously published comparative proteomic studies. These studies were performed with six different plant species, namely *Beta vulgaris* [61], *Cucumis sativus* [62], *Medicago truncatula* [63], *Prunus dulcis* × *Prunus persica* [64], *Solanum lycopersicum* [65] and *A. thaliana* [22,55,66–68]. In order to pinpoint the most consistent proteomic adaptations to iron deficiency across the different studies and species we focused on proteins whose abundance was found significantly changed upon iron deficiency across multiple studies. These include proteins that were similarly regulated in different species as well as proteins that were found regulated in Arabidopsis in different studies. The comparative conditions in each study included either + and − iron and/or WT versus an iron regulation-deficient Arabidopsis *fit* mutant or the ortholog tomato *fer* mutant. We also looked at proteins that were identified within one study in more than one protein spot if the 2-DGE (two-dimensional gel electrophoresis) approach was used. Functional classification and the generation of interaction networks were done by using publicly available tools [69–73].

2. Overview of the responses to iron deficiency

For a general overview on the physiological responses to iron deficiency and the regulation of these responses see the reviews by Colangelo and Guerinot [74], Ivanov et al. [19] and Brumbarova et al. [75]. Here, we first want to provide a brief overview on the proteomic adaptations.

In tomato roots, the proteomic responses to iron deficiency in wild-type and *fer* mutants [65,76] are predominantly related to stress response, miscellaneous peroxidases and redox regulation and among the proteins with differentially altered abundance there are iron-dependent or heme proteins [65]. Proteins regulated in an iron-dependent manner in tomato roots have been assigned to functions in starch metabolism, TCA cycle, amino acid metabolism and protein folding [76]. In *C. sativus* L. roots, the most iron-dependently regulated proteins belong to the glycolytic pathway, N and C metabolism, and a portion is related to stress [62]. In *B. vulgaris*, the most prominent changes upon iron deficiency are involved in carbohydrate catabolism, glycol-

ysis and TCA cycle and to a lesser extent to amino acid and N metabolism as well as riboflavin biosynthesis. The metabolic profiling showed that riboflavin, FAD and riboflavin sulfates were present in yellow root zones [61]. In *M. truncatula*, adaptations to iron deficiency include protein metabolism, N metabolism, glycolysis, TCA cycle, secondary metabolism and stress [63]. *P. dulcis* × *P. persica* roots show changes in redox homeostasis, ascorbic acid synthesis, pathogen response, protein folding, glycolysis, fermentation, electron transport and secondary metabolism [64]. In Arabidopsis roots, TCA cycle, glycolysis, amino acid metabolism and minor CHO metabolism were observed most regulated [55]. In shoots of iron-deficient Arabidopsis plants, regulated protein functions are rather related to protein processing, folding, sorting and degradation and glutathione metabolism [68]. In the iTRAQ-based analyses, the most prominent changes in roots upon iron deficiency have been observed for proteins involved in S-adenosyl methionine (SAM) and nicotianamine (NA) synthesis. Furthermore, proteins involved in the phenylpropanoid pathway were found up-regulated upon iron deficiency [22]. Additionally, RNA and mRNA processing proteins are up-regulated [22]. In the phosphoproteome [66] RNA processing and ribosomal proteins are enriched in iron-deficient roots. Both indicates that post-transcriptional processing and translation may be altered upon iron deficiency in roots.

Taken together, the most commonly regulated functional categories among the plant species investigated in the ten compared studies are C and N metabolism, glycolysis and TCA cycle. Although these previously conducted proteomic studies display results with a very low to no overlap in single proteins regulated under different iron conditions there are still clearly recognizable parallels concerning the regulated functional categories with which the respectively regulated proteins are associated. For more information on the general root and thylakoid proteome responses also see Lopez-Millan et al. [77].

3. Carbon metabolism

3.1. Glycolysis

Accumulation of a number of glycolysis-associated proteins has been observed to be iron-dependent in proteomic studies across the different species. Thus, alteration of glycolysis appears to be an important mechanism for the adaptation of plants to iron deficiency. One of the changes in proteins that were found regulated in at least two studies is related to glycolysis and gluconeogenesis. Two carbohydrate metabolism-related enzymes, namely ENO2 (ENOLASE 2; AT2G36530) and FBA8 (FRUCTOSE BISPHOSPHATE ALDOLASE 8; AT3G52930) were found up-regulated in four of the ten studies (Table 1). ENO2 was found regulated in *P. dulcis* × *persica* [64], *M. truncatula* [63], *B. vulgaris* [61] and *C. sativus* roots [62]. FBA8 was found regulated twice in *A. thaliana* [55,66] and once in *M. truncatula* [63] and *S. lycopersicum* [65], respectively. ENO2 is involved in pyruvate formation. FBAs also act in glycolysis catalyzing the DHAP and BPG forming reaction but are also involved in other pathways. There are two more FBAs whose abundance was found changed similarly across multiple studies: FBA6 (FRUCTOSE-BISPHOSPHATE ALDOLASE 6; AT2G36460) in *C. sativus* [62], *S. lycopersicum* [65] and *M. truncatula* [63] and FBA4 (FRUCTOSE BISPHOSPHATE ALDOLASE 4; AT5G03690) in *A. thaliana* [22] and *B. vulgaris* [61]. Enzymes acting in the glycolytic pathway downstream of the FBAs participating in the C3 compound processing steps were also found up-regulated in at least two of the examined studies. PGK (PHOSPHOGLYCERATE KINASE; AT1G79550) is up-regulated in *B. vulgaris* [61] and *C. sativus* roots [62], IPGAM1 (2,3-biphosphoglycerate-independent phosphoglycerate mutase 1; AT1G09780) in *C. sativus*

Table 1
Proteins which were significantly altered under iron deficiency across multiple studies.

1	2	3	4	5	6	7	8	9	10	AGI Code	Symbol	Metabolic Process
▓									▓	AT3G55440	TPI (TRIOSEPHOSPHATE ISOMERASE)	Glycolysis / Gluconeogenesis
				▓	▓				▓	AT2G36530	ENO2 (ENOLASE 2)	
				▓	▓					AT1G79550	PGK (PHOSPHOGLYCERATE KINASE)	
	▓							▓		AT1G09780	IPGAM1 (2,3-BIPHOSPHOGLYCERATE-INDEPENDENT PHOSPHOGLYCERATE MUTASE 1)	
				▓		▓				AT3G08590	IPGAM2 (2,3-BIPHOSPHOGLYCERATE-INDEPENDENT PHOSPHOGLYCERATE MUTASE 2)	
▓										AT1G77120	ADH1 (ALCOHOL DEHYDROGENASE 1)	
					▓				▓	AT5G03690	FBA4 (FRUCTOSE-BISPHOSPHATE ALDOLASE 4)	Glycolysis/Gluconeogenesis Pentose phosphate pathway
					▓		▓			AT2G36460	FBA6 (FRUCTOSE-BISPHOSPHATE ALDOLASE 6)	
					▓					AT3G52930	FBA8 (FRUCTOSE-BISPHOSPHATE ALDOLASE 8)	
						▓				AT1G04410	C-NAD-MDH1 (CYTOSOLIC-NAD-DEPENDENT MALATE DEHYDROGENASE 1)	Citrate cycle (TCA cycle)
▓										AT1G02500	MAT1 (METHIONINE ADENOSYLTRANSFERASE 1)*	Cysteine and methionine metabolism
					▓					AT2G36880	MAT3 (METHIONINE ADENOSYLTRANSFERASE 3)	
▓										AT3G17390	MAT4 (METHIONINE ADENOSYLTRANSFERASE 4)	
				▓						AT2G15620	NIR1 (NITRITE REDUCTASE 1)	Nitrogen metabolism
		▓								AT5G37600	GLN1;1 (ARABIDOPSIS GLUTAMINE SYNTHASE 1;1)	
▓										AT2G44050	COS1 (COI1 SUPPRESSOR1)	Riboflavin metabolism
▓										AT3G13610	Fe(II)- and 2-oxoglutarate-dependent dioxygenase family gene F6'H1*	Phenylpropanoid biosynthesis
▓									▓	AT4G34050	CCOAOMT1 (CAFFEOYL COENZYME A O-METHYLTRANSFERASE 1)	
						▓				AT4G35090	CAT2 (CATALASE 2)	Tryptophan metabolism
					▓					AT5G06720	PA2 (PEROXIDASE 2)	peroxidase
		▓								AT3G10920	MSD1 (MANGANESE SUPEROXIDE DISMUTASE 1)	Peroxisome
▓										AT1G02930	GST1 (ARABIDOPSIS GLUTATHIONE S-TRANSFERASE 1)*	Glutathione metabolism
						▓				AT3G52880	MDAR1 (MONODEHYDROASCORBATE REDUCTASE 1)	Ascorbate and aldarate metabolism
		▓								AT1G17745	PGDH2 (PHOSPHOGLYCERATE DEHYDROGENASE 2)*	Gly, Ser and Thr metabolism
						▓				AT5G64570	XYL4 (BETA-D-XYLOSIDASE 4)	Starch and sucrose metabolism
	▓									AT1G04820	TUA4 (TUBULIN ALPHA-4 CHAIN)	Phagosome
▓										AT1G13950	ELF5A-1 (EUKARYOTIC ELONGATION FACTOR 5A-1)	translation initiation factor
▓										AT3G60820	Encodes 20S proteasome beta subunit (PBF1)	Proteasome
								▓		AT4G27270	Q-reductase (Quinone reductase family protein)	Ubiquinone and other terpenoid-quinone biosynthesis
	▓									AT4G39090	RD19 (RESPONSIVE TO DEHYDRATION 19)	Plant-pathogen interaction
▓										AT5G37510	Encodes a subunit of the 400 kDa subcomplex of the mitochondrial NADH dehydrogenase (complex I)	Oxidative phosphorylation
▓										AT5G42020	BIP (LUMINAL BINDING PROTEIN)	Protein export
▓										AT1G18970	GLP4 (GERMIN-LIKE PROTEIN 4)*	n/a
	▓									AT2G21660	CCR2 (COLD, CIRCADIAN RHYTHM, AND RNA BINDING 2)	n/a
		▓								AT3G07720	Galactose oxidase/kelch repeat superfamily protein*	n/a
▓										AT3G12900	2-oxoglutarate (2OG) and Fe(II)-dependent oxygenase superfamily protein*	n/a
		▓								AT3G48890	MAPR3 (MEMBRANE-ASSOCIATED PROGESTERONE BINDING PROTEIN 3)*	n/a
▓										AT4G11650	OSM34 (OSMOTIN 34)	n/a
▓										AT5G19510	Translation elongation factor EF1B/ribosomal protein S6 family protein*	n/a

The gray boxes indicate that the proteins were regulated similarly in these compared studies. The numbers in the upper row indicate the study and species in which the protein was found regulated. 1: Mai et al. [55], *Arabidopsis thaliana* root; 2: Lan et al. [22], *Arabidopsis thaliana* root; 3: Lan et al. [66], *Arabidopsis thaliana* root; 4: Zargar et al. [68], *Arabidopsis thaliana* shoot; 5: Sudre et al. [67], *Arabidopsis thaliana* flowers; 6: Rodriguez-Celma et al. [64], *Prunus dulcis* × *Prunus persica* root; 7: Rodriguez-Celma et al. [63], *Medicago truncatula* root; 8. Brumbarova et al. [65], *Solanum lycopersicum* root; 9: Donnini et al. [62], *Cucumis sativus* root; 10: Rellan-Alvarez et al. [61], *Beta vulgaris* root. An asterisk at the end of the symbol/description indicates that the protein was found regulated only in Arabidopsis. We used the Arabidopsis AGI codes and protein annotations from the TAIR database [110] as common base. If the study had been conducted with *A. thaliana*, the AGI codes were obtained directly from the respective study. In a plant species other than Arabidopsis we searched for the closest Arabidopsis orthologs of the proteins whose accumulation changed upon iron deficiency. We retrieved the protein sequence from the NCBI protein database using the protein identifiers provided in the respective publications and performed searches with NCBI BLAST [129] with *Arabidopsis thaliana* (taxid:3702) specified as the organism and retained the result with the lowest E value as the closest homolog. If the lowest E value was above 1E-20 the respective search for an ortholog was considered without result. Proteins identified this way and taken into account for analysis are listed in the Supplementary Dataset. Functional classification was done with the KEGG database [69]. If not otherwise noted some single protein functions were specified using the MetaCyc database [71] or the UniProtKB database [72].

roots [62] and the *A. thaliana* root phosphoproteome [66] and IPGAM2 (2,3-BIPHOSPHOGLYCERATE-INDEPENDENT PHOSPHO-GLYCERATE MUTASE 2; AT3G08590) in *A. thaliana* flowers [67] and *M. truncatula* roots [63]. Among the proteins which similarly changed their abundance in at least two studies, the phosphoglycerate mutases IPGAM1 (2,3-BIPHOSPHOGLYCERATE-INDEPENDENT PHOSPHOGLYCERATE MUTASE 1; AT1G09780) and IPGAM2 (2,3-BIPHOSPHOGLYCERATE-INDEPENDENT PHOSPHO-GLYCERATE MUTASE 2; AT3G08590) are up-regulated under iron deficiency in *C. sativus* roots [62] and in the phosphoproteome in iron-deficient *A. thaliana* roots [22], and in *A. thaliana* flowers [67] and *M. truncatula* roots [63], respectively. The triose phosphate isomerase TPI (AT3G55440) is up-regulated under iron deficiency in *B. vulgaris* [61] and down-regulated in the *fit-3* mutant compared to wild-type in *A. thaliana* irrespective of the iron supply [55].

Taken together, five distinct studies report significant increase of glycolysis-associated proteins in *S. lycopersicum* [76], *C. sativus*

[62], *B. vulgaris* [61], *M. truncatula* [63] and *P. dulcis* × *P. persica* roots [64], respectively. Proteins that were found differentially regulated depending on the iron status in at least two studies also comprise six glycolysis-associated enzymes in *A. thaliana* roots [22,55,66] and flowers [67]. In total, 9 of the 39 proteins that were found similarly regulated in at least two independent studies were associated with glycolysis and gluconeogenesis. It has been suggested that the glycolytic pathway is increased under iron deficiency because of an increased demand for ATP and reducing equivalents to drive the iron uptake machinery, and organic acids in iron deficient roots [77]. This would possibly also require downstream pathways to be adapted accordingly.

3.2. TCA cycle

The tricarboxylic acid (TCA) cycle is a central metabolic pathway that provides energy and reducing equivalents as well as organic acids which can be moved to other catabolic and anabolic pathways or serve other purposes. With regard to iron deficiency, organic acids such as citrate turned out to be important compounds, e.g. for iron mobilization, especially in the xylem root-to-shoot iron transport, and the MATE transporter FRD3 is responsible for loading citrate into the xylem in *A. thaliana* [78–81]. Way over 90% of the xylem sap iron is estimated to be complexed with citrate [82] and an iron–citrate complex was identified in the xylem sap. Under iron deficiency the citrate concentration in *S. lycopersicum* xylem sap and root tips and in *B. vulgaris* xylem sap increases dramatically to enhance root-to-shoot iron mobilization [83,84] and in iron deficient Arabidopsis *nas4x-2* mutants the citrate concentration in the xylem sap is also increased under normal iron supply compared to wild-type [85]. Other carboxylates like malate, oxalate, 2-oxoglutarate and fumarate are also increased in the xylem sap of iron-deficient *S. lycopersicum* and *B. vulgaris* roots [83,84]. The requirement for increased production of such carboxylates under iron deficiency is mirrored by increased PEPC activity [83] and proteomic up-regulation of the TCA cycle [61–63,76]. In *A. thaliana* roots phosphorylated PEPC1 (AT1G53310) peptides are more abundant under iron deficiency [66] indicating increased PEPC activity. PEPC1 phosphorylation increases the affinity to phosphoenol pyruvate, lowers inhibition by malate and aspartate and increases activation by glucose-6-phosphate [86]. The malate dehydrogenase C-NAD-MDH1 (AT1G04410) is up-regulated in iron-deficient *B. vulgaris* roots [61] and in roots of the Arabidopsis *fit-3* mutant compared to wild-type irrespective of iron supply [55]. The abundance of the cytosolic isocitrate dehydrogenase cICDH (AT1G65930) was found altered upon iron deficiency in Arabidopsis [55]. Both, C-NAD-MDH1 and cICDH, are involved in carboxylate interconversion. Hence, up-regulation of PEPC activity as well as C-NAD-MDH1 and cICDH abundance may contribute to increased production of carbonic acids and thereby provide important compounds which are required in the plant's physiologic adaptation to iron deficiency.

3.3. Pentose phosphate pathway

An increased demand for reducing equivalents to drive the iron uptake machinery has been suggested [77]. The pentose phosphate pathway provides such reducing equivalents in the form of NADPH and might be altered to provide more such reducing equivalents. Some pentose phosphate pathway-associated proteins such as fructose bisphosphate aldolases (FBA) were found up-regulated upon iron deficiency in different proteomic studies. Since FBAs do not only act in the oxidative pentose phosphate pathway but also glycolysis, gluconeogenesis and the Calvin–Benson–Bassham cycle, it cannot clearly be stated whether up-regulation of these enzymes happens to increase glycolysis, the pentose phosphate pathway or carbon fixation. In roots, carbon fixation via the Calvin–Benson–Bassham cycle can be excluded. The results of Lan et al. indicate a reduction of the pentose phosphate pathway under iron deficiency in Arabidopsis roots since the two glucose-6-phosphate dehydrogenase proteins G6PD2 (AT5G13110) and G6PD3 (AT1G24280) are down-regulated in iron-deficient Arabidopsis roots [22]. The increased requirement for ATP and reducing equivalents to keep processes such as proton extrusion, FCR activity and other reactions going can be matched by increase of the TCA cycle activity. NADH/NAD$^+$ and NADPH/NAD$^+$ contents are increased in iron-deficient *B. vulgaris* roots [28] and the total NADH/NAD$^+$ and NADPH/NAD$^+$ content is slightly reduced in iron-deficient *S. lycopersicum* roots. However, the NADH-to-NAD$^+$ and NAHPH-to-NADP$^+$ ratios drop dramatically under iron deficiency [83]. Both suggests an increased demand for reducing equivalents in iron-deficient roots. Increased abundance of glycolysis and TCA-related proteins and the drop in G6PD2 and G6PD3 abundance in

iron-deficient Arabidopsis roots indicate down-regulation of the pentose phosphate pathway in favor of glycolysis in *A. thaliana*. At the same time NADPH production in Arabidopsis seems to be shifted toward organic acids-converting reactions catalyzed by enzymes such as MDH or cICDH. In barley, a Strategy II plant, the activity of G6PDH in roots is increased under iron deficiency [87] which leads to the assumption that in Strategy II plants, NADPH production is shifted toward the pentose phosphate pathway. This suggests that Strategy I and Strategy II plants have different preferences of glucose utilization under iron deficiency.

4. Nitrogen and amino acid metabolism

4.1. Yang cycle

Nicotianamine (NA) is a metal chelator that is involved in long distance iron transport and necessary for unloading iron from the phloem in the sink organs [85]. NA is also a precursor of mugineic acids (MA) [88] which are potent phytosiderophores that are excreted from roots of Strategy II plants to enhance iron uptake [7,89]. NA is synthesized from *S*-adenosylmethionine (SAM) [90–92] which also serves a substrate in ethylene biosynthesis and as a methyl group donor in methyl group transferring reactions such as feruloyl-CoA biosynthesis in the phenylpropanoid pathway (see Section 5.2). SAM is synthesized in the Yang cycle (also known as the methionine salvage pathway) by *S*-adenosylmethionine synthetases (SAMS) [93] which are also known as methionine adenosyltransferases (MAT). The proteins whose abundance was found changed similarly across multiple studies also comprise MATs (Table 1). MAT1 (METHIONINE ADENOSYLTRANSFERASE 1; AT1G02500) is up-regulated under iron deficiency in Arabidopsis roots [22,55], MAT3 (METHIONINE ADENOSYLTRANSFERASE 3; AT1G02500) is up-regulated under iron deficiency in Arabidopsis und tomato roots [22,65] and MAT4 (METHIONINE ADENOSYL-TRANSFERASE 4; AT3G17390) is up-regulated under iron deficiency in the *A. thaliana fit-3* mutant [55], down-regulated in *M. truncatula* roots [63] and up-regulated in *C. sativus* roots [62]. MAT2 (METHIONINE ADENOSYLTRANSFERASE 2; AT4G01850) is up-regulated in iron-deficient Arabidopsis roots [22] and in *C. sativus* roots it belongs to the multiply *identified proteins of* which one form is up-regulated and the other form is down-regulated under iron deficiency [62]. However, the overall tendency to up-regulation of MATs under iron deficiency can be explained by the role of SAM as a precursor of ethylene, nicotianamine, feruloyl-CoA and other compounds. Taken together, the proteomic data show activation of the Yang cycle which is an important prerequisite for iron-mobilizing adaptations and for ethylene signaling under iron-deficient conditions [54,94,95].

4.2. Nitrogen assimilation

The role of nitrogen assimilation genes and proteins under iron deficiency is poorly understood. However, there is evidence that the abundance of the nitrate transporter NRT1.1 influences the iron deficiency response via the FIT regulatory pathway [96,97]. Among the nitrogen assimilation-associated proteins NIR1 (NITRITE REDUCTASE 1; AT2G15620) is one of those whose abundance were found regulated similarly across multiple studies. NIR1 is involved in nitrogen assimilation and catalyzes the reduction of nitrite to ammonia which can then be utilized in the downstream reactions. NIR1 was found down-regulated in the Arabidopsis *fit-3* mutant compared to wild-type and the FIT over-expressor only under iron-deficient conditions [55]. In the *P. dulcis × persica* hybrid two close NIR1 homologs were also found down-regulated under iron deficiency. At the transcriptional level, NIR1 showed up-regulation in the *fit-3* mutant irrespective of iron supply. So the protein generally decreases under iron deficiency while the transcript levels increase in the *fit-3* mutant. This suggests that nitrite reduction might be impaired due to reduced protein stability or increased turnover of NIR1 under iron deficiency. On the other hand up-regulation of another protein, GLN1;1 (ARABIDOPSIS GLUTAMINE SYNTHETASE 1 CLONE R1; AT5G37600), has been observed in Arabidopsis [22] and *C. sativus* roots [62]. This up-regulation could happen for two reasons: Enhanced activity of glutamine synthetases might improve assimilation of lower-abundant ammonia or, as suggested by Lopez-Millan et al. [77], increase nitrogen recycling to compensate for decreased nitrogen assimilation. Controversially, the orthologs of another glutamine synthetase, GLN2 (GLUTAMINE SYNTHETASE 2; AT5G35630), are down-regulated in *M. truncatula* roots [63] but up-regulated in *P. dulcis × persica* roots [64] under iron-deficiency. Since, on the one hand, these proteins appear to be regulated in some studies but, on the other hand, lack stable regulation in one direction under iron deficiency and are missing in transcriptome studies, the actual roles of these proteins under iron deficiency as well as the nitrogen status remain elusive.

5. Secondary metabolism

5.1. Riboflavin synthesis

Secondary metabolites like riboflavin have recently gained increasing attention. In *B. vulgaris* the homolog of the DMRL synthase COS1 (AT2G44050) is induced under iron deficiency. In *M. truncatula* the homologs of the DMRL synthase COS1, the dihydroflavonol reductase DFR (AT5G42800) and two distinct protein spots of the GTP cyclohydrolase II RIBA1 (AT5G64300) are up-regulated under iron deficiency. These findings support the assumption that riboflavin synthesis is an important factor in the iron deficiency response at least in these two species. It has been speculated whether flavins are excreted to complex or reduce extracellular iron [26,27] and they have been hypothesized to serve as a redox bridge to ferric chelate reductases [28,29] or to induce changes in the bacterial rhizobiome [98]. Recent results rather point toward a role of flavins in reductive iron (III) mobilization and strengthen the assumption that flavins could play a role in extracellular electron transfer from the ferric chelate reductase to iron (III) oxide [99]. It has also been proposed that some microorganisms use flavins as a carbon source and elevated bacterial H_2CO_3 production might decrease the rhizosphere pH and thereby increase iron solubility [26]. Furthermore, riboflavin is a precursor of FAD (flavin adenine dinucleotide) which, besides NADH, serves as a cofactor of the *A. thaliana* ferric chelate reductase FRO2 [12,63,100]. Interestingly, ectopic expression of the *A. thaliana* bHLH transcription factors bHLH38 and bLHL39, which participate in the regulation of the iron deficiency response in *A. thaliana* [49–51], also lead to ectopic riboflavin accumulation and excretion in tobacco hairy roots and leaves although bHLH38 or bHLH39 over-expressing Arabidopsis roots do not accumulate riboflavin [101]. Transcription of riboflavin synthesis genes is up-regulated in iron-deficient Medicago roots and riboflavin has been detected in roots and found excreted into the nutrient solution only under iron deficiency [63]. This suggests that iron deficiency-induced riboflavin accumulation and excretion is a species-specific competence that seems to lack initiation in species that are not capable of accumulating riboflavin such as *A. thaliana*. However, there are transcriptional regulators that could potentially induce such a reaction. It cannot be excluded that different combinations of the above-mentioned mechanisms with different priorities and riboflavin requirements apply to different plant species with respect to iron homeostasis. Additionally, they do not necessarily have to result in riboflavin accumulation.

5.2. Phenylpropanoid synthesis

Besides carbonic acids, carbohydrates and flavins, phenolic compounds are also known to be excreted by iron-deficient roots [23–25] and there is evidence that phenolics are needed for proper iron uptake: Removal of phenolic compounds from the growth medium results in development of an iron-deficient phenotype in *Trifolium pratense* [24]. Absence of phenolics in the hydroponic medium also leads to decreased leaf iron content in *A. thaliana* [21]. Metabolite analyses in Arabidopsis showed that scopoletin, a simple coumarin, accumulates in roots under iron deficiency [22]. Scopoletin and its derivatives are excreted through the ABC (ATP binding cassette) transporter ABCG37 (PDR9) and excretion of coumarins by Arabidopsis roots is induced by iron deficiency [21]. Among the proteins whose abundance was found altered similarly in at least two distinct studies, two are involved in phenylpropanoid biosynthesis. CCOAOMT1 (CAFFEOYL COENZYME A O-METHYLTRANSFERASE 1; AT4G34050) was found regulated twice in *A. thaliana* [22,55] and down-regulated in *B. vulgaris* [61]. CCOAOMT1 is involved in the formation of feruloyl-CoA for which S-adenosylmethionine serves as the methyl group donor. Feruloyl-CoA is a precursor of scopoletin or scopolin [102]. F6'H1 (FERULOYL-6'-HYDROXYLASE 1; AT3G13610) was found up-regulated in iron-deficient Arabidopsis wild-type roots and down-regulated in the Arabidopsis *fit-3* mutant irrespective of iron supply. F6'H1 gene expression in Arabidopsis roots depends on the transcription factor FIT [17] and down-regulation of the F6'H1 protein in the Arabidopsis *fit-3* mutant [55] confirms this dependence. F6'H1 catalyzes the conversion of feruloyl-CoA to 6-hydroxyferuloyl-CoA [103] which is a precursor of Scopoletin and Scopolin. It has been demonstrated that F6'H1 is necessary to enhance iron mobilization by extrusion of F6'H1-dependent coumarins such as esculin, esculetin and scopoletin of which all three are able to rescue the chlorotic phenotype on high pH soils when added exogenously, whereas only esculetin is able to mobilize iron in vitro [104]. Although found regulated in only one study, three enzymes catalyzing the initial steps of phenylpropanoid biosynthesis, namely PAL2 (PHENYLALANINE AMMONIA LYASE 2, AT3G53260), 4CL1 (4-COUMARATE:COA LIGASE 1; AT1G51680) and 4CL2 (4-COUMARATE:COA LIGASE 2; AT3G21240), are induced under iron deficiency in Arabidopsis roots. Taken together, five enzymes of the phenylpropanoid pathway were found up-regulated under iron deficiency which, together with up-regulation of S-adenosylmethionine synthetases, supports the previously suggested role of coumarins in at least in iron-deficient *Arabidopsis* roots.

6. Redox regulation

Excess free Fe^{2+} and Fe^{3+} ions can cause the formation of reactive oxygen species (ROS) in the Haber-Weiss reaction part of which is also referred to as the Fenton reaction [105–107], and thereby cause oxidative stress. Under iron deficiency, ROS formation is poorly understood but it has been demonstrated that iron-deficient plants also produce ROS [108,109]. The abundance of two ROS-detoxifying proteins is down-regulated upon iron deficiency in at least two studies, respectively: CAT2 (CATALASE 2; AT4G35090) is down-regulated in *S. lycopersicum* [65] and *P. dulcis × persica* [64] roots and PA2 (PEROXIDASE 2; AT5G06720) is decreased in *C. sativus* [62] and *M. truncatula* [63] roots. These enzymes catalyze the conversion of hydrogen peroxide to H_2O, the latter using a phenolic donor. On the other hand, the superoxide dismutase ATMSD1 (ARABIDOPSIS MANGANESE SUPEROXIDE DISMUTASE 1; AT3G10920) displays increased abundance in iron-deficient *M. truncatula* [63] and *P. dulcis × P. persica* [64] roots. Additionally, two enzymes involved in

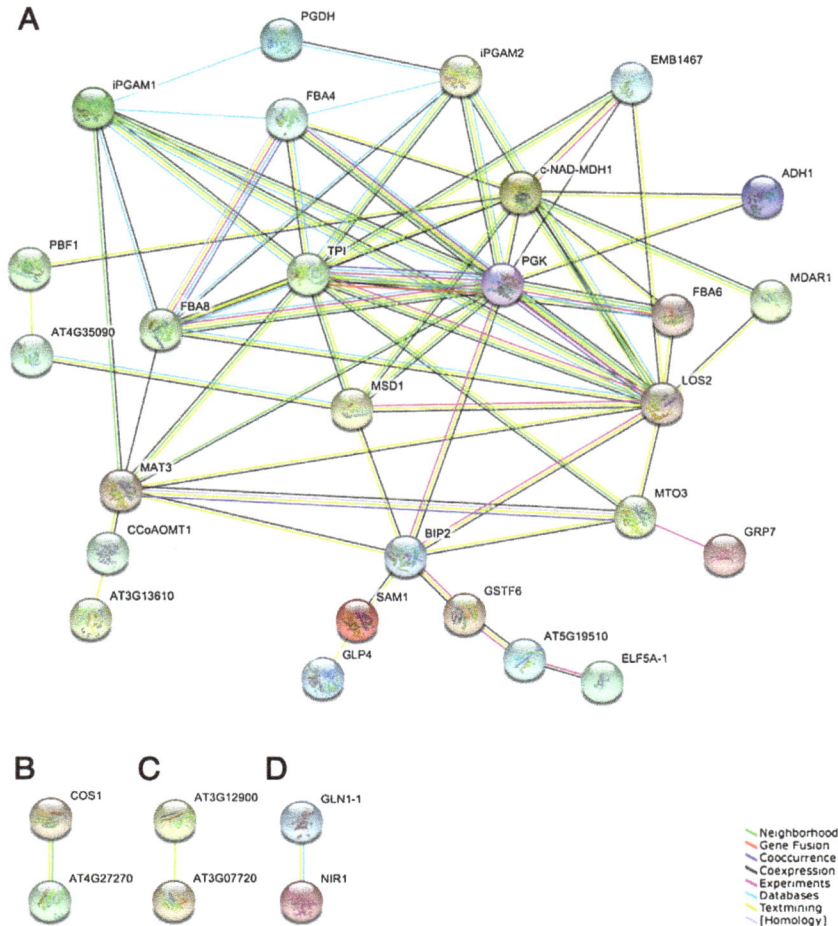

Fig. 1. Protein interaction and gene co-expression networks of the proteins which were significantly altered in their expression during iron stress across multiple studies. A: glycolysis and TCA cycle cluster; B: Riboflavin synthesis pair; C: the Arabidopsis-associated pair; D: nitrogen assimilation pair. For this network we used the AGI codes of the proteins that are listed in Table 1. The network was drawn using the STRING (version 10) protein interaction tool [73].

non-enzymatic ROS scavenging show increased abundance under iron deficiency across multiple studies: GST1 (ARABIDOPSIS GLU-TATHIONE S-TRANSFERASE 1; AT1G02930) is up-regulated under iron deficiency in *A. thaliana* roots [55] and shoots [68] and MDAR1 (MONODEHYDROASCORBATE REDUCTASE 1; AT3G52880) is increased in iron-deficient *A. thaliana* [22] and *M. truncatula* [63] roots.

CAT2 and PA2 are annotated to be heme binding proteins [72,110] and decrease of these peroxidases upon iron deficiency might be a direct consequence of reduced iron availability. Consequently, increased ROS levels might result from the reduced abundance of such enzymes. Up-regulation of ROS-eliminating enzymes such as ATMSD1 might help compensate for the lack of iron-dependent peroxidases and activation of the ascorbate-glutathione cycle by GPX3 (GLUTATHIONE PEROXIDASE 3; AT2G43350), which is induced in iron-deficient leaves and MDAR1 as well as GST1 activity might help detoxify ROS and translocate other oxidatively tagged molecules [111]. Although not all found regulated in multiple studies, the abundance of some distinct glutathione-S-transferases was found changed upon iron deficiency in many of the studies such as GSTF10, GSTL1 and GSTU25 in Arabidopsis roots [22,55], GSTF9 in *P. dulcis* × *P. persica* roots [64], GSTF6, GSTF7 and a putative microsomal glutathione-*S*-transferase in Arabidopsis shoots [68], GSTF8 in Arabidopsis flowers [67] or GSTU19 in *M. truncatula* roots [63]. Also, a total of twelve different peroxidase superfamily proteins was found regulated in one single study, respectively, but distributed among all species.

This shows that redox regulation is immensely influenced by iron deficiency across all the investigated Strategy I species.

7. Arabidopsis-specific proteins

Among the proteins whose abundance changes upon iron deficiency in multiple studies some were found regulated only in Arabidopsis. In this section we focus on these proteins. Apart from F6′H1, which has already been discussed above, another Arabidopsis-associated protein that was found differentially expressed in two proteomic studies, belongs to the 2-oxoglutarate and iron(II)-dependent oxygenase superfamily. Members of this family act in biosynthetic processes like antho-cyanin, flavonoid, coumarin and plant hormone synthesis and in oxidation of other organic substrates using Fe^{2+} as a cofactor [112–115]. Like F6′H1, AT3G12900 has been found regulated upon iron deficiency exclusively in Arabidopsis so far [22,55] (Table 1). The regulatory pattern of the AT3G12900 protein appears to depend on FIT [55] and bHLH039, bHLH100 and bHLH101 [53]. BLAST searches result in F6′H2 (FERULOYL-6′-HYDROXYLASE 2; AT1G55290) and F6′H1 (AT3G13610) as the two closest homologs. This points toward a possible function in coumarin biosynthesis as has been suggested by [55] or, more generally, in phenylpropanoid biosynthesis. Thus, AT3G12900 is an interesting new candidate for further research.

Another hitherto uncharacterized Arabidopsis-associated protein was found regulated in three proteomic studies [55,66,67]: the

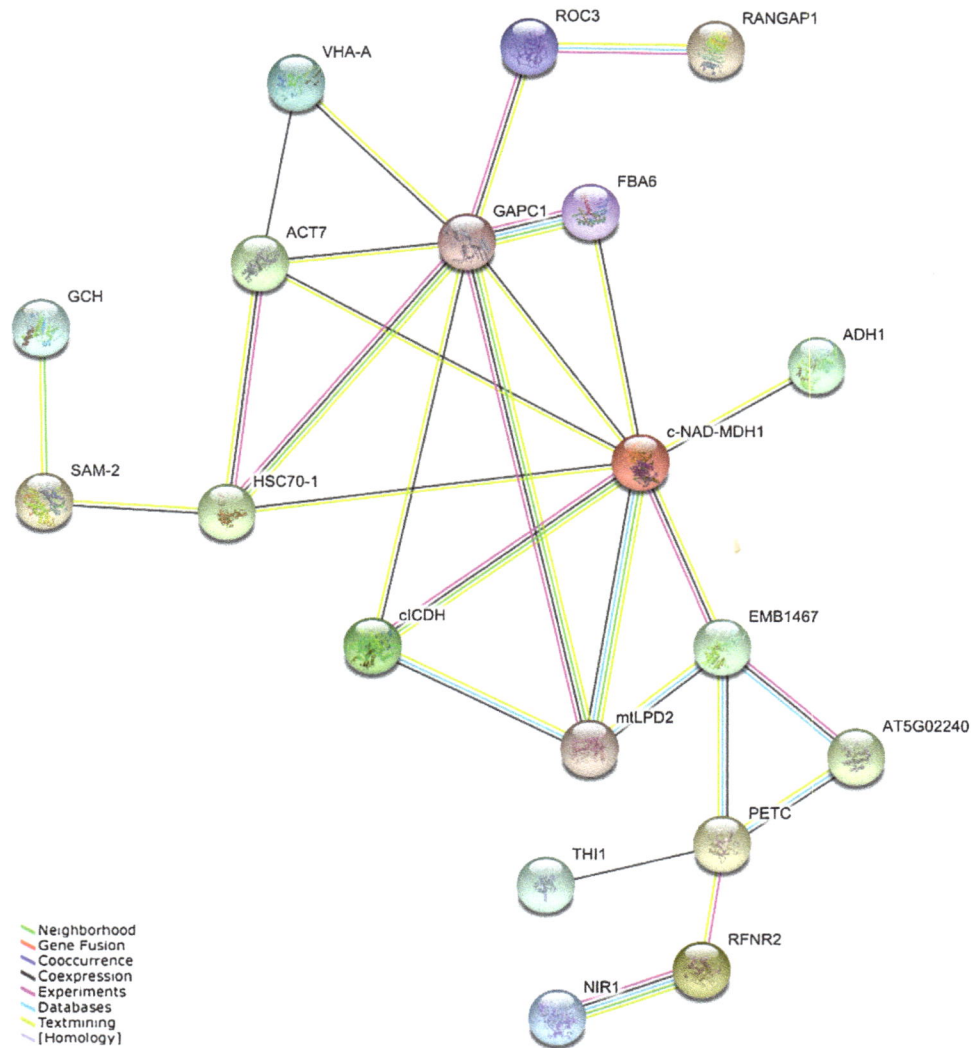

Fig. 2. Protein interaction networks of the multiply identified proteins that had been identified in multiple spots in the 2-DGE-based proteomic studies. For this network we used the AGI codes of the proteins that are listed in the Supplementary Table 1. The network was drawn using the STRING (version 10) protein interaction tool [73].

galactose oxidase/kelch repeat superfamily protein AT3G07720. *AT3G07720* has already been suggested to be a robust marker gene for iron deficiency since the transcript levels are stably regulated upon iron deficiency throughout a number of transcriptomic studies [19,53]. The function of AT3G07720 is yet unknown. Some galactose oxidase/kelch repeat proteins participate in glucosinolate breakdown. These act as specifiers of the fate of the intermediate aglucone which can rearrange to isothiocyanate or, in the presence of and depending on the specifier, yield a nitrile, epitionitrile or thiocyanate. Glucosinolates and their breakdown products have been associated with pathogen response [116]. BLAST searches for the closest homologs and orthologs result in nitrile specifier proteins such as NSP5 (AT5G48180), Nitrile-specifier protein (*M. truncatula*; MTR_2g069030) and nitrile-specifier protein 5-like isoform X1 (*S. lycopersicum*; LOC101257076). Hence, it can be speculated that AT3G07720 might act as a nitrile specifier protein. Like AT3G12900, AT3G07720 was found regulated in Arabidopsis roots [22,55] and shoots [67]. The proteomic profile indicates that AT3G07720 is regulated in a FIT-dependent manner since it is down-regulated in the *fit-3* mutant compared to wild-type and the constitutive FIT over-expressor irrespective of iron supply [55].

The genes of both the above-mentioned Arabidopsis proteins are part of the FIT target network [19]. Expression of the AT3G12900

gene is responsive iron deficiency and excess zinc [17,45,117–121]. The transcriptional profiles of AT3G12900 in wild-type and *fit* mutants with altered FIT activity lead to the conclusion that the AT3G12900 gene is regulated by FIT [17,55]. The gene expression profile of AT3G07720 in the Arabidopsis *fit-3* mutant, wild-type and a constitutive FIT over-expressor leads to the suggestion that AT3G07720 is also regulated by FIT [55]. The fact that these genes are regulated by the central regulator of the iron uptake machinery in Arabidopsis indicates a possibly important role of AT3G12900 and AT3G07720 under iron-deficient conditions although their exact functions in iron homeostasis in *A. thaliana* remain unclear.

Furthermore, the abundances of GST1, MAT1, GLP4, PGDH2, EF1B and MAPR3 were found regulated in multiple studies but only in Arabidopsis. MATs might contribute to NA, ethylene and phenylpropanoid production while GSTs are involved in glutathionylization which appears to be an important mechanism for disposal of damaged molecules under iron deficiency [111]. GLP4 could be related to pathogen response since its wheat and barley orthologs are important for quantitative resistance [122]. These orthologs also display superoxide dismutase activity which implies that AtGLP4 could also be part of the redox regulatory mechanism. In Arabidopsis, GLP4 has been reported to bind auxin and could be involved in developmental processes [123]. PGDH2 is involved in serine biosynthesis [124]. MAPR3 is up-regulated in shoots and in the root phosphoproteome under iron deficiency. MAPR3 is a heme binding cytochrome b5 family protein and is annotated to bind steroids in the TAIR and UniProtKB databases [72,110]. EF1B has been found regulated at the protein level in three studies [55,66,67], which indicates that EF1B is post-transcriptionally regulated. It has been reported that transcriptional efficiency is altered under oxidative stress in yeast [125]. It has also been proposed that in Arabidopsis, the ribosomal composition is altered under iron deficiency, which may lead to biased expression [126]. Both suggest that discrepancy between transcript and protein regulation may occur under iron deficiency.

Taken together, since the genes of AT3G12900 and AT3G07720 are expressed in a FIT-dependent manner and the AT3G12900 and AT3G07720 proteins differentially accumulated under iron deficiency, they appear to take important roles in the response to iron deficiency in Arabidopsis. However, the exact roles of the Arabidopsis-specific proteins AT3G12900, AT3G07720, GLP4, PGDH2 and MAPR3 remain elusive.

8. Post-transcriptional regulation

Post-transcriptional regulation is particularly important in the control of iron acquisition in Arabidopsis. For a review on these specific mechanisms see Brumbarova et al. [75]. A more global approach to investigate post-transcriptional regulation with regard to iron homeostasis in the Arabidopsis proteome has been used by Lan et al. [66]. They have demonstrated that phosphopeptides of 45 proteins are significantly regulated in iron-deficient Arabidopsis roots, 25 of which show decreased and 20 show increased abundance under iron deficiency. Hence, the response to iron deficiency in Arabidopsis roots includes differential phosphorylation of proteins. Among alterations in the carbon flow and proton transport, one of the most interesting findings in the adaptations of the phosphoproteome to iron deficiency is the fact that two of the four regulated protein interaction clusters are related to mRNA processing and translation [66]. This suggests that, besides phosphorylation, transcriptional as well as post-transcriptional regulation might be altered under iron deficiency in Arabidopsis.

Mai et al. [55] focused on the congruence between protein and transcript expression in Arabidopsis roots depending on iron and FIT. 62% of the differentially regulated proteins were not found regulated at the transcriptional level and 12% of the differentially regulated proteins even display opposite regulation at the protein and transcript level in at least one comparison and these miscorrelations largely depend on FIT. In the Arabidopsis root proteome, the overall number of proteins that changes their abundance in the comparisons between wild-type and *fit* mutants with altered FIT activity were larger between the lines at a given iron supply status than in the comparisons between iron-sufficient and iron-deficient root within each of these lines. The proteomic changes were predominantly characterized by down-regulation in the *fit-3* mutant. This indicates that the post-transcriptional regulation of a large subset of the differentially regulated proteins depends on the plant's ability to express FIT. Additionally, 10 proteins were found in at least two distinct protein spots, which we define as 'multiply identified' [55]. These protein forms differ in their *pI* (isoelectric point) rather than 2nd dimension electrophoretic mobility (mass). Phosphorylations cause a mass difference (multiples of 80 Da) which may result in a barely or not at all visible electrophoretic mobility shift but also more acidic isoelectric point (*pI*). Therefore, and based on the multiply identified proteins are partly known or predicted to be phosphorylated, phosphorylation seems to be the most likely reason, although other modifications or mechanisms like alternative splicing cannot be excluded.

Interestingly, in all 2-DGE-based proteomic studies, some proteins were identified in more than one protein spot within the respective study (Supplementary Table 1). No modifications have been detected because identification of proteins rather than modification was focused in the first place. However, for some of these proteins post-translational modifications and alternative splice forms have been reported. The orthologs of three proteins were found in several spots in two studies with different species: c-NAD-MDH1 in *A. thaliana* and *B. vulgaris* roots, ADH1 in *A. thaliana* and *C. sativus* roots and FBA6 in *M. truncatula* and *C. sativus* roots.

9. Protein interaction and gene co-expression

Protein interaction and gene co-expression networks have been shown [22,66]. Here, we have created a network view of protein interactions and co-expression between proteins that were found regulated across multiple studies (Table 1) (Fig. 1). 27 of the 39 proteins form a large cluster, which is mainly related to glycolysis and TCA. The central players in this cluster are FBA6, FBA8, TPI, LOS2 (ENO2) and PGK (Fig. 1A). The edges of this cluster are connected to small sub-clusters related to phenylpropanoid biosynthesis (SAM3, CCoAOMT1 and F6′H1) and to response to stress (AT5G19510, GSTF6 and BIP2). This large cluster reflects the above-discussed observation that glycolysis and TCA cycle as well as biosynthesis of phenylpropanoids and organic acids are crucial pathways that need to be adapted to iron deficiency. Besides the large cluster there are three pairs which are not connected to the large cluster. One is composed of the DMRL synthase COS1 which is involved in riboflavin synthesis, and AT4G27270, a quinone reductase family protein that is predicted to bind FMN (Fig. 1B). Another cluster contains the two 'unknown newcomers' AT3G07720 and AT3G12900 (Fig. 1C) and the third one contains two proteins that synthetize two subsequent reactions in N assimilation, namely NIR1 and the glutamine synthetase GLN1;1 (Fig. 1D). Six proteins do not show interactions with any of the other proteins. Taken together, 27 of 39 proteins (69%) are combined in a cluster, while six more are combined in pairs, which means that ca. 84% of all regulated proteins interact with other proteins of this selection.

We have created a second network view of protein interactions using the proteins that were found in at least two forms in any of the compared 2-DGE-based studies (Fig. 2). 19 of the 26 multiply identified proteins form a cluster of which the central players are GAPC1 and C-NAD-MDH1. Although this cluster is mainly characterized by glycolysis and TCA cycle proteins, it also contains a small edge sub-cluster that is related to N assimilation, namely NIR1 and RFNR2 which maintains the supply of reduced ferredoxin [127]. Seven of 26 proteins do not interact with any of the other proteins. With 73% also the multiply identified proteins show a high degree of interaction and gene co-expression.

Taken together, the high degree of interaction and co-expression among the proteins and genes in the database-generated networks mirror the observed common changes in the abundance of these proteins.

10. Methodological considerations

The 'classical' method to analyze the proteome is two-dimensional gel electrophoresis (2-DGE) in which up to a few thousand protein spots can be detected, quantified and identified. The number of detectable protein spots depends on several factors such as the pH gradient in the 1st dimension, the IPG (immobilized pH gradient) strip length, the amount of total protein analyzed per 2-D gel or the staining method. Protein extraction methods and the further procedure can influence the results of 2-DGE. Therefore, it is not very surprising that the overlap between different

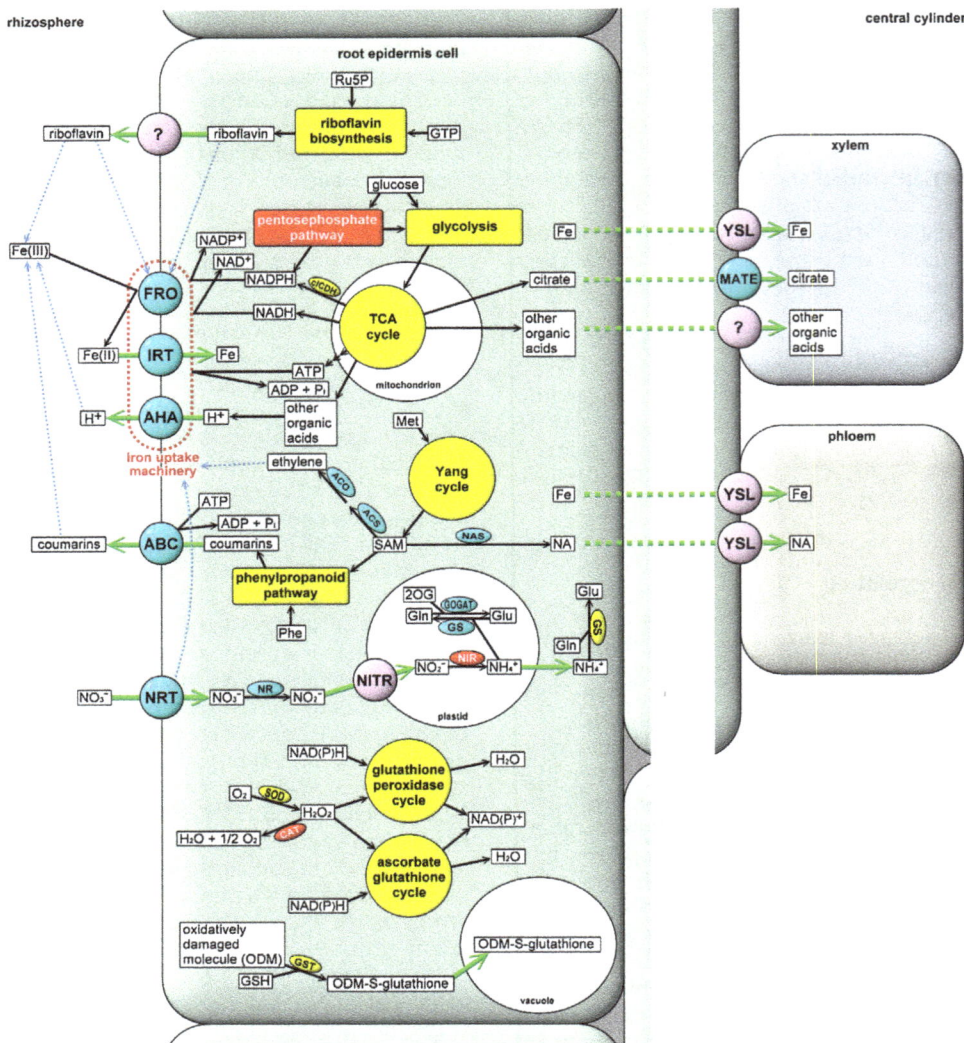

Fig. 3. Hypothetical model of regulated protein functions under iron deficiency in roots of Strategy I plants. Pathways and proteins colored yellow were found up-regulated and pathways and proteins colored in red were found down-regulated. Previously demonstrated important iron homeostasis-related proteins were added to display the links between the observed proteomic adaptations with iron homeostasis. The respective references can be found in the text. These proteins are colored blue. Metabolites and other compounds are displayed in white rectangles. Black arrows indicate biochemical reactions or conversions. Solid green arrows indicate active or passive transport processes. Dotted blue arrows indicate a possible influence of the protein or compound on the respective process. Dotted green arrows indicate transport processes which are still unclear or under discussion. Pink coloration indicates a predicted protein.

2-DGE-based proteomic studies conducted in different laboratories is comparably low. Yet, 2-DGE is an excellent method to find post-translationally modified forms of known and unknown proteins. Immunoblots of 2-D gels (2-D blots) are also very suitable to investigate the dynamics of post-translationally modified protein forms and the detected spots can also be directly analyzed [128].

Recently, quantitative peptide and mass spectrometry-based methods such as iTRAQ (isobaric tags for relative and absolute quantitation) have gained increasing importance. With iTRAQ, it is possible to detect and quantify peptides of many more proteins than with the 2-DGE approach. Besides the sheer number of detectable peptides/proteins which enables more comprehensive investigations of proteomes one of the advantages of this method is that it requires less fresh material per sample and it is less time consuming. Since the whole procedure from handling the extracts up to MS analysis is less complicated than with 2-DGE this approach is less prone to errors and generally, less replicates have been accepted to yield significant results. Analysis can also be done with different parameters such as the specific search for post-translationally modified peptides. Hence, these methods also allow

for investigation of subsets of the proteome such as the phospho-proteome or the ubiquitinome. However, changes in the abundance of single post-translationally modified peptides may also be a result of changes in expression of the respective proteins and the results must be handled with care. Although iTRAQ is less error-prone than 2-DGE, different growth methods, growth media, climatic parameters and age of the investigated plants may also lead to different results.

Due to the above considerations we regard it as crucial to look at the overlaps between different proteomic studies irrespective of the methods used. Some proteins appear as differentially expressed in a more stable manner. To gain a more accurate and condensed overview of the common response to iron deficiency in Strategy I or Strategy II plants, it is also important to investigate overlaps of regulated proteins in different species.

11. Concluding model

Based on the above-described patterns of regulated proteins we present a unifying hypothetical model of the proteomic and metabolic adaptations to iron deficiency in Strategy I plants (Fig. 3). Since so far Strategy II plants are represented in only one proteomic study, an informative model for Strategy II plants could not be created this way. The Strategy I model illustrates how primary and secondary metabolic adaptations may contribute to enhance iron uptake and distribution under iron deficient conditions. Additionally, the model points out many open questions that may have to be addressed by future research. The model can also be used to predict prospective breeding targets for designing iron-efficient crops. Furthermore, we point out proteins of yet unknown function that may take potentially important roles in iron homeostasis and highlight the importance of post-transcriptional control of a large subset of proteins that are regulated under iron deficiency.

References

[1] E. McLean, M. Cogswell, I. Egli, D. Wojdyla, B.d. Benoist, Worldwide Prevalence of Anaemia 1993–2005, WHO Global Database on Anaemia, 2008.

[2] V.M. Römheld Nikolic, Handbook of Plant Nutrition, CRC, Press Taylor & Francis Group, 2006.

[3] A. Tanaka, R. Loe, S.A. Navasero, Some mechanisms involved in the development of iron toxicity symptoms in the rice plant, Soil Sci. Plant Nutr. 12 (1966) 32–38.

[4] M. Yamauchi, Rice bronzing in Nigeria caused by nutrient imbalances and its control by potassium sulfate application, Plant Soil 117 (1989) 275–286.

[5] S. Mohapatra, International Rice Research Institute (IRRI), http://irri.org/index.php?option=com_k2&view=item&id=10762:beware-of-bronzing&lang=en (2011).

[6] V. Romheld, Different strategies for iron acquisition in higher-plants, Physiol. Plantarum 70 (1987) 231–234.

[7] H. Marschner, V. Römheld, M. Kissela, Different strategies in higher plants in mobilization and uptake of iron, J. Plant Nutr. (1986) 9.

[8] V. Römheld, H. Marschner, Mobilization of iron in the rhizophere of different plant species, Adv. Plant Nutr. 2 (1986) 155–204.

[9] M.R. Sussman, Molecular analysis of proteins in the plant plasma-membrane, Ann. Rev. Plant Phys. 45 (1994) 211–234.

[10] S. Santi, W. Schmidt, Dissecting iron deficiency-induced proton extrusion in Arabidopsis roots, New Phytol. 183 (2009) 1072–1084.

[11] E.L. Connolly, N.H. Campbell, N. Grotz, C.L. Prichard, M.L. Guerinot, Overexpression of the FRO2 ferric chelate reductase confers tolerance to growth on low iron and uncovers posttranscriptional control, Plant Physiol. 133 (2003) 1102–1110.

[12] N.J. Robinson, C.M. Procter, E.L. Connolly, M.L. Guerinot, A ferric-chelate reductase for iron uptake from soils, Nature 397 (1999) 694–697.

[13] D. Eide, M. Broderius, J. Fett, M.L. Guerinot, A novel iron-regulated metal transporter from plants identified by functional expression in yeast, Proc. Natl. Acad. Sci. U. S. A 93 (1996) 5624–5628.

[14] R. Henriques, J. Jasik, M. Klein, E. Martinoia, et al., Knock-out of Arabidopsis metal transporter gene IRT1 results in iron deficiency accompanied by cell differentiation defects, Plant Mol. Biol. 50 (2002) 587–597.

[15] C. Varotto, D. Maiwald, P. Pesaresi, P. Jahns, et al., The metal ion transporter IRT1 is necessary for iron homeostasis and efficient photosynthesis in Arabidopsis thaliana, Plant J. 31 (2002) 589–599.

[16] G. Vert, N. Grotz, F. Dedaldechamp, F. Gaymard, et al., IRT1, an Arabidopsis transporter essential for iron uptake from the soil and for plant growth, Plant Cell 14 (2002) 1223–1233.

[17] E.P. Colangelo, M.L. Guerinot, The essential basic helix-loop-helix protein FIT1 is required for the iron deficiency response, Plant Cell 16 (2004) 3400–3412.

[18] M. Jakoby, H.Y. Wang, W. Reidt, B. Weisshaar, P. Bauer, F.R.U. (BHLH029) is required for induction of iron mobilization genes in Arabidopsis thaliana, FEBS Lett. 577 (2004) 528–534.

[19] R. Ivanov, T. Brumbarova, P. Bauer, Fitting into the harsh reality: regulation of iron-deficiency responses in dicotyledonous plants, Mol. Plant 5 (2012) 27–42.

[20] J. Abadia, A. Lopez-Millan, A. Rombola, A. Abadia, Organic acids and Fe deficiency: a review, Plant Soil 24 (2002) 75–86.

[21] P. Fourcroy, P. Siso-Terraza, D. Sudre, M. Saviron, et al., Involvement of the ABCG37 transporter in secretion of scopoletin and derivatives by Arabidopsis roots in response to iron deficiency, New Phytol. 201 (2014) 155–167.

[22] P. Lan, W. Li, T.N. Wen, J.Y. Shiau, et al., iTRAQ protein profile analysis of Arabidopsis roots reveals new aspects critical for iron homeostasis, Plant Physiol. 155 (2011) 821–834.

[23] V. Romheld, H. Marschner, Mechanism of iron uptake by peanut plants: I. Fe reduction, chelate splitting, and release of phenolics, Plant Physiol. 71 (1983) 949–954.

[24] C.W. Jin, G.Y. You, Y.F. He, C. Tang, et al., Iron deficiency-induced secretion of phenolics facilitates the reutilization of root apoplastic iron in red clover, Plant Physiol. 144 (2007) 278–285.

[25] J. Rodriguez-Celma, S. Vazquez-Reina, J. Orduna, A. Abadia, et al., Characterization of flavins in roots of Fe-deficient strategy I plants, with a focus on Medicago truncatula, Plant Cell Physiol. 52 (2011) 2173–2189.

[26] S. Cesco, G. Neumann, N. Tomasi, R. Pinton, L. Weisskopf, Release of plant-borne flavonoids into the rhizosphere and their role in plant nutrition, Plant Soil 329 (2010) 1–25.

[27] E.B. Gonzalez-Vallejo, S. Susin, A. Abadia, J. Abadia, Changes in sugar beet leaf plasma membrane Fe(III)-chelate reductase activities mediated by Fe-deficiency, assay buffer composition, anaerobiosis and the presence of flavins, Protoplasma 205 (1998) 163–168.

[28] A.F. Lopez-Millan, F. Morales, S. Andaluz, Y. Gogorcena, et al., Responses of sugar beet roots to iron deficiency. Changes in carbon assimilation and oxygen use, Plant Physiol. 124 (2000) 885–898.

[29] A. Higa, Y. Mori, Y. Kitamura, Iron deficiency induces changes in riboflavin secretion and the mitochondrial electron transport chain in hairy roots of Hyoscyamus albus, J. Plant Physiol. 167 (2010) 870–878.

[30] E.L. Connolly, J.P. Fett, M.L. Guerinot, Expression of the IRT1 metal transporter is controlled by metals at the levels of transcript and protein accumulation, Plant Cell 14 (2002) 1347–1357.

[31] E. Lombi, K.L. Tearall, J.R. Howarth, F.J. Zhao, et al., Influence of iron status on cadmium and zinc uptake by different ecotypes of the hyperaccumulator Thlaspi caerulescens, Plant Physiol. 128 (2002) 1359–1367.

[32] S. Nishida, C. Tsuzuki, A. Kato, A. Aisu, et al., AtIRT1, the primary iron uptake transporter in the root, mediates excess nickel accumulation in Arabidopsis thaliana, Plant Cell Physiol. 52 (2011) 1433–1442.

[33] S. Arrivault, T. Senger, U. Kramer, The Arabidopsis metal tolerance protein AtMTP3 maintains metal homeostasis by mediating Zn exclusion from the shoot under Fe deficiency and Zn oversupply, Plant J. 46 (2006) 861–879.

[34] T.J. Yang, W.D. Lin, W. Schmidt, Transcriptional profiling of the Arabidopsis iron deficiency response reveals conserved transition metal homeostasis networks, Plant Physiol. 152 (2010) 2130–2141.

[35] P.F. Lindley, Iron in biology: a structural viewpoint, Rep. Prog. Phys. 59 (1996) 867–933.

[36] A.C. Dlouhy, C.E. Outten, The iron metallome in eukaryotic organisms, Met. Ions Life Sci. 12 (2013) 241–278.

[37] C.M. Fraser C. Chapple, The phenylpropanoid pathway in Arabidopsis. Arabidopsis Book, 2011, 9, e0152.

[38] F. Schaller, Enzymes of the biosynthesis of octadecanoid-derived signalling molecules, J. Exp. Bot. 52 (2001) 11–23.

[39] C.A. Helliwell, A. Poole, W.J. Peacock, E.S. Dennis, Arabidopsis ent-kaurene oxidase catalyzes three steps of gibberellin biosynthesis, Plant Physiol. 119 (1999) 507–510.

[40] T. Katsumata, A. Hasegawa, T. Fujiwara, T. Komatsu, et al., Arabidopsis CYP85A2 catalyzes lactonization reactions in the biosynthesis of 2-deoxy-7-oxalactone brassinosteroids, Biosci. Biotechnol. Biochem 72 (2008) 2110–2117.

[41] T.W. Kim, J.Y. Hwang, Y.S. Kim, S.H. Joo, et al., Arabidopsis CYP85A2, a cytochrome P450, mediates the Baeyer–Villiger oxidation of castasterone to brassinolide in brassinosteroid biosynthesis, Plant Cell 17 (2005) 2397–2412.

[42] M. Nakayama, T. Akashi, T. Hase, Plant sulfite reductase: molecular structure, catalytic function and interaction with ferredoxin, J. Inorg. Biochem. 82 (2000) 27–32.

[43] J.R. Lancaster, J.M. Vega, H. Kamin, N.R. Orme-Johnson, et al., Identification of the iron-sulfur center of spinach ferredoxin-nitrite reductase as a tetranuclear center, and preliminary EPR studies of mechanism, J. Biol. Chem. 254 (1979) 1268–1272.

[44] M.J. Murphy, L.M. Siegel, S.R. Tove, H. Kamin, Siroheme: a new prosthetic group participating in six-electron reduction reactions catalyzed by both sulfite and nitrite reductases, Proc. Natl. Acad. Sci. U. S. A 71 (1974) 612–616.

[45] T.A. Long, H. Tsukagoshi, W. Busch, B. Lahner, et al., The bHLH transcription factor POPEYE regulates response to iron deficiency in Arabidopsis roots, Plant Cell 22 (2010) 2219–2236.

[46] J. Zhang, B. Liu, M. Li, D. Feng, et al., The bHLH transcription factor bHLH104 interacts with IAA-LEUCINE RESISTANT3 and modulates iron homeostasis in Arabidopsis, Plant Cell 27 (2015) 787–805.

[47] Y. Zhang, H. Wu, N. Wang, H. Fan, et al., Mediator subunit 16 functions in the regulation of iron uptake gene expression in Arabidopsis, New Phytol. 203 (2014) 770–783.

[48] H. Wu, C. Chen, J. Du, H. Liu, et al., Co-overexpression FIT with AtbHLH38 or AtbHLH39 in Arabidopsis-enhanced cadmium tolerance via increased cadmium sequestration in roots and improved iron homeostasis of shoots, Plant Physiol. 158 (2012) 790–800.

[49] Y. Yuan, H. Wu, N. Wang, J. Li, et al., FIT interacts with AtbHLH38 and AtbHLH39 in regulating iron uptake gene expression for iron homeostasis in Arabidopsis, Cell Res. 18 (2008) 385–397.

[50] N. Wang, Y. Cui, Y. Liu, H. Fan, et al., Requirement and functional redundancy of Ib subgroup bHLH proteins for iron deficiency responses and uptake in Arabidopsis thaliana, Mol. Plant 6 (2013) 503–513.

[51] H.Y. Wang, M. Klatte, M. Jakoby, H. Baumlein, et al., Iron deficiency-mediated stress regulation of four subgroup Ib BHLH genes in Arabidopsis thaliana, Planta 226 (2007) 897–908.

[52] A.B. Sivitz, V. Hermand, C. Curie, G. Vert, Arabidopsis bHLH100 and bHLH101 control iron homeostasis via a FIT-independent pathway, PLoS One 7 (2012) e44843.

[53] F. Maurer, M.A. Naranjo Arcos, P. Bauer, Responses of a triple mutant defective in three iron deficiency-induced Basic Helix-Loop-Helix genes of the subgroup Ib(2) to iron deficiency and salicylic acid, PLoS One 9 (2014) e99234.

[54] S. Lingam, J. Mohrbacher, T. Brumbarova, T. Potuschak, et al., Interaction between the bHLH transcription factor FIT and ETHYLENE INSENSITIVE3/ETHYLENE INSENSITIVE3-LIKE1 reveals molecular linkage between the regulation of iron acquisition and ethylene signaling in Arabidopsis, Plant Cell 23 (2011) 1815–1829.

[55] H.J. Mai, C. Lindermayr, C. von Toerne, C. Fink-Straube, et al., Iron and FER-LIKE IRON DEFICIENCY-INDUCED TRANSCRIPTION FACTOR-dependent regulation of proteins and genes in Arabidopsis thaliana roots, Proteomics (2015).

[56] M. Barberon, E. Zelazny, S. Robert, G. Conejero, et al., Monoubiquitin-dependent endocytosis of the iron-regulated transporter 1 (IRT1) transporter controls iron uptake in plants, Proc. Natl. Acad. Sci. U. S. A 108 (2011) E450–458.

[57] A. Blum, T. Brumbarova, P. Bauer, R. Ivanov, Hormone influence on the spatial regulation of expression in iron-deficient roots, Plant Signal Behav. (2014) 9.

[58] R. Ivanov, T. Brumbarova, A. Blum, A.M. Jantke, et al., SORTING NEXIN1 is required for modulating the trafficking and stability of the Arabidopsis IRON-REGULATED TRANSPORTER1, Plant Cell 26 (2014) 1294–1307.

[59] L. Kerkeb, I. Mukherjee, I. Chatterjee, B. Lahner, et al., Iron-induced turnover of the Arabidopsis IRON-REGULATED TRANSPORTER1 metal transporter requires lysine residues, Plant Physiol. 146 (2008) 1964–1973.

[60] J. Meiser, S. Lingam, P. Bauer, Posttranslational regulation of the iron deficiency basic helix-loop-helix transcription factor FIT is affected by iron and nitric oxide, Plant Physiol. 157 (2011) 2154–2166.

[61] R. Rellan-Alvarez, S. Andaluz, J. Rodriguez-Celma, G. Wohlgemuth, et al., Changes in the proteomic and metabolic profiles of Beta vulgaris root tips in response to iron deficiency and resupply, BMC Plant Biol. 10 (2010) 120.

[62] S. Donnini, B. Prinsi, A.S. Negri, G. Vigani, et al., Proteomic characterization of iron deficiency responses in Cucumis sativus L. roots, BMC Plant Biol. 10 (2010) 268.

[63] J. Rodriguez-Celma, G. Lattanzio, M.A. Grusak, A. Abadia, et al., Root responses of Medicago truncatula plants grown in two different iron deficiency conditions: changes in root protein profile and riboflavin biosynthesis, J. Proteome Res. 10 (2011) 2590–2601.

[64] J. Rodriguez-Celma, G. Lattanzio, S. Jimenez, J.F. Briat, et al., Changes induced by Fe deficiency and Fe resupply in the root protein profile of a peach-almond hybrid rootstock, J. Proteome Res. 12 (2013) 1162–1172.

[65] T. Brumbarova, A. Matros, H.P. Mock, P. Bauer, A proteomic study showing differential regulation of stress, redox regulation and peroxidase proteins by iron supply and the transcription factor FER, Plant J. 54 (2008) 321–334.

[66] P. Lan, W. Li, T.N. Wen, W. Schmidt, Quantitative phosphoproteome profiling of iron-deficient Arabidopsis roots, Plant Physiol. 159 (2012) 403–417.

[67] D. Sudre, E. Gutierrez-Carbonell, G. Lattanzio, R. Rellan-Alvarez, et al., Iron-dependent modifications of the flower transcriptome, proteome, metabolome, and hormonal content in an Arabidopsis ferritin mutant, J. Exp. Bot. 64 (2013) 2665–2688.

[68] S.M. Zargar, R. Kurata, S. Inaba, Y. Fukao, Unraveling the iron deficiency responsive proteome in Arabidopsis shoot by iTRAQ-OFFGEL approach, Plant Signal Behav. 8 (2013), http://dx.doi.org/10.4161/psb.26892.

[69] M. Kanehisa, S. Goto, Y. Sato, M. Kawashima, et al., Data, information, knowledge and principle: back to metabolism in KEGG, Nucleic Acids Res. 42 (2014) D199–205.

[70] M.S. Katari, S.D. Nowicki, F.F. Aceituno, D. Nero, et al., VirtualPlant: a software platform to support systems biology research, Plant Physiol. 152 (2010) 500–515.

[71] R. Caspi, T. Altman, R. Billington, K. Dreher, et al., The MetaCyc database of metabolic pathways and enzymes and the BioCyc collection of pathway/genome databases, Nucleic Acids Res. 42 (2014) D459–471.

[72] C. UniProt, UniProt: a hub for protein information, Nucleic Acids Res. 43 (2015) D204–212.

[73] L.J. Jensen, M. Kuhn, M. Stark, S. Chaffron, et al., STRING 8—a global view on proteins and their functional interactions in 630 organisms, Nucleic Acids Res. 37 (2009) D412–416.

[74] E.P. Colangelo, M.L. Guerinot, Put the metal to the petal: metal uptake and transport throughout plants, Curr. Opin. Plant Biol. 9 (2006) 322–330.

[75] T. Brumbarova, P. Bauer, R. Ivanov, Molecular mechanisms governing Arabidopsis iron uptake, Trends Plant Sci. 20 (2015) 124–133.

[76] J. Li, X.D. Wu, S.T. Hao, X.J. Wang, H.Q. Ling, Proteomic response to iron deficiency in tomato root, Proteomics 8 (2008) 2299–2311.

[77] A.F. Lopez-Millan, M.A. Grusak, A. Abadia, J. Abadia, Iron deficiency in plants: an insight from proteomic approaches, Front Plant Sci. 4 (2013) 254.

[78] T.P. Durrett, W. Gassmann, E.E. Rogers, The FRD3-mediated efflux of citrate into the root vasculature is necessary for efficient iron translocation, Plant Physiol 144 (2007) 197–205.

[79] L.S. Green, E.E. Rogers, FRD3 controls iron localization in Arabidopsis, Plant Physiol. 136 (2004) 2523–2531.

[80] E.E. Rogers, M.L. Guerinot, FRD3 member of the multidrug and toxin efflux family, controls iron deficiency responses in Arabidopsis, Plant Cell 14 (2002) 1787–1799.

[81] K. Yokosho, N. Yamaji, D. Ueno, N. Mitani, J.F. Ma, OsFRDL1 is a citrate transporter required for efficient translocation of iron in rice, Plant Physiol. 149 (2009) 297–305.

[82] M.C. White, Metal complexation in xylem fluid: II. THEORETICAL EQUILIBRIUM MODEL AND COMPUTATIONAL COMPUTER PROGRAM, Plant Physiol. 67 (1981) 301–310.

[83] A.F. Lopez-Millan, F. Morales, Y. Gogorcena, A. Abadia, J. Abadia, Metabolic responses in iron deficient tomato plants, J. Plant Physiol. 166 (2009) 375–384.

[84] A. Larbi, F. Morales, A. Abadia, J. Abadia, Changes in iron and organic acid concentrations in xylem sap and apoplastic fluid of iron-deficient Beta vulgaris plants in response to iron resupply, J. Plant Physiol. 167 (2010) 255–260.

[85] M. Schuler, R. Rellan-Alvarez, C. Fink-Straube, J. Abadia, P. Bauer, Nicotianamine functions in the Phloem-based transport of iron to sink organs, in pollen development and pollen tube growth in Arabidopsis, Plant Cell 24 (2012) 2380–2400.

[86] A.L. Gregory, B.A. Hurley, H.T. Tran, A.J. Valentine, et al., In vivo regulatory phosphorylation of the phosphoenolpyruvate carboxylase AtPPC1 in phosphate-starved Arabidopsis thaliana, Biochem. J. 420 (2009) 57–65.

[87] A.F. Lopez-Millan, M.A. Grusak, J. Abadia, Carboxylate metabolism changes induced by Fe deficiency in barley, a Strategy II plant species, J. Plant Physiol. 169 (2012) 1121–1124.

[88] S. Mori, Iron acquisition by plants, Curr. Opin. Plant Biol. 2 (1999) 250–253.

[89] V. Romheld, H. Marschner, Evidence for a specific uptake system for iron phytosiderophores in roots of grasses, Plant Physiol. 80 (1986) 175–180.

[90] S. Shojima, N.K. Nishizawa, S. Fushiya, S. Nozoe, et al., Biosynthesis of phytosiderophores: in vitro biosynthesis of 2'-deoxymugineic acid from l-methionine and nicotianamine, Plant Physiol. 93 (1990) 1497–1503.

[91] A. Herbik, G. Koch, H.P. Mock, D. Dushkov, et al., Isolation, characterization and cDNA cloning of nicotianamine synthase from barley. A key enzyme for iron homeostasis in plants, Eur. J. Biochem. 265 (1999) 231–239.

[92] K. Higuchi, K. Suzuki, H. Nakanishi, H. Yamaguchi, et al., Cloning of nicotianamine synthase genes, novel genes involved in the biosynthesis of phytosiderophores, Plant Physiol. 119 (1999) 471–480.

[93] T. Kobayashi, M. Suzuki, H. Inoue, R.N. Itai, et al., Expression of iron-acquisition-related genes in iron-deficient rice is co-ordinately induced by partially conserved iron-deficiency-responsive elements, J. Exp. Bot. 56 (2005) 1305–1316.

[94] M.J. Garcia, V. Suarez, F.J. Romera, E. Alcantara, R. Perez-Vicente, A new model involving ethylene, nitric oxide and Fe to explain the regulation of Fe-acquisition genes in Strategy I plants, Plant Physiol. Biochem. 49 (2011) 537–544.

[95] F.J. Romera, M.J. Garcia, E. Alcantara, R. Perez-Vicente, Latest findings about the interplay of auxin, ethylene and nitric oxide in the regulation of Fe deficiency responses by Strategy I plants, Plant Signal. Behav. 6 (2011) 167–170.

[96] X. Liu, H. Cui, A. Li, M. Zhang, Y. Teng, The nitrate transporter NRT1.1 is involved in iron deficiency responses in Arabidopsis, J. Plant Nutr. Soil Sci. 178 (2015) 601–608.

[97] S. Munos, C. Cazettes, C. Fizames, F. Gaymard, et al., Transcript profiling in the chl1-5 mutant of Arabidopsis reveals a role of the nitrate transporter NRT1.1 in the regulation of another nitrate transporter, NRT2.1, Plant Cell 16 (2004) 2433–2447.

[98] S. Susin, J. Abian, M.L. Peleato, F. Sanchezbaeza, et al., FLavin excretion from roots of iron-deficient sugar-beet (Beta-vulgaris L.), Planta 193 (1994) 514–519.

[99] P. Siso-Terraza, J.J. Rios, J. Abadia, A. Abadia, A. Alvarez-Fernandez, Flavins secreted by roots of iron-deficient Beta vulgaris enable mining of ferric oxide via reductive mechanisms, New Phytol. (2015).

[100] U. Schagerlof, G. Wilson, H. Hebert, S. Al-Karadaghi, C. Hagerhall, Transmembrane topology of FRO2, a ferric chelate reductase from Arabidopsis thaliana, Plant Mol. Biol. 62 (2006) 215–221.

[101] A. Vorwieger, C. Gryczka, A. Czihal, D. Douchkov, et al., Iron assimilation and transcription factor controlled synthesis of riboflavin in plants, Planta 226 (2007) 147–158.

[102] B. Shimizu, 2-Oxoglutarate-dependent dioxygenases in the biosynthesis of simple coumarins, Front Plant Sci. 5 (2014) 549.

[103] K. Kai, M. Mizutani, N. Kawamura, R. Yamamoto, et al., Scopoletin is biosynthesized via ortho-hydroxylation of feruloyl CoA by a 2-oxoglutarate-dependent dioxygenase in Arabidopsis thaliana, Plant J. 55 (2008) 989–999.

[104] N.B. Schmid, R.F. Giehl, S. Doll, H.P. Mock, et al., F6'H1-dependent coumarins mediate iron acquisition from alkaline substrates in Arabidopsis thaliana, Plant Physiol. (2016) 2013.

[105] F. Haber, J. Weiss, The catalytic decomposition of hydrogen peroxide by iron salts, Proc. R. Soc. A: Math. Phys. Eng. Sci. 147 (1934) 332–351.

[106] J.P. Kehrer, The Haber–Weiss reaction and mechanisms of toxicity, Toxicology 149 (2000) 43–50.

[107] H.J.H. Fenton, Oxidation of tartaric acid in presence of iron, J. Chem. Soc. Trans. 65 (1894) 899–910.

[108] A. Ranieri, A. Castagna, B. Baldan, G.F. Soldatini, Iron deficiency differently affects peroxidase isoforms in sunflower, J. Exp. Bot. 52 (2001) 25–35.

[109] B. Sun, Y. Jing, K. Chen, L. Song, et al., Protective effect of nitric oxide on iron deficiency-induced oxidative stress in maize (Zea mays), J. Plant Physiol. 164 (2007) 536–543.

[110] P. Lamesch, T.Z. Berardini, D. Li, D. Swarbreck, et al., The Arabidopsis Information Resource (TAIR): improved gene annotation and new tools, Nucleic Acids Res. 40 (2012) D1202–1210.

[111] K. Apel, H. Hirt, Reactive oxygen species: metabolism, oxidative stress, and signal transduction, Annu. Rev. Plant Biol. 55 (2004) 373–399.

[112] R.C. Wilmouth, J.J. Turnbull, R.W. Welford, I.J. Clifton, et al., Structure and mechanism of anthocyanidin synthase from Arabidopsis thaliana, Structure 10 (2002) 93–103.

[113] L. Aravind, E.V. Koonin, The DNA-repair protein AlkB EGL-9, and leprecan define new families of 2-oxoglutarate- and iron-dependent dioxygenases, Genome Biol. 2 (2001), RESEARCH0007.

[114] M.A. McDonough, C. Loenarz, R. Chowdhury, I.J. Clifton, C.J. Schofield, Structural studies on human 2-oxoglutarate dependent oxygenases, Curr. Opin. Struct. Biol. 20 (2010) 659–672.

[115] N.B. Schmid, R.F. Giehl, S. Doll, H.P. Mock, et al., Feruloyl-CoA 6′-hydroxylase1-dependent coumarins mediate iron acquisition from alkaline substrates in Arabidopsis, Plant Physiol. 164 (2014) 160–172.

[116] U. Wittstock, M. Burow, Glucosinolate breakdown in Arabidopsis: mechanism, regulation and biological significance The Arabidopsis Book, 8, American Society of Plant Biologists, 2010, pp. e0134.

[117] J.R. Dinneny, T.A. Long, J.Y. Wang, J.W. Jung, et al., Cell identity mediates the response of Arabidopsis roots to abiotic stress, Science 320 (2008) 942–945.

[118] M.J. Garcia, C. Lucena, F.J. Romera, E. Alcantara, R. Perez-Vicente, Ethylene and nitric oxide involvement in the up-regulation of key genes related to iron acquisition and homeostasis in Arabidopsis, J. Exp. Bot. 61 (2010) 3885–3899.

[119] J.E. van de Mortel, L. Almar Villanueva, H. Schat, J. Kwekkeboom, et al., Large expression differences in genes for iron and zinc homeostasis, stress response, and lignin biosynthesis distinguish roots of Arabidopsis thaliana and the related metal hyperaccumulator Thlaspi caerulescens, Plant Physiol. 142 (2006) 1127–1147.

[120] T.J. Buckhout, T.J. Yang, W. Schmidt, Early iron-deficiency-induced transcriptional changes in Arabidopsis roots as revealed by microarray analyses, BMC Genomics 10 (2009) 147.

[121] A. Perea-Garcia, A. Garcia-Molina, N. Andres-Colas, F. Vera-Sirera, et al., Arabidopsis copper transport protein COPT2 participates in the cross talk between iron deficiency responses and low-phosphate signaling, Plant Physiol. 162 (2013) 180–194.

[122] A.B. Christensen, H. Thordal-Christensen, G. Zimmermann, T. Gjetting, et al., The germinlike protein GLP4 exhibits superoxide dismutase activity and is an important component of quantitative resistance in wheat and barley, Mol. Plant Microbe Interact. 17 (2004) 109–117.

[123] K. Yin, X. Han, Z. Xu, H. Xue, Arabidopsis GLP4 is localized to the Golgi and binds auxin in vitro, Acta Biochim. Biophys. Sin. (Shanghai) 41 (2009) 478–487.

[124] C.L. Ho, M. Noji, M. Saito, K. Saito, Regulation of serine biosynthesis in Arabidopsis: crucial role of plastidic 3-phosphoglycerate dehydrogenase in non-photosynthetic tissues, J. Biol. Chem. 274 (1999) 397–402.

[125] C. Vogel, G.M. Silva, E.M. Marcotte, Protein expression regulation under oxidative stress, Mol. Cell Proteomics 10 (2011), M111 009217.

[126] J. Rodriguez-Celma, I.C. Pan, W. Li, P. Lan, et al., The transcriptional response of Arabidopsis leaves to Fe deficiency, Front. Plant Sci. 4 (2013) 276.

[127] G.T. Hanke, S. Okutani, Y. Satomi, T. Takao, et al., Multiple iso-proteins of FNR in Arabidopsis: evidence for different contributions to chloroplast function and nitrogen assimilation, Plant Cell Environ. 28 (2005) 1146–1157.

[128] J.L. Luque-Garcia, G. Zhou, D.S. Spellman, T.T. Sun, T.A. Neubert, Analysis of electroblotted proteins by mass spectrometry: protein identification after Western blotting, Mol. Cell Proteomics 7 (2008) 308–314.

[129] S.F. Altschul, W. Gish, W. Miller, E.W. Myers, D.J. Lipman, Basic local alignment search tool, J. Mol. Biol. 215 (1990) 403–410.

15

The KnownLeaf literature curation system captures knowledge about *Arabidopsis* leaf growth and development and facilitates integrated data mining

Dóra Szakonyi [a,***], Sofie Van Landeghem [a,****], Katja Baerenfaller [b], Lieven Baeyens [a],
Jonas Blomme [a], Rubén Casanova-Sáez [c], Stefanie De Bodt [a], David Esteve-Bruna [c],
Fabio Fiorani [a,1], Nathalie Gonzalez [a], Jesper Grønlund [d], Richard G.H. Immink [e],
Sara Jover-Gil [c], Asuka Kuwabara [b], Tamara Muñoz-Nortes [c], Aalt D.J. van Dijk [e],
David Wilson-Sánchez [c], Vicky Buchanan-Wollaston [d], Gerco C. Angenent [e],
Yves Van de Peer [a,f], Dirk Inzé [a], José Luis Micol [c], Wilhelm Gruissem [b],
Sean Walsh [b,**], Pierre Hilson [a,g,*]

[a] Department of Plant Systems Biology, VIB, and Department of Plant Biotechnology and Bioinformatics, Ghent University, B-9052 Ghent, Belgium
[b] Department of Biology, ETH Zurich, CH-8093 Zurich, Switzerland
[c] Instituto de Bioingeniería, Universidad Miguel Hernández, 03202 Elche, Alicante, Spain
[d] Warwick Systems Biology Centre, and School of Life Sciences, University of Warwick, Coventry CV4 7AL, United Kingdom
[e] Plant Research International, Bioscience, 6708 PB Wageningen, The Netherlands
[f] Genomics Research Institute (GRI), University of Pretoria, Private Bag X20, Pretoria 0028, South Africa
[g] INRA, UMR1318, and AgroParisTech, Institut Jean-Pierre Bourgin, RD10, F-78000 Versailles, France

ARTICLE INFO

Keywords:
Arabidopsis
Leaf growth
Literature curation
Data integration

ABSTRACT

The information that connects genotypes and phenotypes is essentially embedded in research articles written in natural language. To facilitate access to this knowledge, we constructed a framework for the curation of the scientific literature studying the molecular mechanisms that control leaf growth and development in *Arabidopsis thaliana* (*Arabidopsis*). Standard structured statements, called relations, were designed to capture diverse data types, including phenotypes and gene expression linked to genotype description, growth conditions, genetic and molecular interactions, and details about molecular entities. Relations were then annotated from the literature, defining the relevant terms according to standard biomedical ontologies. This curation process was supported by a dedicated graphical user interface, called Leaf Knowtator. A total of 283 primary research articles were curated by a community of annotators, yielding 9947 relations monitored for consistency and over 12,500 references to *Arabidopsis* genes. This information was converted into a relational database (KnownLeaf) and merged with other public *Arabidopsis* resources relative to transcriptional networks, protein–protein interaction, gene co-expression, and additional molecular annotations. Within KnownLeaf, leaf phenotype data can be searched together with molecular data originating either from this curation initiative or from external public resources. Finally, we built a network (LeafNet) with a portion of the KnownLeaf database content to graphically represent the leaf phenotype relations in a molecular context, offering an intuitive starting point for knowledge mining. Literature curation efforts such as ours provide high quality structured information accessible to computational analysis, and thereby to a wide range of applications.

* Corresponding author at: Institut Jean-Pierre Bourgin, INRA Centre de Versailles-Grignon, Route de St. Cyr (RD10), F-78026 Versailles Cedex, France.
** Corresponding author at: Albert-Ludwigs-University of Freiburg, Center for BioSystems Analysis, Faculty of Biology, Habsburgerstr. 49, D-79104 Freiburg, Germany
*** Corresponding author at: Instituto Gulbenkian de Ciencia, Rua da Quinta Grande 6, 2780-156 Oeiras, Portugal.
**** Corresponding author at: VIB-UGent, Department of Plant Systems Biology, Technologiepark 927, B-9052 Zwijnaarde, Belgium.
E-mail addresses: dszakonyi@igc.gulbenkian.pt (D. Szakonyi), sofie.van.landeghem@gmail.com (S. Van Landeghem), sean.walsh@biologie.uni-freiburg.de (S. Walsh), pierre.hilson@versailles.inra.fr (P. Hilson).
[1] Present address: Institute of Bio- and Geosciences, IBG2: Plant Sciences, Forschungszentrum Jülich, Wilhelm-Johnen-Straße, 52428 Jülich, Germany.

DATA: The presented work was performed in the framework of the AGRON-OMICS project (Arabidopsis GRO wth Network integrating OMICS technologies) supported by European Commission 6th Framework Programme project (Grant number LSHG-CT-2006-037704). This is a data integration and data sharing portal collecting all the all the major results from the consortium. All data presented in our paper is available here. https://agronomics.ethz.ch/.

1. Introduction

An increasing number of research papers are published every year containing vast amounts of scientific data. It is therefore difficult to rigorously monitor even a subset of these publications for a topic of particular interest. With the publication process becoming essentially digital, the scientific community now develops tools to capture aspects of published information into databases that can be queried by researchers to reveal previously veiled gene or protein functions [1].

Much effort has been devoted to the development of automatic text mining, a field supported by a growing community, with some initiatives focused on the processing of plant-related textual data [2–7]. However, most advanced text mining methods are not generic enough to be applied to a novel domain without additional work or to provide the level of detail required for specific studies.

A complimentary approach to text mining is the creation of high-quality datasets through the manual curation of primary research articles. In this context, specific nuggets of information are extracted with a rich and controlled vocabulary, compatible with algorithmic processing to identify valued relationships [8,9]. One of the most challenging tasks is to transform free text descriptions of complex phenotypes into structured statements linked to corresponding genotypes. Phenotype/genotype datasets are extremely valuable because they summarize current knowledge, they may reveal unknown biological mechanisms, and they facilitate the comparative analysis of functional studies across species [1,10,11].

As part of the AGRON-OMICS project, we created a standard framework to collect various types of information about *Arabidopsis* leaf growth (Fig. 1). Special attention was paid to gather phenotype data with the corresponding genotypes and additional parameters that characterize them as described in primary research papers. The components of our integrated system include (1) lists of selected ontology terms, (2) relationships between different types of entities expressed in a constrained structure, (3) a customized curation interface, (4) a semi-automated quality control pipeline, (5) a relational database capturing the collected data, and (6) an integrated network summarizing data curated within this project together with pre-existing knowledge. Information from 283 articles was compiled by multiple curators and the quality of the dataset was assessed for consistency. The workflow was designed in such a way as to fully support future text mining methods to be built on top of this data collection. All computer programs and data are publicly available at http://www.agronomics.ethz.ch in the "Knowtator, KnownLeaf, LeafNet" section.

2. Materials and methods

2.1. Annotation software and files

The Leaf Knowtator annotation interface was built with the software Protégé version 3.3.1 (http://protege.cim3.net/download/old-releases/3.3.1/full/) and the Knowtator plug-in version 1.9 beta (http://sourceforge.net/projects/knowtator/files/Knowtator/knowtator-1.9-beta2/). The program is available for various operating systems including Windows, Mac OS X, AIX, Solaris, Linux, HP-UX, any Unix platform and other Java-enabled platforms. Protégé

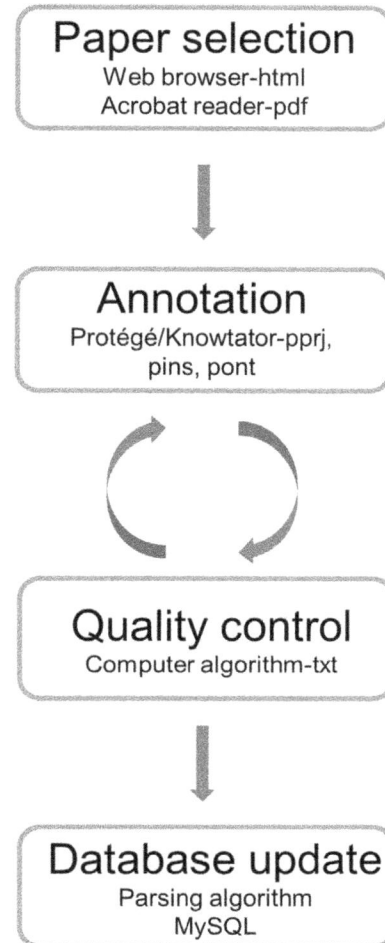

Fig. 1. Annotation workflow. Primary research papers were selected based on phenotypes, genes or interactions of interest. The full text html and pdf files were converted into a text format and imported into the Leaf Knowtator platform for manual annotation (.pont and .pins files contain domain classes and instances, respectively; the Protégé pprj project file identifies these files) The resulting data were exported from the Protégé software into individual XML files for each paper. These files were processed with dedicated in-house scripts for two distinct purposes. First, a quality control algorithm automatically detected predefined errors made during curation. Based on the output analytical text logs, the annotators corrected common inconsistencies within the Leaf Knowtator environment. The quality assessment and correction steps could be repeated until properly amended data files were obtained. Second, the corrected XMLs were parsed and flattened into the MySQL table embodying the KnownLeaf database.

requires Java 1.5 or a later version installed (http://www.java.com/en/download/index.jsp). Relations and corresponding slots were manually implemented with the Protégé/Knowtator tools. Ontology libraries were imported from publicly available onlice resources (Table S2).

The software was deployed as a standalone desktop application on each curator's computer and the Leaf Knowtator files (Leaf Knowtator.pprj, Leaf Knowtator.pins, Leaf Knowtator.pont) were shared. The annotations resulting from the curation of each individual paper were exported as an XML file from Protégé/Knowtator.

2.2. Databases

The MySQL database server (v5.x series) was used throughout together with the MySQL WorkBench graphical client tool. The annotations created by each curator were exported from Leaf Knowtator to produce one XML file per article, with a self-contained representation of the annotations consistent with the Protégé-Knowtator data-model, together with meta-data. Annotated phrases link to annotated classes, themselves compositions of slots associated with values (the ontology terms) through internally generated string identifiers. This complex network of classes and pointers was transformed with Perl ("knowtator2table.pl") into (1) tables in which each row represents a straight link between an annotated class, the corresponding text phrase and the assigned ontology terms or collection of terms, and (2) records to view each annotation as a collection of key value pairs. These tables and records were queried and viewed to identify missing or inconsistent annotations and to track progress.

The annotation tables were parsed into a MySQL relational database form with a Perl script ("PhenotypesEtc.pl"). The table **knowtator** represents the input tabular format with an auto-increment numeric row identifier ("id"). The table **knowtator_agi** holds references to AGI codes and has the id as a foreign key. The table **knowtator_papers** holds bibliographic information resolved by PubMed identifier with NCBI's efetch tool (http://eutils.ncbi.nlm.nih.gov/entrez/eutils/efetch.fcgi), together with the id as a foreign key, and the curator names per paper. In addition to data mining, this MySQL database served to generate summaries about content and to track the curation progress and the incremental growth of the data at a fine-grained level.

Additional publicly available sources of knowledge, such as gene and protein information (TAIR), protein–protein interactions, the *Arabidopsis* regulatory network (AGRIS AtRegNet) and the gene co-expression network (ATTED-II), were downloaded and parsed with Perl into a database format for uploading. These data and resources can be conveniently joined, filtered and summarized with structured query language (SQL). The database stored routines *user_get_knowtator_edges_for_cytoscape* and *user_get_knowtator_nodes_for_cytoscape* encode the filtering and joining functionality to develop the Cytoscape network files. Cytoscape version 2.8.3 was used throughout. The database and Perl scripts can be downloaded at www.agronomics.ethz.ch together with instructions on how to install the tables and data on an instance of the MySQL database server.

2.3. Quality control

The quality control script, applied to assist the expert curators to enhance data consistency and completeness, has been implemented as a Java (JDK 5) standalone program in Netbeans IDE 7.2. As input, the program receives a set of Knowtator-exported XML files, which are subsequently parsed and checked for completeness. Next, a quality check pipeline is run consisting of three different modules.

First, the program imposes the relation-slot structures as detailed in the annotation guidelines (Table S3), and reports any plausible violation of these guidelines, such as a missing required slot. These results are written to a human-readable log file, which can be opened with any text editor. Second, a quality module investigated the consistent usage of ontology terms. (i) The many text spans assigned to the same ontology term were listed together. In this way, the term "leaf_PO:0025034" was found to correctly relate to text spans such as "leaf" and "leaves", while the text fragment

"rosette leaves" was reassigned to the more specific term "rosette leaf_PO:0000014". (ii) Instances where the same text referred to multiple ontology terms were also reported. For example, the word "irregular" was found to refer to the ontology term "abnormal" (PATO:0000460), but also "variant" (PATO:0001227). It may be acceptable to assign a different ontology term depending on context, but in most cases it is desirable to review synonymous annotations to increase consistency in the ontological assignments. (iii) Additional error logs reported the usage of undefined ontology terms as well as manually entered information not included in a relation (e.g. a highlighted gene name with no other linked data).

All quality logs produced by these fully automated scripts include the article identifier and the original textual information, enabling a fast look-up by the expert annotators in the Leaf Knowtator program. The annotator may choose to adjust the annotations or to ignore the log output when the annotations are deemed correct. After adjustments are made, the new XML file can be exported from Leaf Knowtator, and the process repeated until no more changes are necessary.

2.4. Training

The recruitment of community curators was supported with manuals and reference documents for the KnownLeaf annotation scheme and the Leaf Knowtator curation interface. It is recommended to first learn about the annotation scheme in the 'Annotation structures' document that provides a description of the ten relation categories and the corresponding table-like slot system. The 'Knowtator manual' is a hands-on guide to the Knowtator plug-in, including installation instructions, an introduction to the major curation functions, and general instructions on how to build an annotation project in the Knowtator framework. A 'Training document' with annotation solutions presents various examples that illustrate the different aspects of the curation process, useful for understanding the annotation practices before working on unannotated texts. All KnownLeaf project files can be downloaded at http://www.agronomics.ethz.ch, including the described annotation scheme, the original training document and a copy of the fully annotated training document.

3. Results

3.1. Data collection

Our first objective was to design and implement a framework to collect information about genes reported to be involved in leaf development (Fig. 1). We focused on phenotypes resulting from genetic alterations, but additional relevant relations were also captured, such as genetic interactions and protein–protein interactions (Table 1). Our annotation effort was focused on a specific subdomain of the available knowledge by imposing the following restrictions. (1) Data were acquired solely on the model organism *Arabidopsis thaliana*, because it offers the richest body of literature describing the molecular and genetic control of plant development. This initial choice does not preclude the later inclusion of articles describing gene functions and leaf phenotypes in other plant species, in particular major crops. (2) Statements were recorded if they referred to leaves, cotyledons, meristems, or the apical part of an embryo. (3) Text curation was limited to the Results sections of primary research articles, excluding the Introduction, Discussion and Supplemental data sections, to include actual data but avoid repetitions. For the same reasons, review articles were not taken into consideration.

The annotation of research articles was completed in two successive phases. The KnownLeaf system was initially developed on the basis of 174 publications curated by the reference annotator,

Table 1
Relation categories.

Relation	Example	References[a]
Phenotype	The rot3-2 allele causes enlarged leaf blades	Kim et al. (1999)
Gene expression	1-h BR treatments resulted in increased EXO transcript levels in...wild-type...plants	Coll-Garcia et al. (2004)
Feature	AtCPL2 contains one dsRNA-binding domain	Koiwa et al. (2002)
DNA–protein interaction	ARF2...bound to the promoter region of GH3.1	Wang et al. (2011)
Protein–protein interaction	AN3 interacted strongly with...AtGRF9	Horiguchi et al. (2005)
Genetic interaction	hyl1...appeared to suppress the as2 phenotypes	Xu et al. (2006)
Process	RHL2...involved during endocycles	Sugimoto-Shirasu et al. (2002)
Regulation of gene expression	AtCPL1...negative regulators of RD29A expression	Koiwa et al. (2002)
Regulation of process	AN3...promoting...cell proliferation	Horiguchi et al. (2005)
Regulation of phenotype	PHABULOSA...influence leaf shape	Garcia et al. (2006)

[a] Full references in Table S1.

to define the main principles of the process and to establish the annotation structure (Table S1). This initial set of articles was selected because they reported notable progress in the field of leaf growth and development, and described the function of key genes in relation with relevant mutant phenotypes. In the second phase, twelve community annotators, recruited within five laboratories part of the AGRON-OMICS consortium, were trained to work with the established tools, and curated an additional 109 articles, chosen because these were of particular interest to the respective contributors. The quality of the resulting composite data set was monitored through manual inspections, reiterative feedback between the reference and community annotators, and semi-automated tests as described below.

3.2. Leaf Knowtator, a custom-made curation tool

Powerful and adaptable tools are required to capture complex information and relationships involving leaf development in a structured and detailed framework. To this end, we chose the open-source software Protégé that was specifically created for ontology development and knowledge acquisition, and supported by an international user community [12]. Among the add-ons that expand the functions of this platform, the Knowtator plugin was designed to annotate text [13]. In combination, Protégé and Knowtator provide a flexible tool to build customized curation projects.

Ontology libraries or structured vocabularies were imported together with the hierarchical organization of their terms (Table S2). Ontologies and relations were entered as classes and each recorded statement defined a separate instance (according to Protégé definitions). Relations consisted of multiple slots in a tabular structure. Constraints (facets, in Protégé) were linked to the values contained in the slots.

Within the Leaf Knowtator interface (Fig. 2), the full text of the original research article is displayed in the center. There, the annotator selects a portion of the text that carries semantic knowledge and creates novel relation annotations accordingly by selecting from a list of relation categories. Each such relation is built up with a set of predefined information "slots" which appear automatically after selecting a relation type. These slots are filled in by the annotator who selects the relevant parts of the original text and tags them with an ontology term when appropriate. These textual fragments do not need to be contiguous nor originating from the same sentence, although this is most often the case. Additionally, the system automatically logs the name of the annotator, the date when the given relation was created, and the exact location of the tagged text span within the curated document.

As a result, both the original text and the attached ontology terms were linked together via the slot name. Often, the information implied in the original text was not clearly spelled out, in which case the appropriate ontology term was entered in the relevant slot without an explicit link to specific words in the original article. Alternatively, non-required slots could be left empty. A full example of a phenotype relation annotation is shown in Table 2 and its structure is explained in more detail in the next sections.

3.3. Definition of relation categories

Ten different categories of structured statements or "relations" were created to capture the data published on leaf growth and development (Tables 1 and 3). The relation names, in bold and italicized hereafter, are defined as follows. *Phenotype* entries contain morphological descriptions about wild-type, mutant or transgenic plants. *Gene expression* relations encode observations about RNA or protein levels corresponding to a given gene, in wild-type, mutant or transgenic background. They relate to molecular experiments such as (q)RT-PCR assays, microarray analyses, Northern and Western blots or capture information about gene expression patterns studied with a range of microscopy methods. The *Feature* category includes records of structural elements in DNA, RNA or protein molecules, e.g. promoter elements bound by transcription factors, miRNA target sites or protein domains. The *DNA–protein*

Table 2
Example of a *Phenotype* annotation table. *Annotated sentence*: The reduced leaf area in the *hub1-1* mutant was confirmed by morphological measurements of the fully expanded leaves 1 and 2 [57].

Information type	Slot	Example	
		Annotated text	Entry
Genotype	Genotype	hub1-1	mutated gene_MI:0804
	Gene	hub1	RDO4 HUB1_AT2G44950
	Genotype Zygosity		homozygous diploid _APO:0000229
	Mutant LOF_GOF[a] Mutant type		loss of function_APO:0000011
Phenotype	Plant part	leaf	leaf_PO:0025034
	Localization	leaves 1 and 2	juvenile leaf_PO:0006339
	Property Process	area	area_PATO:0001323
	Value	reduced	decreased area_PATO:0002058
	Factuality Developmental stage	fully expanded leaves	3 leaf fully expanded_PO:0001053
Environment	Growth condition		
Experiment	Methodology		

[a] LOF, loss of function; GOF, gain of function.

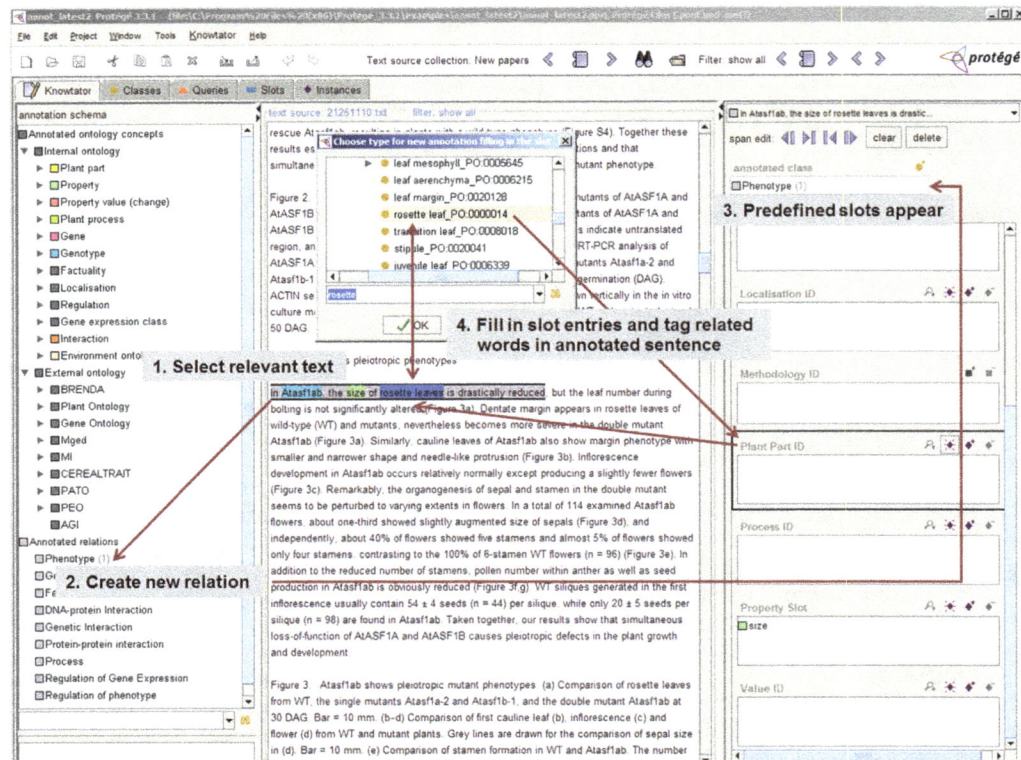

Fig. 2. Leaf Knowtator interface. The Protégé/Knowtator interface consists of three main windows. The middle window displays the full text of a research paper. The relation categories and ontology collections are listed in the window on the left. Relation slots appear on the right side window as a series of smaller panels. When annotating an article, the curated sentence or sentence fragments are first selected and highlighted (gray background, black upper and lower lines) in the large middle window. To start recording a new statement, the relevant relation category is selected in the left window, resulting in the presentation of the corresponding slot panels. Each slot entry is then typed in or selected from ontology menus if applicable and tagged to the corresponding words in the middle window (color highlights).

Table 3
Summary of the KnownLeaf database content.

Relation category	# Annotations	# Unique AGI	Ratio
Phenotype	5608	381	14.7
Gene expression	4767	704	6.8
Genetic interaction	658	186	3.5
Feature	462	175	2.6
Protein–protein interaction	310	121	2.6
Process	235	140	1.7
Regulation of gene expression	204	70	2.9
Regulation of process	178	85	2.1
DNA–protein interaction	92	47	2.0
Regulation of phenotype	20	17	1.2
Total	12,534	883	14.2

interaction relations report the direct interaction between a protein and a target DNA molecule as established experimentally, for example with mobility shift or yeast one-hybrid assays, or chromatin immunoprecipitation. Similarly, **Protein–protein interaction** reports a direct interaction between two protein molecules, as determined by yeast two-hybrid assays, *in vitro* affinity enrichment experiments, co-immunoprecipitation, FRET assays, or split molecular tag studies. **Genetic interactions** report those relations. **Process** relations correspond to sentences with information about the biological or molecular function of a given gene or the corresponding gene product (RNA, protein). **Regulation of gene expression** reflects the functional activation or repression of genes rather than a direct mechanistic binding, which is comprised in the category **DNA–protein interaction**. In this context, the regulation can take place at the DNA, RNA or protein level, for example via the action of transcription factors, epigenetic marks, small RNAs

guiding transcript cleavage, or ubiquitin labels targeting proteins for degradation. The remaining two relation categories include general statements describing the involvement of genes or gene products in either biological processes, **Regulation of process**, or phenotype, **Regulation of phenotype** without additional information. While not exhaustive, the factual results recorded with these ten distinct categories are sufficient for this scope.

3.4. Structure of the relations

Relations are defined in terms of specific parcels of data ("slots"), each containing a different type of information. A slot has a self-explanatory name such as 'Plant part' or 'Growth condition', can be linked to the corresponding words in the article text, and contains a single value to enable a seamless import of recorded relations into relational databases. In most cases, the data in a slot is a structured ontology term (see next section). However, some slots allow for free-text entries when a relevant ontology is not available or not detailed or extensive enough, such as the description of the experimental methodology. The structure of the ten relation categories is provided in Table S3. For clarity, slot names are hereafter italicized and underlined.

The principle of the annotation method is illustrated with a particular **Phenotype** relation in Table 2. The different slots (second column) of a relation can be logically grouped into a few high-level categories or "information units" such as Genotype, Phenotype and Environment (first column).

1. **Genotype.** This information unit consists of several slots specifying and identifying the relevant mutation and corresponding gene and mutant type. First, the required *Genotype* slot holds

one of the following values: (a) "Wild type_SO:0000817" when wild-type plants were studied (numbers here and below refer to the unique identifier in the listed ontology). (b) The "mutated gene_MI:0804" term for mutants with a known defective allele, irrespective of the exact nature of the mutation, except plants stably transformed with constructs that resulted in decreased gene expression that were labeled with the "knock down_MI:0789" term. (c) Overexpressor plants were labeled with the "over expressed level_MI:0506" term when they carried constructs for ectopic expression resulting in elevated RNA or protein levels. (d) A "Complex genotype" label captures transgenes or genetic configurations that cannot be described with any of the above terms, such as heterologous promoter-gene constructs or overexpressed transgenes in mutant backgrounds.

A *Gene* slot captures the appropriate AGI code for mutants, overexpressors and complex genotype entries, or a "no AGI" mark when there is no applicable code. Through the *Mutant LOF_GOF* slot, mutants can be further characterized by the nature of the mutation (e.g. LOF, "loss of function_APO:0000011" or GOF, "gain of function_APO:0000010") and the zygosity is stored in the *Genotype Zygosity* slot (e.g. "homozygous diploid_APO:0000229" or "heterozygous diploid_APO:0000230"). Finally, additional details about a mutation such as mutagen, allele, site of the lesion and exact change, can be recorded as free text in the slot *Mutant type*.

2. **Phenotype**. The phenotypic information unit is documented in a format reminiscent of the entity-attribute-value (EAV) model [14]. First, the entity is recorded in the *Plant part* slot (e.g. "leaf_PO:0025034"). Then, the attribute under consideration is subsequently filled into the *Property* slot (e.g. "area_PATO:0001323") to indicate the plant feature that was studied. Finally, the *Value* slot indicates the change of that specific feature (e.g. "decreased area_PATO:0002058").

More detailed and flexible annotations can be introduced with a few additional slots. The *Localization* slot further specifies the plant part, for example to define in which organ a given cell type was located or when the subject is a subset of the original plant part term. The *Developmental stage* of the plant part can be recorded with the corresponding ontology term. GO terms can be entered in the *Process* slot. Finally, the *Factuality* slot qualifies certain statements with *ad hoc* labels such as 'negation' and 'speculation', to respectively mark negative statements or capture recorded statements that are suggested by the experimental data but not fully supported by additional evidence.

3. **Additional information types**. The details of plant growth conditions, special treatments or stress circumstances ("Environment") were entered in the *Growth condition* slot only if they were specifically stated in the Results section of the paper. These records relate to conditions that differ from the standard environment or have a direct effect on the phenotype. They included specific Plant Environment Ontology (EO) terms or in a few cases the CHEBI identifier of the chemical with which plants were treated. The *Methodology slot* (information type "Experiment") further allows for free text entries to describe the experimental method used.

The slots in the other nine relation categories are organized according to a similar scheme as detailed in the 'Annotation structure' section in Supplemental Information (Table S3).

3.5. Ontologies for standardized statements

Importantly, we adopted well-developed and widely accepted biological ontologies to build our annotations. The inclusion of such structured vocabularies enhances the interoperability of the resulting data, facilitating data integration and providing the proper basis for complex queries and computer reasoning. Concretely, whenever possible, the relevant words in the original articles were tagged with ontology terms from authoritative resources (Tables S2 and S4).

Plant organs, tissues and cell types were described with terms defined in the Plant Ontology (PO) and, in rare cases, in the BRENDA Tissue Ontology (BTO) [15,16]. Subcellular components were marked with Gene Ontology (GO) identifiers [17]. GO entries also provided information about biological processes. Plant traits and features, and changes that affect them, were described with terms from the Phenotype, Attribute and Trait Ontology (PATO) [12], Plant Trait Ontology (TO) [18], and BTO [15].

Relevant growth conditions or treatments were labeled with Plant Environment Ontology (EO) when specifically mentioned in the annotated Results section [18,19]. Unfortunately, the description of growth conditions and experimental treatments is generally not standardized [20,21], details describing experimental growth conditions vary widely between articles and are often cursory, and public ontologies addressing this semantic field are still under development. Therefore, it remains problematic to undertake coherent comparative or integrative analysis on the basis of fragmentary data, despite the basic relationship between phenotypes and the environment in which they are expressed [22,23]. Thus, exhaustive records of environmental conditions were not included.

Records about molecular events, modifications and interactions were defined with the Molecular Interaction (MI) standard assembled by the Proteomics Standards Initiative [19]. Finally, genes were mapped to their unique identifiers based on the *Arabidopsis* Genome Initiative (AGI) format and extracted from the latest TAIR10 genome annotation published by The *Arabidopsis* Information Resource [24]. While these common ontology libraries provided most necessary terms, they lacked some of the concepts important in leaf development biology. In such cases, we introduced *ad hoc* terms or included terms from alternative ontology systems, always ensuring consistency throughout the framework.

3.6. Monitoring relation consistency

Literature curation is prone to occasional errors and inconsistencies, especially when carried out by multiple contributors. To deliver a coherent data set, the annotation effort was designed with rigorous guidelines and community curators were trained during hands-on sessions backed up by documentation explaining the details of the annotation scheme and customized functions within Leaf Knowtator (see Supplemental Files, including an annotation manual and training documents). In addition, the quality of the records was monitored throughout the project with scripts designed to detect different types of errors that were subsequently corrected. Fully automated scripts first validated the completeness of the relation annotations, i.e. whether all required slots had been filled. Next, the consistency of ontology terms was automatically verified. Finally, orphan annotations or seemingly undefined ontology terms were also reported. Details of the quality control scripts are provided in Section 2.

The curators examined the resulting logs and the relations were adjusted when necessary. In rare cases, it was impossible to enter data in all required slots because textual descriptions were ambiguous or incomplete. While such annotations violated the initial guidelines, their information value often justified their inclusion in the final version of our database. As expected, the relations produced initially by the sole reference annotator – and main developer of Leaf Knowtator – were highly consistent and complete. Based on 174 curated articles, the quality control script reported on average only 3.1 missing required slots per article (19,267 required

External resources   LeafNet   Leaf Knowtator curated

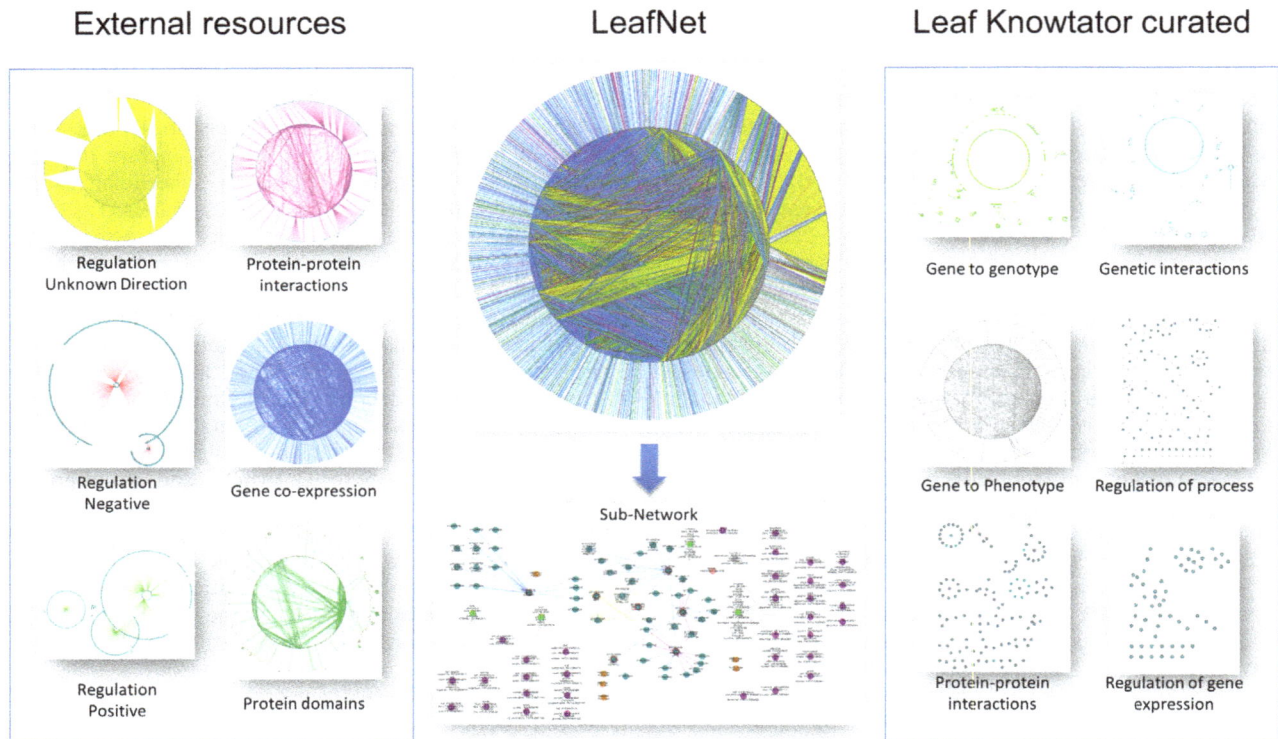

Fig. 3. Cytoscape representation of LeafNet. LeafNet is a composite assembly of publicly available knowledge resources (left panel) and Knowtator curated (right panel). The large central network represents the merged information from which sub-networks, including genes of interest, are derived using standard Cytoscape functionality. The sub-networks are practical to inspect the connectivity between genes of interest, the corresponding mutants and their phenotypes within the multi-faceted knowledge network landscape. See Fig. 4 for the details of the pictured subnetwork and the color code.

slots in total or an average of 111 slots per article; 2.8% missing slots). Note that this number is not expected to reach zero due to incomplete textual information. In contrast, the relations encoded originally by twelve community annotators contained on average 4.9 missing slots per article (369 missing slots among 12,122 slots, for 75 articles). After two rounds of automated evaluation with the quality control script followed by corrections, this average number dropped to 2.8, thus matching the quality of the initial set. Taking into account the total number of relations and corresponding mandatory slots, less than 2% of the required information could not be identified in the text.

3.7. The KnownLeaf database

XML files resulting from the curation of each annotated article were converted, flattened and parsed into database tables. The resulting KnownLeaf relational database consists of three tables: **knowtator** contains the annotated relations and has database foreign keys to **knowtator_papers** with bibliographic information and to **knowtator_agi** with AGI code references. The final database contains 9947 relations in the **knowtator** table with a total of 12,534 references to AGI codes, corresponding to 883 unique genes (3.2% of the TAIR10 protein coding genes) with on average 14.2 references per gene. Table 3 shows a breakdown of AGI references by annotated relation type in the completed data set. **Phenotype** statements form the largest category and refer to 381 distinct *Arabidopsis* genes, on average 14.7 statements per gene across the text corpus, highlighting the dense phenotype knowledge extracted from the literature. Statements on ***Gene expression*** are fewer but refer to more genes because the transcription of multiple genes was often described in a single mutant background or plant part.

A wide range of queries can be performed within the KnownLeaf database. For example, the annotated relations can be looked up

for a given *Arabidopsis* gene or protein; genes can be searched that are linked to a particular plant part or to specific phenotype alterations (see example SQL within https://www.agronomics.ethz.ch/knowtator/code_repository.zip). To provide molecular context to our records, publicly available *Arabidopsis* resources were merged into the database, including transcriptional networks (AtRegNet from AGRIS) [25,26], *Arabidopsis* protein interactome (AI-1) [27] gene co-expression measures (ATTED-II) [28], and TAIR gene and protein annotations [24]. Thus, complex queries within KnownLeaf may combine information embedded in the Knowtator-curated annotations together with molecular data from additional public resources (Fig. 3).

3.8. Graphical representation of leaf phenotype records in a molecular context

As an alternative to structured command line queries, an interactive graph was created with the Cytoscape tool [29], in which Knowtator relations are part of a larger network of objects. The data residing in KnownLeaf were joined and simplified in several ways to produce a graph with information pertinent to interpret and further expand these relations.

Co-expression edges were only represented below a threshold value (ATTED-II co-expression mutual rank score < 25) to reduce the noise in the co-expression dataset. Next, the network was seeded with a list of AGI codes including (i) all 883 genes curated via Knowtator (Table 3) and (ii) 111 genes coding for proteins whose levels vary significantly across leaf development as determined by iTRAQ (according to more stringent cut-off criteria than previously reported in Baerenfaller, et al. [22]; global fold change > 2.8, p-value < 0.05) (Table S5). This combined set counted 977 non-redundant AGIs. Finally, these initial gene nodes were enriched with neighboring gene nodes through the connectivity

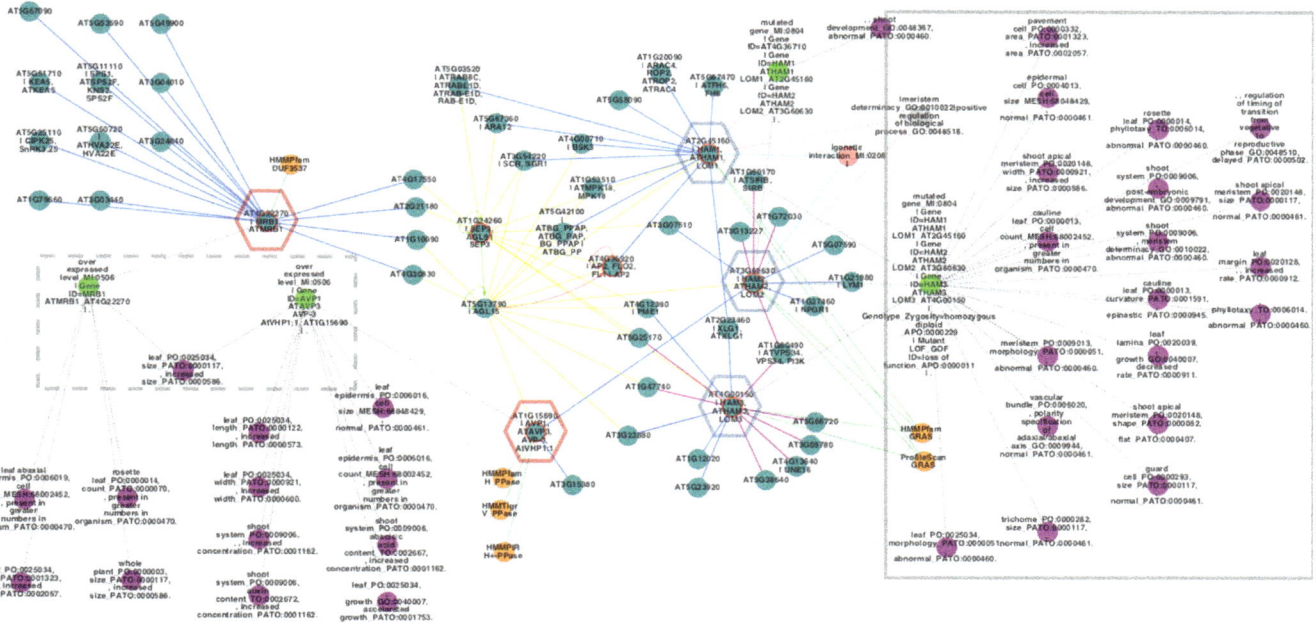

Fig. 4. LeafNet neighborhood around MRB1 and AVP1. The sub-network was derived by querying for MRB1, AVP1 and LOM* with Cytoscape's Extended Search. The color code is as follows. Nodes: AGI referring to gene or protein, teal; genotype, bright green; phenotype, purple; protein domain, orange; regulation of process, gray. Edges: AGRIS AtRegNet transcriptional regulations; yellow, red and green, representing regulation with unknown, negative and positive direction, respectively; protein–protein interactions, light purple; co-expression, blue; between AGI nodes and protein domains, light green; between AGI and genotype, orange; genetic interactions between AGIs, light blue; between AGI and phenotype, gray; between AGI and regulation of process, pink; protein–protein interaction, dark purple; indicating regulation of gene expression, olive green. Edges from external resources are solid lines, whilst Knowtator curated edges are dashed lines. The gray box highlights the nodes and edges collectively describing the phenotype of the *lom1 lom2 lom3* triple mutant. The LOM1, LOM2 and LOM3 AGI nodes are framed in blue hexagons. The dashed gray box highlights the nodes and edges indicating that *MRB1* and *AVP1* mutant plants have common leaf phenotypes. The MRB1 and AVP1 AGI nodes are framed in red hexagons.

defined by the public molecular resources incorporated within the KnownLeaf database (above).

The corresponding network, known as LeafNet, contains 19,055 nodes connected by a total of 39,649 edges, combining gene/protein–phenotype relationships with molecular information (Fig. 3). In LeafNet, the information about molecular functions, collected from the primary literature via Knowtator, complements that from the public resources. For example, 123 non-redundant protein–protein interactions were annotated involving 121 proteins. Of these, 41 (33%) overlap with those found in the AI-1 interactome, in line with the intersection commonly found between literature-curated datasets and high-throughput yeast-two hybrid datasets [7]. All the components of the software system have been made available (see Materials & Methods for details) and can be modified to create alternative network versions by adjusting threshold values or by seeding with different AGI code sets.

LeafNet is a starting point for knowledge mining. Within Cytoscape, genes, proteins or mutants can be searched with AGI codes or synonymous names (Enhanced Search plugin), their network neighborhood visualized (Select > First Neighbors of Selected Nodes), and new sub-networks created (File > New > Network > From Selected Nodes, All Edges). Combining automated and manual layout, the network context of genes/proteins of interest can be inspected and help formulate novel hypotheses. The following use-case illustrates this process with a specific example.

In *Arabidopsis*, the overexpression of both *MEMBRANE RELATED BIGGER1* (*MRB1*) and *ARABIDOPSIS VACUOLAR PYROPHOSPHATASE1* (*AVP1*) results in large leaves producing more cells, although of equal size, compared to wild type [30,31] (in red hexagons; Fig. 4). Their common phenotype is represented by connections in their LeafNet neighborhood (purple phenotype nodes and green mutant nodes in the gray box). Searching for potential regulatory relationships involving *MRB1* and *AVP1*, we noticed that they are linked

in LeafNet through a path including *AGAMOUS-like 15* (*AGL15*) and *LOST MERISTEMS 2* (*LOM2*). AGL15 is a MADS-domain transcription factor that controls somatic growth [32–34]. Plants that ectopically express *AGL15* have pleiotropic mutant phenotypes, including a defect in leaf morphology [33]. LOM2 is a GRAS transcription factor [35] with two close homologs in *Arabidopsis*, LOM1 and LOM3. The *lom1 lom2 lom3* triple mutant shows an abnormal leaf morphology, indicating that all three *LOM* genes regulate cell division and cell differentiation (in dashed gray box) [36,37] (nodes in blue hexagons). This subnetwork suggests that the possible co-regulation of *MRB1* and *AVP1* by AGL15 and LOM transcription factors could be considered. While this simplified graph built with data from diverse origins may not completely or faultlessly represent actual functional links, it is useful to visualize plausible connections that warrant additional investigation.

4. Discussion

Our workflow was developed to record, among other data types, anatomical details in phenotype description (entity), what changes in that plant part or cells (attribute), and in simple terms how it changes (value). Compared to other biocuration and text mining efforts, Leaf Knowtator captures more detailed information about leaf growth and development than, for example, the more general TAIR workflow [8] or the generic large-scale text mining resource EVEX [38]. Additionally, this project was not restricted to information available in the abstract of the curated articles, but targeted any relevant sentences found in the results section of full-text articles. While the rich format of our resource resulted in a more time-consuming project, the resulting high-quality data will be a strong asset in future leaf development studies due to the high complementarity to other relevant resources.

The single most time-consuming step during annotation was the tagging of text spans in the original text of the curated article with the elements (slots) recorded in the relations. We included this constraint in the workflow to allow for semi-automated error checks, as described, and, importantly, to assist future text mining efforts. Indeed, the text mining research field is currently dominated by machine learning methods, in which lexical and grammatical patterns relevant to the problem domain are derived from manually curated training sets [39–42]. The machine learning algorithm subsequently identifies these patterns in unseen texts to predict novel annotations. The quality of such predictions relies heavily on the size and quality of the training sets. Throughout our project, we have ensured full compatibility of our methods and annotation scheme to future text mining efforts, by storing the exact offsets of the textual annotations within the original article files, thereby enabling the automatic retrieval of the specific sentence(s) and paragraph from which the information was deduced. The dimensions of our dataset are comparable to recent general-purpose annotation efforts [43], but it is unique in size and scope in the plant domain.

The specific slot-value structure defined for each relation category was also designed to facilitate future text mining efforts. By providing formal semantics and ontology terms, computer reasoning can be enhanced and well-structured annotations produced in a more time-efficient manner. For instance, the combination of manual curation, as described here, with partially automated extraction of textual information [44–46] would dramatically speed up literature curation projects. We view these opportunities as interesting follow-up work to this study.

The text corpus at the basis of this work is not exhaustive and can be expanded in several ways. Considering all primary research articles in which detailed leaf phenotypes can be linked to specific alleles of identified *Arabidopsis* genes (i.e. AGI codes), we estimate that the 283 papers we curated represent a third to a quarter of the relevant published literature. While the current dataset demonstrates the usefulness of high quality manual annotation, it would be even more valuable if it encompassed all targeted research results. Our system provides solid grounds to build up the resource by drawing in a larger community. The Leaf Knowtator interface can be adopted by any willing researcher, with documentation and training examples available to guide the first steps. Downstream software is available to monitor the consistency of the recorded relations, to transfer them into the KnownLeaf relational database, and to represent them graphically in an increased Cytoscape version of the LeafNet network.

Alternatively, Leaf Knowtator relations can be merged with large inventories of *Arabidopsis* mutations that are precious for their exhaustive coverage but that provide little detail about associated phenotypes [47,48]. Leaf Knowtator relations can also be imported into other *Arabidopsis* databases and online web tools designed to query large-scale datasets, for example TAIR [24], BAR [49], Genevestigator [11], VirtualPlant [50] or CORNET [51]. However, it is worthwhile emphasizing that such integrative systems do not yet include advanced functions to probe – beyond free text search or display – the connections between mutations in specific genes and corresponding phenotypes, stressing the usefulness of structured phenotype data enriched with ontology terms as presented here.

Through LeafNet, the integration of curated and reference knowledge resources showed that genes of interest can be placed into an informative molecular and phenotypic network landscape. LeafNet recapitulates aspects of what is already known, which is *per se* quite valuable as it joins together information dispersed in literature. In addition, it suggests new leads and close proximal associations that can be leveraged for hypothesis generation and testing. Moreover, the approach we developed could be extended and enhanced to help describe gene function in *Arabidopsis*, since a vast number of unknown or partially described genes have been placed into a molecular network landscape that goes beyond the usual GO descriptors or homology reports from sequence analysis based annotations.

An additional benefit of ontology-based phenotype descriptions is that they facilitate the comparison of phenomena between related species. For example, the Plant Ontology (PO) community is continuously improving its structured term lists to include anatomical entities that reflect the organizing principle of the plant body, thereby enabling interspecific comparisons of gene expression, phenotypes and gene functions [52]. At the other end of the research spectrum, ecologists, agronomists and breeders are also codifying trait descriptions with the implementation of dedicated ontologies to integrate field observations and measurements across experimental sites and for different species, including crops [53,54]. As our understanding of the functional modules that govern plant growth and development improves, information formatted through studies such as this one, focusing on mutant phenotypes in one plant species, could eventually assist trait development efforts in another.

To conclude, we have mustered the good will of about 15 biologists and distilled a sizable portion of the published information describing the molecular control of leaf growth and development. As the *Arabidopsis* community builds up its international bioinformatics infrastructure [55,56], we suggest that small initiatives similar to ours focusing on complementary biological domains could together contribute significantly to the inclusion of phenotype information in reference resources.

Acknowledgments

We thank Annick Bleys for help with the manuscript. We thank Dr Georgios Gkoutos for advise on the potential for machine reasoning with the curated ontological relations. This work was supported by the Integrated Project AGRON-OMICS, in the Sixth Framework Programme of the European Commission (LSHG-CT-2006-037704). S.W. was supported by TiMet - Linking the Clock to Metabolism, a Collaborative Project (Grant Agreement 245143) funded by the European Commission FP7, in response to call FP7-KBBE-2009-3. S.V.L. thanks the Research Foundation Flanders (FWO) for funding her research. A.D.J.v.D. was supported by the transPLANT project (funded by the European Commission within its 7th Framework Programme under the thematic area 'Infrastructures', contract number 283496). The predoctoral fellowship of J.B. was sponsored by the Agency for Innovation by Science and Technology in Flanders (IWT-Vlaanderen), a predoctoral fellowship (project no. 111164).

References

[1] R. Hoehndorf, P.N. Schofield, G.V. Gkoutos, PhenomeNET: a whole-phenome approach to disease gene discovery, Nucleic Acids Res. 39 (2011) e119.

[2] V. Exner, P. Taranto, N. Schönrock, W. Gruissem, L. Hennig, Chromatin assembly factor CAF-1 is required for cellular differentiation during plant development, Development 133 (2006) 4163–4172.

[3] M. Krallinger, C. Rodriguez-Penagos, A. Tendulkar, A. Valencia, PLAN2L: a web tool for integrated text mining and literature-based bioentity relation extraction, Nucleic Acids Res. 37 (2009) W160–W165.

[4] A. Loyola, G. Almouzni, Histone chaperones, a supporting role in the limelight, Biochim. Biophys. Acta 1677 (2004) 3–11.

[5] H.-M. Müller, E.E. Kenny, P.W. Sternberg, Textpresso: an ontology-based information retrieval and extraction system for biological literature, PLoS Biol. 2 (2004) e309.

[6] S. Van Landeghem, S. De Bodt, Z.J. Drebert, D. Inzé, Y. Van de Peer, The potential of text mining in data integration and network biology for plant research: a case study on *Arabidopsis*, Plant Cell 25 (2013) 794–807.

[7] K. Venkatesan, J.-F. Rual, A. Vazquez, U. Stelzl, I. Lemmens, T. Hirozane-Kishikawa, T. Hao, M. Zenkner, X. Xin, K.-I. Goh, M.A. Yildirim, N. Simonis, K. Heinzmann, F. Gebreab, J.M. Sahalie, S. Cevik, C. Simon, A.-S. de Smet, E. Dann, A. Smolyar, A. Vinayagam, H. Yu, D. Szeto, H. Borick, A. Dricot, N. Klitgord, R.R. Murray, C. Lin, M. Lalowski, J. Timm, K. Rau, C. Boone, P. Braun, M.E. Cusick, F.P. Roth, D.E. Hill, J. Tavernier, E.E. Wanker, A.-L. Barabási, M. Vidal, An empirical framework for binary interactome mapping, Nat. Methods 6 (2009) 83–90.

[8] D. Li, T.Z. Berardini, R.J. Muller, E. Huala, Building an efficient curation workflow for the *Arabidopsis* literature corpus, Database 2012 (2012) bas047.

[9] N. Tsesmetzis, M. Couchman, J. Higgins, A. Smith, J.H. Doonan, G.J. Seifert, E.E. Schmidt, I. Vastrik, E. Birney, G. Wu, P. D'Eustachio, L.D. Stein, R.J. Morris, M.W. Bevan, S.V. Walsh, *Arabidopsis* reactome: a foundation knowledgebase for plant systems biology, Plant Cell 20 (2008) 1426–1436.

[10] J.N. Cobb, G. DeClerck, A. Greenberg, R. Clark, S. McCouch, Next-generation phenotyping: requirements and strategies for enhancing our understanding of genotype–phenotype relationships and its relevance to crop improvement, Theor. Appl. Genet. 126 (2013) 867–887.

[11] T. Hruz, O. Laule, G. Szabo, F. Wessendorp, S. Bleuler, L. Oertle, P. Widmayer, W. Gruissem, P. Zimmermann, Genevestigator v3: a reference expression database for the meta-analysis of transcriptomes, Adv. Bioinform. 2008 (2008) 420747.

[12] B. Smith, M. Ashburner, C. Rosse, J. Bard, W. Bug, W. Ceusters, L.J. Goldberg, K. Eilbeck, A. Ireland, C.J. Mungall, The OBI Consortium, N. Leontis, P. Rocca-Serra, A. Ruttenberg, S.-A. Sansone, R.H. Scheuermann, N. Shah, P.L. Whetzel, S. Lewis, The OBO Foundry: coordinated evolution of ontologies to support biomedical data integration, Nat. Biotechnol. 25 (2007) 1251–1255.

[13] P.V. Ogren, Knowtator: a protégé plug-in for annotated corpus construction, in: Proceedings of the Conference of the North American Chapter of the Association for Computational Linguistics: Human Language Technology, Companion Volume: Demonstrations, Association for Computational Linguistics, New York City, USA, 2006, pp. 273–275.

[14] N.S.B. Miyoshi, D.G. Pinheiro, W.A. Silva Jr., J.C. Felipe, Computational framework to support integration of biomolecular and clinical data within a translational approach, BMC Bioinform. 14 (2013) 180.

[15] M. Gremse, A. Chang, I. Schomburg, A. Grote, M. Scheer, C. Ebeling, D. Schomburg, The BRENDA Tissue Ontology (BTO): the first all-integrating ontology of all organisms for enzyme sources, Nucleic Acids Res. 39 (2011) D507–D513.

[16] P. Jaiswal, S. Avraham, K. Ilic, E.A. Kellogg, S. McCouch, A. Pujar, L. Reiser, S.Y. Rhee, M.M. Sachs, M. Schaeffer, L. Stein, P. Stevens, L. Vincent, D. Ware, F. Zapata, Plant Ontology (PO): a controlled vocabulary of plant structures and growth stages, Comp. Funct. Genomics 6 (2005) 388–397.

[17] M. Ashburner, C.A. Ball, J.A. Blake, D. Botstein, H. Butler, J.M. Cherry, A.P. Davis, K. Dolinski, S.S. Dwight, J.T. Eppig, M.A. Harris, D.P. Hill, L. Issel-Tarver, A. Kasarskis, S. Lewis, J.C. Matese, J.E. Richardson, M. Ringwald, G.M. Rubin, G. Sherlock, Gene ontology: tool for the unification of biology, Nat. Genet. 25 (2000) 25–29.

[18] C. Liang, P. Jaiswal, C. Hebbard, S. Avraham, E.S. Buckler, T. Casstevens, B. Hurwitz, S. McCouch, J. Ni, A. Pujar, D. Ravenscroft, L. Ren, W. Spooner, I. Tecle, J. Thomason, C.-W. Tung, X. Wei, I. Yap, K. Youens-Clark, D. Ware, L. Stein, Gramene: a growing plant comparative genomics resource, Nucleic Acids Res. 36 (2008) D947–D953.

[19] H. Hermjakob, L. Montecchi-Palazzi, G. Bader, R. Wojcik, L. Salwinski, A. Ceol, S. Moore, S. Orchard, U. Sarkans, C. von Mering, B. Roechert, S. Poux, E. Jung, H. Mersch, P. Kersey, M. Lappe, Y. Li, R. Zeng, D. Rana, M. Nikolski, H. Husi, C. Brun, K. Shanker, S.G.N. Grant, C. Sander, P. Bork, W. Zhu, A. Pandey, A. Brazma, B. Jacq, M. Vidal, D. Sherman, P. Legrain, G. Cesareni, L. Xenarios, D. Eisenberg, B. Steipe, C. Hogue, R. Apweiler, The HUPO PSI's molecular interaction format – a community standard for the representation of protein interaction data, Nat. Biotechnol. 22 (2004) 177–183.

[20] J. Hannemann, H. Poorter, B. Usadel, O.E. Bläsing, A. Finck, F. Tardieu, O.K. Atkin, T. Pons, M. Stitt, Y. Gibon, Xeml Lab: a tool that supports the design of experiments at a graphical interface and generates computer-readable metadata files, which capture information about genotypes, growth conditions, environmental perturbations and sampling strategy, Plant Cell Environ. 32 (2009) 1185–1200.

[21] H. Poorter, F. Fiorani, M. Stitt, U. Schurr, A. Finck, Y. Gibon, B. Usadel, R. Munns, O.K. Atkin, F. Tardieu, T.L. Ponsi, The art of growing plants for experimental purposes: a practical guide for the plant biologist, Funct. Plant Biol. 39 (2012) 821–838.

[22] K. Baerenfaller, C. Massonnet, S. Walsh, S. Baginsky, P. Bühlmann, L. Hennig, M. Hirsch-Hoffmann, K.A. Howell, S. Kahlau, A. Radziejwoski, D. Russenberger, D. Rutishauser, I. Small, D. Stekhoven, R. Sulpice, J. Svozil, N. Wuyts, M. Stitt, P. Hilson, C. Granier, W. Gruissem, Systems-based analysis of *Arabidopsis* leaf growth reveals adaptation to water deficit, Mol. Syst. Biol. 8 (2012) 606.

[23] C. Massonnet, D. Vile, J. Fabre, M.A. Hannah, C. Caldana, J. Lisec, G.T.S. Beemster, R.C. Meyer, G. Messerli, J.T. Gronlund, J. Perkovic, E. Wigmore, S. May, M.W. Bevan, C. Meyer, S. Rubio-Díaz, D. Weigel, J.L. Micol, V. Buchanan-Wollaston, F. Fiorani, S. Walsh, B. Rinn, W. Gruissem, P. Hilson, L. Hennig, L. Willmitzer, C. Granier, Probing the reproducibility of leaf growth and molecular phenotypes: a comparison of three *Arabidopsis* accessions cultivated in ten laboratories, Plant Physiol. 152 (2010) 2142–2157.

[24] P. Lamesch, T.Z. Berardini, D. Li, D. Swarbreck, C. Wilks, R. Sasidharan, R. Muller, K. Dreher, D.L. Alexander, M. Garcia-Hernandez, A.S. Karthikeyan, C.H. Lee, W.D. Nelson, L. Ploetz, S. Singh, A. Wensel, E. Huala, The *Arabidopsis* Information Resource (TAIR): improved gene annotation and new tools, Nucleic Acids Res. 40 (2012) D1202–D1210.

[25] S.K. Palaniswamy, S. James, H. Sun, R.S. Lamb, R.V. Davuluri, E. Grotewold, AGRIS and AtRegNet: a platform to link cis-regulatory elements and transcription factors into regulatory networks, Plant Physiol. 140 (2006) 818–829.

[26] A. Yilmaz, M.K. Mejia-Guerra, K. Kurz, X. Liang, L. Welch, E. Grotewold, AGRIS: the *Arabidopsis* Gene Regulatory Information Server, an update, Nucleic Acids Res. 39 (2011) D1118–D1122.

[27] Arabidopsis Interactome Mapping Consortium, Evidence for network evolution in an *Arabidopsis* interactome map, Science 333 (2011) 601–607.

[28] T. Obayashi, S. Hayashi, M. Saeki, H. Ohta, K. Kinoshita, ATTED-II provides coexpressed gene networks for *Arabidopsis*, Nucleic Acids Res. 37 (2009) D987–D991.

[29] P. Shannon, A. Markiel, O. Ozier, N.S. Baliga, J.T. Wang, D. Ramage, N. Amin, B. Schwikowski, T. Ideker, Cytoscape: a software environment for integrated models of biomolecular interaction networks, Genome Res. 13 (2003) 2498–2504.

[30] N. Gonzalez, S. De Bodt, R. Sulpice, Y. Jikumaru, E. Chae, S. Dhondt, T. Van Daele, L. De Milde, D. Weigel, Y. Kamiya, M. Stitt, G.T.S. Beemster, D. Inzé, Increased leaf size: different means to an end, Plant Physiol. 153 (2010) 1261–1279.

[31] H. Guan, D. Kang, M. Fan, Z. Chen, L.-J. Qu, Overexpression of a new putative membrane protein gene *AtMRB1* results in organ size enlargement in *Arabidopsis*, J. Integr. Plant Biol. 51 (2009) 130–139.

[32] B.J. Adamczyk, M.D. Lehti-Shiu, D.E. Fernandez, The MADS domain factors AGL15 and AGL18 act redundantly as repressors of the floral transition in *Arabidopsis*, Plant J. 50 (2007) 1007–1019.

[33] D.E. Fernandez, G.R. Heck, S.E. Perry, S.E. Patterson, A.B. Bleecker, S.-C. Fang, The embryo MADS domain factor AGL15 acts postembryonically: inhibition of perianth senescence and abscission via constitutive expression, Plant Cell 12 (2000) 183–197.

[34] E.W. Harding, W. Tang, K.W. Nichols, D.E. Fernandez, S.E. Perry, Expression and maintenance of embryogenic potential is enhanced through constitutive expression of *AGAMOUS-Like 15*, Plant Physiol. 133 (2003) 653–663.

[35] C. Bolle, The role of GRAS proteins in plant signal transduction and development, Planta 218 (2004) 683–692.

[36] E.M. Engstrom, C.M. Andersen, J. Gumulak-Smith, J. Hu, E. Orlova, R. Sozzani, J.L. Bowman, *Arabidopsis* homologs of the *Petunia HAIRY MERISTEM* gene are required for maintenance of shoot and root indeterminacy, Plant Physiol. 155 (2011) 735–750.

[37] S. Schulze, B.N. Schäfer, E.A. Parizotto, O. Voinnet, K. Theres, LOST MERISTEMS genes regulate cell differentiation of central zone descendants in *Arabidopsis* shoot meristems, Plant J. 64 (2010) 668–678.

[38] S. Van Landeghem, J. Björne, C.-H. Wei, K. Hakala, S. Pyysalo, S. Ananiadou, H.-Y. Kao, Z. Lu, T. Salakoski, S. Van de Peer, F. Ginter, Large-scale event extraction from literature with multi-level gene normalization, PLOS ONE 8 (2013) e55814.

[39] J. Björne, F. Ginter, T. Salakoski, University of Turku in the BioNLP'11 Shared Task, BMC Bioinform. 13 (2012) S4.

[40] D. McClosky, S. Riedel, M. Surdeanu, A. McCallum, C.D. Manning, Combining joint models for biomedical event extraction, BMC Bioinform. 13 (2012) S9.

[41] R. Sætre, K. Yoshida, M. Miwa, T. Matsuzaki, Y. Kano, J. Tsujii, Extracting protein interactions from text with the unified AkaneRE event extraction system, IEEE-ACM Trans. Comput. Biol. Bioinform. 7 (2010) 442–453.

[42] Y. Zhang, H. Lin, Z. Yang, J. Wang, Y. Li, A single kernel-based approach to extract drug–drug interactions from biomedical literature, PLOS ONE 7 (2012) e48901.

[43] C. Nédellec, R. Bossy, J.-D. Kim, J.-J. Kim, T. Ohta, S. Pyysalo, P. Zweigenbaum, Overview of BioNLP Shared Task 2013, in: Proceedings of the BioNLP Workshop, Sofia, Bulgaria, 9 August 2013, 2013, pp. 1–7.

[44] C.N. Arighi, P.M. Roberts, S. Agarwal, S. Bhattacharya, G. Cesareni, A. Chatr-Aryamontri, S. Clematide, P. Gaudet, M.G. Giglio, I. Harrow, E. Huala, M. Krallinger, U. Leser, D. Li, F. Liu, Z. Lu, L.J. Maltais, N. Okazaki, L. Perfetto, F. Rinaldi, R. Sætre, D. Salgado, P. Srinivasan, P.E. Thomas, L. Toldo, L. Hirschman, C.H. Wu, BioCreative III interactive task: an overview, BMC Bioinform. 12 (2011) S4.

[45] L. Hirschman, G.A.P.C. Burns, M. Krallinger, C. Arighi, K.B. Cohen, A. Valencia, C.H. Wu, A. Chatr-Aryamontri, K.G. Dowell, E. Huala, A. Lourenço, R. Nash, A.-L. Veuthey, T. Wiegers, A.G. Winter, Text mining for the biocuration workflow, Database 2012 (2012) bas020.

[46] C.-H. Wei, H.-Y. Kao, Z. Lu, PubTator: a web-based text mining tool for assisting biocuration, Nucleic Acids Res. 41 (2013) W518–W522.

[47] J. Lloyd, D. Meinke, A comprehensive dataset of genes with a loss-of-function mutant phenotype in *Arabidopsis*, Plant Physiol. 158 (2012) 1115–1129.

[48] D.W. Meinke, A survey of dominant mutations in *Arabidopsis thaliana*, Trends Plant Sci. 18 (2013) 84–91.

[49] S.M. Brady, N.J. Provart, Web-queryable large-scale data sets for hypothesis generation in plant biology, Plant Cell 21 (2009) 1034–1051.

[50] K. Toufighi, S.M. Brady, R. Austin, E. Ly, N.J. Provart, The botany array resource: e-Northerns, expression angling, and promoter analyses, Plant J. 43 (2005) 153–163.

[51] S. De Bodt, J. Hollunder, H. Nelissen, N. Meulemeester, D. Inzé, CORNET 2.0: integrating plant coexpression, protein-protein interactions, regulatory interactions, gene associations and functional annotations, New Phytol. 195 (2012) 707–720.

[52] L. Cooper, R.L. Walls, J. Elser, M.A. Gandolfo, D.W. Stevenson, B. Smith, J. Preece, B. Athreya, C.J. Mungall, S. Rensing, M. Hiss, D. Lang, R. Reski, T.Z. Berardini, D. Li, E. Huala, M. Schaeffer, N. Menda, E. Arnaud, R. Shrestha, Y. Yamazaki, P. Jaiswal, The Plant Ontology as a tool for comparative plant anatomy and genomic analyses, Plant Cell Physiol. 54 (2013) e1 (1–23).

[53] R. Shrestha, L. Matteis, M. Skofic, A. Portugal, G. McLaren, G. Hyman, E. Arnaud, Bridging the phenotypic and genetic data useful for integrated breeding through a data annotation using the Crop Ontology developed by the crop communities of practice, Front. Physiol. 3 (2012) 326.

[54] N. Pérez-Harguindeguy, S. Díaz, E. Garnier, S. Lavorel, H. Poorter, P. Jau-reguiberry, M.S. Bret-Harte, W.K. Cornwell, J.M. Craine, D.E. Gurvich, C. Urcelay, E.J. Veneklaas, P.B. Reich, L. Poorter, I.J. Wright, P. Ray, L. Enrico, J.G. Pausas, A.C. de Vos, N. Buchmann, G. Funes, F. Quétier, J.G. Hodgson, K. Thompson, H.D. Morgan, H. ter Steege, M.G.A. van der Heijden, L. Sack, B. Blonder, P. Poschlod, M.V. Vaieretti, G. Conti, A.C. Staver, S. Aquino, J.H.C. Cornelissen, New handbook for standardised measurement of plant functional traits worldwide, Aust. J. Bot. 61 (2013) 167–234.

[55] International Arabidopsis Informatics Consortium, An international bioinformatics infrastructure to underpin the *Arabidopsis* community, Plant Cell 22 (2010) 2530–2536.

[56] The International Arabidopsis Informatics Consortium, Taking the next step: building an *Arabidopsis* Information Portal, Plant Cell 24 (2012) 2248–2256.

[57] D. Fleury, K. Himanen, G. Cnops, H. Nelissen, T.M. Boccardi, S. Maere, G.T. Beemster, P. Neyt, S. Anami, P. Robles, J.L. Micol, D. Inzé, M. Van Lijsebettens, The Arabidopsis thaliana homolog of yeast BRE1 has a function in cell cycle regulation during early leaf and root growth, Plant Cell 19 (2) (2007) 417–432.

16

Computational approaches to identify regulators of plant stress response using high-throughput gene expression data

Alexandr Koryachko[a], Anna Matthiadis[b], Joel J. Ducoste[c], James Tuck[a], Terri A. Long[b,*], Cranos Williams[a,**]

[a] *Electrical and Computer Engineering, North Carolina State University, Raleigh, NC, USA*
[b] *Plant and Microbial Biology, North Carolina State University, Raleigh, NC, USA*
[c] *Civil, Construction, and Environmental Engineering, North Carolina State University, Raleigh, NC, USA*

ARTICLE INFO

ABSTRACT

Keywords:
Stress response
Transcription factors
Gene regulatory networks
Algorithms
Arabidopsis thaliana

Insight into biological stress regulatory pathways can be derived from high-throughput transcriptomic data using computational algorithms. These algorithms can be integrated into a computational approach to provide specific testable predictions that answer biological questions of interest. This review conceptually organizes a wide variety of developed algorithms into a classification system based on desired type of output predictions. This classification is then used as a structure to describe completed approaches in the literature, with a focus on project goals, overall path of implemented algorithms, and biological insight gained. These algorithms and approaches are introduced mainly in the context of research on the model plant species *Arabidopsis thaliana* under stress conditions, though the nature of computational techniques makes these approaches easily applicable to a wide range of species, data types, and conditions.

1. Introduction

Plants are sessile organisms subject to constantly changing environments. The ability to respond to these environmental changes, therefore, is key to plant adaptation and survival. An overall goal of plant abiotic stress research is to develop an understanding of the molecular components of a single or combinatorial stress response and show how these components interact, enabling directed genetic manipulations that can enhance stress tolerance [1]. Transcription factors are one of the first categories of genes activated in response to a stress [2]. Transcription factor activity can lead to alterations in activity and accumulation of downstream transcription factors and proteins that modulate plant morphology and molecular composition. In this way, manipulation of the activity of just one transcription factor or a small family of transcription factors can lead to alterations in a transcriptional cascade with dramatic outcomes. This strategy is the basis for both evolutionary adaptation as well as genetic manipulation of stress responses [3,2,4]. Despite a widespread focus on stress-induced transcription factors and a recent breadth of high throughput data, successful

genetic manipulations of transcription factors in crop plants that improve stress tolerance are limited. Identification of transcriptional regulators in the model species *Arabidopsis thaliana* is a first step to the search for candidates for genetic modification in crop species, yet a number of limitations exist in this identification. First, with 5–10% of plant genomes reported to code for transcription factors, the number of candidate genes to study is extensive [5–7]. Second, it is difficult to predict the effects of one transcription factor in isolation and even more difficult to predict combinatorial effects of transcription factors acting on the same targets or acting in complexes. Finally, the ways in which transcription factor activity is modulated in response to one stress or a combination of stresses are complex. In other words, a huge number of possible experiments exist to test the effect of combinations of transcription factors under combinations of stresses. Computational approaches play a critical role in the research process by producing a set of testable predictions, thus limiting the space of experiments needed to yield a better understanding of the cascading responses resulting from stress. These predictions range from involvement of a transcription factor in a stress response to detailed descriptions of transcription factor and target gene interaction dynamics. An increasing abundance of computational approaches necessitates a careful evaluation of the utility and application of these tools.

In this review, we summarize and organize algorithms involved in current and promising computational approaches into 5 categories ("Types") based on the type of inference an algorithm aims

* Corresponding author.
** Co-corresponding author.
   E-mail addresses: terri_long@ncsu.edu (T.A. Long), cmwilli5@ncsu.edu (C. Williams).

to obtain from a biological dataset. We focus on algorithms that can operate on gene expression data, as this is currently the most available form of high throughput data. We then demonstrate how algorithms from different categories have been combined in the scope of computational approaches to achieve specific objectives associated with plant stress response. Additional examples of algorithms that fall into the proposed classification system can be found either in plant related [8–10] or general computational approach [11–15] review articles. We build on these reviews by organizing algorithms based on their utility and highlighting how inferences achieved by algorithms in multiple categories can be systematically combined to achieve a better understanding of the transcriptional cascade involved in stress response. We then discuss how these algorithms have been utilized in recent studies to achieve a certain objective. In this way, researchers aiming to acquire specific predictions from an expression dataset can more efficiently choose appropriate algorithms.

2. Background

Environmental conditions in agricultural settings are highly variable, leading to suboptimal crop yields and survival rates [16]. The frequency and intensity of environmental extremes, particular drought, heat, and pests are expected to increase with climate change [1,17,16]. A large number of stress response studies are focused on elucidating transcriptional cascades regulating responses to individual and combined stresses. Transcription factors that play important roles in modulating such cascades are candidates for genetic engineering approaches and are worthy of intensive study.

Gene expression analysis is a widely proposed means of bringing a greater understanding to all abiotic stress responses for several reasons. A large number of genes have altered expression in response to stress and these alterations play an important role in adaptation [18,19]. Expression data is also relatively cheap. Because of this, high-throughput gene expression datasets have been generated and are publicly available for a multitude of stresses, both biotic and abiotic, with examples in *A. thaliana* including but not limited to pathogen infection [20–22], cold [23], pH [24], salt [25], light [26], and nutrient [25,27–30] stress. Though these studies are comparable in theory, a few large studies have attempted to mitigate the effects of variations in experimental setup by collecting expression data under different stresses imposed with otherwise identical growth conditions [18] or under combinations of stresses [31–34]. Analyses of these concurrent and combinatorial experiments in particular have revealed distinct patterns of different stress responses along with some common features, inferring that both general and specific stress response pathways exist. For example, analysis of the AtGenExpress database of concurrent stress application indicates that some abiotic stresses result in sustained gene expression alterations and others in transient alterations [18]. A set of early- and commonly induced genes, representing the so called Plant Core Environmental Stress Response (PCESR), includes transcription factors, indicating that a general stress response may be transcription factor mediated and likely occurs early in stress response cascades [18,35]. Combinatorial studies indicate that genes responding to combined stresses are often distinct from those responding to individual stresses, highlighting a need for both more studies of this type as well as computational methods to attempt to predict these emergent behaviors [31–34]. Despite these extensive analyses, limited direct predictions concerning stress pathways have been made and validated. The majority of detailed characterizations of transcription factors, including direct promoter binding and influence on target gene expression, are the result of traditional studies. These

studies are time and cost intensive. Furthermore, since many key regulators have been found through phenotypic mutant screens, subtle yet important phenotypes and genes can easily be missed. Redundancy is expected in critical regulatory mechanisms [36], and predictions concerning which regulators or mutants to combine in a genetic engineering strategy would be extremely valuable. A recent increase in algorithm development and utilization will help to increase the predictive power in available datasets so that regulators and combinatorial regulatory mechanisms beyond the "low hanging fruit" can be identified. In the following sections, we describe and organize sets of algorithms and implementations thereof in experimental approaches, aiming to bring attention to the benefit of these approaches and facilitate future increases in frequency and strength of computational biology studies.

3. Classification of inference algorithms

Many computational algorithms have been developed for analyzing gene expression data. We focus here on algorithms capable of identifying stress related genes, grouping genes by function, inferring connections between genes, estimating gene interaction direction and type, and predicting gene expression states and values in interconnected regulatory networks. These algorithms differ in complexity and implied assumptions, but can be classified based on functionality. We categorize these algorithms in 5 distinct groups based on the type of insight they provide to a biological process of interest. Depending on research objectives, these algorithms can either be used separately or as a part of a systematic computational approach where inferences from algorithms of one type can be used as input for algorithms of another type. For example, a computational approach designed to predict gene interactions and their type based on time course microarray data can be comprised of 3 algorithms of different types that sequentially process input data to obtain a desired output (Fig. 1).

The algorithms described can be applied to transcriptomic data obtained at M time points or treatments (t_j, $j = 1, \ldots, M$) for a set of N genes (g_i, $i = 1, \ldots, N$). Examples of such datasets include the global abiotic stress expression database AtGenExpress [18]. This database includes datasets for multiple abiotic stress treatments that are obtained for $N \approx 24,000$ genes at $M = 7$ time points using Affymetrix ATH1 GeneChip microarray analysis. Hence, the activity of each gene can be represented by a set of numbers $g_i(t_1)$, $g_i(t_2)$, $\ldots$, $g_i(t_M)$, forming a pattern that is used by algorithms to make inferences.

Type 1 algorithms attempt to capture genes that are relevant to a particular condition. Techniques for determining differentially expressed genes are an example of algorithms falling into this category [37]. Differential expression techniques work by assuming that significant change in transcript levels of a given gene under stress condition relative to its activity under normal conditions indicates that the gene plays a role in the stress response. This assumption disregards posttranscriptional modifications as alternate means of gene product regulation. Since transcript measurement precision can vary from one experimental approach to another, statistical tests are often applied to determine the significance of the change in transcript levels. Student's t-test for 2 treatments or ANOVA for a set of treatments are commonly applied to deduce statistical significance. Other differential expression inference algorithms were developed for large scale experimental techniques such as microarrays, for which the correlation between within-array replicates can be taken into consideration [38], or RNA-Seq, for which count based statistics are more appropriate [39,40].

Type 2 algorithms aim to identify relationships between genes. These algorithms work by assuming that genes with "similar"

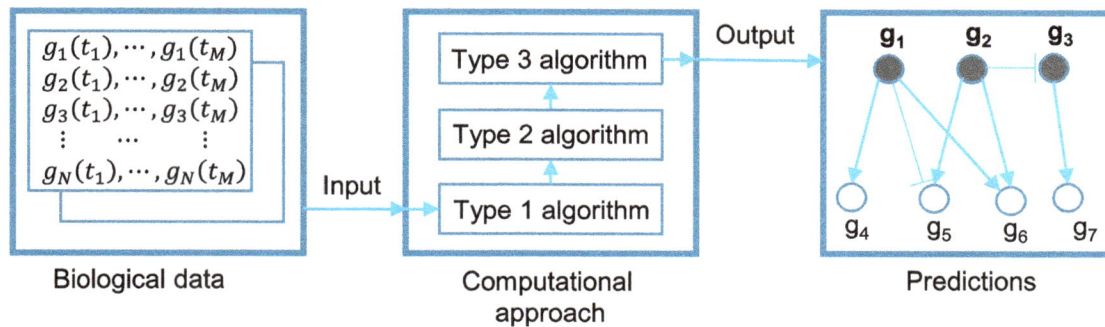

Fig. 1. Conceptual view of the information flow in a computational approach. Biological data is used to identify genes of interest (Type 1 algorithm), infer connections between these genes (Type 2 algorithm), and predict types of these connections (Type 3 algorithm).

expression patterns are co-regulated or are part of the same regulatory pathway [13]. Techniques like co-expression analysis [41–45] fall into this category.

Common metrics that have been used to assess similarities between genes based on their expression patterns include Pearson correlation coefficient [42,46–48], Spearman correlation coefficient [49–51], partial correlation coefficient [52–54], Euclidean distance [55,56], and mutual information [57–59]. These metrics typically represent a quantified measure that establishes a pairwise comparison between the expression levels of two genes, g_1 and g_2, across time points or experimental treatments. Kumari et al. [60] presented a study that evaluated the utility of Spearman rank correlation, Weighted Rank Correlation, Kendall, Hoeffdings D measure, Theil-Sen, Rank Theil-Sen, Distance Covariance, and Pearson correlation coefficient on transcriptional data for determining gene association. The authors found that Spearman, Hoeffding, and Kendall correlation coefficients were more effective in identifying related pathway genes than others. In contrast, Ma et al. [61] claim that based on manual inspection of the expression patterns of several pairs of TF-target genes, the Gini correlation coefficient can compensate for the shortcomings of the Pearson, Spearman, Kendall, and Tukeys biweight correlations in detecting transient regulatory relationships between transcription factors and their targets. Metrics such as area between expression curves [62], Z-score [63], and others appear in the literature but have not been extensively evaluated.

Relationships between individual genes or across established groups of genes can be generated based on these similarity metrics. A typical procedure for estimating relationships between individual genes is to set a threshold value and assign connections between genes whose pairwise similarity value is higher than a selected threshold [62,58]. The statistical significance of the similarity can also be taken into consideration when establishing a connection [57]. Groups of similarly behaving genes are in most cases identified using clustering algorithms. Clustering algorithms apply similarity metrics to isolate groups of co-expressed genes. k-means clustering [64], the Markov Cluster algorithm [65,66], biclustering [67], self-organizing maps [68], hierarchical clustering [69], and affinity propagation [70] are examples of clustering algorithms applied to transcriptomic data. Martin et al. [64] applied k-means clustering, hierarchical clustering, and self-organizing maps to time series transcriptomic data from mice. The results suggested that k-means was able to convey comparable grouping to hierarchical clustering, and self-organizing maps (more than 80% agreement) while maintaining less of a computational load than other approaches. Frey and Dueck [70] showed that the affinity propagation algorithm yields more compact clusters compared to k-means in terms of the sum of intercluster distances which might imply tighter relationships between genes in the same cluster.

Clustering has also been used to reduce the complexity of building transcriptional networks by reducing high dimensional networks with many genes to lower dimensional networks of clusters of genes or "metagenes", which represent groups of genes with similar expression activity. The expression pattern of a metagene may be defined as the cluster average or the expression pattern of the gene with the highest sum of similarities with its cluster members. Some algorithms have extracted metagene expression patterns first by applying principal component analysis (PCA) or singular value decomposition (SVD) to the overall expression dataset. The clusters are then assembled based on similarities between gene and metagene expression patterns [71,72].

Type 3 algorithms aim to infer causal relationships between genes. Causal inference procedures are often based on the assumption that a change in one gene (g_1) will result in a subsequent change in another gene (g_2) at some later time if g_1 activates or inhibits g_2 [73–77]. Thus, the approach is similar to co-expression analysis in that it aims to find genes with similar temporal expression patterns. The key difference distinguishing this approach from those in **Type 2** is the assumption that these similarities will occur at a delay, allowing for inference on the direction of regulation (which gene comes first in a regulatory cascade) in addition to a relationship connection. The equation for Pearson correlation coefficient, for example, can be modified to assess this temporal characteristic by incorporating a time delay. Eq. (1) reflects similarity at the delay of one time unit. The algorithms capture the regulation delay for a pair of genes by selecting the time unit duration that maximizes the correlation coefficient [75].

$$\rho_{g_1 \to g_2} = \frac{\sum_{j=1}^{M-1}(g_1(t_j) - \bar{g}_1)(g_2(t_{j+1}) - \bar{g}_2)}{\sqrt{\sum_{j=1}^{M-1}(g_1(t_j) - \bar{g}_1)^2}\sqrt{\sum_{j=1}^{M-1}(g_2(t_{j+1}) - \bar{g}_2)^2}},$$

$$\text{where} \quad \bar{g}_1 = \frac{1}{M-1}\sum_{j=1}^{M-1}g_1(t_j), \quad \bar{g}_2 = \frac{1}{M-1}\sum_{j=1}^{M-1}g_2(t_{j+1}) \quad (1)$$

Two sets of similarity values, each corresponding to a range of delays for a certain direction of shift, are calculated to assess the strength and directionality of connection in each pair of genes. Small similarity values, corresponding to a low probability of regulation, can be removed, leaving the remaining high confidence connections to characterize genes that have potential causal relationships. Approaches that use modifications of the metric in (1) have been effective for single datasets with 50 and 27 time points and sampling intervals of 20 min [75] and for a collection of 18 datasets with 7 time points in each and sampling intervals ranging from 0.5 to 12 h [78]. Other sample times may be relevant depending on the features that exist in the data.

Another class of algorithms that infer regulatory interactions between genes is Bayesian networks [79,80]. Bayesian networks are capable of inferring regulatory connections from time course and non-time course data. These algorithms attempt to find causal connections based on Bayes' rule by explicitly choosing a network structure that best describes experimental data. The algorithm considers a network of gene regulations as a set of dependencies where the probability of expression of a target is conditioned on the expression of its regulator. These regulations are described as conditional probabilities. Algorithms then try to find a network structure that best describes the data based on a scoring function. Identification of the network structure is a computationally intensive problem. Complexity grows exponentially with an increasing number of nodes [11]. For example, around 10^{18} different topologies arise for a network of only 10 genes [12]. Thus, most of the approaches using Bayesian networks concentrate on a small subset of genes (typically when some portion of a gene regulatory network is already known) or employ sub-optimal but less computationally intense solutions to handle larger networks [11].

Dynamic Bayesian Networks (DBN) [81,82] incorporate ordering information in time course data to allow for feedback loops (not allowed in standard Bayesian networks). These feedback loops are allowed by treating expression of the same gene at different time points as different nodes. Nodes corresponding to the same gene are combined after the structure inference procedure. This algorithm leads to an increase in complexity since the number of nodes involved in structure inference routine is a product of the number of genes and the number of time points.

Type 4 algorithms aim to infer combinations of regulator expression states that are necessary to result in a particular state of target. These algorithms can be conceptualized as a search for a functional relationship between a target and its regulator(s) ($g_i = f(g_1, g_2, ..., g_N)$). In this case, a qualitative measure of gene behavior can be used, with gene expression values represented as either high or low, active or inactive, or "ON" or "OFF" to simplify the problem. An "ON" state of only a couple regulators may suffice to upregulate the expression of the target. This qualitative assumption allows the use of Boolean networks [83] in **Type 4** inference problems. Expression values in Boolean network inference approaches are discretized mostly in two states, representing an activity level at each time point [84–86]. Regulatory connection inference algorithms try to find a binary function that computes the next state of a gene based on a combination of other genes' states using simple Boolean operations, e.g. AND (&) if more than one regulator should have a certain state to influence a common target, OR (|) if any of the regulator states suffice for the same purpose, and NOT (¬) in the case of repression (Fig. 2). The goal of this approach is to find the simplest function for each gene, which is the function that depends on the fewest regulator genes possible.

A direct approach to find the simplest Boolean function that satisfies a given data set is to compare all possible functions capable of generating the observed expression pattern. The number of Boolean functions that can represent the expression activity of a gene regulated by as many as n transcription factors is 2^{2^n} [87], making the problem computationally infeasible for a large (more than 10) number of genes. Some algorithms use prior knowledge to confine the number of genes to analyze. Others rely on network structures inferred by other types of algorithms to confine the number, type and directionality of possible regulatory relationships between individual genes or groups of genes. Another factor constraining the use of Boolean networks in whole genome dataset analysis is the small number of samples (time points) associated with most datasets. These small sample sizes typically do not provide the diversity needed to uniquely define relationships across a large number of individual genes. For example, for 5 time points, which is the median number in typical gene expression

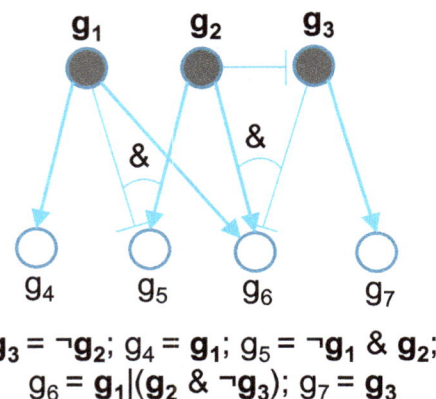

Fig. 2. Boolean network representation in graphical and functional forms. Combinations of transcription factors g_1, g_2, and g_3 influence expression of each other and target genes. The state of g_6, for example, is influenced by a combination of g_2 and g_3 or by g_1 alone.

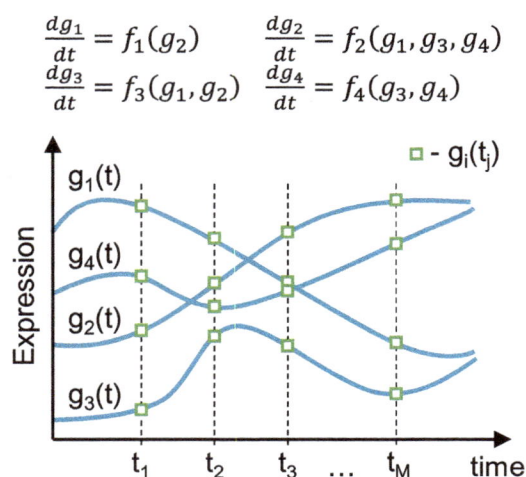

Fig. 3. Type 5 algorithms output in terms of the system of ODEs and predicted gene expression dynamics ($g_i(t)$) based on experimental values ($g_i(t_j)$). In this example, the expression pattern of each gene is influenced by the expression of at least one other gene, with some genes (g_4) influenced by their own expression (feedback loop).

datasets [88], the number of genes with distinct Boolean expression patterns is limited to only $2^5 = 32$. Any attempted analysis of more than 32 genes with such a dataset would result in at least 2 genes with identical behavior which would limit resolution to groups of genes (e.g. metagenes) as opposed to individual genes.

Type 5 algorithms aim to describe dynamic behavior in a transcriptional network. The resulting network representation allows for the reconstruction of continuous changes in transcripts over time (Fig. 3). Ordinary differential equations are commonly used to capture the dynamics associated with gene expression changes [89]. These equations allow for the estimation of gene expression values at any given time point either between samples (interpolation) or beyond the last collected sample (extrapolation) [29]. When a gene regulatory network is represented in terms of linear differential equations, the instantaneous change in expression of a gene is related to the sum of weighted expression values of influencing genes:

$$\frac{dg_i}{dt} = \sum_{k=1}^{N} a_{ik}g_k, \tag{2}$$

where a_{ik} represent influence coefficients. Coefficients for linear differential equations are often inferred using the Least Absolute Shrinkage and Selection Operator (LASSO) algorithm [90], a modification of the linear regression approach. When LASSO is used for ODE inference purposes, the changes in expression, i.e. differences between expression values at consecutive time points, are approximated by a linear combination of other genes' expression values. Expression patterns for target genes are replaced with patterns of changes in expression [91–93] to infer influence coefficients. Given that biological processes are assumed to be inherently nonlinear, linear Ordinary Differential Equation (ODE) inference algorithms for transcriptional networks rely on the assumption that the system operates close to a stability point [93]. The system may not stay close to a stability point in the case of stress induced responses, where a plant may transition from one stable steady state to another. Nonlinear ODEs, though potentially more biologically relevant because they do not rely on the steady state assumption, typically require the estimation of more coefficients associated with nonlinear terms [94]. Coefficient estimation routines for inference algorithms search the parameter space to find coefficients that yield solutions closest to measured expression values [95,96].

All of the described algorithms require implementation and validation in biological systems in order to assess their utility. A number of validation techniques exist, depending on the type of algorithm [97–113]. These validation techniques are visualized with key references in Fig. 4. Validation for algorithms of **Types 1** and **2**, which predict associations between a gene and a process or a gene and a group of genes, are limited to analysis of Gene Ontology (GO) enrichment or phenotypes in mutants of transcriptional regulators. These phenotypes range widely depending on the stress response in question, and could involve extensive experimentation to search for a phenotype of interest. A wider range of techniques exist for algorithms of **Types 3–5**, algorithms that predict relationships between transcription factors and target genes. These relationships can be tested indirectly through expression profiling, computationally through promoter analysis, or directly through binding interactions. Given that no "gold standard" validation technique exists [114], convincing support often involves the combination of multiple validation techniques, such as expression analysis and binding activity for a regulator and target of interest. Similarly, complex predictions such as those derived from **Type 4** and **Type 5** algorithms require a combination of static and dynamic validation techniques – including expression profiling at multiple time points, preferably along with determination of binding activity.

4. Computational approaches

Computational approaches are used widely to gain insight into processes underlying plant response to stress conditions. These approaches have a similar structure in terms of the types of algorithms they use and differ in the combination of and order in which these algorithms are applied. In the following examples, we describe how algorithms of different types have been combined in particular computational approaches to answer research specific questions.

4.1. Relevant gene identification

A large number of current computational approaches are focused on identifying genes that play a key role in a process of interest. The importance of these genes is then typically tested through mutant phenotypic analysis. Ma et al. [56] analyzed a set of A. thaliana abiotic stress response transcriptome datasets with

6 time points to identify stress related genes. The computational approach started by partitioning each stress dataset into "informative" and "noninformative" genes using differential network analysis (**Type 1** algorithm). The authors stated that differential network analysis that involves machine learning and training based on a priori information is more sensitive than differential expression analysis, which is statistics oriented. The Gini correlation coefficient was then calculated for pairs of "informative" genes to establish significant connections (**Type 2** algorithm). Stress related genes were identified from the resulting network based on the combination of 33 topology scores obtained from the network of significant connections (**Type 1** algorithm). The authors validated their algorithm by performing a phenotypic screen for 89 candidates identified as salt stress related. Mutants of 2 previously unreported salt stress-related genes showed phenotypes.

Dinneny et al. [25] conducted DNA microarray experiments on A. thaliana root response to iron deficient media with 7 time points spanning 72 h to identify common stress response behavior patterns. The authors applied differential expression analysis [115] to identify genes having at least a 1.5-fold change in expression with a false discovery rate value less than 10^{-4} at a sampling time point compared to no treatment (**Type 1** algorithm). The analysis showed that the strongest transcriptional response occurred after 24 h of treatment. Dinneny et al. [25] then applied the affinity propagation clustering algorithm [70] to form groups of similarly expressed genes and thus identify general patterns of gene expression (**Type 2** algorithm). Long et al. [28] used the results of this analysis and screened through mutants of 38 identified genes coding for coexpressed transcription factors. The screens led to identification of important iron homeostasis regulators POPEYE (PYE) and BRUTUS (BTS).

Lin et al. [30] investigated the effect of phosphate starvation on A. thaliana root gene signaling using a DNA microarray time course with 3 time points to infer functional modules in early transcriptional responses. The authors used differential expression analysis with the requirement of a 2-fold change in expression with a p-value cutoff of 0.05 to identify stress related genes (**Type 1** algorithm). Additional information from 2671 experimental datasets, 300 of which are root specific, was used to select 187 root specific genes (**Type 1** algorithm). The authors used the Multi-Array Correlation Computation Utility (MACCU) toolbox based on thresholding pairwise Pearson correlation coefficients to obtain 3 functional modules of stress specific genes (**Type 2** algorithm). To validate the results, Lin et al. [30] conducted mutant screens on 31 members of a cluster where most of the genes are known to participate in root development. Only 5 tested lines did not show a statistically significant root hair length phenotype.

4.2. Gene function elucidation

Another group of computational approaches aim to associate genes with a specific function during a process of interest. The guilt-by-association heuristic [43] is often used to assign a function to an unknown gene based on known functions of co-regulated genes (Gene Ontology enrichment). Polanski et al. [48] analyzed six A. thaliana stress response transcriptome datasets to identify gene modules showing evidence for co-regulation. The computational approach revealed 78 modules of co-regulated genes, 71 of which were overrepresented in Gene Ontology categories and 51 of which were enriched in transcription factor binding motifs (compared to 24 and 6 of 78 randomly assigned modules, respectively). The approach used information about which genes were differentially expressed in each stress response as an input (previously determined in other publications using **Type 1** algorithms). For each gene differentially expressed under at least 2 conditions, the algorithm assembled a set of correlated genes for each condition (**Type**

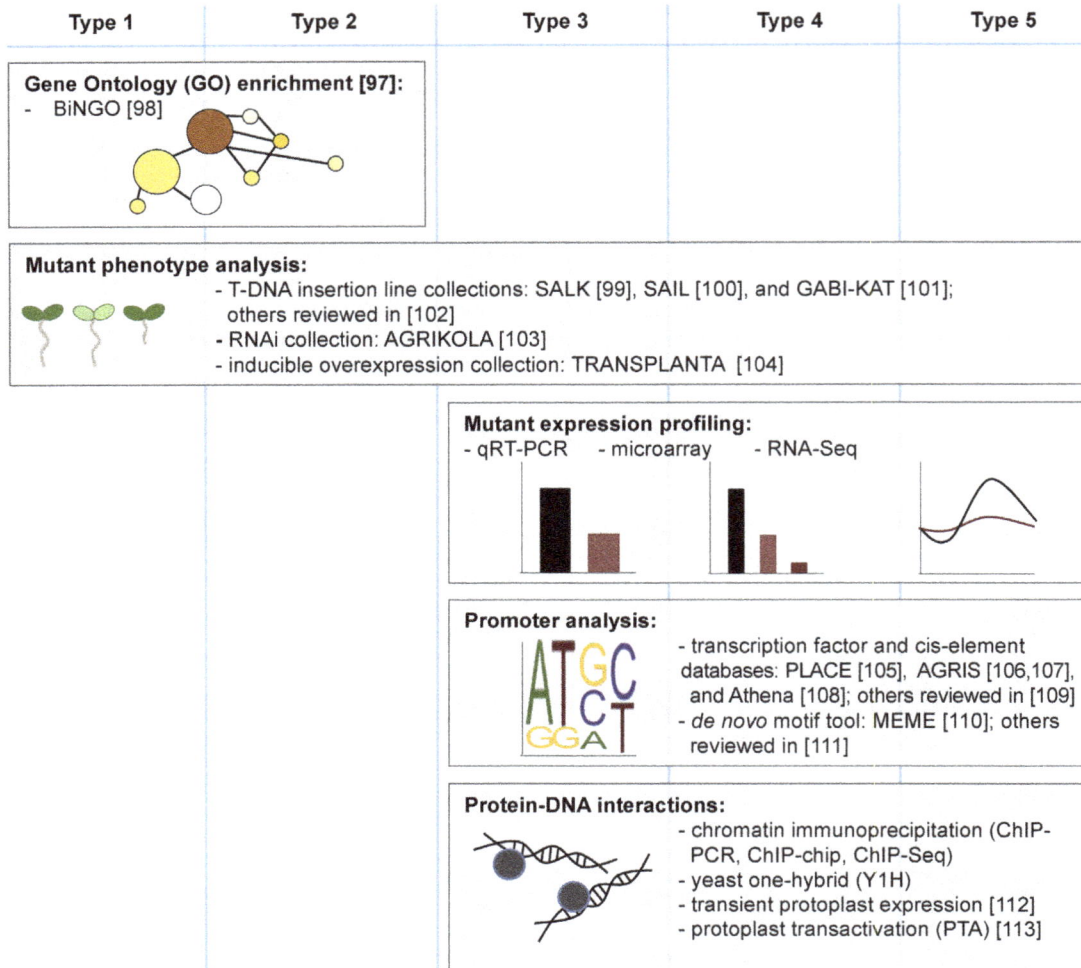

Fig. 4. Validation techniques for algorithm types with key references. Examples shown are those typically seen in current computational research approaches, specifically for research projects in *A. thaliana*.

2 algorithm). A co-regulation relationship in a pair of genes was established if these genes had shared a significant number of correlated genes across stress conditions (**Type 2** algorithm). The authors used Gene Ontology enrichment, promoter analysis, and yeast one-hybrid protein–DNA interactions to validate the resulting modules of co-regulated genes.

Ma and Bohnert [116] integrated time course and cell specific transcriptomics data with gene promoter structures to identify stress related *cis*-elements in *A. thaliana*. The computational approach used in this work detected known stress related *cis*-elements and identified secondary motifs. The authors combined abiotic and biotic stress, hormone and chemical treatment time courses and different light condition samples to create one combined expression pattern of 145 values per gene. Differentially expressed genes were identified by combining the results from fuzzy *k*-means clustering [117] applied to all gene probes and the 'limma' statistical program [118] which identified genes differentially expressed in at least one condition (**Type 1** algorithm). Fuzzy *k*-means clustering was again applied to the resulting set to identify stress related clusters of genes (**Type 2** algorithm). The authors assigned functions to clusters based on GO enrichment. Binding motif analysis using Plant Cis-acting Regulatory DNA Elements (PLACE) database [105] revealed motifs significantly overexpressed in the function related clusters. Further analysis of 22 major clusters resulted in the identification of new DNA regulatory motifs [119].

4.3. Gene relationship inference

Computational approaches that aim to unravel influential relationships between regulators and their targets are less common but are increasing in frequency. Windram et al. [22] applied a computational approach to identify transcription factor families operating at different stages of *A. thaliana* pathogen defense response. The authors analyzed transcriptional profiles at 24 time points with 4 replicates per time point. The computational approach predicted gene regulatory interactions, confirmed experimentally or by binding motif enrichment. The analysis started with assessment of differentially expressed genes based on a combination of MAANOVA (MicroArray ANalysis Of VAriance) [120], approximate F tests, GP2S (Gaussian process 2 sample) test [121], and Hotelling statistic (T^2) [122] (**Type 1** algorithm). Next, a SplineCluster [123] algorithm separated differentially expressed genes into clusters associated with different stages of stress response (**Type 2** algorithm). The clusters were validated by GO enrichment analysis. Nonparametric modification of Bayesian network inference algorithm [124] was applied to cluster representatives to infer regulatory connections between clusters (**Type 3** algorithm). The authors validated the regulatory effect of one of the clusters through experiments with a knockout mutant line for the transcription factor TGA3. Experimental data showed altered gene expression in predicted TGA3 target clusters in the *tga3-2* mutant,

Active | Significant similarity

g₁ g₂ g₃
g₄ g₅ g₆ g₇

g₁ g₂ g₃
& &
g₄ g₅ g₆ g₇

Expression — g₁, g₆, g₅, g₂, g₄, g₃, g₇ — time

Not active | Clusters

○ - Gene ● - TF

→ - Transitions used in computational approaches

⇒ - Perspective transitions

Type 1 algorithms	**Type 2 algorithms**	**Type 3 algorithms**	**Type 4 algorithms**	**Type 5 algorithms**
Differentiate between stress related and not stress related	Identify co-expressed genes or groups of genes	Infer regulatory connections	Establish qualitative functional relationships	Describe dynamic behavior in transcriptional network

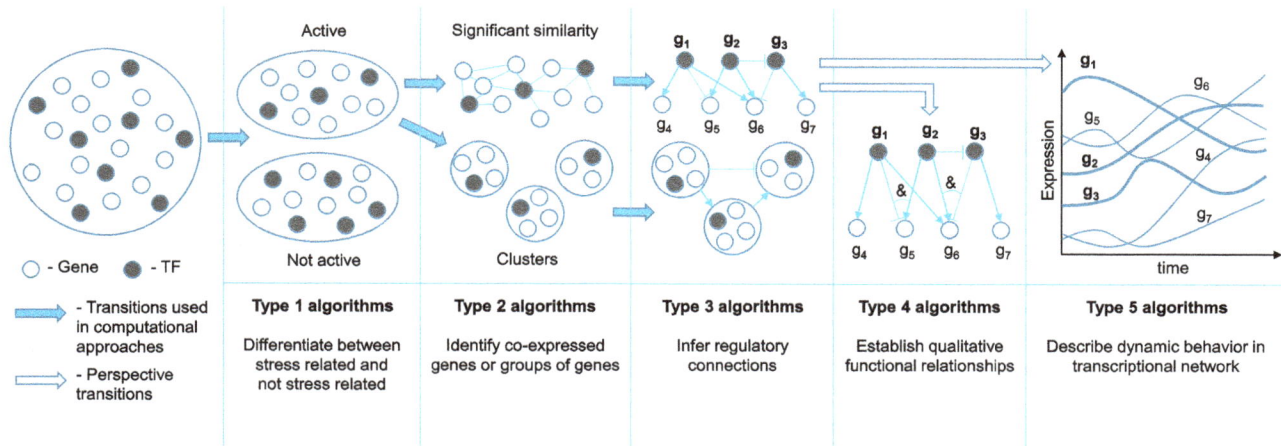

Fig. 5. Transitions between algorithms of different types. Typical experimental transitions between algorithms are indicated with blue arrows and perspective future transitions, less common but possible with more reliable supporting algorithms, are indicated with white arrows.

whereas targets regulated by non TGA3 clusters were less affected. The effect of another transcription factor, ANAC055, was validated by binding motif enrichment in target clusters.

Redestig et al. [78] analyzed a set of 18 DNA microarray time series corresponding to nine different abiotic stresses with seven time points obtained from root and shoot of *A. thaliana* seedlings with the aim of associating stress responsive transcription factors with their targets. The authors concluded that their computational approach delivered a usable number of high-confidence target genes (12–59% of identified true targets) for stress related transcription factors. The computational approach identified stress related transcription factors by selecting ones with maximum overall response and maximum change in response satisfying a specific threshold criteria (**Type 1** algorithm). Covariance values between a transcription factor and other genes over a set of delays were calculated for a set of conditions (**Type 3** algorithm). High scores corresponded to a high probability of regulation.

Krouk et al. [29] conducted DNA Microarray experiments on *A. thaliana* nitrate response with six time points spanning 20 minutes to capture a gene regulatory network underlying plant adaptation to nitrate provision. The inferred temporal model of the reaction process built for 20 cluster representatives resulted in 70% correct predictions of expression value direction change after the last time point in the time course. The computational approach started with ANOVA to identify nitrogen regulated genes (**Type 1** algorithm). Next, MeV software [125] was used to separate the nitrogen regulated genes into 20 clusters, eight of which appeared to have over-represented biological functions (**Type 2** algorithm). The application of LASSO based algorithm to cluster representatives provided coefficients for a system of linear ODEs describing the dynamics of each cluster (**Type 5** algorithm). Predictions on the direction of change obtained from ODEs were tested by comparing them with expression values from a time point that was not used for inference purposes.

4.4. Summary

As can be surmised from the examples given, algorithms from **Type 1**, **Type 2**, and **Type 3** are more common in current experimental approaches applied to plants. The problem of dimensionality prevents the extensive use of **Type 4** and **Type 5** algorithms for individual genes based on whole genome datasets due to data requirements for such type of inference [13]. Thus, the dimension of the problem is typically reduced by limiting a set of genes to ones known to interact or participate in the same biological process. Recent non-stress related approaches in *A. thaliana* have

employed such techniques. Espinosa et al. [126] used experimentally obtained knowledge about relationships of 15 genes in *A. thaliana* flower development process to predict development scenarios using Boolean networks approach (**Type 4** algorithm). Sankar et al. [127] built a model to predict states of the components from auxin and brassinosteroid signaling networks in *A. thaliana* by applying Boolean logic approach (**Type 4** algorithm) and then transformed the resulting discrete network representation to a set of ordinary differential equations (**Type 5** algorithm) to obtain quantitative predictions. Cruz-Ramirez et al. [128] investigated the dynamics of asymmetric cell division within the *A. thaliana* root by analyzing a system of nonlinear differential equations for 7 interacting complexes (**Type 5** output). The analysis predicted a bistable behavior of the process. Finally, Pokhilko et al. [129] refined the interaction model describing circadian rhythms in *A. thaliana* by modeling the process with a system of nonlinear ODEs (**Type 5** output).

Similarities in regulatory processes on a genomic level allow for the application of computational approaches that were developed for non-plant species. Some computational approaches are available in software packages. An extensive use of these packages shows that even if a technique was developed and tested for one species, it can be applied to a similar dataset from another species. Examples of these approaches are briefly described here. Vermeirssen et al. [130] combined the Learning Module Networks algorithm [131] developed for yeast, Context Likelihood of Relatedness algorithm [132] tested on *Escherichia coli*, and Double Two-way *t*-tests algorithm tested on *E. coli* to identify oxidative stress regulatory transcription factors in *A. thaliana* (**Type 3** output). The Algorithm for the Reconstruction of Gene Regulatory Networks (ARACNE) [58] was developed to infer transcriptional regulations in human B cells, but then used for other applications including the inference of transcriptional interactions underlying root development and physiological processes in *A. thaliana* [133] (**Type 3** output). Other software packages that showed the ability to recover gene regulatory networks from transcriptomic data include CLR [132], MRNET [134], C3NET [57], and ARTIVA [135]. The Dialogue on Reverse Engineering Assessment and Methods (DREAM) project attempted to compare such GRN inference methods applied to *E. coli*, *Staphylococcus aureus*, *Saccharomyces cerevisiae* and *in silico* microarray data [136]. The authors discovered that these methods have complementary advantages and limitations under different contexts. In the case of multicellular organisms, the performance of techniques has so far been measured based on goals achieved for a specific application. Such performance is difficult to compare between methods since goals and applications are often diverse.

5. Conclusions

We presented a classification of computational algorithms based on the type of information they aim to infer. This structure was used to describe approaches in the literature that have been used to gain insight into biological processes of interest based on transcriptomics data. Examples of existing computational approaches applied to plant stress transcriptional datasets demonstrated a pattern of transition between algorithms of different types (displayed graphically in Fig. 5). This progression demonstrates that the quality of predictions made by an algorithm in the scope of a computational approach often depends on the quality of predictions made by a preceding algorithm as well as on the quality of the original biological data. Based on available algorithms and example implementations, we can state that even though both stress related gene identification and grouping algorithms (**Type 1** and **Type 2**) are still evolving, confidence in **Type 2** algorithm predictions is sufficient to allow for a transition to causality inference (**Type 3**). **Type 3** algorithms have the potential to supply **Type 4** and **Type 5** algorithms with information about the structure of gene regulatory networks. This information will reduce the number of possible functional relationships to consider for these types of algorithms dramatically and thus allow for the increase in scope and predictive power. Therefore, the perspective transitions shown in Fig. 5 will likely appear more often in future computational approaches as reliability of **Type 3** algorithms predictions increase.

Acknowledgements

This material is based upon work supported by the National Science Foundation under Grant No. 1247427 and by the National Science Foundation Graduate Research Fellowship under Grant No. 1252376. We thank Rosangela Sozzani and Robert Franks for critical reading of this manuscript.

References

[1] R. Mittler, E. Blumwald, Genetic engineering for modern agriculture: challenges and perspectives, Annu. Rev. Plant Biol. 61 (2010) 443–462.

[2] L. Vaahtera, M. Brosché, More than the sum of its parts – how to achieve a specific transcriptional response to abiotic stress, Plant Sci. 180 (3) (2011) 421–430.

[3] W. Wang, B. Vinocur, A. Altman, Plant responses to drought, salinity and extreme temperatures: towards genetic engineering for stress tolerance, Planta 218 (1) (2003) 1–14.

[4] A.E. Valdés, Forced adaptation: plant proteins to fight climate change, Front. Plant Sci. 5 (2014) 762.

[5] J. Riechmann, J. Heard, G. Martin, L. Reuber, J. Keddie, L. Adam, O. Pineda, O. Ratcliffe, R. Samaha, R. Creelman, et al., Arabidopsis transcription factors: genome-wide comparative analysis among eukaryotes, Science 290 (5499) (2000) 2105–2110.

[6] M.K. Udvardi, K. Kakar, M. Wandrey, O. Montanari, J. Murray, A. Andriankaja, J.-Y. Zhang, V. Benedito, J.M. Hofer, F. Chueng, et al., Legume transcription factors: global regulators of plant development and response to the environment, Plant Physiol. 144 (2) (2007) 538–549.

[7] R. Melzer, G. Theißen, MADS and more: transcription factors that shape the plant, in: Plant Transcription Factors, Springer, 2011, pp. 3–18.

[8] G.R. Cramer, K. Urano, S. Delrot, M. Pezzotti, K. Shinozaki, Effects of abiotic stress on plants: a systems biology perspective, BMC Plant Biol. 11 (1) (2011) 163.

[9] S. Friedel, B. Usadel, N. Von Wirén, N. Sreenivasulu, Reverse engineering: a key component of systems biology to unravel global abiotic stress cross-talk, Front. Plant Sci. 3 (2012) 294.

[10] G. Krouk, J. Lingeman, A.M. Colon, G. Coruzzi, D. Shasha, et al., Gene regulatory networks in plants: learning causality from time and perturbation, Genome Biol. 14 (6) (2013) 123.

[11] C. Sima, J. Hua, S. Jung, Inference of gene regulatory networks using time-series data: a survey, Curr. Genomics 10 (6) (2009) 416.

[12] K. Cho, S. Choo, S. Jung, J. Kim, H. Choi, J. Kim, Reverse engineering of gene regulatory networks, Syst. Biol. IET 1 (3) (2007) 149–163.

[13] M. Hecker, S. Lambeck, S. Toepfer, E. van Someren, R. Guthke, Gene regulatory network inference: data integration in dynamic models – a review, Biosystems 96 (1) (2009) 86.

[14] G. Karlebach, R. Shamir, Modelling and analysis of gene regulatory networks, Nat. Rev. Mol. Cell Biol. 9 (10) (2008) 770–780.

[15] A.M. Middleton, E. Farcot, M.R. Owen, T. Vernoux, Modeling regulatory networks to understand plant development: small is beautiful, Plant Cell 24 (10) (2012) 3876–3891.

[16] N.J. Atkinson, P.E. Urwin, The interaction of plant biotic and abiotic stresses: from genes to the field, J. Exp. Bot. 63 (10) (2012) 3523–3543.

[17] C. de Sassi, J.M. Tylianakis, Climate change disproportionately increases herbivore over plant or parasitoid biomass, PLOS ONE 7 (7) (2012) e40557.

[18] J. Kilian, D. Whitehead, J. Horak, D. Wanke, S. Weinl, O. Batistic, C. D'Angelo, E. Bornberg-Bauer, J. Kudla, K. Harter, The AtGenExpress global stress expression data set: protocols, evaluation and model data analysis of UV-B light, drought and cold stress responses, Plant J. 50 (2) (2007) 347–363.

[19] L. López-Maury, S. Marguerat, J. Bähler, Tuning gene expression to changing environments: from rapid responses to evolutionary adaptation, Nat. Rev. Genet. 9 (8) (2008) 583–593.

[20] R.F. Ditt, K.F. Kerr, P. de Figueiredo, J. Delrow, L. Comai, E.W. Nester, The Arabidopsis thaliana transcriptome in response to Agrobacterium tumefaciens, Mol. Plant Microbe Interact. 19 (6) (2006) 665–681.

[21] R.J. O'Connell, M.R. Thon, S. Hacquard, S.G. Amyotte, J. Kleemann, M.F. Torres, U. Damm, E.A. Buiate, L. Epstein, N. Alkan, et al., Lifestyle transitions in plant pathogenic colletotrichum fungi deciphered by genome and transcriptome analyses, Nat. Genet. 44 (9) (2012) 1060–1065.

[22] O. Windram, P. Madhou, S. McHattie, C. Hill, R. Hickman, E. Cooke, D.J. Jenkins, C.A. Penfold, L. Baxter, E. Breeze, et al., Arabidopsis defense against Botrytis cinerea: chronology and regulation deciphered by high-resolution temporal transcriptomic analysis, Plant Cell 24 (9) (2012) 3530–3557.

[23] B.-h. Lee, D.A. Henderson, J.-K. Zhu, The Arabidopsis cold-responsive transcriptome and its regulation by ICE1, Plant Cell 17 (11) (2005) 3155–3175.

[24] A.S. Iyer-Pascuzzi, T. Jackson, H. Cui, J.J. Petricka, W. Busch, H. Tsukagoshi, P.N. Benfey, Cell identity regulators link development and stress responses in the Arabidopsis root, Dev. Cell 21 (4) (2011) 770–782.

[25] J.R. Dinneny, T.A. Long, J.Y. Wang, J.W. Jung, D. Mace, S. Pointer, C. Barron, S.M. Brady, J. Schiefelbein, P.N. Benfey, Cell identity mediates the response of Arabidopsis roots to abiotic stress, Science 320 (5878) (2008) 942–945.

[26] S. González-Pérez, J. Gutiérrez, F. Garcí a-Garcí a, D. Osuna, J. Dopazo, Ó. Lorenzo, Ó. Lorenzo, J.L. Revuelta, J.B. Arellano, Early transcriptional defense responses in Arabidopsis cell suspension culture under high-light conditions, Plant Physiol. 156 (3) (2011) 1439–1456.

[27] T.J. Buckhout, T.J. Yang, W. Schmidt, Early iron-deficiency-induced transcriptional changes in Arabidopsis roots as revealed by microarray analyses, BMC Genomics 10 (1) (2009) 147.

[28] T.A. Long, H. Tsukagoshi, W. Busch, B. Lahner, D.E. Salt, P.N. Benfey, The bHLH transcription factor POPEYE regulates response to iron deficiency in Arabidopsis roots, Plant Cell 22 (7) (2010) 2219–2236.

[29] G. Krouk, P. Mirowski, Y. LeCun, D.E. Shasha, G.M. Coruzzi, et al., Predictive network modeling of the high-resolution dynamic plant transcriptome in response to nitrate, Genome Biol. 11 (12) (2010) R123.

[30] W.-D. Lin, Y.-Y. Liao, T.J. Yang, C.-Y. Pan, T.J. Buckhout, W. Schmidt, Coexpression-based clustering of Arabidopsis root genes predicts functional modules in early phosphate deficiency signaling, Plant Physiol. (2011) 110.

[31] L. Rizhsky, H. Liang, J. Shuman, V. Shulaev, S. Davletova, R. Mittler, When defense pathways collide: the response of Arabidopsis to a combination of drought and heat stress, Plant Physiol. 134 (4) (2004) 1683–1696.

[32] S. Rasmussen, P. Barah, M.C. Suarez-Rodriguez, S. Bressendorff, P. Friis, P. Costantino, A.M. Bones, H.B. Nielsen, J. Mundy, Transcriptome responses to combinations of stresses in Arabidopsis, Plant Physiol. 161 (4) (2013) 1783–1794.

[33] C.M. Prasch, U. Sonnewald, Simultaneous application of heat, drought, and virus to Arabidopsis plants reveals significant shifts in signaling networks, Plant Physiol. 162 (4) (2013) 1849–1866.

[34] N. Sewelam, Y. Oshima, N. Mitsuda, M. Ohme-Takagi, A step towards understanding plant responses to multiple environmental stresses: a genome-wide study, Plant Cell Environ. 37 (9) (2014) 2024–2035.

[35] A. Hahn, J. Kilian, A. Mohrholz, F. Ladwig, F. Peschke, R. Dautel, K. Harter, K.W. Berendzen, D. Wanke, Plant core environmental stress response genes are systemically coordinated during abiotic stresses, Int. J. Mol. Sci. 14 (4) (2013) 7617–7641.

[36] J.L. Riechmann, Transcriptional regulation: a genomic overview, The Arabidopsis Book 16 (1) (2002) 1.

[37] X. Cui, G.A. Churchill, et al., Statistical tests for differential expression in cDNA microarray experiments, Genome Biol. 4 (4) (2003) 210.

[38] G.K. Smyth, J. Michaud, H.S. Scott, Use of within-array replicate spots for assessing differential expression in microarray experiments, Bioinformatics 21 (9) (2005) 2067–2075.

[39] S. Anders, W. Huber, Differential expression analysis for sequence count data, Genome Biol. 11 (10) (2010) R106.

[40] M.D. Robinson, D.J. McCarthy, G.K. Smyth, edgeR: a bioconductor package for differential expression analysis of digital gene expression data, Bioinformatics 26 (1) (2010) 139–140.

[41] K. Aoki, Y. Ogata, D. Shibata, Approaches for extracting practical information from gene co-expression networks in plant biology, Plant Cell Physiol. 48 (3) (2007) 381–390.

[42] B. Zhang, S. Horvath, et al., A general framework for weighted gene co-expression network analysis, Stat. Appl. Genet. Mol. Biol. 4 (1) (2005) 1128.

[43] C.J. Wolfe, I.S. Kohane, A.J. Butte, Systematic survey reveals general applicability of "guilt-by-association" within gene coexpression networks, BMC Bioinform. 6 (1) (2005) 227.

[44] B. Usadel, T. Obayashi, M. Mutwil, F.M. Giorgi, G.W. Bassel, M. Tanimoto, A. Chow, D. Steinhauser, S. Persson, N.J. Provart, Co-expression tools for plant biology: opportunities for hypothesis generation and caveats, Plant Cell Environ. 32 (12) (2009) 1633–1651.

[45] I. Lee, B. Ambaru, P. Thakkar, E.M. Marcotte, S.Y. Rhee, Rational association of genes with traits using a genome-scale gene network for Arabidopsis thaliana, Nat. Biotechnol. 28 (2) (2010) 149–156.

[46] A. Gupta, C.D. Maranas, R. Albert, Elucidation of directionality for co-expressed genes: predicting intra-operon termination sites, Bioinformatics 22 (2) (2006) 209–214.

[47] J. Ehlting, V. Sauveplane, A. Olry, J.-F. Ginglinger, N.J. Provart, D. Werck-Reichhart, An extensive (co-) expression analysis tool for the cytochrome P450 superfamily in Arabidopsis thaliana, BMC Plant Biol. 8 (1) (2008) 47.

[48] K. Polanski, J. Rhodes, C. Hill, P. Zhang, D.J. Jenkins, S.J. Kiddle, A. Jironkin, J. Beynon, V. Buchanan-Wollaston, S. Ott, et al., Wigwams: identifying gene modules co-regulated across multiple biological conditions, Bioinformatics 30 (7) (2014) 962–970.

[49] R. Balasubramaniyan, E. Hüllermeier, N. Weskamp, J. Kämper, Clustering of gene expression data using a local shape-based similarity measure, Bioinformatics 21 (7) (2005) 1069–1077.

[50] J. Nie, R. Stewart, H. Zhang, J. Thomson, F. Ruan, X. Cui, H. Wei, TF-Cluster: a pipeline for identifying functionally coordinated transcription factors via network decomposition of the shared coexpression connectivity matrix (SCCM), BMC Syst. Biol. 5 (1) (2011) 53.

[51] X. Cui, T. Wang, H.-S. Chen, V. Busov, H. Wei, Tf-finder: a software package for identifying transcription factors involved in biological processes using microarray data and existing knowledge base, BMC Bioinform. 11 (1) (2010) 425.

[52] H. Kishino, P.J. Waddell, Correspondence analysis of genes and tissue types and finding genetic links from microarray data, Genome Inform. 11 (2000) 83–95.

[53] A. Wille, P. Zimmermann, E. Vranová, A. Fürholz, O. Laule, S. Bleuler, L. Hennig, A. Prelic, P. von Rohr, L. Thiele, et al., Sparse graphical Gaussian modeling of the isoprenoid gene network in Arabidopsis thaliana, Genome Biol. 5 (11) (2004) R92.

[54] J. Schäfer, K. Strimmer, An empirical bayes approach to inferring large-scale gene association networks, Bioinformatics 21 (6) (2005) 754–764.

[55] P. D'haeseleer, et al., How does gene expression clustering work? Nat. Biotechnol. 23 (12) (2005) 1499–1502.

[56] C. Ma, M. Xin, K.A. Feldmann, X. Wang, Machine learning-based differential network analysis: a study of stress-responsive transcriptomes in Arabidopsis, Plant Cell 26 (2) (2014) 520–537.

[57] G. Altay, F. Emmert-Streib, Inferring the conservative causal core of gene regulatory networks, BMC Syst. Biol. 4 (1) (2010) 132.

[58] A.A. Margolin, I. Nemenman, K. Basso, C. Wiggins, G. Stolovitzky, R.D. Favera, A. Califano, ARACNE: an algorithm for the reconstruction of gene regulatory networks in a mammalian cellular context, BMC Bioinform. 7 (Suppl 1) (2006) S7.

[59] R. Steuer, J. Kurths, C.O. Daub, J. Weise, J. Selbig, The mutual information: detecting and evaluating dependencies between variables, Bioinformatics 18 (suppl 2) (2002) S231–S240.

[60] S. Kumari, J. Nie, H.-S. Chen, H. Ma, R. Stewart, X. Li, M.-Z. Lu, W.M. Taylor, H. Wei, Evaluation of gene association methods for coexpression network construction and biological knowledge discovery, PLOS ONE 7 (11) (2012) e50411.

[61] C. Ma, X. Wang, Application of the gini correlation coefficient to infer regulatory relationships in transcriptome analysis, Plant Physiol. 160 (1) (2012) 192–203.

[62] L. Rueda, A. Bari, A. Ngom, Clustering time-series gene expression data with unequal time intervals, in: Transactions on Computational Systems Biology X, Springer, 2008, pp. 100–123.

[63] M. Triska, D. Grocutt, J. Southern, D.J. Murphy, T. Tatarinova, cisExpress: motif detection in DNA sequences, Bioinformatics 29 (17) (2013) 2203–2205.

[64] S. Martin, Z. Zhang, A. Martino, J. Faulon, Boolean dynamics of genetic regulatory networks inferred from microarray time series data, Bioinformatics 23 (7) (2007) 866–874.

[65] S. Van Dongen, A cluster algorithm for graphs, Rep. Inf. Syst. (10) (2000) 1–40.

[66] W.I. Mentzen, E.S. Wurtele, Regulon organization of Arabidopsis, BMC Plant Biol. 8 (1) (2008) 99.

[67] Y. Zhang, H. Zha, C.-H. Chu, A time-series biclustering algorithm for revealing co-regulated genes, in: Information Technology: Coding and Computing, 2005. ITCC 2005, vol. 1, IEEE, 2005, pp. 32–37.

[68] P. Tamayo, D. Slonim, J. Mesirov, Q. Zhu, S. Kitareewan, E. Dmitrovsky, E.S. Lander, T.R. Golub, Interpreting patterns of gene expression with self-organizing maps: methods and application to hematopoietic differentiation, Proc. Natl. Acad. Sci. U. S. A. 96 (6) (1999) 2907–2912.

[69] M.B. Eisen, P.T. Spellman, P.O. Brown, D. Botstein, Cluster analysis and display of genome-wide expression patterns, Proc. Natl. Acad. Sci. U. S. A. 95 (25) (1998) 14863–14868.

[70] B.J. Frey, D. Dueck, Clustering by passing messages between data points, Science 315 (2007) 972–976.

[71] X. Li, Y. Ye, M. Ng, Q. Wu, MultiFacTV: module detection from higher-order time series biological data, BMC Genomics 14 (Suppl 4) (2013) S2.

[72] J.-X. Liu, C.-H. Zheng, Y. Xu, Extracting plants core genes responding to abiotic stresses by penalized matrix decomposition, Comput. Biol. Med. 42 (5) (2012) 582–589.

[73] T. Chen, V. Filkov, S.S. Skiena, Identifying gene regulatory networks from experimental data, in: Proceedings of the Third Annual International Conference on Computational Molecular Biology, ACM, 1999, pp. 94–103.

[74] A.T. Kwon, H.H. Hoos, R. Ng, Inference of transcriptional regulation relationships from gene expression data, Bioinformatics 19 (8) (2003) 905–912.

[75] W.A. Schmitt, R.M. Raab, G. Stephanopoulos, Elucidation of gene interaction networks through time-lagged correlation analysis of transcriptional data, Genome Res. 14 (8) (2004) 1654–1663.

[76] W. Zhao, E. Serpedin, E.R. Dougherty, Inferring gene regulatory networks from time series data using the minimum description length principle, Bioinformatics 22 (17) (2006) 2129–2135.

[77] P.C. Ma, K.C. Chan, Inferring gene regulatory networks from expression data by discovering fuzzy dependency relationships, IEEE Trans. Fuzzy Syst. 16 (2) (2008) 455–465.

[78] H. Redestig, D. Weicht, J. Selbig, M.A. Hannah, Transcription factor target prediction using multiple short expression time series from Arabidopsis thaliana, BMC Bioinform. 8 (1) (2007) 454.

[79] D. Heckerman, A tutorial on learning with Bayesian networks, Innov. Bayesian Netw. (2008) 33–82.

[80] N. Friedman, M. Linial, I. Nachman, D. Pe'er, Using Bayesian networks to analyze expression data, J. Comput. Biol. 7 (3–4) (2000) 601–620.

[81] K. Murphy, S. Mian, et al., Modelling gene expression data using dynamic Bayesian networks, Tech. rep., Technical Report, Computer Science Division, University of California, Berkeley, CA, 1999.

[82] N. Dojer, A. Gambin, A. Mizera, B. Wilczyński, J. Tiuryn, Applying dynamic Bayesian networks to perturbed gene expression data, BMC Bioinform. 7 (1) (2006) 249.

[83] S. Liang, S. Fuhrman, R. Somogyi, et al., REVEAL, a general reverse engineering algorithm for inference of genetic network architectures, in: Pacific Symposium on Biocomputing, vol. 3, 1998, p. 2.

[84] R. Albert, Boolean modeling of genetic regulatory networks, in: Complex Networks, Springer, 2004, pp. 459–481.

[85] E. Dimitrova, L. Garcí a-Puente, F. Hinkelmann, A. Jarrah, R. Laubenbacher, B. Stigler, M. Stillman, P. Vera-Licona, Parameter estimation for Boolean models of biological networks, Theor. Comput. Sci. 412 (26) (2011) 2816–2826.

[86] R. Laubenbacher, B. Stigler, A computational algebra approach to the reverse engineering of gene regulatory networks, J. Theor. Biol. 229 (4) (2004) 523–537.

[87] T. Akutsu, S. Miyano, S. Kuhara, et al., Identification of genetic networks from a small number of gene expression patterns under the boolean network model, in: Pacific Symposium on Biocomputing, vol. 4, World Scientific Maui, Hawaii, 1999, pp. 17–28.

[88] B.A. Rosa, J. Zhang, I.T. Major, W. Qin, J. Chen, Optimal timepoint sampling in high-throughput gene expression experiments, Bioinformatics 28 (21) (2012) 2773–2781.

[89] G. Bernot, J.-P. Comet, A. Richard, M. Chaves, J.-L. Gouzé, F. Dayan, Modeling and analysis of gene regulatory networks, in: Modeling in Computational Biology and Biomedicine, Springer, 2013, pp. 47–80.

[90] R. Tibshirani, Regression shrinkage and selection via the lasso, J. R. Stat. Soc. B: Methodological (1996) 267–288.

[91] R. Guthke, U. Möller, M. Hoffmann, F. Thies, S. Töpfer, Dynamic network reconstruction from gene expression data applied to immune response during bacterial infection, Bioinformatics 21 (8) (2005) 1626–1634.

[92] M. Gustafsson, M. Hornquist, A. Lombardi, Large-scale reverse engineering by the lasso, arXiv preprint q-bio/0403012.

[93] M.S. Yeung, J. Tegnér, J.J. Collins, Reverse engineering gene networks using singular value decomposition and robust regression, Proc. Natl. Acad. Sci. U. S. A. 99 (9) (2002) 6163–6168.

[94] M. Gustafsson, M. Hörnquist, J. Lundström, J. Björkegren, J. Tegnér, Reverse engineering of gene networks with LASSO and nonlinear basis functions, Ann. N. Y. Acad. Sci. 1158 (1) (2009) 265–275.

[95] L. Palafox, N. Noman, H. Iba, Reverse engineering of gene regulatory networks using dissipative particle swarm optimization, IEEE Trans. Evol. Comput. 17 (4) (2013) 577–587.

[96] M. Kabir, N. Noman, H. Iba, Reverse engineering gene regulatory network from microarray data using linear time-variant model, BMC Bioinform. 11 (Suppl 1) (2010) S56.

[97] M. Ashburner, C.A. Ball, J.A. Blake, D. Botstein, H. Butler, J.M. Cherry, A.P. Davis, K. Dolinski, S.S. Dwight, J.T. Eppig, et al., Gene Ontology: tool for the unification of biology, Nat. Genet. 25 (1) (2000) 25–29.

[98] S. Maere, K. Heymans, M. Kuiper, BiNGO: a Cytoscape plugin to assess overrepresentation of gene ontology categories in biological networks, Bioinformatics 21 (16) (2005) 3448–3449.

[99] J.M. Alonso, A.N. Stepanova, T.J. Leisse, C.J. Kim, H. Chen, P. Shinn, D.K. Stevenson, J. Zimmerman, P. Barajas, R. Cheuk, et al., Genome-wide insertional mutagenesis of Arabidopsis thaliana, Science 301 (5633) (2003) 653–657.

[100] A. Sessions, E. Burke, G. Presting, G. Aux, J. McElver, D. Patton, B. Dietrich, P. Ho, J. Bacwaden, C. Ko, et al., A high-throughput Arabidopsis reverse genetics system, Plant Cell 14 (12) (2002) 2985–2994.

[101] M.G. Rosso, Y. Li, N. Strizhov, B. Reiss, K. Dekker, B. Weisshaar, An Arabidopsis thaliana T-DNA mutagenized population (GABI-Kat) for flanking sequence tag-based reverse genetics, Plant Mol. Biol. 53 (1–2) (2003) 247–259.

[102] B. Ülker, E. Peiter, D.P. Dixon, C. Moffat, R. Capper, N. Bouché, R. Edwards, D. Sanders, H. Knight, M.R. Knight, Getting the most out of publicly available T-DNA insertion lines, Plant J. 56 (4) (2008) 665–677.

[103] P. Hilson, J. Allemeersch, T. Altmann, S. Aubourg, A. Avon, J. Beynon, R.P. Bhalerao, F. Bitton, M. Caboche, B. Cannoot, et al., Versatile gene-specific sequence tags for *Arabidopsis* functional genomics: transcript profiling and reverse genetics applications, Genome Res. 14 (10b) (2004) 2176–2189.

[104] A. Coego, E. Brizuela, P. Castillejo, S. Ruí z, C. Koncz, J.C. del Pozo, M. Pi neiro, J.A. Jarillo, J. Paz-Ares, J. León, The TRANSPLANTA collection of *Arabidopsis* lines: a resource for functional analysis of transcription factors based on their conditional overexpression, Plant J. 77 (6) (2014) 944–953.

[105] K. Higo, Y. Ugawa, M. Iwamoto, T. Korenaga, Plant cis-acting regulatory DNA elements (PLACE) database: 1999, Nucleic Acids Res. 27 (1) (1999) 297–300.

[106] R.V. Davuluri, H. Sun, S.K. Palaniswamy, N. Matthews, C. Molina, M. Kurtz, E. Grotewold, AGRIS. *Arabidopsis* gene regulatory information server, an information resource of *Arabidopsis* cis-regulatory elements and transcription factors, BMC Bioinform. 4 (1) (2003) 25.

[107] S.K. Palaniswamy, S. James, H. Sun, R.S. Lamb, R.V. Davuluri, E. Grotewold, AGRIS and AtRegNet. A platform to link cis-regulatory elements and transcription factors into regulatory networks, Plant Physiol. 140 (3) (2006) 818–829.

[108] T.R. O'Connor, C. Dyreson, J.J. Wyrick, Athena: a resource for rapid visualization and systematic analysis of *Arabidopsis* promoter sequences, Bioinformatics 21 (24) (2005) 4411–4413.

[109] S.M. Brady, N.J. Provart, Web-queryable large-scale data sets for hypothesis generation in plant biology, Plant Cell 21 (4) (2009) 1034–1051.

[110] T.L. Bailey, M. Boden, F.A. Buske, M. Frith, C.E. Grant, L. Clementi, J. Ren, W.W. Li, W.S. Noble, MEME SUITE: tools for motif discovery and searching, Nucleic Acids Res. (2009) gkp335.

[111] M.K. Das, H.-K. Dai, A survey of DNA motif finding algorithms, BMC Bioinform. 8 (Suppl. 7) (2007) S21.

[112] K.W. Berendzen, K. Harter, D. Wanke, Analysis of plant regulatory dna sequences by transient protoplast assays and computer aided sequence evaluation, in: Plant Signal Transduction, Springer, 2009, pp. 311–335.

[113] N. Wehner, L. Hartmann, A. Ehlert, S. Böttner, L. O nate-Sánchez, W. Dröge-Laser, High-throughput protoplast transactivation (PTA) system for the analysis of *Arabidopsis* transcription factor function, Plant J. 68 (3) (2011) 560–569.

[114] C. Olsen, K. Fleming, N. Prendergast, R. Rubio, F. Emmert-Streib, G. Bontempi, B. Haibe-Kains, J. Quackenbush, Inference and validation of predictive gene networks from biomedical literature and gene expression data, Genomics 103 (5) (2014) 329–336.

[115] T.-M. Chu, B. Weir, R. Wolfinger, A systematic statistical linear modeling approach to oligonucleotide array experiments, Math. Biosci. 176 (1) (2002) 35–51.

[116] S. Ma, H.J. Bohnert, Integration of *Arabidopsis thaliana* stress-related transcript profiles, promoter structures, and cell-specific expression, Genome Biol. 8 (4) (2007) R49.

[117] A.P. Gasch, M.B. Eisen, Exploring the conditional coregulation of yeast gene expression through fuzzy k-means clustering, Genome Biol. 3 (11) (2002) 1–22.

[118] G.K. Smyth, Linear models and empirical bayes methods for assessing differential expression in microarray experiments, Stat. Appl. Genet. Mol. Biol. 3 (1) (2004) 1–25.

[119] S. Ma, S. Bachan, M. Porto, H.J. Bohnert, M. Snyder, S.P. Dinesh-Kumar, Discovery of stress responsive DNA regulatory motifs in *Arabidopsis*, PLOS ONE 7 (8) (2012) e43198.

[120] H. Wu, M.K. Kerr, X. Cui, G.A. Churchill, MAANOVA: a software package for the analysis of spotted cDNA microarray experiments, in: The Analysis of Gene Expression Data, Springer, 2003, pp. 313–341.

[121] O. Stegle, K.J. Denby, E.J. Cooke, D.L. Wild, Z. Ghahramani, K.M. Borgwardt, A robust bayesian two-sample test for detecting intervals of differential gene expression in microarray time series, J. Comput. Biol. 17 (3) (2010) 355–367.

[122] Y.C. Tai, T.P. Speed, et al., A multivariate empirical Bayes statistic for replicated microarray time course data, Ann. Stat. 34 (5) (2006) 2387–2412.

[123] N.A. Heard, C.C. Holmes, D.A. Stephens, D.J. Hand, G. Dimopoulos, Bayesian coclustering of anopheles gene expression time series: study of immune defense response to multiple experimental challenges, Proc. Natl. Acad. Sci. U. S. A. 102 (47) (2005) 16939–16944.

[124] S. Klemm, et al., Causal Structure Identification in Nonlinear Dynamical Systems, Dept. of Eng., Univ. of Cambridge, United Kingdom, 2008 (Master's thesis).

[125] E. Howe, K. Holton, S. Nair, D. Schlauch, R. Sinha, J. Quackenbush, MeV: multi-experiment viewer, in: Biomedical Informatics for Cancer Research, Springer, 2010, pp. 267–277.

[126] C. Espinosa-Soto, P. Padilla-Longoria, E.R. Alvarez-Buylla, A gene regulatory network model for cell-fate determination during *Arabidopsis thaliana* flower development that is robust and recovers experimental gene expression profiles, Plant Cell 16 (11) (2004) 2923–2939.

[127] M. Sankar, K.S. Osmont, J. Rolcik, B. Gujas, D. Tarkowska, M. Strnad, I. Xenarios, C.S. Hardtke, A qualitative continuous model of cellular auxin and brassinosteroid signaling and their crosstalk, Bioinformatics 27 (10) (2011) 1404–1412.

[128] A. Cruz-Ramí rez, S. Dí az-Trivi no, I. Blilou, V.A. Grieneisen, R. Sozzani, C. Zamioudis, P. Miskolczi, J. Nieuwland, R. Benjamins, P. Dhonukshe, et al., A bistable circuit involving scarecrow-retinoblastoma integrates cues to inform asymmetric stem cell division, Cell 150 (5) (2012) 1002–1015.

[129] A. Pokhilko, A.P. Fernández, K.D. Edwards, M.M. Southern, K.J. Halliday, A.J. Millar, The clock gene circuit in *Arabidopsis* includes a repressilator with additional feedback loops, Mol. Syst. Biol. 8 (1) (2012) 574.

[130] V. Vermeirssen, I. De Clercq, T. Van Parys, F. Van Breusegem, Y. Van de Peer, *Arabidopsis* ensemble reverse-engineered gene regulatory network discloses interconnected transcription factors in oxidative stress, Plant Cell (2014), tpc-114.

[131] A. Joshi, R. De Smet, K. Marchal, Y. Van de Peer, T. Michoel, Module networks revisited: computational assessment and prioritization of model predictions, Bioinformatics 25 (4) (2009) 490–496.

[132] J.J. Faith, B. Hayete, J.T. Thaden, I. Mogno, J. Wierzbowski, G. Cottarel, S. Kasif, J.J. Collins, T.S. Gardner, Large-scale mapping and validation of *Escherichia coli* transcriptional regulation from a compendium of expression profiles, PLoS Biol. 5 (1) (2007) e8.

[133] R.A. Montes, G. Coello, K.L. González-Aguilera, N. Marsch-Martí nez, S. de Folter, E.R. Alvarez-Buylla, ARACNe-based inference, using curated microarray data, of *Arabidopsis* root transcriptional regulatory networks, BMC Plant Biol. 14 (1) (2014) 97.

[134] P.E. Meyer, K. Kontos, F. Lafitte, G. Bontempi, Information-theoretic inference of large transcriptional regulatory networks, EURASIP J. Bioinform. Syst. Biol. 2007 (2007) 79879.

[135] S. Lebre, J. Becq, F. Devaux, M.P. Stumpf, G. Lelandais, Statistical inference of the time-varying structure of gene-regulation networks, BMC Syst. Biol. 4 (1) (2010) 130.

[136] D. Marbach, J.C. Costello, R. Küffner, N.M. Vega, R.J. Prill, D.M. Camacho, K.R. Allison, M. Kellis, J.J. Collins, G. Stolovitzky, et al., Wisdom of crowds for robust gene network inference, Nat. Methods 9 (8) (2012) 796–804.

17

Transcriptional networks governing plant metabolism

Allison Gaudinier[a,1], Michelle Tang[a,b,1], Daniel J. Kliebenstein[b,c,*]

[a] Department of Plant Biology, College of Biological Sciences, University of California, Davis One Shields Avenue, Davis, CA 95616, USA
[b] Department of Plant Sciences, College of Agriculture and Environmental Sciences, University of California, Davis One Shields Avenue Davis, CA 95616, USA
[c] DynaMo Center of Excellence, University of Copenhagen, Thorvaldsensvej 40, DK-1871, Frederiksberg C., Denmark

ARTICLE INFO

ABSTRACT

Keywords:
Metabolism
Transcription
Feed-forward
Feed-back
Regulation
Coordination
Plant

Efficiently obtaining and utilizing energy and elements is critical for an organism to maximize its fitness. Optimizing these processes requires precise regulation and coordination of an organism's metabolic networks in response to diverse environmental conditions and developmental stages. Metabolic regulation is often considered to largely occur by allosteric feedback where the metabolites directly influence the enzymes function. Recent work is showing that there is also an extensive role for transcriptional control of the enzyme encoding genes to construct the metabolic network in response to developmental and environmental stimuli. Within this review, we go through the extensive evidence of how transcription can coordinate the necessary metabolic shifts required to coordinate a plants metabolism with its development and environment. Additionally, we discuss evidence that the metabolites not only feed-back regulate the enzymes but also the upstream transcriptional processes, possibly to stabilize the system.

1. Introduction

Efficiently obtaining and utilizing energy and elements is critical for an organism to maximize its fitness. Optimizing these processes requires precise regulation and coordination of an organism's metabolic networks in response to diverse environmental conditions and developmental stages [1,2]. These metabolic networks are the key avenues by which an organism obtains and produces all of the necessary building blocks for cells and the resulting biomass and they must be fine-tuned to make the most efficient use of resources available. This precise coordination is a foundational hypothesis for numerous fields of biology including predicting organismal growth and analyzing how a plant interacts with its environment [3]. This essential need to coordinate organismal metabolism has led to strong interest in understanding how metabolism is regulated [3,4].

A dominant feature of metabolic regulation is allosteric feedback whereby metabolites bind the enzymes within a biochemical pathway to control the activity of the enzymes and pathway flux [5,6]. This often leads to the argument that flux based measurements of metabolism are critical to understand metabolism and there is little reason to study the transcriptional control of metabolism [7–9].

This argument, however, is in stark contrast to the rapidly growing body of literature that shows that transcriptional control of metabolism is also a key component of plant metabolic regulation. One way we suggest to reconcile the transcriptional and allosteric views of metabolic regulation is that transcription may establish the base patterns of metabolism within cells in response to developmental and environmental cues [10–12] (Fig. 1). Flux regulation then works within the base rules that transcription establishes to precisely coordinate metabolic regulation. Within this review, we will summarize the evidence of the importance of metabolic transcriptional control by focusing on the intersection of metabolism with development and environmental stresses2. (Fig. 1). Then finally, we will discuss emerging evidence that metabolic outputs can feedback regulate the upstream transcriptional networks.

2. Metabolism and development are mutually dependent

Metabolism and metabolites are spatially organized across all levels of developmental organization for a breadth of reasons. A key example of metabolic spatial distribution is the presence of photosynthates in the shoots and their absence in roots that coordinates with the location of photosynthesis. Modern metabolomics platforms are greatly expanding our understanding of the precise developmental patterns displayed by metabolites by showing that they can be targeted to a diversity of different cell and tissue types in the plant [10,13–15]. In addition to metabolites being developmentally localized because of their function, the metabolites function

* Corresponding author.
  E-mail address: kliebenstein@ucdavis.edu (D.J. Kliebenstein).
[1] A.G. and M.T. are co-first authors.

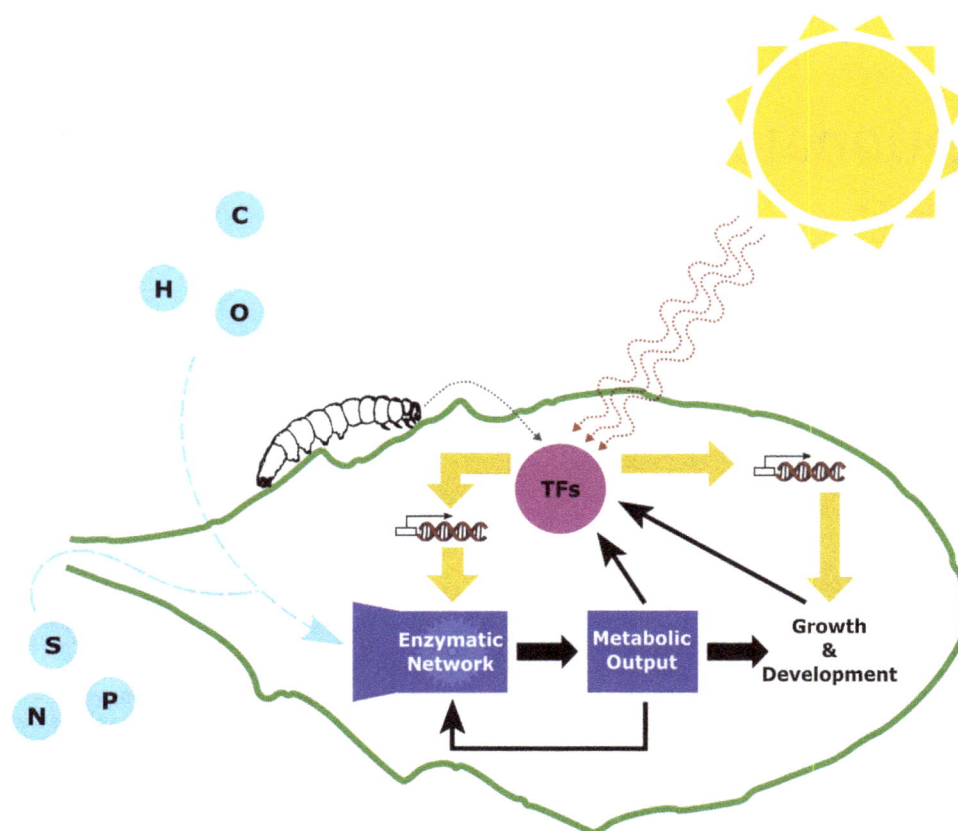

Fig. 1. A central role for TFs in plant metabolism. TFs differentially regulate expression of enzyme encoding genes by interacting with promoters to shape the potential enzymatic network as well as to control plant growth and development. The enzymatic network takes energy and chemical elements from the environment to generate the metabolites necessary for building plant cells, tissues and organs. The metabolites can modulate the enzymatic network via allosteric feedback. Additionally, TFs can coordinate metabolic shifts by integrating feedback signals from metabolic outputs and developmental processes, as well as integrating abiotic (example: UV-B radiation) and biotic stress (example: lepitdopteran herbivory) signals from the environment.

often plays a role in regulating specific developmental patterns. For example, auxin is transported and accumulates in very distinct domains in the root, and these auxin maxima specify meristematic regions and root cell identities [16]. Thus, regulation of auxin biosynthesis, transport and regulation must be strictly coordinated with the growth and development of the root, from cell type differentiation to cell elongation and maturation to ensure proper development of the plant. Essential to this coordination between development and metabolism are transcription factors (TFs) that regulate both developmental programs and metabolic pathways. To illustrate how recent research is discovering the coordination of development and metabolism, we will discuss three vignettes (i) Developmental TFs control metabolism essential to build the root vasculature, (ii) lipids accumulation in reproductive tissues and (iii) specialized defense metabolites localized in myrosin idioblasts. We were unable to describe all examples of the link between development and metabolism for spatial reasons.

2.1. Developmental TFs control metabolism essential to build the root vasculature

Central to the xylems' ability to transport water and nutrients from the roots to the aboveground parts of the plants are the fibers and lignin of the secondary cell walls that enable the formation of tracheid elements. The creation of these tracheid elements requires the precise timing of secondary cell wall deposition, lignification and finally programmed cell death. This requires proper timing of production of the requisite lignin monomers as any deviation in the pattern results in aberrant xylem vessels lacking structural

integrity or subjected to embolisms later in developmental age. Thus, the formation of functional xylem requires the coordinated regulation of development and metabolism.

Recent work has begun to show how key TFs responsible for the initiation and development of xylem vessels also coordinate the production of metabolites required for secondary cell walls formation. The NAC domain transcription factors VASCULAR NAC DOMAIN 6 (VND6), VND7 and SECONDARY WALL ASSOCIATED NAC DOMAIN 1 (SND1) are key TFs that directly regulate xylem cell differentiation [17–20]. These TFs also regulate cellulose and lignin biosynthetic genes necessary for secondary cell wall deposition via the intermediary TFs MYB46 and MYB83 [21]. Overexpression of VND6, VND7 and MYB83 resulted in increased levels of secondary cell wall precursors and ectopic secondary cell wall formation, while knockout mutants of MYB46 and MYB83 displayed severe reduction of secondary cell wall thickening and loss of secondary cell wall precursors [17–21]. This hierarchical model of NAC to MYB to metabolic gene was recently changed when it was found that VND6 and VND7 can also directly interact with the promoters of enzyme encoding genes critical for the formation of metabolites required for secondary cell wall synthesis [22]. Thus, VND6 and VND7 create a regulatory feed-forward loop that likely helps to coordinate lignin metabolism with xylem development. Supporting this concept, when these VND6 and VND7 are mis-regulated, the coordination between root patterning in development and biosynthesis and delivery of metabolites is disrupted. Thus, the VND TFs play a role in bridging the initiation and final maturation of xylem vessel formation by linking regulatory control of development and the requisite metabolism.

2.2. TFs governing both morphogenesis and lipid metabolism programs in seed embryo development

Embryo development is divided into two temporally distinct phases of morphogenesis and maturation. Accumulation of lipids during the maturation phase provides energy stores for seed dormancy and germination when photosynthetic machinery is inactive. While lipid metabolism occurs on the whole plant level, lipid accumulation in the seeds is specifically controlled by several TFs involved in mid to late stages of embryogenesis and embryo maturation. Transcripts of LEAFY COTYLEDON (LEC1, LEC2 and FUS3), ABSCISIC ACID-INSENTIVE3 (ABI3) and WRINKLED1 (WRI1) TFs are found to overlap in the morphogenesis and maturation phase of embryo development, indicating their potential involvement in controlling embryo patterning and lipid accumulation in seeds [23–30].

LEC TFs and ABI3 are critical regulators of embryogenesis. Single and higher order mutants of these TFs resulted in early arrest of embryogenesis and switch to seedling development [25,26,28,29]. Ectopic expression of LEC TFs in vegetative tissues was sufficient to give rise to somatic embryos, demonstrating their roles in cell fate specification and totipotency [25,31]. Concurrently, while the LEC and ABI3 TFs regulate embryo development, they coordinate with each other to properly time lipid metabolism and accumulation in seeds after arrest in embryogenesis and at the beginning of the maturation phase. Studies measuring gene induction in response to LEC2 and WRI1 showed enzymes involved in fatty acid synthesis, including oleosin genes, are targets of these TFs [30,32,33]. In line with these findings, fatty acid analysis of wri1 revealed decreases in total fatty acid accumulation in seeds and ectopic expression of LEC2 caused oil accumulation in leaves [24,33]. While LEC2 and WRI1 can regulate target lipid biosynthetic genes, LEC2 also creates a feed-forward loop wherein it also regulate lipid metabolism as an upstream regulator [24,33]. Thus, as with xylem formation, key developmental regulators also create feed-forward loops to directly regulate the necessary metabolic shifts in coordination with the development. While studies provided evidence of the dual roles these TFs play in both embryogenesis and lipid metabolism in seeds, the mechanism governing the switch between the two temporally distinct phases of seed embryo remain unclear [34,35].

2.3. Novel role of the conserved stomatal development TF, FAMA in specialized defense metabolism

In addition to primary metabolism, plant specialized defense metabolism also displays precise coordination of regulation to target specific metabolites and/or enzymes to explicit cells. For example, laticifers and glandular trichomes are unique repositories of defense metabolites in numerous species [36,37]. These defensive metabolites and their associated structures are highly evolutionarily labile with numerous independent events recreating similar developmental structures but it is not known how this independent evolution occurs [36,37]. Recent work on the development of myrosin idioblast (MI) fate and patterning is beginning to show how conserved TFs can be modified to evolve new tissues. MIs are specialized cells that contain myrosinase, an enzyme that activates defensive glucosinolates and in combination with other proteins turns them into toxic isothiocyanates, thiocyanates and nitriles to deter pathogens and herbivores [38–41]. Glucosinolates are unique to the Capparales and their accumulation varies in different plant organs as well as in different life cycle phases. The greatest accumulation of glucosinolates is found in seeds and lowest in leaves, though amounts of glucosinolate increase with leaf expansion [42,43]. The development of MIs is linked to the regulation of tissue- and organ-specific glucosinolate accumulation as a way for plants to strategize activating these toxic defense compounds without unnecessary and costly production of secondary metabolites.

In contrast to their importance for defense of *Arabidopsis* and other Capparales, it was not known how MI were initiated or patterned. Recent work has shown that the bHLH TF FAMA was required for the formation of the MI. Moreover, FAMA is required for the activation of myrosinase genes TRANSPARENT TESTA GALABRA 1 (*TGG1*) and *TGG2*, further evidence of FAMA's role in MI development [38]. Intriguingly, FAMA is a key regulator controlling cell division and cell differentiation in stomatal development, particularly specifying guard cell fate [38,44,45].

Moreover, FAMA and other bHLH TFs regulating stomatal development are conserved across seed plants [46]. This suggests that the specialized metabolism of glucosinolates in the Brassicaceae have co-opted the conserved function of FAMA to also function in regulating myrosin idioblast differentiation. This suggests the potential for a general pattern wherein new developmental structures are generated by co-opting conserved developmental TFs [36,37]. As with the NACs and LEC2, FAMA co-ordinately regulates both the developmental structure, MI, and the requisite enzyme, myrosinase. This raises the intriguing question of what stomatal metabolism might also be regulated by FAMA or if this metabolic function of FAMA is unique to the MIs [38,44,45].

2.4. Concluding remarks on transcriptional coordination of metabolism and development

Metabolites frequently intersect with developmental processes. While the TFs discussed in this section play a critical role in cell and tissue differentiation and regulating in the metabolism of those cells and tissues, they also highlight the beginning appearance of what may be a recurrent theme. In all instances, the developmental regulator also directly regulated the cell- or tissue-specific metabolism often via a feed-forward regulatory loop. Regulatory loops provide an enhanced ability to precisely coordinate different processes and as such may be a key feature of how TFs regulate both development and metabolism. This further suggests that as the two processes appear to be intrinsically linked, they are not mutually exclusive events and their study should be more coordinated in the future.

3. Stress and nutrient alteration of metabolism via transcription

In addition to development, metabolites are also central to the plants' response to a diverse array of environmental inputs. Examples include the accumulation of proline and polyamines in response to drought stress and altered lipid composition in response to freezing that are considered to be key changes facilitating the ability of the plant to survive these stresses [47,48]. These metabolic changes in response to stress are largely mediated via direct transcriptional shifts in the expression of enzyme-encoding genes that may enable shifts in the metabolic steady state to optimize growth within the new environmental or stressful condition. Recent work is beginning to highlight how TF networks modulate metabolism both to express key stress resistance metabolites and also to restructure the entire primary metabolic network to facilitate these shifts. We outline the transcriptional stress response to UV-B perception to illustrate a well-described regulatory pathway linking input to transcriptional control of metabolic output. In addition to stress regulation of metabolism, nutrient status is also a key component of metabolic regulation with plants having well characterized networks controlling the response to different carbon sources, nitrogen, phosphorous, iron, sulfur and a myriad of other nutrients. A full discussion of these pathways is beyond our available space. Thus, we have decided to focus on nitrogen and sulfur to compare the considerably studied transcriptional control of nitrogen metabolism with the well modeled metabolic research but limited transcriptional studies of sulfur metabolism to provide an illustration of how combining these approaches is necessary for future research. Finally, we describe how these nutrient and environmental signals are being integrated as can be seen through the clock affecting both primary and secondary metabolism.

3.1. Direct pathway of UV-B photoperception to resistance metabolite expression

Phenylpropanoids, especially flavonoids, are critical metabolites to allow plants to live in the wild by absorbing ultraviolet-B radiation [49]. As a part of this function, the plant induces flavonoid biosynthesis in response to sensing UV-B [49,50]. Recent studies have developed the model of UV-B perception and the subsequent metabolic resistance response into what may be the best understood pathway linking environmental signal perception to transcriptional change in metabolic output to resist the stress (Fig. 2). UV-B is directly perceived by the UVR8 photoreceptor, which then transmits a signal to activate the TFs ELONGATED HYPOCOTYL 5 (HY5) and MYB12 [50–53]. The HY5 and MYB12 TFs create a feed-forward loop in which HY5 binds the MYB12 promoter and upregulates MYB12 and they both bind the promoters and upregulate the enzyme genes involved in flavonoid biosynthesis [54–57]. This transcriptional network then directly regulates the accumulation of the phenylpropanoids allowing for increased resistance to UV-B irradiation and survival in the wild [50,51,58]. This system provides a unique model whereby all of the direct molecular steps between the perception of a stress (UV-B/UVR8) and downstream transcriptional control of the metabolic enzyme genes to provide resistance to that stress are known. Interestingly, even in this largely linear pathway there are regulatory loops whose function in modulating the response remains to be determined (Fig. 2).

3.2. Remodulation of primary metabolism: nitrogen

A key component of any plant transcriptional response to stress is to properly control how the plant assimilates macronutrients and uses them in downstream reactions. For example, numerous studies have shown that many of the genes involved in the sensing, acquisition and downstream metabolic processes using nitrogen are transcriptionally responsive to diverse and stressful environmental conditions [59]. In accordance with this known transcriptional control for nitrogen metabolism, many studies have focused on discovering the involved TFs. These studies have identified at least a dozen TFs that control various aspects of nitrogen metabolism [60,61]. The best studied of these TFs is NIN-LIKE PROTEIN 7 (NLP7), which is retained in the nucleus quickly under nitrate sensing via an unknown signal. Once in the nucleus, NLP7 directly binds to the promoters of many core nitrogen responsive genes to regulate their expression [62]. NLP7 mutants are small and have a high shoot to root ratio, likely because the plants are suffering from limited nitrogen availability due to the altered nitrogen metabolic network [63]. However, most of the metabolic consequences of these nitrogen associated TFs are inferred based on the growth of mutant plants on different nutrients with minimal efforts to measure the broader impacts on metabolism, besides amino acid level, of these TF mutants. Thus, it is currently unknown how the transcriptional changes caused by these TFs actually affect the metabolic network and/or nitrogen metabolism within the plant. Developing this understanding of how TFs targeted to the primary metabolic network shift the actual metabolites in the network will

be key to our understanding of the transcriptional link between stress and the resulting metabolic changes.

3.3. Remodulation of primary metabolism: sulfur

In contrast to the studies performed in regards to nitrogen metabolism, there have been extensive metabolic studies profiling and modeling sulfur metabolism that have been reviewed elsewhere [64]. The emphasis on modeling in this field has allowed researchers to predict accurately how changes in the plants genotype affect the accumulation of sulfurous metabolites [65]. In contrast, only a single transcription factor has been found to control transcription in response to altered sulfur status: SULFUR LIMITATION1 (SLIM1) also known as ETHYLENE INSENSITIVE LIKE 3 (EIL3). Mutant slim1 plants suffer from severe limitations in sulfur availability and have a decrease in the content of sulfur-containing metabolite glutathione in their shoots [66]. No direct targets of SLIM1 have been identified; therefore its role in sulfur metabolism although clear, is linked indirectly to metabolic control. Incorporating the transcriptional control provided by genes like SLIM1 and other unidentified TFs into the extensive flux models has the potential to provide a more accurate model involving transcriptional and metabolic control of the system that could inform our understanding of how plants control their metabolic networks.

3.4. Integration of diverse stress transcriptional signals to coordinate metabolism: CCA1

Studies on the regulation of plant metabolism typically focused on the study of an individual pathway, environmental impact or nutrient source. While this singular focus has led to abundant knowledge about how individual metabolic pathways are regulated, genomic and systems biology tools are showing the importance of links across pathways and metabolites to coordinate responses [67–74].

An example of a gene that integrates information to coordinate the regulation of diverse metabolic networks is the phytochrome-activated, Myb-related TF CIRCADIAN CLOCK ASSOCIATED 1 (CCA1) [75]. CCA1 potentially affects a majority of plant metabolism directly, as a TF, and indirectly, via its role as part of the circadian clock oscillator [73]. CCA1 directly binds the light-harvesting chlorophyll a/b protein (CAB1) promoter allowing it to regulate the expression of photosynthetic machinery [76]. Altered expression of CCA1 also leads to perturbation in the circadian rhythms that affect the expression of genes in the carbon cycle including starch accumulation, starch conversion to sugars, and downstream metabolic pathways that use the carbon backbones [77]. Metabolic profiling of the CCA1 overexpression line, however, showed that few metabolites had significantly altered abundance. This indicates that there are likely other regulatory factors that can counter misexpression of CCA1 to maintain the stability of central metabolism in arrhythmic growth [78].

In addition to controlling the carbon availability, CCA1 has been linked to mineral nutrient regulation. CCA1 directly targets the promoters and regulates the expression of the nitrogen metabolism genes glutamine synthase (GLN1;3) and glutamate dehydrogenase (GDH1) [79]. Moreover, CCA1 can indirectly affect the expression of nitrogen and sulfur assimilation via altered circadian regulation [80]. The CCA1 protein interacts with the HY5 protein and together can bind and regulate the promoters of CAB1 and CAB2 [81,82]. Therefore, CCA1 links together circadian regulation, nutrient metabolism, multiple light responses, and potentially pathways not yet analyzed. This ability of CCA1 to integrate diverse signals and influence numerous downstream pathways begins to illustrate how the hierarchical linear model of single pathway regulation

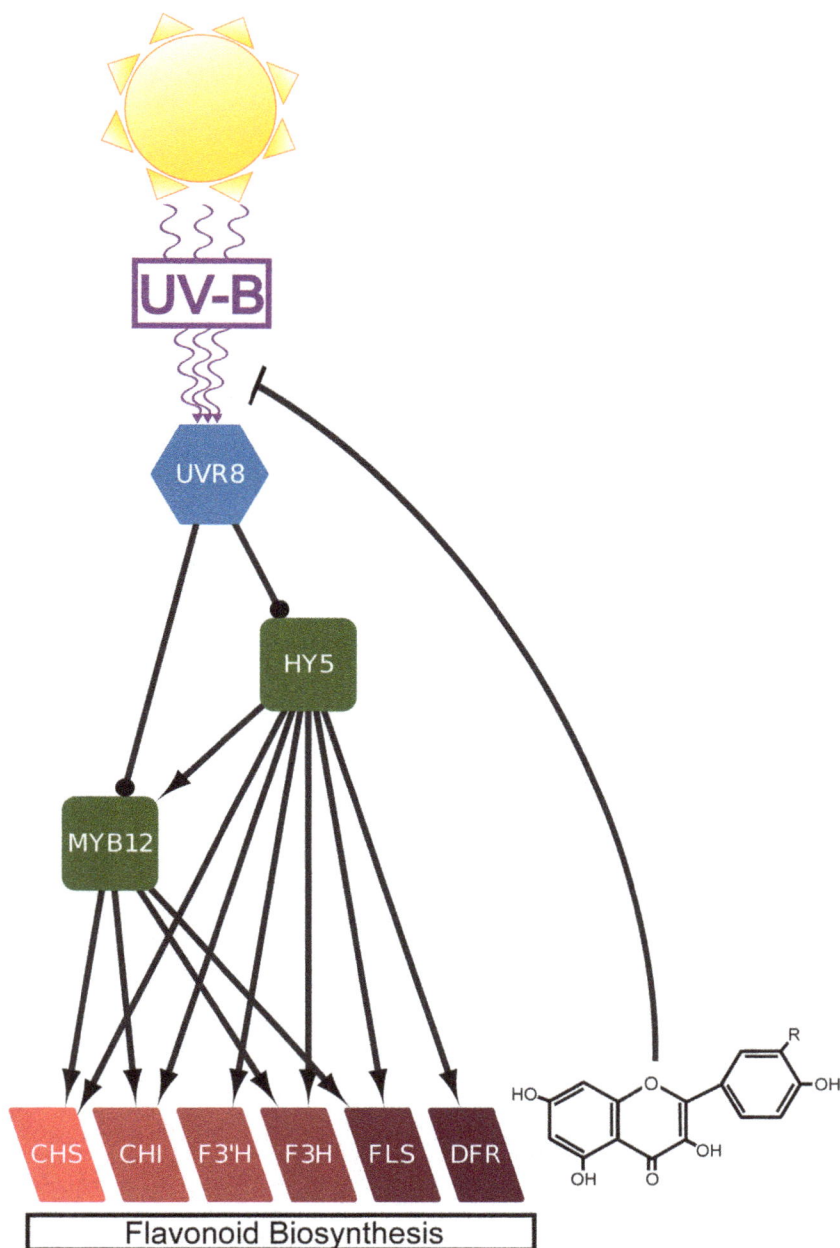

Fig. 2. Signalling pathway from UV-B perception to flavonoid biosynthesis. UV-B wavelengths from sunlight is directly perceived by the UVR8 photoreceptor (purple photons). UVR8 then induces HY5 and MYB12 expression (lines ending in circles). HY5 binds to the promoter of MYB12 and activates its expression creating a feed-forward transcriptional loop (line ending in arrow). HY5 and MYB12 bind to promoters of genes encoding enzymes in the flavonoid biosynthesis pathway (Rhombi) and activate their expression (lines ending in arrows). Finally, the flavonoids absorb the UV-B to prevent damage and possibly cause decreased UV-B photoperception.

breaks down when viewing the transcriptional regulatory mechanisms and their regulation of metabolic systems [69].

3.5. Concluding remarks on transcriptional coordination of metabolism and stress

Having a thorough understanding of the regulation of a metabolic system allows for prediction of how it will respond to a change in the state of the system. An altered state could be varying environmental conditions, a genetic perturbation, or an interaction of the two. Developing a predictive metabolic model will require combining flux models of metabolism with dynamic analysis of the transcriptional control of multiple metabolic pathways. Combining

this knowledge is the next step to understand how these two levels of metabolism coordinate to change the plant's metabolic state.

4. Metabolites alter transcription to provide feedback and stabilize regulatory networks

The current model of transcriptional regulation within plants is largely based on the concept of a hierarchical regulatory system. This hierarchy posits that environmental signals stimulate a regulatory network that integrates these signals and then generates an output supposed to optimize the plants fitness. While this base model may include numerous circuits and interconnections, it is inherently linear because there are no connections between the output and the signaling network. However, systems engineering theory shows that these purely hierarchical/linear systems are highly unstable [83–85]. Stabilizing these hierarchical/linear systems requires the introduction of feedback whereby the outputs of the systems are linked to the inputs to allow the system to simultaneously coordinate both.

In the modeling of plant development, this link of output to input is directly built into the model by the inclusion of spatial constraints (i.e. tissue size and/or cell position) as a consequence of development [86,87]. Thus, the necessity of links between the output and input to stabilize these models has not always been widely recognized. In contrast to developmental models, models involving the transcriptional regulation of metabolic pathways or transcriptional responses do not allow for the inclusion of feedback by the metabolite or other outputs. This is in contrast to growing evidence that metabolites can provide this feedback regulation in plants and there is the potential that metabolite feedback of transcriptional pathways may be a ubiquitous albeit underappreciated facet of plant biology [70,88–90]. In this section, we will discuss several lines of evidence supporting this argument.

4.1. Metabolites and retrograde signaling

A key for any organism to modulate its metabolism is the ability to coordinate the functioning of organelles with the transcription of nuclear genes whose proteins are targeted to the organelles. This coordination involves retrograde signaling in which multiple signals from the organelle somehow migrate to the nucleus to control nuclear transcription [91–93]. Several pathways have been identified that enable retrograde signaling with all involving an organellar metabolite influencing the generation of the signal. The GUN (Genomes Uncoupled) pathway of retrograde signaling involves the likely detection of some metabolic component of the Heme biosynthetic pathway to affect nuclear expression of photosynthetic genes [93–95]. Another metabolite, methylerythritol cyclodiphosphate (MEcPP) derived from the plastid methylerythritol phosphate (MEP) pathway appears to conduct retrograde signaling by targeting a pool of stress responsive genes that largely focus on plant interactions with their biotic environment [96]. 3'-phosphoadenosine 5'-phosphate (PAP) also provides a retrograde signal that targets stress responsive nuclear genes largely linked to the abiotic environment such as drought and high light [97]. While these retrograde studies suggest that there is likely a breadth of metabolites that can influence gene transcription in plants, the underlying perception and signaling mechanisms have yet to be uncovered.

4.2. metabolic feedback loop involving phenylpropanoids and the Mediator transcriptional complex

Key evidence that metabolites and/or their direct consequences have the capacity to generate a feedback loop to regulate transcription and properly coordinate a regulatory network has come from the study of metabolic mutants affecting plant phenylpropanoid metabolism [70,98,99]. Mutating genes within the lignin/phenylpropanoid network frequently lead to plants with diminished growth, which had been presumed to be a side effect of the altered lignin [98]. This hypothesis has begun to be overturned by the application of a suppressor screen wherein Arabidopsis ref8 mutants deficient in p-coumaroylshikimate 3'-hydroxylase (C3'H) was screened for second site mutations that rescued growth of this genotype. This screen found several second site suppressor mutations that alleviated the growth defect while maintaining the biochemical defect thus indicating that the biochemical deficiency was not causing the diminished growth. Cloning of these suppressors showed that they were mutated in components of Mediator, a multi-subunit complex that is required for eukaryotic transcrip-

tion and influences numerous plant phenotypes [100]. were. The disruption of two Mediator subunits, MED5a/5b was able to rescue the growth phenotype and restore to near wild-type level of lignin albeit of a different biochemical form [70]. This suggests that the growth and lignin deficiencies were not a direct result of the biochemical mutation but instead caused by a regulatory effect triggered by the Mediator complex in response to the biochemical mutation and the plants altered metabolism [70].

This hypothesis was supported in a separate suppressor screen of another phenylpropanoid mutant, fah1, that had an unexplained repressed accumulation of hydroxycinnamate ester (HCE) and anthocyanins. Mutations in the MED5a/5b genes within this fah1 mutant lead to a recovered ability to synthesis and accumulate HCEs and anthocyanins [99]. Thus, the decreased accumulation of HCEs and anthocyanins was not due to the biochemical deficiency in the fah1 enzyme but was instead caused by a regulatory response that works via Mediator and is triggered by the fah1 mutation. A similar link between Mediator based regulation of the phenylpropanoid pathway and biosynthetic deficiency was found using a mutant in the indole glucosinolate biosynthesis pathway that altered phenylpropanoid biosynthesis [101]. Similarly, mutations in Mediator again recovered the proper expression of the phenylpropanoid pathway suggesting that Mediator is not just involved in regulating the phenylpropanoid pathway in response to phenylpropanoid mutations but also for other biochemical deficiencies. However, in all cases, the immediate signal stimulating this transcriptional effect remains to be identified.

In agreement with the concept that the above three examples outlined may be examples of metabolite-facilitated feedback, the Mediator complex is a direct regulator of transcription for the phenylpropanoid genes [71]. Thus, the Mediator complex regulates the phenylpropanoid pathway and is sensitive to perturbations in the output of this complex establishing a feedback loop whereby the output can influence the input. Interestingly, this link between lignin and anthocyanin metabolism and transcriptional regulation is not the only observed link for phenlypropanoids. There has also been a link between flavonoid flux and potential regulatory consequences independent of the above system [88]. Thus, it is likely that multiple aspects of the phenylpropanoid biosynthetic pathways can influence the plant transcriptional networks.

4.3. Rapid evolution of metabolic signals

In addition to the above molecularly characterized examples where conserved primary metabolites cause what appear to be conserved transcriptional signaling events, there is a rapid growth in the observation of plant secondary metabolites also having potential transcriptional regulatory effects upon the plant. In Arabidopsis thaliana, there is evidence that an indolic glucosinolate defense compound can lead to altered defense responses like altered callose formation [102]. Interestingly, DIMBOA, a monocot limited indolic secondary metabolite, has also been linked to altered regulation of callose formation in plant/pathogen interactions in maize [103]. Thus, two separate plant lineages have evolved a regulatory link between evolutionarily distinct indolic metabolites and callose formation [102,103]. One possible solution to how divergent metabolites can lead to conserved transcriptional responses was recently suggested by the possibility that regulatory proteins that perceive plant defense compounds can coevolve with the structural variation in distinct plant lineages [104]. This allows the metabolites structure to change while maintaining the same regulatory linkage.

There have also been examples of plant secondary metabolite having regulatory influence over non-defense mechanisms. In Raphanus sativus, a glucosinolate hydrolysis product, raphanusanin, can directly modulate plant growth by interacting with the

TIR1 auxin receptor to alter the downstream transcription of auxin responsive genes and the resulting hypocotyl response [105,106]. In *Arabidopsis thaliana*, variation in the glucosinolate structure has been linked to altered circadian clock transcriptional patterns [89]. In oat, beta-amyrin has been associated with altered cell patterning in the roots [107]. While the molecular mechanism of most of these regulatory connections remains to be determined, there is a recurrent pattern where plant metabolites have the ability to provide feedback linkages to broad aspects of plant physiology that are under transcriptional control. Only by cloning all of the underlying genes for a number of these different connections will we be able to understand how these connections have developed and how many other metabolites may have similar but unmeasured links.

5. Conclusions

As shown above, recent work is beginning to highlight the role of transcription factors in controlling plant metabolism to potentially optimize growth under diverse conditions and tissues. A common feature of all these studies is the frequent observation of regulatory loops. These can be either feed-forward loops involving TFs regulating each other and a common enzyme-encoding gene (i.e. LEC2/WRI1/Lipids) or feedback loops wherein the metabolic output of a network influences the activity of TFs within the plant (i.e. phenylpropanoids/Mediator). This leads us to propose that the largely hierarchical modeling of plant metabolic regulation is insufficient to the task at hand and instead we need to move into alternative approaches that can better integrate these regulatory loops. Similarly, this suggests that plant metabolism, due to its ease of manipulation and measurement, may provide the optimal platform to conduct these integrative studies on linking TFs to their downstream consequences in transcription, metabolism and resulting growth phenotypes. It will be exciting to see the outcomes of these future integrative systems experiments studying how TFs and metabolic flux interconnect in plants.

Acknowledgements

This effort was funded by the NSF DBI grant 0820580 to DJK, the NSF MCB grant 1330337 to DJK, the USDA National Institute of Food and Agriculture, Hatch project number CA-D-PLS-7033-H to DJK, the Danish National Research Foundation (DNRF99) grant to DJK, the UC Davis Department of Plant Sciences Jastro Scholarship to MT, the NSF GRFP to MT via NSF DGE 1148897 to Jeffery C. Gibeling, Dean and Vice Provost of Office of Graduate Studies at UC Davis.

References

[1] U. Sauer, D.R. Lasko, J. Fiaux, M. Hochuli, R. Glaser, et al., Metabolic Flux Ratio Analysis of Genetic and Environmental Modulations of *E. coli* Central Carbon, Metabol. J. Bacteriol. 181 (1999) 6679–6688.

[2] P.D. Keightley, Models of quantitative variation of flux in metabolic pathways, Genetics 121 (1989) 869–876.

[3] Karban, R., Baldwin, I.T. 1997. Induced Responses to Herbivory. Chicago, IL, USA. University of Chicago Press.

[4] A.M. Smith, M. Stitt, Coordination of carbon supply and plant growth, Plant Cell Environ. 30 (2007) 1126–1149.

[5] J. Li, R.L. Last, The *Arabidopsis thaliana* trp5 mutant has a feedback-resistant anthranilate synthase and elevated soluble tryptophan, Plant Physiol. 110 (1996) 51–59.

[6] V.-R. Chellamuthu, E. Ermilova, T. Lapina, J. Lueddecke, E. Minaeva, et al., A widespread glutamine-sensing mechanism in the plant kingdom, Cell 159 (2014) 1188–1199.

[7] Y. Shachar-Hill, Metabolic network flux analysis for engineering plant systems, Curr. Opin. Biotechnol. 24 (2013) 247–255.

[8] D.K. Allen, I.G.L. Libourel, Y. Shachar-Hill, Metabolic flux analysis in plants: coping with complexity, Plant Cell Environ. 32 (2009) 1241–1257.

[9] I.G.L. Libourel, Y. Shachar-Hill, Metabolic flux analysis in plants: from intelligent design to rational engineering, Ann. Rev. Plant Biol. (2008) 625–650.

[10] A. Moussaieff, I. Rogachev, L. Brodsky, S. Malitsky, T.W. Toal, et al., High-resolution metabolic mapping of cell types in plant roots, Proc. Natl. Acad. Sci. U. S. A 110 (2013) E1232–E1241.

[11] J.R. Dinneny, T.A. Long, J.Y. Wang, J.W. Jung, D. Mace, et al., Cell identity mediates the response of *Arabidopsis* roots to abiotic stress, Science 320 (2008) 942–945.

[12] S.M. Brady, D.A. Orlando, J.Y. Lee, J.Y. Wang, J. Koch, et al., A high-resolution root spatiotemporal map reveals dominant expression patterns, Science 318 (2007) 801–806.

[13] R. Shroff, F. Vergara, A. Muck, A. Svatos, J. Gershenzon, Nonuniform distribution of glucosinolates in *Arabidopsis thaliana* leaves has important consequences for plant defense, Proc. Natl. Acad. Sci. U. S. A. 105 (2008) 6196–6201.

[14] A. Maruyama-Nakashita, Y. Nakamura, T. Tohge, K. Saito, H. Takahashi, *Arabidopsis* SLIM1 is a central transcriptional regulator of plant sulfur response and metabolism *Arabidopsis* SLIM1 is a central transcriptional regulator of plant sulfur response and metabolism, Plant Cell 18 (2006) 3235–3251.

[15] S. Krueger, P. Giavalisco, L. Krall, M.C. Steinhauser, D. Bussis, et al., A Topological map of the compartmentalized *Arabidopsis thaliana* leaf metabolome, Plos One 6 (2011).

[16] S. Sabatini, D. Beis, H. Wolkenfelt, J. Murfett, T. Guilfoyle, et al., An auxin-dependent distal organizer of pattern and polarity in the *Arabidopsis* root, Cell (1999) 463–472.

[17] K. Ohashi-Ito, Y. Oda, H. Fukuda, *Arabidopsis* VASCULAR-RELATED NAC-DOMAIN6 directly regulates the genes that govern programmed death and secondary wall formation during xylem differentiation *Arabidopsis* VASCULAR-RELATED NAC-DOMAIN6 directly regulates the genes that govern programmed death and secondary wall formation during xylem differentiation, Plant Cell (2010) 3461–3473.

[18] M. Yamaguchi, N. Goue, H. Igarashi, M. Ohtani, Y. Nakano, et al., VASCULAR-RELATED NAC-DOMAIN6 and VASCULAR-RELATED NAC-DOMAIN7 effectively induce transdifferentiation into xylem vessel elements under control of an induction system, Plant Physiol. 153 (2010) 906–914.

[19] M. Yamaguchi, M. Kubo, H. Fukuda, T. Demura, VASCULAR-RELATED NAC-DOMAIN7 is Involved in the Differentiation of All Types of Xylem Vessels in *Arabidopsis* Roots and Shoots, Plant J: Blackwell Publishing Ltd., 2008, pp. 652–664.

[20] M. Yamaguchi, N. Mitsuda, M. Ohtani, M. Ohme-Takagi, K. Kato, et al., VASCULAR-RELATED NAC-DOMAIN 7 Directly Regulates the Expression of a Broad Range of Genes for Xylem Vessel Formation, Plant J: Blackwell Publishing Ltd., 2011, pp. 579–590.

[21] R.L. McCarthy, R. Zhong, Z.-H. Ye, MYB83 Is a Direct Target of SND1 and acts redundantly with myb 46 in the regulation of secondary cell wall biosynthesis in *Arabidopsis*, Plant Cell Physiol (2009) 1950–1964, Oxford University Press.

[22] M. Taylor-Teeples, L. Lin, M. de Lucas, G. Turco, T.W. Toal, et al., An *Arabidopsis* gene regulatory network for secondary cell wall synthesis, Nature (2014): Nature Publishing Group.

[23] J.G. Angeles-Núñez, A. Tiessen, Mutation of the transcription factor LEAFY COTYLEDON 2 alters the chemical composition of *Arabidopsis* seeds, decreasing oil and protein content, while maintaining high levels of starch and sucrose in mature seeds, J. Plant Physiol. (2011) 1891–1900.

[24] S. Baud, M.S. Mendoza, A. To, E. Harscoët, L. Lepiniec, et al., WRINKLED1 specifies the regulatory action of LEAFY COTYLEDON2 towards fatty acid metabolism during seed maturation in *Arabidopsis*, Plant J. (2007) 825–838.

[25] J.J. Harada, Role of *Arabidopsis* LEAFY COTYLEDON genes in seed development, J. Plant Physiol. (2001).

[26] F. Parcy, C. Valon, A. Kohara, S. Miséra, J. Giraudat, The ABSCISIC ACID-INSENSITIVE3, FUSCA3, and LEAFY COTYLEDON1 loci act in concert to control multiple aspects of *Arabidopsis* seed development, Plant Cell: Am. Soc. Plant Biol. (1997) 1265–1277.

[27] M. Santos-Mendoza, B. Dubreucq, M. Miquel, M. Caboche, M.L. Lepiniec, LEAFY COTYLEDON 2 activation is sufficient to trigger the accumulation of oil and seed specific mRNAs in *Arabidopsis* leaves, FEBS Lett. (2005) 4666–4670.

[28] S.L. Stone, S.A. Braybrook, S.L. Paula, L.W. Kwong, J. Meuser, et al., Arabidopsis LEAFY COTYLEDON2 induces maturation traits and auxin activity: implications for somatic embryogenesis, Proc. Natl. Acad. Sci. U.S.A. 105 (2008) 3151–3156.

[29] M. West, K.M. Yee, J. Danao, J.L. Zimmerman, R.L. Fischer, et al., LEAFY COTYLEDON1 Is an Essential Regulator of Late Embryogenesis and Cotyledon Identity in *Arabidopsis*, Plant Cell, Am. Soc. Plant Biol. (1994) 1731–1745.

[30] K. Maeo, T. Tokuda, A. Ayame, N. Mitsui, T. Kawai, et al., An AP2-type transcription factor, WRINKLED1, of *Arabidopsis thaliana* binds to the AW-box sequence conserved among proximal upstream regions of genes involved in fatty acid synthesis, Plant J. (2009) 476–487, Blackwell Publishing Ltd.

[31] S.L. Stone, L.W. Kwong, K.M. Yee, J. Pelletier, L. Lepiniec, et al., LEAFY COTYLEDON2 encodes a B3 domain transcription factor that induces embryo development, Proc. Natl. Acad. Sci. U. S. A. 98 (2001) 11806–11811.

[32] S.A. Braybrook, S.L. Stone, S. Park, A.Q. Bui, B.H. Le, et al., Genes directly regulated by LEAFY COTYLEDON2 provide insight into the control of embryo maturation and somatic embryogenesis, Proc. Natl. Acad. Sci. U. S. A. 103 (2006) 3468–3473.

[33] A. To, J. Joubès, G. Barthole, A. Lécureuil, A. Scagnelli, et al., WRINKLED transcription factors orchestrate tissue-specific regulation of fatty acid biosynthesis in *Arabidopsis*, The Plant Cell (2012) 5007–5023, American Society of Plant Biologists.

[34] C.S. Johnson, B. Kolevski, D.R. Smyth, TRANSPARENT TESTA GLABRA2, a trichome and seed coat development gene of *Arabidopsis* encodes a WRKY transcription factor, Plant Cell (2002) 1359–1375.

[35] T. Ishida, S. Hattori, R. Sano, K. Inoue, Y. Shirano, et al., *Arabidopsis* TRANSPARENT TESTA GLABRA2 is directly regulated by R2R3 MYB transcription factors and is involved in regulation of GLABRA2 transcription in epidermal differentiation *Arabidopsis* TRANSPARENT TESTA GLABRA2 is directly regulated by R2R3 MYB transcription factors and is involved in regulation of GLABRA2 transcription in epidermal differentiation, Plant Cell (2007) 2531–2543.

[36] D.J. Kliebenstein, New synthesis-regulatory evolution, the veiled world of chemical diversification, J. Chem. Ecol. 39 (2013) 349.

[37] Kliebenstein D.J. 2013. Making new molecules – evolution of structures for novel metabolites in plants. Curr Opin Plant Biol Online.

[38] M. Li, F.D. Sack, Myrosin idioblast cell fate and development are regulated by the arabidopsis transcription factor FAMA the auxin pathway, and vesicular trafficking, Plant Cell Am. Soc. Plant Biol. (2014) 4053–4066.

[39] M. Burow, A. Losansky, R. Muller, A. Plock, D.J. Kliebenstein, et al., The genetic basis of constitutive and herbivore-induced ESP-independent nitrile formation in *Arabidopsis*, Plant Physiol. 149 (2009) 561–574.

[40] M. Burow, Z.Y. Zhang, J.A. Ober, V.M. Lambrix, U. Wittstock, et al., ESP and ESM1 mediate indol-3-acetonitrile production from indol-3-ylmethyl glucosinolate in *Arabidopsis*, Phytochemistry 69 (2008) 663–671.

[41] M. Burow, M. Rice, B. Hause, U. Wittstock, J. Gershenzon, Cell- and tissue-specific localization and regulation of the epithiospecifier protein in *Arabidopsis thaliana*, Plant Mol.Biol. 64 (2007) 173–185.

[42] P.D. Brown, J.G. Tokuhisa, M. Reichelt, J. Gershenzon, Variation of glucosinolate accumulation among different organs and developmental stages of *Arabidopsis thaliana*, Phytochem 62 (2003) 471–781.

[43] I.E. Sønderby, F. Geu-Flores, B.A. Halkier, Biosynthesis of glucosinolates – gene discovery and beyond, Trends Plant Sci. 15 (2010) 283–290.

[44] D.C. Bergmann, W. Lukowitz, C.R. Somerville, Stomatal development and pattern controlled by a MAPKK kinase, Sci.: Am. Assoc. Adv. Sci. (2004) 1494–1497.

[45] K. Ohashi-Ito, D.C. Bergmann, *Arabidopsis* FAMA controls the final proliferation/differentiation switch during stomatal development *Arabidopsis* FAMA controls the final proliferation/differentiation switch during stomatal development, Plant Cell: Am. Soc. Plant Biol. (2006) 2493–2505.

[46] C.A. MacAlister, D.C. Bergmann, Sequence and function of basic helix-loop-helix proteins required for stomatal development in *Arabidopsis* are deeply conserved in land plants, Evol. Dev. (2011) 182–192.

[47] M. Seki, T. Umezawa, K. Urano, K. Shinozaki, Regulatory metabolic networks in drought stress responses, Curr. Opin. Plant. Biol. 10 (2007) 296–302.

[48] R. Welti, W. Li, M. Li, Y. Sang, H. Biesiada, et al., Profiling membrane lipids in plant stress responses Role of phospholipase D-alpha in freezing-induced lipid changes in *Arabidopsis*, J. Biol. Chem. 277 (31) (2002) 994–32002.

[49] L.G. Landry, C.C.S. Chapple, R.L. Last, *Arabidopsis* mutants lacking phenolic sunscreens exhibit enhanced Ultraviolet-B injury and oxidative damage *Arabidopsis* mutants lacking phenolic sunscreens exhibit enhanced Ultraviolet-B injury and oxidative damage, Plant Physiol. 109 (1995) 1159–1166.

[50] D.J. Kliebenstein, J.E. Lim, L.G. Landry, R.L. Last, *Arabidopsis* UVR8 regulates ultraviolet-B signal transduction and tolerance and contains sequence similarity to human Regulator of Chromatin Condensation 1 *Arabidopsis* UVR8 regulates ultraviolet-B signal transduction and tolerance and contains sequence similarity to human Regulator of Chromatin Condensation 1, Plant Physiol. 130 (2002) 234–243.

[51] B.A. Brown, C. Cloix, G.H. Jiang, E. Kaiserli, P. Herzyk, et al., A UV-B-specific signaling component orchestrates plant UV protection, Proc. Natl. Acad. Sci. U. S. A. 102 (2005) 18225–18230.

[52] C. Cloix, G.I. Jenkins, Interaction of the *Arabidopsis* UV-B-specific signaling component UVR8 with chromatin, Mol. Plant 1 (2008) 118–128.

[53] C. Cloix, G.I. Jenkins, Interaction of the *Arabidopsis* UV-B-specific signaling component UVR8 with chromatin, Mol. Plant 1 (2008) 118–128.

[54] F. Mehrtens, H. Kranz, P. Bednarek, B. Weisshaar, The *Arabidopsis* transcription factor MYB12 is a flavonol-specific regulator of phenylpropanoid biosynthesis, Plant Physiol. 138 (2005) 1083–1096.

[55] R. Stracke, J.J. Favory, H. Gruber, L. Bartelniewoehner, S. Bartels, et al., The *Arabidopsis* bzip transcription factor hy5 regulates expression of the PFG1/MYB12 gene in response to light and ultraviolet-b radiation, Plant Cell Environ. 33 (2010) 88–103.

[56] BINKERTM, L. Kozma-Bognár, K. Terecskei, L. De Veylder, F. Nagy, et al., UV-B-responsive association of the *Arabidopsis* bZIP transcription factor ELONGATED HYPOCOTYL5 with target genes, including its own promoter, Plant Cell 26 (2014) 4200–4213.

[57] T. Tohge, M. Kusano, A. Fukushima, K. Saito, A.R. Fernie, Transcriptional and metabolic programs following exposure of plants to UV-B irradiation, Plant Signal. Behav. 6 (2011) 1987–1992.

[58] H. Jin, E. Cominelli, P. Bailey, A. Parr, F. Mehrtens, et al., Transcriptional repression by AtMYB4 controls production of UV-protecting sunscreens in *Arabidopsis*, EMBO J. 19 (2000) 6150–6161.

[59] N.M. Crawford, A.D.M. Glass, Molecular and physiological aspects of nitrate uptake in plants, Trends Plant Sci. 3 (1998) 389–395.

[60] R.A. Gutierrez, Systems biology for enhanced plant nitrogen nutrition, Science 336 (2012) 1673–1675.

[61] P. Guan, R. Wang, P. Nacry, G. Breton, S.A. Kay, et al., Nitrate foraging by *Arabidopsis* roots is mediated by the transcription factor TCP 20 through the systemic signaling pathway, Proc. Natl. Acad. Sci. U. S. A. (2014).

[62] C. Marchive, F. Roudier, L. Castaings, V. Brehaut, E. Blondet, et al., Nuclear retention of the transcription factor NLP7 orchestrates the early response to nitrate in plants, Nat. Commun. 4 (2013) 1713.

[63] L. Castaings, A. Camargo, D. Pocholle, V. Gaudon, Y. Texier, et al., The nodule inception-like protein 7 modulates nitrate sensing and metabolism in *Arabidopsis*, Plant J. 57 (2009) 426–435.

[64] A. Calderwood, R.J. Morris, S. Kopriva, Predictive sulfur metabolism – a field in flux, Frontiers in Plant Science 5 646 (2014).

[65] G. Curien, S. Ravanel, R. Dumas, A kinetic model of the branch-point between the methionine and threonine biosynthesis pathways in *Arabidopsis* thaliana, Eur. J. Biochem. 270 (2003) 4615–4627.

[66] A. Maruyama-Nakashita, Y. Nakamura, T. Tohge, K. Saito, H. Takahashi, *Arabidopsis* SLIM1 is a central transcriptional regulator of plant sulfur response and metabolism *Arabidopsis* SLIM1 is a central transcriptional regulator of plant sulfur response and metabolism, Plant Cell Online 18 (2006) 3235–3251.

[67] B. Li, D.J. Kliebenstein, The AT-hook Motif-encoding Gene METABOLIC NETWORK MODULATOR 1 underlies natural variation in *Arabidopsis* primary metabolism, Front. Plant Sci. 5 (2014).

[68] M. Taylor-Teeples, L. Lin, M. de Lucas, G. Turco, C. Doherty, et al., Environmental developmental and genotype-dependent regulation of xylem cell specification and secondary cell wall biosynthesis in *Arabidopsis thaliana*, Nature Accepted (2014).

[69] B. Li, A. Gaudinier, M. Taylor-Teeples, N.T. Nham, C. Ghaffari, et al., Promoter based integration in plant defense regulation, Plant Physiol. 166 (2014) 1803–1820.

[70] N.D. Bonawitz, J.I. Kim, Y. Tobimatsu, P.N. Ciesielski, N.A. Anderson, et al., Disruption of mediator rescues the stunted growth of a lignin-deficient *Arabidopsis mutant*, Nature 509 (2014) 376–380.

[71] N.D. Bonawitz, W.L. Soltau, M.R. Blatchley, B.L. Powers, A.K. Hurlock, et al., REF4 and RFR1, subunits of the transcriptional coregulatory complex mediator, are required for phenylpropanoid homeostasis in *Arabidopsis*, J. Biol. Chem. 287 (2012) 5434–5445.

[72] M.F. Covington, J.N. Maloof, M. Straume, S.A. Kay, S.L. Harmer, Global transcriptome analysis reveals circadian regulation of key pathways in plant growth and development, Genome Biol. 9 (2008) R130.

[73] S.L. Harmer, The circadian system in higher plants, Ann. Rev. Plant Biol. 60 (2009) 357–377.

[74] S.L. Harmer, S.A. Kay, Positive and negative factors confer phase-specific circadian regulation of transcription in *Arabidopsis*, Plant Cell 17 (2005) 1926–1940.

[75] Z.Y. Wang, E.M. Tobin, Constitutive Expression of the CIRCADIAN CLOCK ASSOCIATED 1(CCA1) gene disrupts circadian rhythms and suppresses Its Own . . ., Cell 93 (1998) 1207–1217.

[76] Z.Y. Wang, D. Kenigsbuch, L. Sun, E. Harel, M.S. Ong, et al., A Myb-related transcription factor is involved in the phytochrome regulation of an *Arabidopsis Lhcb* gene, Plant Cell Online 9 (1997) 491–507.

[77] S.M. Smith, D.C. Fulton, T. Chia, D. Thorneycroft, A. Chapple, et al., Diurnal changes in the transcriptome encoding enzymes of starch metabolism provide evidence for both transcriptional and posttranscriptional regulation of starch metabolism in *Arabidopsis* leaves, Plant Physiol. 136 (2004) 2687–2699.

[78] A. Fukushima, M. Kusano, N. Nakamichi, M. Kobayashi, N. Hayashi, et al., Impact of clock-associated *Arabidopsis* pseudo-response regulators in metabolic coordination, Proc. Natl. Acad. Sci. 106 (2009) 7251–7256.

[79] R.A. Gutierrez, T.L. Stokes, K. Thum, X. Xu, M. Obertello, et al., Systems approach identifies an organic nitrogen-responsive gene network that is regulated by the master clock control gene CCA1, Proc. Natl. Acad. Sci. U. S. A. 105 (2008) 4939–4944.

[80] S.L. Harmer, Orchestrated transcription of key pathways in *Arabidopsis* by the circadian Clock, Science 290 (2000) 2110–2113.

[81] S. Chattopadhyay, L.H. Ang, P. Puente, X.W. Deng, N. Wei, *Arabidopsis* bZIP protein HY5 directly interacts with light-responsive promoters in mediating light control of gene expression *Arabidopsis* bZIP protein HY5 directly interacts with light-responsive promoters in mediating light control of gene expression, Plant Cell Online 10 (1998) 673–683.

[82] C. Andronis, S. Barak, S.M. Knowles, S. Sugano, E.M. Tobin, The clock protein CCA1 and the bZIP transcription factor HY5 physically interact to regulate gene expression in *Arabidopsis*, Mol. Plant 1 (2008) 58–67.

[83] E.J. Chikofsky, J.H. Cross, Reverse engineering and design recovery – A taxonomy, Ieee Software 7 (1990) 13–17.

[84] I. Dobson, B.A. Carreras, V.E. Lynch, D.E. Newman, Complex systems analysis of series of blackouts: Cascading failure, critical points, and self-organization, Chaos 17 (2007).

[85] E. Eilam, Reversing: Secrets of Reverse Engineering, Wiley Publishing, 595, 2005.

[86] G.D. Bilsborough, A. Runions, M. Barkoulas, H.W. Jenkins, A. Hasson, et al., Model for the regulation of *Arabidopsis thaliana* leaf margin development, Proc. Natl. Acad. Sci. U. S. A. 108 (2011) 3424–3429.

[87] O. Hamant, M.G. Heisler, H. Jonsson, P. Krupinski, M. Uyttewaal, et al., Developmental Patterning by Mechanical Signals in *Arabidopsis*, Science 322 (2008) 1650–1655.

[88] L. Pourcel, N.G. Irani, A.J.K. Koo, A. Bohorquez-Restrepo, G.A. Howe, et al., A chemical complementation approach reveals genes and interactions of flavonoids with other pathways, Plant J 74 (2013) 383–397.

[89] R.E. Kerwin, J.M. Jiménez-Gómez, D. Fulop, S.L. Harmer, J.N. Maloof, et al., Network quantitative trait loci mapping of circadian clock outputs identifies metabolic pathway-to-clock linkages in *Arabidopsis*, Plant Cell 23 (2011) 471–485.

[90] W.L. Araujo, A. Nunes-Nesi, Z. Nikoloski, L.J. Sweetlove, A.R. Fernie, Metabolic control and regulation of the tricarboxylic acid cycle in photosynthetic and heterotrophic plant tissues, Plant Cell Environ. 35 (2012) 1–21.

[91] A. Nott, H.-S. Jung, S. Koussevitzky, J. Chory, Plastid-to-nucleus retrograde signaling, Ann. Rev. Plant Biol. (2006) 739–759.

[92] S. Koussevitzky, A. Nott, T.C. Mockler, F. Hong, G. Sachetto-Martins, et al., Signals from chloroplasts converge to regulate nuclear gene expression, Science 316 (2007) 715–719.

[93] J.D. Woodson, J. Chory, Coordination of gene expression between organellar and nuclear genomes, Nature Rev. Gen. 9 (2008) 383–395.

[94] G. Vinti, A. Hills, S. Campbell, J.R. Bowyer, N. Mochizuki, et al., Interactions between hy1 and gun mutants of *Arabidopsis*, and their implications for plastid/nuclear signalling, Plant J. 24 (2000) 883–894.

[95] R.M. Larkin, J.M. Alonso, J.R. Ecker, J. Chory, GUN4, a regulator of chlorophyll synthesis and intracellular signaling, Science 299 (2003) 902–906.

[96] Y. Xiao, T. Savchenko, E.E.K. Baidoo, W.E. Chehab, D.M. Hayden, et al., Retrograde signaling by the plastidial metabolite MEcPP regulates expression of nuclear stress-response genes, Cell 149 (2012) 1525–1535.

[97] G.M. Estavillo, P.A. Crisp, W. Pornsiriwong, M. Wirtz, D. Collinge, et al., Evidence for a SAL1-PAP chloroplast retrograde pathway that functions in drought and high light signaling in *Arabidopsis*, Plant Cell 23 (2011) 3992–4012.

[98] J.I. Kim, P.N. Ciesielski, B.S. Donohoe, C. Chapple, X. Li, Chemically induced conditional rescue of the reduced epidermal fluorescence8 mutant of *Arabidopsis* reveals rapid restoration of growth and selective turnover of secondary metabolite pools, Plant Physiol. 164 (2014) 584–595.

[99] N. Anderson, N.D. Bonawitz, K.E. Nyffeler, C. Chapple, Loss of ferulate 5-hydroxylase leads to Mediator-dependent inhibition of soluble phenylpropanoid biosynthesis in *Arabidopsis*, Plant Physiol. (2015).

[100] R.D. Kornberg, The molecular basis of eukaryotic transcription, Proc. Natl. Acad. Sci. 104 (2007) 12955–12961.

[101] J.I. Kim, W.L. Dolan, N.A. Anderson, C. Chapple, Indole glucosinolate biosynthesis limits phenylpropanoid accumulation in *Arabidopsis thaliana*, Plant Cell 27 (2015) 1529–1546.

[102] N.K. Clay, A.M. Adio, C. Denoux, G. Jander, F.M. Ausubel, Glucosinolate metabolites required for an *Arabidopsis* innate immune response, Science 323 (2009) 95–101.

[103] L.N. Meihls, V. Handrick, G. Glauser, H. Barbier, H. Kaur, et al., Natural variation in maize aphid resistance is associated with 2,4-Dihydroxy-7-Methoxy-1,4-Benzoxazin-3-One glucoside methyltransferase activity, Plant Cell 25 (2013) 2341–2355.

[104] Heinze, M., Brandt, W., Marillonnet, S., Roos, W., 2015, Self and Non-Self in the Control of Phytoalexin Biosynthesis: Plant Phospholipases A2 with Alkaloid-Specific Molecular Fingerprints. In Press.

[105] T. Hasegawa, K. Yamada, S. Kosemura, S. Yamamura, K. Hasegawa, Phototropic stimulation induces the conversion of glucosinolate to phototropism-regulating substances of radish hypocotyls, Phytochemistry 54 (2000) 275–279.

[106] K. Yamada, T. Hasegawa, E. Minami, N. Shibuya, S. Kosemura, et al., Induction of myrosinase gene expression and myrosinase activity in radish hypocotyls by phototropic stimulation, J. Plant Physiol 160 (2003) 255–259.

[107] A.C. Kemen, S. Honkanen, R.E. Melton, K.C. Findlay, S.T. Mugford, et al., Investigation of triterpene synthesis and regulation in oats reveals a role for beta-amyrin in determining root epidermal cell patterning, Proc, Nat, Acad, Sci. U. S. A. 111 (2014) 8679–8684.

18

Towards a comprehensive and dynamic gynoecium gene regulatory network

Ricardo A. Chávez Montes, Humberto Herrera-Ubaldo, Joanna Serwatowska, Stefan de Folter*

Unidad de Genómica Avanzada (LANGEBIO), Centro de Investigación y de Estudios Avanzados del Instituto Politécnico Nacional (CINVESTAV-IPN), Km. 9.6 Libramiento Norte, Carretera Irapuato-León, CP 36821 Irapuato, Guanajuato, Mexico

ARTICLE INFO

Keywords:
Gynoecium development
Gene regulatory network
Transcription factor
Protein complex

ABSTRACT

The *Arabidopsis thaliana* gynoecium arises at the center of the flower as a simple structure, which will successively develop novel cell types and tissues, resulting in a complex organ. Genetic and hormonal factors involved in this process have been identified, but we are still far from understanding how these elements interact, and how these interactions rearrange according to spatial and temporal cues. In this work we propose the first steps in a roadmap to attain an ambitious goal: to obtain a comprehensive and dynamic gene regulatory network that will help us elucidate the patterning events leading to the formation of a fully developed gynoecium.

1. Introduction

In flowering plants the gynoecium is the female reproductive organ, which is composed either of a single carpel, or multiple carpels that are usually fused. The gynoecium of *Arabidopsis thaliana* is composed of two carpels fused along their margins. This fusion zone, called medial ridge or carpel margin meristem (CMM), acts as meristematic tissue giving rise to internal structures such as placenta, ovules, replum, septum and transmitting tract [1,2]. According to the floral development stages described by Smyth and collaborators [3], the gynoecium is clearly established at stage 6, when the spatial domains of the gynoecial tube become visible. These include the medial domain versus the lateral domains, and the inner (adaxial) and outer (abaxial) regions. Later, during stages 7–8, two meristematic outgrowths (CMMs) form at the inner medial domain, which subsequently (stages 9–12) will produce key reproductive tissues: placenta, ovules and transmitting tract. A schematic representation of gynoecium inner tissues development from stages 8 to 12 is shown in Fig. 1a. At anthesis, by stage 12, all tissues required for fertilization, and the following fruit development are present. The former are fully developed,

while the latter will develop after fruit set [4,5]. The fully developed, stage 12 gynoecium is composed of three distinct structures along the apical-basal axis: stigma, style and ovary (Fig. 1b).

Several genetic and hormonal factors that participate in gynoecium development have been identified (reviewed in [2]), but we are still far from understanding how all these elements interact to give rise to the morphogenetic patterns we observe. Available genome-wide analyses in other organs, such as roots or the shoot apical meristem, have revealed that developmental processes are controlled by the coordinated action of regulators, that is gene regulatory networks (GRNs), that control gene expression according to spatial and temporal cues [6–8]. These networks are composed of transcription factors, hormones, microRNAs, peptides and chromatin-modifying proteins, together with their interactions, in particular protein-DNA and protein–protein interactions. For gynoecia, a few, small GRNs have been proposed for specific cell types (Fig. 2). However, we are still far from obtaining a definitive, comprehensive gynoecium GRN, and understanding how it evolves in a spatio-temporal context. In this review we propose the first steps in a roadmap to attain such goal, with a particular emphasis on transcription factors, and their protein-DNA and protein–protein interactions.

2. Identifying gynoecium expressed genes

The first cloned gene involved in gynoecium development, *AGAMOUS (AG)*, was described in 1990, 25 years ago [9]. Since then, dozens of genes have been identified and, in a recent review, Reyes-

---

Abbreviations: GRN, gene regulatory network; TF, transcription factor; ChIP, chromatin immunoprecipitation; TRAP, translating ribosome affinity purification; PBM, protein binding microarray.

* Corresponding author.

E-mail address: sdfolter@langebio.cinvestav.mx (S. de Folter).

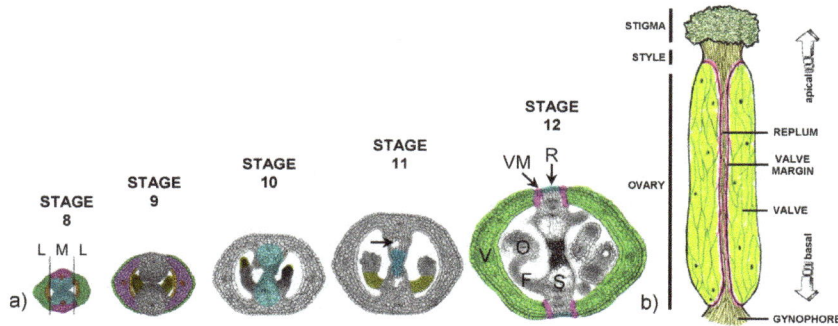

Fig. 1. *Arabidopsis thaliana* carpel tissues development. (a) False-colored histological cross sections of Arabidopsis gynoecia from stages 8 to 12 (accession Col-0). Stage 8 gynoecium: the medial domain (M) is observable, composed of the carpel margin meristem (blue) and the abaxial margin or replum zone (pink). Main vascular bundles differentiate at this stage (red). Lateral domains (L) are divided morphologically into inner (adaxial, orange) and outer (abaxial, green) regions. Stage 9: carpel wall differentiates into three tissues: exocarp (green), mesocarp (purple) and endocarp (orange). Lateral vascular bundles differentiate (red) and ovule primordia begin to form (yellow). Stage 10: the medial ridges meet and give rise to the septum (blue). Ovules differentiate (yellow). Stage 11: the transmitting tract differentiates (blue), and cell death begins in the septum (arrow). Stage 12 (anthesis): valve margins (VM, pink) and replum (R, blue) become morphologically distinct from the valves (V, green). Funiculus (F) and ovules (O) are fully developed. (b) Cartoon representation of a stage 12 wild type *Arabidopsis* gynoecium showing the different tissues that can be distinguished along the apico-basal axis: stigma, style, ovary and gynophore.

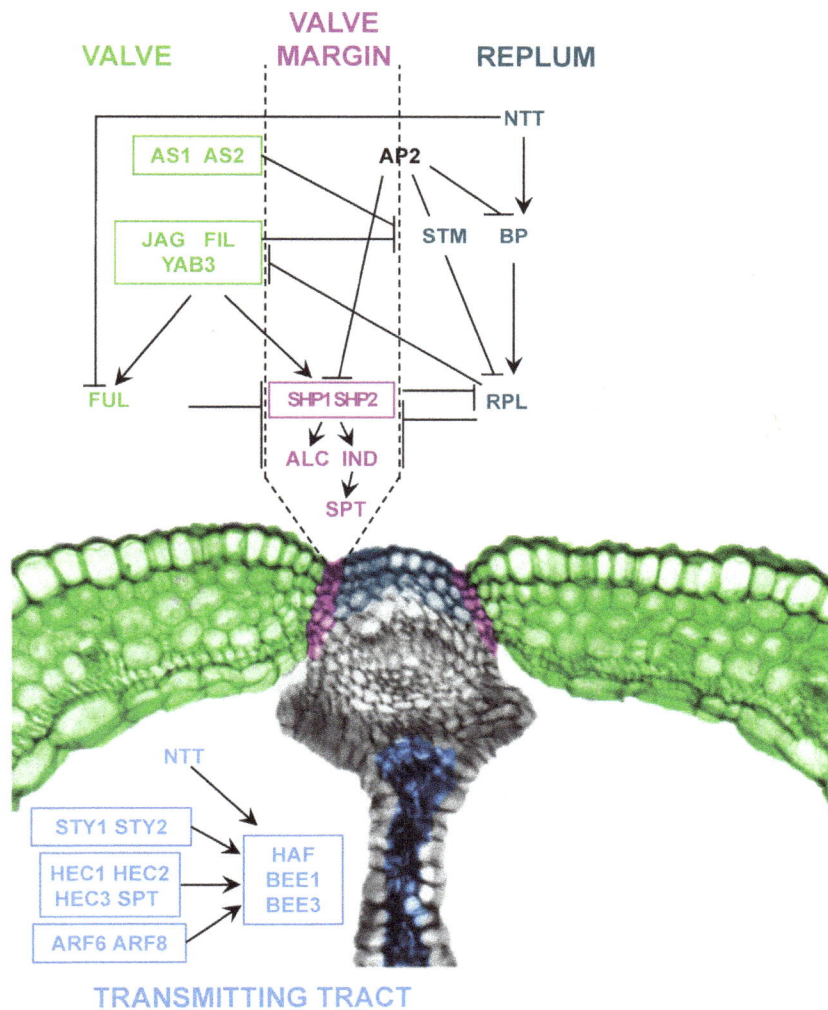

Fig. 2. Known gynoecium GRNs for valve, valve margin, replum and transmitting tract. *Arabidopsis thaliana* (ecotype Col-0) fruit false-colored histological cross section showing the valve, valve margin, replum and transmitting tract region. The corresponding GRNs are displayed in the same color as the tissues they specify. Functionally redundant genes are boxed. Specification along the medial-lateral axis is the result of the antagonistic activities of lateral factors (*ASYMMETRIC LEAVES 1* and *2*, *AS1,2*; *JAGGED, JAG*; *FILAMENTOUS FLOWER, FIL*; *YABBY3, YAB3*; *FRUITFULL, FUL*) and medial factors (*NO TRANSMITTING TRACT, NTT*; *SHOOT MERISTEMLESS, STM*; *BREVIPEDICELLUS, BP*; *REPLUMLESS, RPL*). Differentiation of the valve margin, controlled by *SHATTERPROOF 1* and *2* (*SHP1, 2*), *ALCATRAZ* (*ALC*), *INDEHISCENT* (*IND*) and *SPATULA* (*SPT*), is regulated by both lateral and medial factors. *APETALA 2* (*AP2*) is involved in valve margin and replum specification, and is expressed in both tissues. Several genes are known to participate in transmitting tract development (*STYLISH 1* and *2*, *STYs*; *HECATE 1, 2* and *3*, *HEC1,2,3*; *SPT*; *AUXIN RESPONSE FACTOR 6* and *8*, *ARF6, 8*; *HALF FILLED, HAF*; *BR ENHANCED EXPRESSION 1* and *3*, *BEE1, 3*; *NTT*), but the underlying GRN has not been fully elucidated. *NTT* was first described as an essential gene for transmitting tract formation and, more recently, was also shown to promote replum development. The presented GRNs were reconstructed according to [5,16,55,86–89].

Olalde and collaborators [2] compiled a list of 86 genes, including 62 transcription factors involved in this process. While the information obtained so far has been crucial to our understanding of gynoecium development, we can arguably predict that a final, comprehensive gynoecium GRN will include more than 80 genes. With over 30,000 genes present in the Arabidopsis genome, a priori knowledge can help us focus our search for novel genes, and high-throughput technologies are best suited to provide such information.

2.1. Gene co-expression clusters and networks

Systematic, large scale searches for genes involved in floral organ development were undertaken over ten years ago [10–13]. All these studies performed co-expression analyses on gene expression data, and identified hundreds of organ or stage-specific genes. In particular, de Folter and collaborators [10] used high-density filter arrays in order to obtain the expression profiles of 1100 Arabidopsis TFs across five stages of gynoecium and fruit development, starting at anthesis. Genes identified in this screen, such as *JAIBA* (*JAB*) or *NO TRANSMITTING TRACT* (*NTT*), were later proven to be important for floral determination or transmitting tract and replum development [14–16], with hundreds still unexplored. Moreover, their data shows that over 500 TFs are expressed in stage 12 gynoecia. Although it is probable that not all of them are essential for gynoecium development, this result is a first indication that the list of 62 TFs in [2] will certainly need to be expanded in the future. Nowadays, databases such as NCBI's Gene Expression Omnibus [17], or EBI's ArrayExpress [18] contain impressive amounts of microarray-based gene expression information. For example, ArrayExpress currently contains over 1140 experiments that used the Affymetrix ATH1-121501 complete genome chip. Carefully curated microarray data can be fed to inference algorithms, which allow for the identification of co-expression interactions in a massive scale [19]. The resulting co-expression analyses are a vast source for gene correlation hypothesis, and have been proven to contain valuable biological information in other organs, such as seedlings and roots [20,21]. Using as a starting point known gynoecium genes, these inference-based methodologies can help us identify new genes putatively involved in gynoecium development.

2.2. Cell type and stage-specific transcriptomes

Co-expression analysis from microarray data is a valuable tool for the proposal of novel genes likely to be involved in gynoecium development. However, samples such as "inflorescence" or "flower buds", which are frequently the source material for microarray data, contain more than one cell type, usually from more than one developmental stage. In such samples, spatial and temporal information is scrambled. In order to overcome this limitation, high-throughput techniques can be preceded by: (1) protocols that isolate specific cell types, or messenger RNAs from specific cell types, which will provide spatial information, and (2) flower and floral tissues synchronization protocols [11,12], which allow for the collection of large quantities of floral tissues at a specific developmental stage, thus providing temporal information. Two cell or mRNA isolation protocols have been applied to gynoecium and fruit tissues: laser assisted microdissection (LAM) and translating ribosome affinity purification (TRAP). A third technique, fluorescence-activated cell sorting (FACS) of protoplasts, which has been successfully used in roots for the isolation of specific cell types [22,23], is still being developed for gynoecia. LAM followed by microarray hybridization was used to obtain a list of 1539 differentially expressed probe sets likely to be involved in replum development [24]. More recently, TRAP followed by massively parallel sequencing (TRAP-seq) of synchronized flowers was used by Jiao and collaborators [25] in order

to obtain the translatomes of the AGAMOUS-expression domain at stages 4 and 6–7. Their dataset contains evidence of expression (at least 0.5 reads per kilobase of transcript per million mapped reads in both replicates) for 76 of the 86 genes (including 50 TFs) listed in [2] at stage 4, and 79 genes (including 53 TFs) at stage 6. It would therefore appear that most genes later involved in gynoecium development are already expressed at stage 4, when the gynoecium primordium is not yet apparent. The data also shows that 18301 genes are expressed at stage 4 and 18189 at stage 6–7 but, more importantly, that 17790 (97%) of these are expressed at both stages. Very similar numbers of expressed genes and overlap have been observed in another cell type-specific translatome [26], and it is likely that further TRAP-seq datasets will confirm these observations. This important overlap strongly suggests that the establishment of the gynoecium primordium, rather than being defined by the novel or differential expression of a handful of genes, is the result of GRN dynamics, that is an ongoing rearrangement of interactions between genes. Therefore, identifying these interactions and their dynamics will be essential to our understanding of gynoecium development.

3. Connecting the gene expression atlas

Data from Jiao and collaborators [25] has demonstrated that techniques such as TRAP-seq are an ideal source for very detailed information on gene expression and abundance. On the other hand, with tens of thousands of genes expressed in a particular cell type, the complexity of the underlying GRNs has been multiplied dramatically. While it is tempting to embrace a reductionist approach, and postulate that one or a few genes are actually essential for the establishment of a particular tissue or cell type, a far more interesting challenge will be to identify the interactions among all expressed genes, and to understand how they in turn underlie morphology and development. In particular, it will be essential to identify TF–DNA binding events, which are the main drivers of gene expression. Available techniques for the identification of protein-DNA interactions have been previously reviewed [27]. In this section we will focus on those techniques that have generated data suitable to be applied to the reconstruction of a gynoecium GRN.

3.1. Chromatin immunoprecipitation (ChIP)

ChIP [28,29] followed by whole-genome microarray hybridization (ChIP-chip) or massively parallel sequencing (ChIP-seq) has been used to identify the genomic binding regions of 31 TFs [30,31], and data was obtained from flower material for seven of them (Table 1). ChIP-chip or ChIP-seq data provides detailed information on TF–DNA interactions for a particular TF that can help us identify the transcriptional regulatory pathways in which this TF is involved, and we expect these pathways to be organ-specific. AG, APETALA 2 (AP2) and SEPALLATA 3 (SEP3), participate in gynoecium development, and ChIP-seq data for these three TFs is available from either whole inflorescences (AP2), or stage 5 (AG and SEP3) and stage 9 (SEP3) flowers. Since the gynoecium primordium is visible at stage 6, it can arguably be proposed that the corresponding GRN should be established by the end of stage 5. Therefore, it is possible to filter non-expressed genes from ChIP-seq data in order to reconstruct a putative gynoecium GRN at this stage. Such a network is presented in Fig. 3.

ChIP-seq data for APETALA 1 (AP1) and SEP3 obtained at different time points shows that, while there is a significant overlap of targets[1] at different stages for either SEP3 or AP1, there is also

---

[1] Overlap of target genes has been observed across ChIP experiments for numerous *Arabidopsis* TFs (Pajoro et al. (2014) J. Exp. Bot. 65: 473-4745). These results

Table 1
TFs for which ChIP data was obtained from floral tissues.

Gene	Sample	Stage	References
AP2	Inflorescence	1–12	[90]
AMS	Inflorescence	10–12 (0.6–1.1 mm buds [3])	[91]
AP3	Synchronized flowers	5	[92]
PI	Synchronized flowers	5	[92]
SVP	Inflorescence	1–11	[93]
AG	Synchronized flowers	5	[94]
SEP3	Inflorescence	1–12	[95]
	Synchronized flowers	2, 5 and 9 (2, 4 and 8 days post-flower induction [3])	[32]
AP1	Inflorescence	1–13	[96]
	Synchronized flowers	2, 5 and 9 (2, 4 and 8 days post-flower induction [3])	[32]

an important number of targets whose occupancy varies during development [32]. For example, SEP3 binds 6843 genomic regions at day 4 of flower induction, 7816 at day 8, with 5592 common to both. Using ChIP-seq peak score ratios as a measure of the relative binding level of a TF [32], binding goes up or down between both time points for 625 (11%) of these common targets. Differences in the number of target genes, and the corresponding target occupancies, imply that protein-DNA interactions are dynamic, and we expect such dynamics to underlie the developmental processes that occur during gynoecium development. Unfortunately, considering the limitations of ChIP [33], and the number of TFs likely to be involved in gynoecium development, ChIP-chip or -seq might not be an ideal technique for the massive scale-identification of time-resolved transcriptional regulatory pathways.

3.2. Yeast one-hybrid (Y1H)

A second possible technique that can identify TF–DNA binding events is Y1H [34]. This technique involves the cloning of both the TF and the candidate DNA target sequence, and thus interactions identified using Y1H can be applied to any organ or cell type. In [35–37], there is Y1H information for 8 of the TFs listed in [2], and for 59 TFs that are ChIP-identified targets of AG, AP2 or SEP3. This data can be used to complement a ChIP-derived gynoecium GRN (Fig. 3), with two major drawbacks. First, no spatial or temporal data for Y1H-identified interactions is available, although experimental validation in roots shows that most interactions have biological relevance, and are in fact able to recover a xylem development GRN [37]. We therefore expect the available information on gynoecium-expressed TFs to be equally relevant. Second, Y1H requires prior selection of the target sequences. For some biological processes, such as xylem and secondary wall formation, most participants have been identified, which allows for the straightforward reconstruction of the corresponding GRNs. For other processes however, for example transmitting tract development, only a few genes are known to be involved. In such cases, novel putative TFs and target genes could be proposed through other methods, such as gene co-expression analyses.

3.3. Protein-binding microarrays (PBMs)

A third methodology that can be used to identify protein-DNA interactions, and does not require a priori knowledge of target sequences, involves the use of PBMs. PBMs are microarrays spot-

ted with all possible combinations of DNA n-mers (where n usually equals 8 or 11), and thus it is possible to identify both the sequence and affinity of all possible genomic binding motifs for as many TFs as can be expressed. Franco-Zorrilla et al. [38] identified the binding motifs, and the corresponding affinities, for 69 Arabidopsis TFs, 31 of which are part of the network presented in Fig. 3, including six listed in [2]. These binding motifs can now be used to identify all putative targets for these TFs, and complement available datasets of protein-DNA interactions, with caution: although most binding sites are overrepresented in differentially expressed genes in plants with altered expression of the corresponding TF, binding of a DNA n-mer might not necessarily reflect binding in planta. Scanning, using RSAT's [39] matrix-scan, of the 1000 bp upstream sequences for the 931 genes in Fig. 3 using the 31 available position weight matrices results in a network containing 6790 edges or putative TF–DNA interactions (Supplementary data S1). TFs have an average of 219 targets, and targets are bound by an average of 7 TFs. Even if only a fraction of these interactions actually occur in gynoecium tissues and lead to changes in target expression levels, this data indicates a comprehensive map of TF–TF regulatory loops will be extremely complex. Finally, the results presented by Franco-Zorrilla and collaborators [38] provide new insights into the dynamics of GRNs: first, TFs can bind different n-mers with varying affinities, and many TFs have more than one high-affinity binding site. Second, even closely related TFs can bind similar, but not identical n-mers. Third, for 33 TFs analyzed, the second high-affinity site differs partially or completely from the highest affinity binding site. This variability in both binding motifs and binding affinity, which in turn is different for each TF, points to the complex molecular events at play in TF–DNA interactions.

4. Enriching the GRN

Understanding the dynamics of TF–DNA interactions is an essential step in the reconstruction of GRNs [40,41]. However, there are other factors that also contribute to the regulatory mechanisms of gene expression. In the following section we will focus on two of these factors in the context of GRNs: TF protein–protein interactions, and auxin and cytokinin-responsive TFs.

4.1. Protein–protein interactions

ChIP-seq, Y1H or PBMs allow for the identification of TF–DNA interactions, which in turn serve as a basis for the reconstruction of GRNs. In such networks, nodes represent individual proteins, yet we know that TFs from several families, including MADS-box [42,43], ARR [44], TCP [45] or ARF [46] form protein–protein interactions, and that their molecular and biological function is intrinsically linked to this property. Finally, since protein–protein interactions occur between TFs, ChIP data is most likely a compendium of all sites for all DNA-binding protein complexes in which the particular TF under study participates. It is therefore

should be interpreted with some caution. The ChIP technique presents a bias which results in an over-representation of sites of high transcriptional activity. These "hyper-chipable" sites are consistently recovered across ChIP experiments, which could then be considered to represent highly occupied target (HOT) regions (Teytelman et al. (2013) PNAS 110: 18,602-18607; Park et al. (2013) PLoS One 8: e83506). However, it has been proposed that, in plants, HOT regions could have biological relevance (Heyndrickx et al. (2014) Plant Cell 26: 3894-3910).

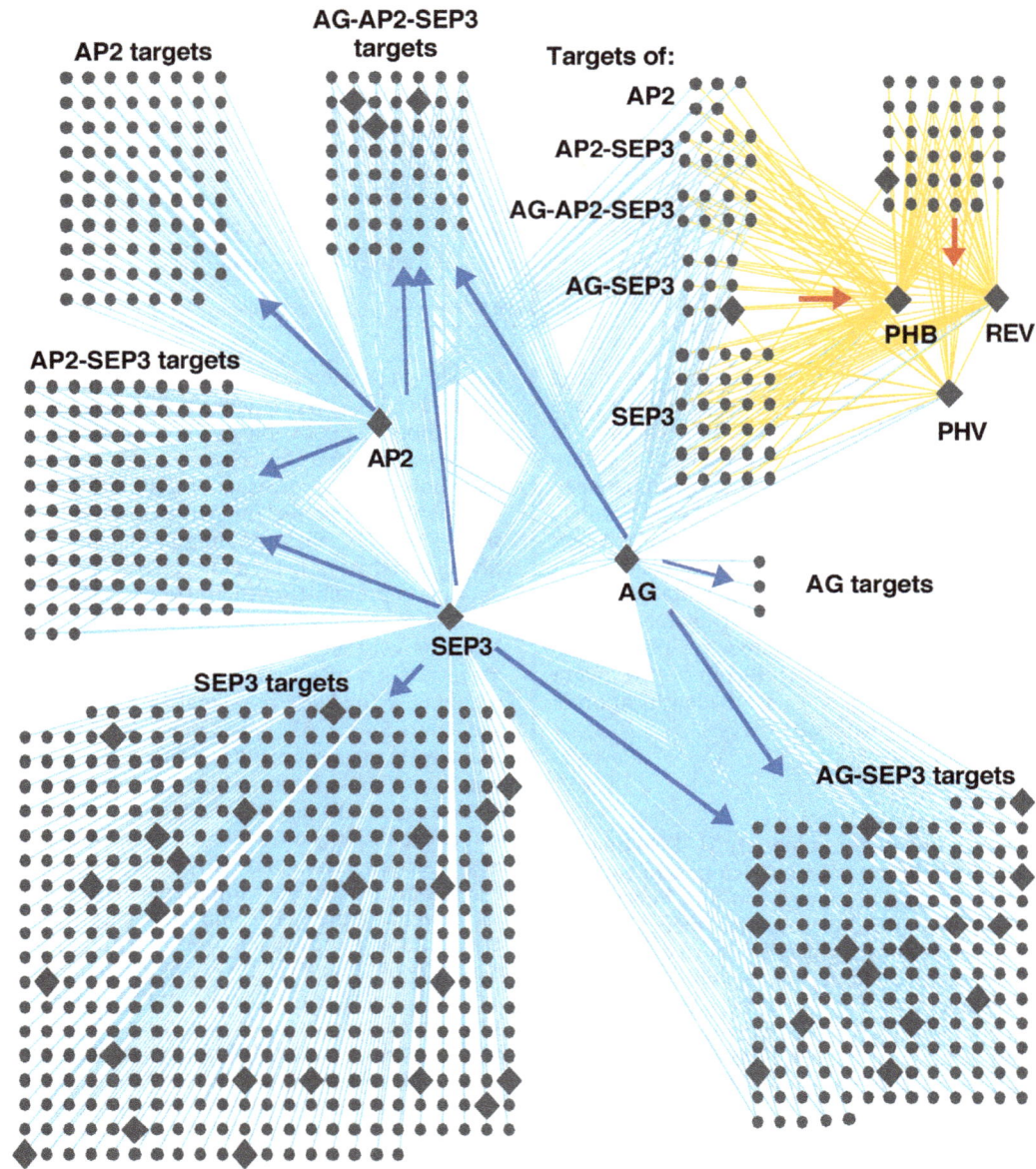

Fig. 3. Stage 5 gynoecium GRN. In this network, the 934 nodes represent transcription factors (TFs), and the 1451 edges represent protein-DNA interactions from ChIP data (blue edges) or Y1H data (orange edges). ChIP data includes interactions identified in inflorescences for AP2 [90], and in stage 5 synchronized flowers for AG and SEP3 [32]. Y1H data includes interactions listed in [35–37]. Interactions were further restricted to those involving at least one TF listed in [2], which are shown as larger, diamond-shaped nodes. Finally, the network only includes TFs expressed at stage 4 in the AGAMOUS domain [25]. Since there is a strong overlap in gene expression between stages 4 and 6–7 [25], we expect that there is also a strong overlap between stages 4 and 5. Arrows show the direction of transcriptional information flows, from protein to DNA. Blue arrows overlay ChIP interactions and red arrows overlay Y1H interactions. PHABULOSA (PHB), REVOLUTA (REV) and PHAVOLUTA (PHV) are targets of AG and SEP3, but not AP2. They are also targets of two groups of TFs, one of which contains targets for AG and/or AP2 and/or SEP3. Additionally, REV is a target of PHV. This suggests: (1) the existence of complex feed-forward loops linking homeotic genes (AG, AP2 and SEP3) to abaxial/adaxial patterning genes (PHB, REV and PHV), and (2) the existence of a second, AG/AP2/SEP3-independent transcriptional regulatory pathway for abaxial/adaxial patterning. Genes participating in the feed-forward loop linking AG, AP2 and SEP3 to PHB, REV and PHV are: AT1G03840 (MGP), AT1G04100 (IAA10), AT1G07640 (OBP2), AT1G13960 (WRKY4), AT1G21910 (DREB26), AT1G24625 (ZFP7), AT1G51700 (DOF1), AT1G53910 (RAP2.12), AT1G64620, AT1G66140 (ZFP4), AT1G68360, AT1G71930 (VND7), AT1G75540 (BBX21), AT1G76880, AT1G77450 (NAC032), AT1G78600 (LZF1), AT2G01570 (RGA1), AT2G22840 (GRF1), AT2G23320 (WRKY15), AT2G33880 (WOX9), AT2G36080 (ABS2), AT2G37590 (DOF2.4), AT2G41940 (ZFP8), AT2G44730, AT3G07650 (COL9), AT3G11280, AT3G14230 (RAP2.2), AT3G15210 (ERF4), AT3G15510 (ANAC056), AT3G16770 (EBP), AT3G19290 (ABF4), AT3G19580 (ZF2), AT3G23690, AT3G28920 (HB34), AT3G50410 (OBP1), AT3G61850 (DAG1), AT3G62100 (IAA30), AT4G14770 (TCX2), AT4G23980 (ARF9), AT4G27240, AT4G28140, AT4G34610 (BLH6), AT4G37260 (MYB73), AT4G39070 (BZS1), AT5G01200, AT5G05410 (DREB2A), AT5G08520, AT5G10510 (AIL6), AT5G13180 (NAC083), AT5G14000 (NAC084), AT5G15210 (HB30), AT5G43270 (SPL2), AT5G43700 (IAA4), AT5G47640 (NF-YB2), AT5G60120 (TOE2), AT5G60200 (TMO6), AT5G60850 (OBP4), AT5G62610, AT5G63790 (NAC102), AT5G66730 (ENY). Source data for this network in Cytoscape-compatible format is available as Supplementary data S1.

clear that the study of gynoecium development is also a story about protein–protein interactions.

4.1.1. Protein–protein interactions in gynoecium development

Protein–protein interactions play a fundamental role in gynoecium development (reviewed in [47]), starting with the

establishment of carpel identity. Two key regulators of this event are the MADS-box TFs AG and SEP3. The existence of the AG-SEP3 protein dimer has been described both in vitro [42] and in planta [48,49], strongly suggesting that the roles of AG and SEP3 are dependent on this interaction. A second, AG-independent carpel development pathway involves the *SHATTERPROOF1* and *2* (*SHP1,*

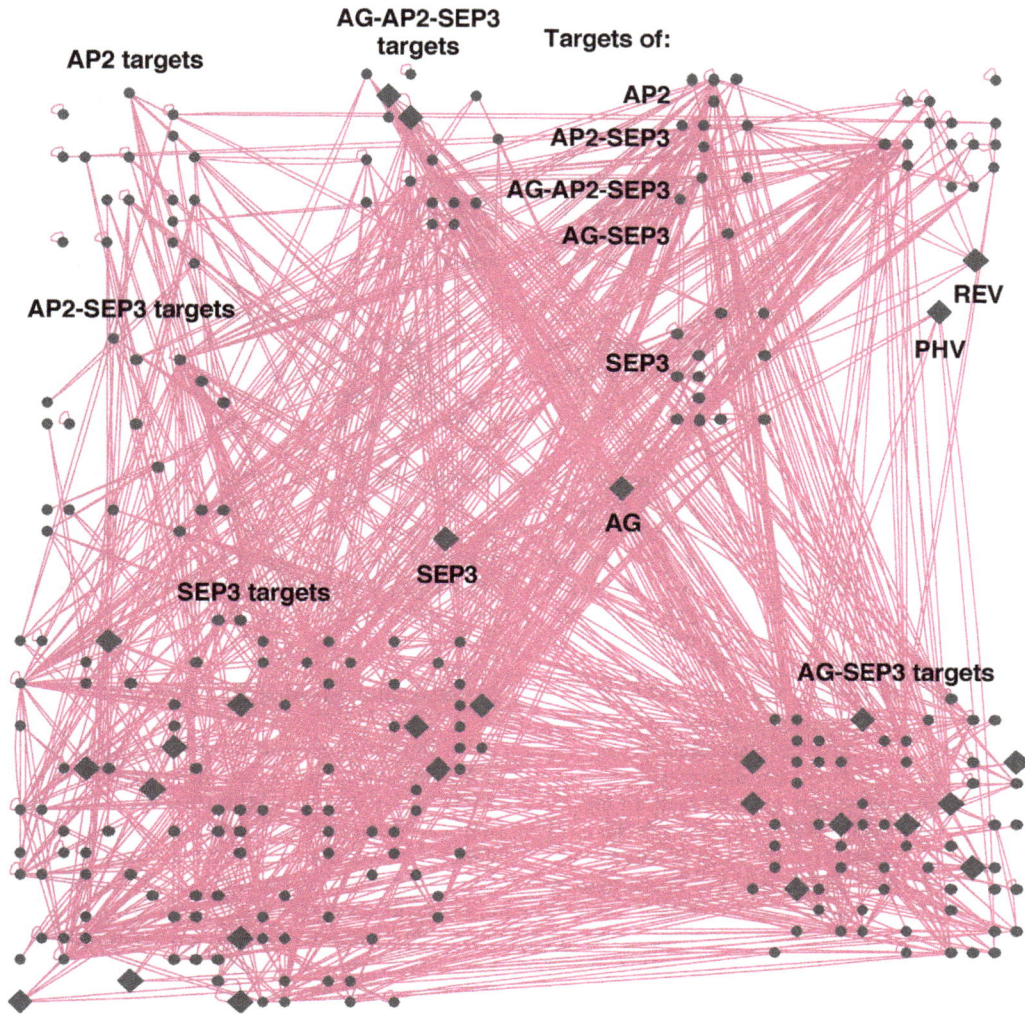

Fig. 4. Stage 5 gynoecium protein–protein interactions network. In this network, the 262 nodes represent transcription factors (TFs) and the 1259 edges represent protein–protein interactions. The network only includes interactions from the IntAct database between the TFs presented in Figure 3. Node positions and labels are the same as in Figure 3. This network illustrates how protein–protein interactions add a second layer of complexity to GRNs. It also shows that some nodes in GRNs should in fact be a representation of protein complexes and not individual proteins. For this, it will be necessary to obtain information on the composition and size of TF complexes present at different stages of gynoecium development. Source data for this network in Cytoscape-compatible format is available as Supplementary data S1.

2) genes [50]. Again, there is evidence that SHP1, 2 and SEP3 can form dimers [42] and higher-order complexes [43]. A third, parallel pathway for the establishment of gynoecium identity involves the *SPATULA* (*SPT*) gene [51], but the molecular mechanism of this pathway has not been determined. After carpel identity has been established, tens of transcription factors continue sculpting the gynoecium, and an interesting observation is that most alterations of gynoecium development occur in double or higher-order mutants [2]. It would therefore appear that carpel development is distributed across multiple protein complexes (Fig. 4), and that redundancy in protein–protein interactions could underlie redundancy in gene function. A first example concerns the AG and SEP1, 2, 3 and 4 proteins. While the *ag* single mutant completely lacks a gynoecium, it is necessary to obtain a triple *sep* mutant to observe a similar phenotype [52,53]. As all four SEP proteins can interact with AG, the observed genetic redundancy can be explained by the redundancy of SEP proteins as AG interactors. A second example involves HALF FILLED (HAF), BR ENHANCED EXPRESSION 1 (BEE1) and BEE3. HAF-BEE1 and HAF-BEE3 interactions have been described [54], and the triple mutant *haf bee1 bee3*, but not single or double mutants, presents severe defects in transmitting tract formation [55], which suggests that BEE1 and BEE3 can replace

each other in a complex with HAF. A third example involves the *HECATE1, 2* and *3* (*HEC1, 2, 3*) genes together with *SPT*. Extreme phenotypes of *HEC2-RNAi hec1 hec3* gynoecia include lack of transmitting tract and defects in septum and stigma development and fusion [56], which are reminiscent of the phenotypes presented by the *spt* mutant. Additionally, HEC proteins do not interact with each other, but they do form heterodimers with SPT. These results suggest that HEC-SPT protein–protein interactions are important for septum and stigma development.

4.1.2. Dynamics of protein complexes

Information on TF protein complexes in gynoecia, and in Arabidopsis in general, is painfully scarce. Most of the available information on protein–protein interactions has been obtained through the use of the yeast two-hybrid (Y2H) technique [57]: the IntAct database [58,59] contains 15000 interactions for Arabidopsis proteins, of which only 3600 (24%) have been identified through methods other than Y2H. Still, available data on MADS-box proteins shows that they can form higher-order protein complexes between themselves, but also with TFs from other families [43,49]. Also, the number of interactors recovered for a particular MADS protein strongly suggests that protein complexes are dynamic, and their formation is likely to be cell type-dependent [49]. This can be easily visualized for SEP proteins: AG can interact with SEP1, 2 and 3 *in planta* [49], and we know that individual SEP proteins have distinct affinities for DNA motifs and for the distance between motifs [60]. Complexes involving AG and different SEP proteins will thus have overlapping, but not identical binding sites and target genes. Additionally, as individual SEP proteins have distinct protein–protein interactions [59], the resulting complexes will have different properties in the formation of the complex itself, further contributing to the dynamics of the corresponding GRNs. In the future, Y2H data complemented with information on protein complexes, derived from techniques such as affinity purification or protein chips [61], will generate a complete protein–protein interactions dataset, and will provide new insights on the dynamics of protein complexes, and their role in gynoecium development.

4.1.3. Protein–protein interactions and chromatin remodeling

Chromatin remodelling also plays a role in gynoecium development, as evidenced by the phenotypes observed in mutants for the *SEUSS* (*SEU*), *LEUNIG* (*LUG*), or *ULTRAPETALA1* (*ULT1*) genes [62,63]. The SEU-LUG complex can associate with histone deacetylase HDA19 and with the Mediator complex components STRUWWELPETER (SWP) and HUA ENHANCER 3 (HEN3) [64], while ULT1, a trithorax group protein, can modify histone methylation status and interact with AG and KANADI (KAN) [65,66]. Additionally, several lines of evidence suggest that TFs from the MADS-box family are able to regulate gene expression through the modification of chromatin structure. Nucleosome-associated factors and transcriptional coregulators are present in immunoprecipitated AG and SEP3 complexes [49]. Interestingly, the double mutant for the nucleosome remodelers CHROMATIN-REMODELING PROTEIN 11 (CHR11) and CHR17 presents pleiotropic phenotypic alterations, including carpel fusion defects. Also, SEP3 binding to DNA alters the DNAse I accessibility of chromatin, suggesting that TF binding alters chromatin structure [32]. Finally, short-range chromatin loops are important for target gene expression of the SEEDSTICK (STK)-SEP3 dimer in ovules [67], suggesting that this same mechanism could be used by MADS-box protein dimers present in carpels. All this fragmented information shows that complexes involving TFs and chromatin-remodelling factors play a role in the regulation of gene expression during gynoecium development.

4.2. TFs involved in auxin and cytokinin responses

Phytohormone regulatory and signaling pathways are involved in nearly all plant growth processes. Therefore, a comprehensive understanding of these pathways is essential for decoding whole plant biology. There is ample information on the role of auxins and cytokinins in gynoecium development and patterning (recently reviewed in [2,68–70]), and the available TF expression data [25] underscores the importance of both hormones: 15 *AUXIN RESPONSE FACTORS* (*ARFs*), 25 *Aux/IAAs*, *KANADI* (*KAN*) 1 to 4, *NGATHA1* to 4 and *STYLISH 1* and 2, all of which participate in auxin responses and/or auxin-mediated gynoecium patterning, are expressed in the AG domain at stage 6–7, together with 7 *ARABIDOPSIS RESPONSE REGULATORS* (*ARRs*) type B, and 10 members of the *CYTOKININ RESPONSE FACTOR* (*CRF*) families, which mediate cytokinins responses [69–72]. Yet despite this, it is curious to note that only Y1H and PBM data is available for 18 TFs of the above mentioned families [35–38]. Furthermore, although the role of polar auxin transport in gynoecium patterning has been clearly established [73–76], only two auxin-responsive genes (*HEC1* and *PIN7*) [26], and six cytokinin-responsive genes (*BEE1*; *GIANT KILLER,GIK; REBELOTE, RBL*; *TRYPTOPHAN AMINOTRANSFERASE OF ARABIDOPSIS 1, TAA1*; *CYTOKININ OXIDASE 3, CKX3* and *CKX5*) [72], whose role in the gynoecium has also been clearly demonstrated [77–79], can be found in the list of 86 genes from Reyes-Olalde et al. [2]. This indicates that we are still far from having a comprehensive view of the genes and transcriptional events involved in auxin and cytokinin responses. This is unfortunate, since hormones are key mediators of information flows within the GRN of a particular cell, but also between GRNs in connecting cells. For example, IND interacts with SPT and regulates polar auxin transport through the induction and repression of *WAG2* and *PINOID*, respectively [74,80]. Both IND and SPT do not belong to any of the families mentioned above, and it is unknown if they are direct mediators of auxin responses. This would imply that TFs (including those not part of the primary response) can adjust hormone concentrations across a series of cells, some of which are probably located, spatially or temporally, beyond their expression sites. This information will then be fed back to the GRN of each cell in the gradient, adjusting gene expression accordingly. Thus, the identification of the transcriptional pathways involved in auxin and cytokinin response will be indispensable for our understanding of gynoecium patterning. A novel tool that can be used for this purpose is the HRGRN database [81]. HRGRN is a hormone-specific database of protein–protein, protein-DNA, and regulatory small RNA-mRNA interactions, and should provide useful hypothesis for novel hormone-related genes and interactions.

5. Final thoughts on gynoecium GRNs

In flowers, organ identity is established by the expression of a few homeotic genes, and absence of a particular gene directly leads to the absence of the corresponding organ. This conceptual framework has guided research on gynoecium and fruit development, which has endeavored to search for "the", or "the few", genes that establish the different cell identities observed in developing gynoecia and fruits. Twenty five years after the identification of the first gene involved in gynoecium development [9], tens of genes have been identified, the majority of which are not homeotic [2]. Most mutants for these genes present defects in more than one cell type (Fig. 5), and current models for the establishment of gynoecium cell identities involve several genes, together with their interactions (Fig. 2). This distribution of roles across tens of genes suggests that many biological processes are involved beyond, though certainly triggered by, the establishment of carpel identity. Gynoecium development is therefore a perfect illustration of the current paradigm transition, from assigning genes to biological processes, to understanding how an ensemble of genes plus their interactions, that is a GRN, is the driving force behind organ development.

The network presented in Fig. 3 illustrates how TF–DNA interactions can serve as a starting point for the reconstruction of a gynoecium GRN, which we hope will allow us in the future to understand the molecular mechanisms that lead to the establishment of gynoecium cell identities. It should be noted that gene expression data, and not differential gene expression data, was used to reconstruct this network. It is usual to find in the literature the assumption that a "relevant" binding event is the one that leads to differential expression of the corresponding target. TF–DNA interactions, in particular those identified through ChIP experiments, are likely to be a collection of events that occur for a number of reasons [27,82]. Regulation of target gene expression is certainly one of them, and the one we seem most interested in. Yet experimental data also shows that TF binding to DNA can alter chromatin structure [32,49,67], which is also relevant as to how gene expression is regulated [83]. But perhaps more intriguing in this context are the experiments of single molecule tracking of (animal) TFs, which show how TF molecules move across chromatin in their search for their target genes [84,85]. These TF search events

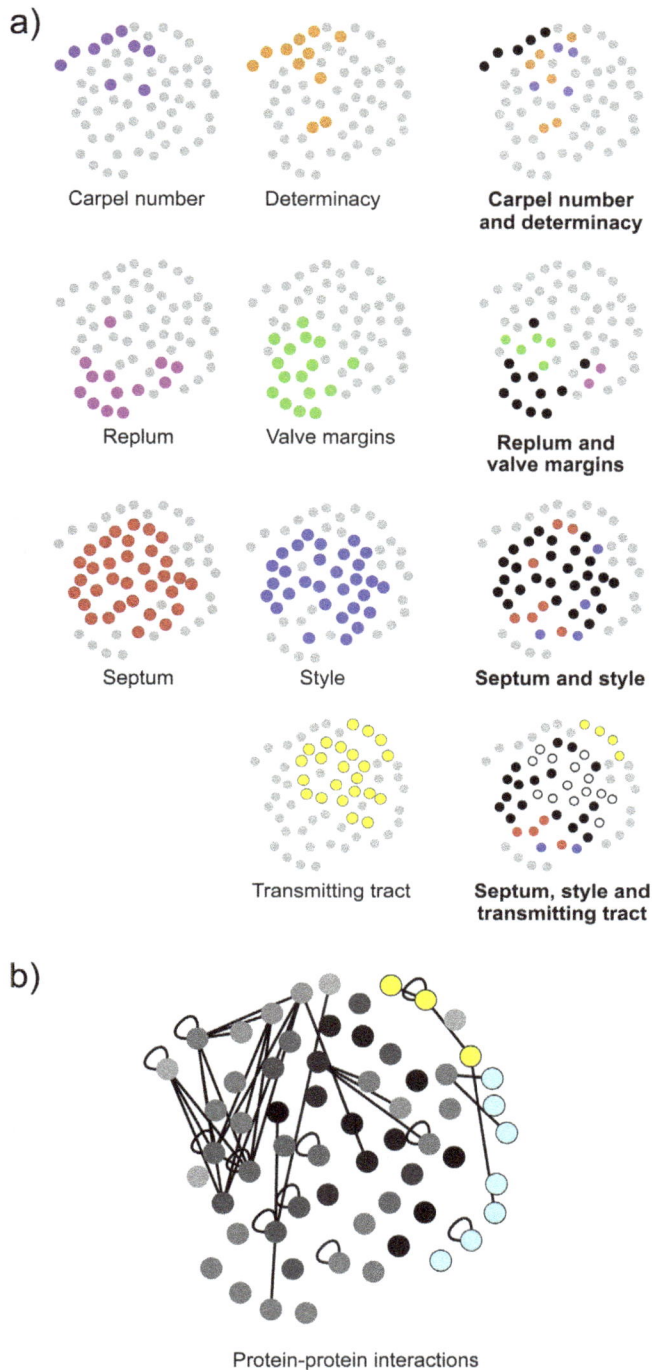

a)

Carpel number | Determinacy | **Carpel number and determinacy**

Replum | Valve margins | **Replum and valve margins**

Septum | Style | **Septum and style**

Transmitting tract | **Septum, style and transmitting tract**

b)

Protein-protein interactions

Fig. 5. High redundancy of TF function during gynoecium development. In all networks, nodes represent the transcription factors listed in [2]. (a) Most gynoecium TFs participate in two or more processes. Colors indicate TFs that participate in the same developmental process, and black and white nodes indicate TFs that participate in two or three processes, respectively. Side by side comparisons show that processes (carpel number establishment and determinacy) or tissue development (replum and valve margins, or septum, style and transmitting tract) share a common set of TFs. (b) Protein-protein interactions network of gynoecium TFs. Protein-protein interactions are those listed in the IntAct database. Most nodes remain unconnected, strongly suggesting that information on interactions is missing, nodes have yet to be discovered, or both. Most TFs participate in two (gray) to six processes (black) indicating that, beyond carpel identity establishment by AG and SEP3, there are no master regulators of gynoecium development. Although some genes are specific for transmitting tract (yellow) or polarity (adaxial–abaxial and apical-basal) establishment (blue), no phenotypes can be observed in the corresponding single mutants. This again indicates that the coordinated action among TFs, rather than the expression of a particular one, gives rise to the different morphogenetic patterns we observe.

could be related to the different affinities for distinct binding sites observed in PBMs data, and ChIP data likely captures TF molecules along their search path, and not just at their destination (presumably their target genes). The question as to how TF target search strategies contribute to the regulation of gene expression is an open one.

Accession numbers

The Arabidopsis Genome Initiative identification numbers of the genes mentioned throughout the manuscript are the following: AG, AT4G18960; ALC, AT5G67110; AMS, AT2G16910; AP1, AT1G69120; AP2, AT4G36920; AP3, AT3G54340; ARF6, AT1G30330; ARF8, AT5G37020; AS1, AT2G37630; AS2, AT1G65620; BEE1, AT1G18400; BP, AT4G08150; CHR11, AT3G06400; CHR17, AT5G18620; CKX3, AT5G56970; CKX5, AT1G75450; FIL, AT2G45190; FUL, AT5G60910; GIK, AT2G35270; HAF, AT1G25330; HDA19, AT4G38130; HEC1, AT5G67060; HEC2, AT3G50330; HEC3, AT5G09750; HEN3, AT5G63610; IND, AT4G00120; JAB, AT4G17460; JAG, AT1G68480; KAN, AT5G16560; LUG, AT4G32551; NTT, AT3G57670; PHB, AT2G34710; PHV, AT1G30490; PI, AT5G20240; PID, AT2G34650; PIN7, AT1G23080; RBL, AT3G55510; REV, AT5G60690; RPL, AT5G02030; SEP1, AT5G15800; SEP2, AT3G02310; SEP3, AT1G24260; SEP4, AT2G03710; SEU, AT1G43850; SHP1, AT3G58780; SHP2, AT2G42830; SPT, AT4G36930; STK, AT4G09960; STM, AT1G62360; STY1, AT3G51060; STY2, AT4G36260; SVP, AT2G22540; SWP, AT3G04740; TAA1, AT1G70560; ULT1, AT4G28190; WAG2, AT3G14370; YAB3, AT4G00180.

References

[1] A.N. Wynn, E.E. Rueschhoff, R.G. Franks, Transcriptomic characterization of a synergistic genetic interaction during carpel margin meristem development in *Arabidopsis thaliana*, PLoS One 6 (2011) e26231.
[2] J.I. Reyes-Olalde, V.M. Zuniga-Mayo, R.A. Chávez Montes, N. Marsch-Martínez, S. de Folter, Inside the gynoecium: at the carpel margin, Trends Plant Sci. 18 (2013) 644–655.
[3] D.R. Smyth, J.L. Bowman, E.M. Meyerowitz, Early flower development in *Arabidopsis*, Plant Cell 2 (1990) 755–767.
[4] E.R. Alvarez-Buylla, M. Benitez, A. Corvera-Poire, C. Chaos, A. ador, F. de, S. olter, Gamboa, B. de, A. uen, A. Garay-Arroyo, B. Garcia-Ponce, F. Jaimes-Miranda, R.V. Perez-Ruiz, et al., Flower Development, 8, Arabidopsis Book, 2010, pp. e0127.
[5] C. Ferrandiz, C. Fourquin, N. Prunet, C.P. Scutt, E. Sundberg, C. Trehin, A.C.M. Vialette-Guiraud, Carpel Development, Adv. Bot. Res. 55 (2010) 1–73.
[6] M. De Lucas, S.M. Brady, Gene regulatory networks in the *Arabidopsis* root, Curr. Opin. Plant Biol. 16 (2013) 50–55.
[7] R.K. Yadav, M. Tavakkoli, M. Xie, T. Girke, G.V. Reddy, A high-resolution gene expression map of the Arabidopsis shoot meristem stem cell niche, Development 141 (2014) 2735–2744.
[8] E.R. Alvarez-Buylla, M. Benítez, E.B. Dávila, A. Chaos, C. Espinosa-Soto, P. Padilla-Longoria, Gene regulatory network models for plant development, Curr. Opin. Plant Biol. 10 (2007) 83–91.
[9] M.F. Yanofsky, H. Ma, J.L. Bowman, G.N. Drews, K.A. Feldmann, E.M. Meyerowitz, The protein encoded by the *Arabidopsis* homeotic gene agamous resembles transcription factors, Nature 346 (1990) 35–39.
[10] S. Gomez-Mena, J. Busscher, L. Colombo, A. Losa, G.C. Angenent, Transcript profiling of transcription factor genes during silique development in *Arabidopsis*, Plant Mol. Biol. 56 (2004) 351–366.
[11] C. Gomez-Mena, S. de Folter, M.M. Costa, G.C. Angenent, R. Sablowski, Transcriptional program controlled by the floral homeotic gene AGAMOUS during early organogenesis, Development 132 (2005) 429–438.
[12] F. Wellmer, M. Alves-Ferreira, A. Dubois, J.L. Riechmann, E.M. Meyerowitz, Genome-wide analysis of gene expression during early *Arabidopsis* flower development, PLoS Genet. 2 (2006) e117.
[13] F. Wellmer, J.L. Riechmann, M. Alves-Ferreira, E.M. Meyerowitz, Genome-wide analysis of spatial gene expression in Arabidopsis flowers, Plant Cell 16 (2004) 1314–1326.
[14] V.M. Zúñiga-Mayo, N. Marsch-Martínez, S. de Folter, JAIBA, a class-II HD-ZIP transcription factor involved in the regulation of meristematic activity, and important for correct gynoecium and fruit development in Arabidopsis, Plant J. 71 (2012) 314–326.
[15] B.C. Crawford, G. Ditta, M.F. Yanofsky, The NTT gene is required for transmitting-tract development in carpels of *Arabidopsis thaliana*, Curr Biol. 17 (2007) 1101–1108.

[16] N. Marsch-Martínez, V.M. Zúñiga-Mayo, H. Herrera-Ubaldo, P.B. Ouwerkerk, J. Pablo-Villa, P. Lozano-Sotomayor, R. Greco, P. Ballester, V. Balanza, S.J. Kuijt, et al., The NTT transcription factor promotes replum development in Arabidopsis fruits, Plant J. 80 (2014) 69–81.

[17] [http://www.ncbi.nlm.nih.gov/geo/] last (accessed 08.05.15).

[18] [https://www.ebi.ac.uk/arrayexpress/] last (accessed 08.05.15).

[19] S.Y. Rhee, M. Mutwil, Towards revealing the functions of all genes in plants, Trends Plant Sci. 19 (2014) 212–221.

[20] R.A. Chávez Montes, G. Coello, K.L. González-Aguilera, N. Marsch-Martínez, S. de Folter, E.R. Alvarez-Buylla, ARACNe-based inference, using curated microarray data, of Arabidopsis thaliana root transcriptional regulatory networks, BMC Plant Biol. 14 (2014) 97.

[21] X. Yu, L. Li, J. Zola, M. Aluru, H. Ye, A. Foudree, H. Guo, S. Anderson, S. Aluru, P. Liu, et al., A brassinosteroid transcriptional network revealed by genome-wide identification of BESI target genes in Arabidopsis thaliana, Plant J. 65 (2011) 634–646.

[22] K. Birnbaum, J.W. Jung, J.Y. Wang, G.M. Lambert, J.A. Hirst, D.W. Galbraith, P.N. Benfey, Cell type-specific expression profiling in plants via cell sorting of protoplasts from fluorescent reporter lines, Nat. Methods 2 (2005) 615–619.

[23] A.S. Iyer-Pascuzzi, P.N. Benfey, Fluorescence-activated cell sorting in plant developmental biology, Methods Mol. Biol. 655 (2010) 313–319.

[24] S. Cai, C.C. Lashbrook, Laser capture microdissection of plant cells from tape-transferred paraffin sections promotes recovery of structurally intact RNA for global gene profiling, Plant J. 48 (2006) 628–637.

[25] Y. Jiao, E.M. Meyerowitz, Cell-type specific analysis of translating RNAs in developing flowers reveals new levels of control, Mol. Syst. Biol. 6 (2010) 419.

[26] C. Tian, X. Zhang, J. He, H. Yu, Y. Wang, B. Shi, Y. Han, G. Wang, X. Feng, C. Zhang, et al., An organ boundary-enriched gene regulatory network uncovers regulatory hierarchies underlying axillary meristem initiation, Mol. Syst. Biol. 10 (2014) 755.

[27] M. Slattery, T. Zhou, L. Yang, M. Dantas, A.C. achado, R. Gordan, R. Rohs, Absence of a simple code: how transcription factors read the genome, Trends Biochem. Sci. 39 (2014) 381–399.

[28] M.H. Kuo, C.D. Allis, In vivo cross-linking and immunoprecipitation for studying dynamic Protein: DNA associations in a chromatin environment, Methods 19 (1999) 425–433.

[29] S. de Folter, Protein tagging for chromatin immunoprecipitation from Arabidopsis, Methods Mol. Biol. 678 (2011) 199–210.

[30] K.S. Heyndrickx, J. Van de Velde, C. Wang, D. Weigel, K. Vandepoele, A functional and evolutionary perspective on transcription factor binding in Arabidopsis thaliana, Plant Cell 26 (2014) 3894–3910.

[31] A. Pajoro, S. Biewers, E. Dougali, F. Leal Valentim, M.A. Mendes, A. Porri, G. Coupland, Y. Van de Peer, A.D. van Dijk, L. Colombo, et al., The (r) evolution of gene regulatory networks controlling Arabidopsis plant reproduction: a two-decade history, J. Exp. Bot. 65 (2014) 4731–4745.

[32] A. Pajoro, P. Madrigal, J.M. Muino, J.T. Matus, J. Jin, M.A. Mecchia, J.M. Debernardi, J.F. Palatnik, S. Balazadeh, M. Arif, et al., Dynamics of chromatin accessibility and gene regulation by MADS-domain transcription factors in flower development, Genome Biol. 15 (2014) R41.

[33] S. de Bruijn, G.C. Angenent, K. Kaufmann, Plant 'evo-devo' goes genomic: from candidate genes to regulatory networks, Trends Plant Sci. 17 (2012) 441–447.

[34] J.S. Reece-Hoyes, A. Diallo, B. Lajoie, A. Kent, S. Shrestha, S. Kadreppa, C. Pesyna, J. Dekker, C.L. Myers, A.J. Walhout, Enhanced yeast one-hybrid assays for high-throughput gene-centered regulatory network mapping, Nat. Methods 8 (2011) 1059–1064.

[35] S.M. Brady, L. Zhang, M. Megraw, N.J. Martinez, E. Jiang, C.S. Yi, W. Liu, A. Zeng, M. Taylor-Teeples, D. Kim, et al., A stele-enriched gene regulatory network in the Arabidopsis root, Mol. Syst. Biol. 7 (2011) 459.

[36] A. Gaudinier, L. Zhang, J.S. Reece-Hoyes, M. Taylor-Teeples, L. Pu, Z. Liu, G. Breton, J.L. Pruneda-Paz, D. Kim, S.A. Kay, et al., Enhanced Y1H assays for Arabidopsis, Nat. Methods 8 (2011) 1053–1055.

[37] M. Taylor-Teeples, L. Lin, L. de, M. russa, G. Turco, T.W. Toal, A. Gaudinier, N.F. Young, G.M. Trabucco, M.T. Veling, R. Lamothe, et al., An Arabidopsis gene regulatory network for secondary cell wall synthesis, Nature 517 (2015) 571–575.

[38] J.M. Franco-Zorrilla, I. Lopez-Vidriero, J.L. Carrasco, M. Godoy, P. Vera, R. Solano, DNA-binding specificities of plant transcription factors and their potential to define target genes, Proc. Natl. Acad. Sci. U.S.A. 111 (2014) 2367–2372.

[39] [http://www.rsat.eu/] last accessed (08. 05.15).

[40] A.J. Walhout, Unraveling transcription regulatory networks by protein-DNA and protein–protein interaction mapping, Genome Res. 16 (2006) 1445–1454.

[41] E.H. Davidson, Emerging properties of animal gene regulatory networks, Nature 468 (2010) 911–920.

[42] S. de Folter, R.G. Immink, M. Kieffer, L. Parenicova, S.R. Henz, D. Weigel, M. Busscher, L. Kooiker, L. Colombo, M.M. Kater, et al., Comprehensive interaction map of the Arabidopsis MADS Box transcription factors, Plant Cell 17 (2005) 1424–1433.

[43] R.G. Immink, I.A. Tonaco, S. de Folter, A. Shchennikova, A.D. van Dijk, J. Busscher-Lange, J.W. Borst, G.C. Angenent, SEPALLATA3: the 'glue' for MADS box transcription factor complex formation, Genome Biol. 10 (2009) R24.

[44] H. Dortay, N. Gruhn, A. Pfeifer, M. Schwerdtner, T. Schmulling, A. Heyl, Toward an interaction map of the two-component signaling pathway of Arabidopsis thaliana J. Proteome Res. 7 (2008) 3649–3660.

[45] S. Danisman, A.D. van Dijk, A. Bimbo, F. van der Wal, L. Hennig, S. de Folter, G.C. Angenent, R.G. Immink, Analysis of functional redundancies within the Arabidopsis TCP transcription factor family, J. Exp. Bot. 64 (2013) 5673–5685.

[46] M.H. Nanao, T. Vinos-Poyo, G. Brunoud, E. Thevenon, M. Mazzoleni, D. Mast, S. Laine, S. Wang, G. Hagen, H. Li, et al., Structural basis for oligomerization of auxin transcriptional regulators, Nat. Commun. 5 (2014) 3617.

[47] H. Herrera-Ubaldo, E. Zanchetti, L. Colombo, F. de, S. olter, Protein interactions guiding carpel and fruit development in Arabidopsis, Plant Biosyst. 148 (2014) 169–175.

[48] T. Honma, K. Goto, Complexes of MADS-box proteins are sufficient to convert leaves into floral organs, Nature 409 (2001) 525–529.

[49] C. Smaczniak, R.G. Immink, J.M. Muino, R. Blanvillain, M. Busscher, J. Busscher-Lange, Q.D. Dinh, S. Liu, A.H. Westphal, S. Boeren, et al., Characterization of MADS-domain transcription factor complexes in Arabidopsis flower development, Proc. Natl. Acad. Sci. U.S.A. 109 (2012) 1560–1565.

[50] A. Pinyopich, G.S. Ditta, B. Savidge, S.J. Liljegren, E. Baumann, E. Wisman, M.F. Yanofsky, Assessing the redundancy of MADS-box genes during carpel and ovule development, Nature 424 (2003) 85–88.

[51] J. Alvarez, D.R. Smyth, CRABS CLAW and SPATULA, two Arabidopsis genes that control carpel development in parallel with AGAMOUS, Development 126 (1999) 2377–2386.

[52] G. Ditta, A. Pinyopich, P. Robles, S. Pelaz, M.F. Yanofsky, The SEP4 gene of Arabidopsis thaliana functions in floral organ and meristem identity, Curr. Biol. 14 (2004) 1935–1940.

[53] S. Pelaz, G.S. Ditta, E. Baumann, E. Wisman, M.F. Yanofsky, B and C floral organ identity functions require SEPALLATA MADS-box genes, Nature 405 (2000) 200–203.

[54] B. Poppenberger, W. Rozhon, M. Khan, S. Husar, G. Adam, C. Luschnig, S. Fujioka, T. Sieberer, CESTA, a positive regulator of brassinosteroid biosynthesis, EMBO J. 30 (2011) 1149–1161.

[55] B.C. Crawford, M.F. Yanofsky, HALF FILLED promotes reproductive tract development and fertilization efficiency in Arabidopsis thaliana, Development 138 (2011) 2999–3009.

[56] K. Gremski, G. Ditta, M.F. Yanofsky, The HECATE genes regulate female reproductive tract development in Arabidopsis thaliana, Development 134 (2007) 3593–3601.

[57] S. de Folter, R.G. Immink, Yeast protein–protein interaction assays and screens, Methods Mol. Biol. 754 (2011) 145–165.

[58] S. Orchard, M. Ammari, B. Aranda, L. Breuza, L. Briganti, F. Broackes-Carter, N.H. Campbell, G. Chavali, C. Chen, N. del-Toro, et al., The MIntAct project—IntAct as a common curation platform for 11 molecular interaction databases, Nucleic Acids Res. 42 (2014) D358–363.

[59] [http://www.ebi.ac.uk/intact/] last accessed (08. 05.15).

[60] K. Jetha, G. Theissen, R. Melzer, Arabidopsis SEPALLATA proteins differ in cooperative DNA-binding during the formation of floral quartet-like complexes, Nucleic Acids Res. 42 (2014) 10927–10942.

[61] M. Morsy, S. Gouthu, S. Orchard, D. Thorneycroft, J.F. Harper, R. Mittler, J.C. Cushman, Charting plant interactomes: possibilities and challenges, Trends Plant Sci 13 (2008) 183–191.

[62] R.G. Franks, C. Wang, J.Z. Levin, Z. Liu, SEUSS, a member of a novel family of plant regulatory proteins, represses floral homeotic gene expression with LEUNIG, Development 129 (2002) 253–263.

[63] V.V. Sridhar, A. Surendrarao, Z. Liu, APETALA1 and SEPALLATA3 interact with SEUSS to mediate transcription repression during flower development, Development 133 (2006) 3159–3166.

[64] D. Gonzalez, A.J. Bowen, T.S. Carroll, R.S. Conlan, The transcription corepressor LEUNIG interacts with the histone deacetylase HDA19 and mediator components MED14 (SWP) and CDK8 (HEN3) to repress transcription, Mol. Cell. Biol. 27 (2007) 5306–5315.

[65] C.C. Carles, J.C. Fletcher, The SAND domain protein ULTRAPETALA1 acts as a trithorax group factor to regulate cell fate in plants, Genes Dev. 23 (2009) 2723–2728.

[66] H.R. Pires, M.M. Monfared, E.A. Shemyakina, J.C. Fletcher, ULTRAPETALA trxG genes interact with KANADI transcription factor genes to regulate Arabidopsis gynoecium patterning, Plant Cell 26 (2014) 4345–4361.

[67] M.A. Mendes, R.F. Guerra, M.C. Berns, C. Manzo, S. Masiero, L. Finzi, M.M. Kater, L. Colombo, MADS domain transcription factors mediate short-range DNA looping that is essential for target gene expression in Arabidopsis, Plant Cell 25 (2013) 2560–2572.

[68] N. Marsch-Martínez, J.I. Reyes-Olalde, D. Ramos-Cruz, P. Lozano-Sotomayor, V.M. Zúñiga-Mayo, S. de Folter, Hormones talking: does hormonal cross-talk shape the Arabidopsis gynoecium? Plant Signal Behav. 7 (2012) 1698–1701.

[69] E. Larsson, R.G. Franks, E. Sundberg, Auxin and the Arabidopsis thaliana gynoecium, J. Exp. Bot. 64 (2013) 2619–2627.

[70] G.E. Schaller, A. Bishopp, J.J. Kieber, The Yin-Yang of hormones: cytokinin and auxin Interactions in plant development, Plant Cell 27 (2015) 44–63.

[71] A.M. Rashotte, M.G. Mason, C.E. Hutchison, F.J. Ferreira, G.E. Schaller, J.J. Kieber, A subset of Arabidopsis AP2 transcription factors mediates cytokinin responses in concert with a two-component pathway, Proc. Natl. Acad. Sci. U.S.A. 103 (2006) 11081–11085.

[72] W.G. Brenner, T. Schmülling, Summarizing and exploring data of a decade of cytokinin-related transcriptomics, Front Plant Sci. 6 (2015) 29.

[73] E. Larsson, C.J. Roberts, A.R. Claes, R.G. Franks, E. Sundberg, Polar auxin transport is essential for medial versus lateral tissue specification and vascular-mediated valve outgrowth in *Arabidopsis* gynoecia, Plant Physiol. 166 (2014) 1998–2012.

[74] L. Moubayidin, L. Ostergaard, Dynamic control of auxin distribution imposes a bilateral-to-radial symmetry switch during gynoecium development, Curr. Biol. 24 (2014) 2743–2748.

[75] J.L. Nemhauser, L.J. Feldman, P.C. Zambryski, Auxin and ETTIN in *Arabidopsis* gynoecium morphogenesis, Development 127 (2000) 3877–3888.

[76] J.J. Sohlberg, M. Myrenas, S. Kuusk, U. Lagercrantz, M. Kowalczyk, G. Sandberg, E. Sundberg, STY1 regulates auxin homeostasis and affects apical-basal patterning of the *Arabidopsis* gynoecium, Plant J. 47 (2006) 112–123.

[77] I. Bartrina, E. Otto, M. Strnad, T. Werner, T. Schmulling, Cytokinin regulates the activity of reproductive meristems, flower organ size, ovule formation, and thus seed yield in *Arabidopsis thaliana*, Plant Cell 23 (2011) 69–80.

[78] N. Marsch-Martínez, D. Ramos-Cruz, R. Irepan, J. eyes-Olalde, P. Lozano-Sotomayor, V.M. Zúñiga-Mayo, F. de, S. olter, The role of cytokinin during *Arabidopsis* gynoecia and fruit morphogenesis and patterning, Plant J. 72 (2012) 222–234.

[79] V.M. Zúñiga-Mayo, J.I. Reyes-Olalde, N. Marsch-Martínez, S. de Folter, Cytokinin treatments affect the apical-basal patterning of the *Arabidopsis* gynoecium and resemble the effects of polar auxin transport inhibition, Front Plant Sci. 5 (2014) 191.

[80] T. Girin, T. Paicu, P. Stephenson, S. Fuentes, E. Korner, M. O'Brien, K. Sorefan, T.A. Wood, V. Balanza, C. Ferrandiz, et al., INDEHISCENT and SPATULA interact to specify carpel and valve margin tissue and thus promote seed dispersal in *Arabidopsis*, Plant Cell 23 (2011) 3641–3653.

[81] [http://plantgrn.noble.org/hrgrn/] last accessed (08. 05.15).

[82] N. Marsch-Martínez, W. Wu, S. de Folter, The MADS symphonies of transcriptional regulation, Front Plant Sci. 2 (2011) 26.

[83] K.L. MacQuarrie, A.P. Fong, R.H. Morse, S.J. Tapscott, Genome-wide transcription factor binding: beyond direct target regulation, Trends Genet. 27 (2011) 141–148.

[84] J. Chen, Z. Zhang, L. Li, B.C. Chen, A. Revyakin, B. Hajj, W. Legant, M. Dahan, T. Lionnet, E. Betzig, et al., Single-molecule dynamics of enhanceosome assembly in embryonic stem cells, Cell 156 (2014) 1274–1285.

[85] I. Izeddin, V. Recamier, L. Bosanac, I.I. Cisse, L. Boudarene, C. Dugast-Darzacq, F. Proux, O. Benichou, R. Voituriez, O. Bensaude, et al., Single-molecule tracking in live cells reveals distinct target-search strategies of transcription factors in the nucleus, eLife 3 (2014).

[86] S. Gonzalez-Reig, J.J. Ripoll, A. Vera, M.F. Yanofsky, A. Martinez-Laborda, Antagonistic gene activities determine the formation of pattern elements along the mediolateral axis of the *Arabidopsis* fruit, PLoS Genet. 8 (2012) e1003020.

[87] P. Kay, M. Groszmann, J.J. Ross, R.W. Parish, S.M. Swain, Modifications of a conserved regulatory network involving INDEHISCENT controls multiple aspects of reproductive tissue development in *Arabidopsis*, New Phytol. 197 (2013) 73–87.

[88] S.J. Liljegren, A.H. Roeder, S.A. Kempin, K. Gremski, L. Ostergaard, S. Guimil, D.K. Reyes, M.F. Yanofsky, Control of fruit patterning in *Arabidopsis* by INDEHISCENT, Cell 116 (2004) 843–853.

[89] J.J. Ripoll, A.H. Roeder, G.S. Ditta, M.F. Yanofsky, A novel role for the floral homeotic gene APETALA2 during *Arabidopsis* fruit development, Development 138 (2011) 5167–5176.

[90] L. Yant, J. Mathieu, T.T. Dinh, F. Ott, C. Lanz, H. Wollmann, X. Chen, M. Schmid, Orchestration of the floral transition and floral development in *Arabidopsis* by the bifunctional transcription factor APETALA2, Plant Cell 22 (2010) 2156–2170.

[91] C. Wang, J. Xu, D. Zhang, Z.A. Wilson, D. Zhang, An effective approach for identification of in vivo protein-DNA binding sites from paired-end ChIP-Seq data, BMC Bioinf. 11 (2010) 81.

[92] S.E. Wuest, D.S. O'Maoileidigh, L. Rae, K. Kwasniewska, A. Raganelli, K. Hanczaryk, A.J. Lohan, B. Loftus, E. Graciet, F. Wellmer, Molecular basis for the specification of floral organs by APETALA3 and PISTILLATA, Proc. Natl. Acad. Sci. U.S.A. 109 (2012) 13452–13457.

[93] V. Gregis, F. Andres, A. Sessa, R.F. Guerra, S. Simonini, J.L. Mateos, F. Torti, F. Zambelli, G.M. Prazzoli, K.N. Bjerkan, et al., Identification of pathways directly regulated by SHORT VEGETATIVE PHASE during vegetative and reproductive development in *Arabidopsis*, Genome Biol. 14 (2013) R56.

[94] D.S. O'Maoiléidigh, S.E. Wuest, L. Rae, A. Raganelli, P.T. Ryan, K. Kwasniewska, P. Das, A.J. Lohan, B. Loftus, E. Graciet, F. Wellmer, Control of reproductive floral organ identity specification in *Arabidopsis* by the C function regulator AGAMOUS, Plant Cell 25 (2013) 2482–2503.

[95] K. Kaufmann, J.M. Muino, R. Jauregui, C.A. Airoldi, C. Smaczniak, P. Krajewski, G.C. Angenent, Target genes of the MADS transcription factor SEPALLATA3: integration of developmental and hormonal pathways in the *Arabidopsis* flower, PLoS Biol. 7 (2009) e1000090.

[96] K. Kaufmann, F. Wellmer, J.M. Muino, T. Ferrier, S.E. Wuest, V. Kumar, A. Serrano-Mislata, F. Madueno, P. Krajewski, E.M. Meyerowitz, et al., Orchestration of floral initiation by APETALA1, Science 328 (2010) 85–89.

19

Biological process annotation of proteins across the plant kingdom

Joachim W. Bargsten [a,c,e], Edouard I. Severing [b], Jan-Peter Nap [a,c], Gabino F. Sanchez-Perez [a,d], Aalt D.J. van Dijk [a,f,*]

[a] Applied Bioinformatics, Bioscience, Plant Sciences Group, Wageningen University and Research Centre, Wageningen, The Netherlands
[b] Laboratory of Genetics, Plant Sciences Group, Wageningen University and Research Centre, Wageningen, The Netherlands
[c] Netherlands Bioinformatics Centre (NBIC), Nijmegen, The Netherlands
[d] Laboratory of Bioinformatics, Plant Sciences Group, Wageningen University and Research Centre, Wageningen, The Netherlands
[e] Laboratory for Plant Breeding, Plant Sciences Group, Wageningen University and Research Centre, The Netherlands
[f] Biometris, Wageningen University and Research Centre, Wageningen, The Netherlands

ARTICLE INFO

Keywords:
Gene function prediction
Gene function divergence

ABSTRACT

Accurate annotation of protein function is key to understanding life at the molecular level, but automated annotation of functions is challenging. We here demonstrate the combination of a method for protein function annotation that uses network information to predict the biological processes a protein is involved in, with a sequence-based prediction method. The combined function prediction is based on co-expression networks and combines the network-based prediction method BMRF with the sequence-based prediction method Argot2. The combination shows significantly improved performance compared to each of the methods separately, as well as compared to Blast2GO. The approach was applied to predict biological processes for the proteomes of rice, barrel clover, poplar, soybean and tomato. The novel function predictions are available at www.ab.wur.nl/bmrf. Analysis of the relationships between sequence similarity and predicted function similarity identifies numerous cases of divergence of biological processes in which proteins are involved, in spite of sequence similarity. This indicates that the integration of network-based and sequence-based function prediction is helpful towards the analysis of evolutionary relationships. Examples of potential divergence are identified for various biological processes, notably for processes related to cell development, regulation, and response to chemical stimulus. Such divergence in biological process annotation for proteins with similar sequences should be taken into account when analyzing plant gene and genome evolution.

DATA: All gene functions predictions are available online (http://www.ab.wur.nl/bmrf/). The online resource can be queried for predictions of proteins or for Gene Ontology terms of interest, and the results can be downloaded in bulk. Queries can be based on protein identifiers, biological process Gene Ontology identifiers, or text descriptors of biological processes.

1. Introduction

The amount of plant genome data grows disproportional to the amount of available experimental data on these genomes [1–5]. To connect this ever increasing amount of genome data to plant biology, structural gene annotation followed by function annotation is imperative. For example, the identification of candidate genes involved in a trait of interest greatly benefits from gene function annotation [6]. In the context of the study of genome evolution,

gene function annotations are necessary in order to enable comparison between sets of genes with different evolutionary histories, e.g. those retained vs. those lost after duplication [7]. To annotate gene or protein function, experimental data, if available, can be used to annotate gene or protein function. However, the scarcity of experimental data highlights the attractiveness of computational approaches to assist in gene function annotation [8]. Indeed, newly sequenced genomes are in general accompanied by a function annotation which heavily relies on computational predictions. Such automated annotations are delivered by a variety of approaches, often without much knowledge about their reliability. For studying plant genomes and plant genome evolution, reliable function annotation is therefore a major challenge.

One way to annotate proteins without experimental data is to infer function from sequence data [3]. The *de facto* standard

* Corresponding author at: Applied Bioinformatics, Bioscience, Plant Sciences Group, Wageningen University and Research Centre, Wageningen, The Netherlands.

E-mail address: aaltjan.vandijk@wur.nl (A.D.J. van Dijk).

to capture function annotation today is the Gene Ontology (GO), in particular, the Molecular Function (MF) and Biological Process (BP) sub-ontologies [9]. MF describes activities, such as catalytic or binding activities, that occur at the molecular level, whereas BP describes a series of events accomplished by one or more ordered assemblies of molecular functions [9]. Compared to MF, terms in the BP ontology are generally associated with more conceptual levels of function; BP terms describe the execution of one or more molecular function instances working together to accomplish a certain biological objective. The prediction of BP terms can depend on the cellular and organismal context [10]. Therefore, BP terms tend to be poorly predicted by methods based on sequence similarity only, such as BLAST [10,11]. The reliability of BP predictions increases with advanced approaches that employ, e.g., phylogenetic frameworks [12,13] or network data such as protein–protein interactions [14].

We recently developed a protein function prediction method for BP terms called Bayesian Markov Random Field (BMRF) [15], which uses network data as input. In BMRF, each protein is represented as a node in the network, and connections in the network indicate functional relationships between proteins. Networks can be based on, e.g., protein–protein interactions or co-expression data. BMRF uses existing BP annotations for proteins in the network to infer biological processes for unannotated proteins in that network. To do so, BMRF uses a statistical model describing how likely neighbors are to participate in the same BP; this constitutes the Markov Random Field. Existing BP annotations are used as "seed" or "training" data, providing a set of initial labels for the Markov Random Field. Parameters in the statistical model are trained using a Bayesian approach by performing simultaneous estimation of the model parameters and prediction of protein functions. Importantly, BMRF can transfer functional information beyond direct interactions. Therefore, it is able to generate function predictions for proteins that are only linked with other proteins with unknown function.

In the Critical Assessment of Function Annotations (CAFA) protein function prediction challenge [10] BMRF obtained particularly good performance in human (first place) and Arabidopsis (second place) for BP term prediction [10]. In these species, BMRF performance benefits from the wealth of existing function annotation, i.e. experimental data. Because of its dependence on training data, function annotation for species with more sparse function annotation is challenging for BMRF. To improve the prediction performance in sparsely annotated species, we present here a strategy to combine BMRF with the sequence-based function prediction method Argot2 [16]. Argot2 was among the top performing sequence-based algorithms in the CAFA category "eukaryotic BP". In its computational approach Argot2 is complementary to BMRF, because it is purely sequence-based.

We demonstrate that a combination of Argot2 and BMRF has a markedly better function prediction performance than each method separately. This integrated method was applied to predict BP terms for proteins in five plant species, *Medicago truncatula* (barrel clover), *Oryza sativa* (rice), *Populus trichocarpa* (poplar), *Glycine max* (soybean) and *Solanum lycopersicum* (tomato), using microarray co-expression networks as input. Numerous new proteins were associated with specific biological processes, such as seed development in rice or nitrogen fixation in Medicago. By comparison between sequence divergence and predicted function divergence, numerous cases of putative neo-functionalization involving various biological processes were identified. This new method and the resulting set of predicted gene functions will be of great value in capitalizing on the large amount of plant genome data that is currently being generated for the study of the evolution of genome and gene function.

2. Results

2.1. Method development and evaluation

We previously developed the protein function prediction method BMRF and used it to annotate protein function in *Arabidopsis thaliana* [17]. This method relies, besides on network data, on existing function annotation as input. For Arabidopsis, we demonstrated that the amount of available annotation (training) data was sufficient to achieve a good prediction performance [17]. However, for crop species, much less annotation data is available as input. To increase the overall function prediction performance for plants with sparse experimental data, we explored combining BMRF with the sequence-based method Argot2.

Argot2 and BMRF were tested separately (standalone setting) or in two combinations (Fig. 1). Performance assessment focussed on rice, the crop with the largest amount of annotation data available: 415 proteins with experimental evidence for a biological process. The rice network used as input for BMRF was obtained from a combination of microarray-based co-expression data, data from STRING [18] and FunctionalNet [19] (Table S1). Of the 415 proteins with experimental evidence, 394 were present in the network, and were used for validation of predicted functions.

Function prediction performance was assessed on the basis of cross-validation, leaving out randomly selected proteins with known function and comparing the predictions with those data. The area under the receiver operator characteristic curve (AUC) was used to compare the performance of the predictions that come as ordered lists of predicted proteins per biological process. In the standalone setting (Fig. 1A and B) with rice sequence and network data, BMRF and Argot2 both have a low performance, with AUC (average ± standard deviation) of 0.6 ± 0.12 and 0.67 ± 0.11, respectively (Tables 1 and S2). These values are considerably lower than the AUC previously obtained with BMRF for Arabidopsis (0.75) [17] due to the small amount of training data (annotated gene functions) that is available for rice. Assuming information from Arabidopsis would improve the performance of rice protein function predictions in BMRF, we connected proteins in an available Arabidopsis network (Table S1) to proteins in the rice network based on sequence similarity using BLAST. With this rice-Arabidopsis interspecies network in addition to the networks of both species separately (Fig. 1C), BMRF performed slightly better than Argot2 (AUC 0.70 ± 0.12). The precise value of the BLAST E-value cut-off used to create the interspecies network did not influence the performance of BMRF (data not shown).

Both methods use complimentary information about biological processes (network input for BMRF, sequence input for Argot2). Therefore, we tested combining the two. Argot2 and BMRF can be combined in multiple ways. We used a simple rank-based approach

Table 1
Prediction performance for rice protein function of various combinations of methods and input datasets.

	Network	Method[a]	AUC[b]
(i)	Rice only	BMRF	0.60 (0.12)
(ii)	Rice only	Argot2	0.67 (0.11)
(iii)	Arabidopsis and rice combined	BMRF	0.70 (0.12)
(iv)	Arabidopsis and rice combined	Blast2GO	0.72 (0.13)
(v)	Arabidopsis and rice combined	Argot2 + BMRF	0.71 (0.12)
(vi)	Arabidopsis and rice combined	Argot2 → BMRF	0.83 (0.15)

[a] Methods analyzed were BMRF, Argot2, Blast2GO, Argot2 + BMRF (rank sum) and Argot2 → BMRF (seeding). Rice network was used separately (rice only), or it was connected to an Arabidopsis network based on sequence similarity (combined).
[b] Area under the curve; mean (standard deviation).

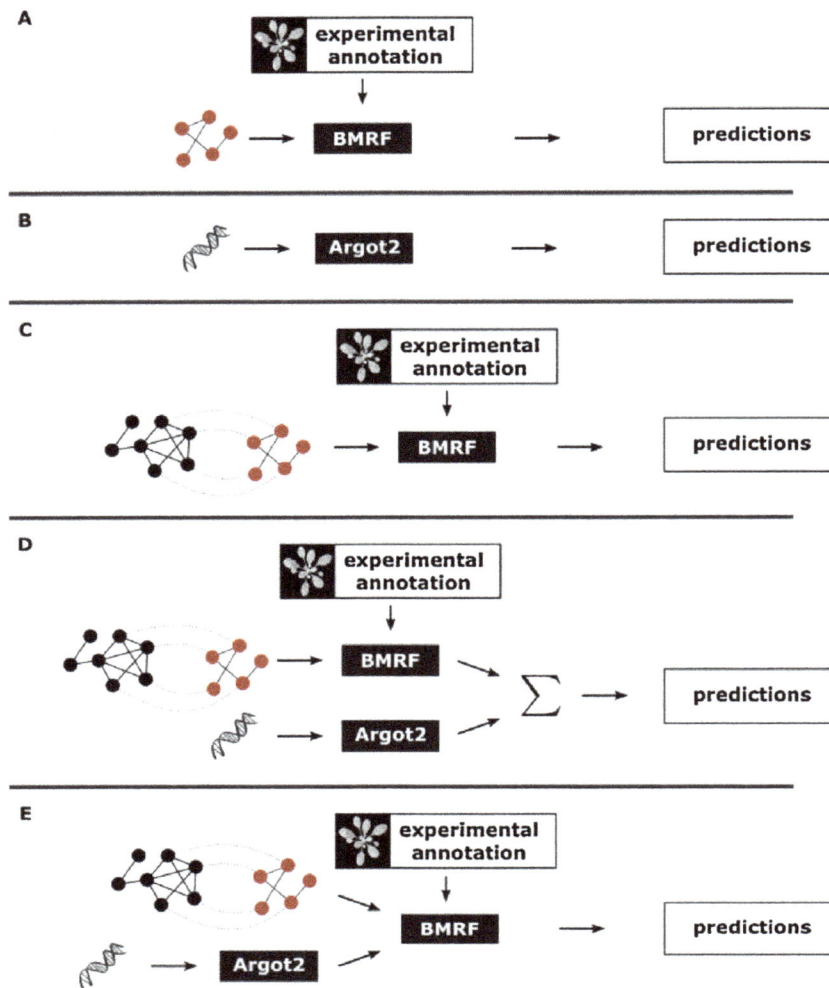

Fig. 1. Strategies for predicting protein function. BMRF (A and C) and Argot2 (B) were used in a standalone setting or in two different combinations (D and E). Combining BMRF and Argot2 was done by combining the results of each of the two methods (D), and by using Argot2 predictions as input for BMRF (E). The rice network is indicated in red, the Arabidopsis network in black and interspecies connections in grey dashed lines. Sequence-based input is indicated by a DNA-helix symbol. (For interpretation of the references to color in this figure legend, the reader is referred to the web version of this article.)

to predict biological processes by ordering Argot2 and BMRF results separately and then combining their ranks to produce a final rank (Fig. 1D). This integration was performed for each biological process separately by sorting the proteins based on their score for that process and using the sum of the ranks induced by this ordering for BMRF and for Argot2. This integration of Argot2 and BMRF did not improve results compared to standalone BMRF (Table 1). Performance was markedly improved, however, by generating initial predictions with Argot2 and supplying these to BMRF as training data (seed data; Fig. 1E). In this integration method, the initial labelling of proteins in the network (i.e. the seed data for BMRF) was based on the Argot2 predictions. Argot2 uses an algorithm-specific score to rank its results and requires a threshold for such a score. To assess the influence of different thresholds on the performance of BMRF, BMRF was seeded with 5 different output sets of Argot2 (Table S3). The best performance was achieved with the default threshold of 5.

The results above indicate that our integrated method performed markedly better than each of the two methods separately. As additional assessment of performance, we predicted annotations with the often-used method Blast2GO [21]. The resulting AUC of Blast2GO was 0.72 ± 0.13, and the AUC of the combined Argot2-BMRF predictions was 0.83 ± 0.15 which is significantly ($p < 10^{-15}$;

Mann–Whitney U) better than Blast2GO (Fig. 2A). The small number of experimentally verified annotations (true positives) and high number of unannotated proteins (true negatives) could introduce a skew in the cross-validation sets, leading to a bias in the AUC performance assessment [22]. The F-score (harmonic mean of precision and recall) does not suffer from this skew and the final prediction performance was therefore also assessed with the maximum F-score (F_{max}-score). In agreement with the AUC evaluation, the F_{max}-scores of Argot2-seeded BMRF (0.56 ± 0.24) were significantly better ($p < 10^{-15}$; Mann–Whitney U) than Blast2GO (0.51 ± 0.23). Visual inspection of a histogram of AUC values and of F_{max}-score values for different BP terms in different cross-validation runs confirms the performance difference between the combined Argot2-BMRF predictions and Blast2GO (Fig. 2B and C).

To obtain independent validation in addition to the cross-validation performed above, the Argot2-seeded BMRF predictions were compared to annotations available in the Oryzabase database [23] which were not present in our input data (71 proteins). The AUC of 0.88 ± 0.13 we obtained was similar to the AUC obtained in the cross-validation, confirming the performance assessment. Overall, the performance evaluation demonstrates that Argot2-seeded BMRF is an effective way to predict BP protein function in sparsely annotated plant genomes.

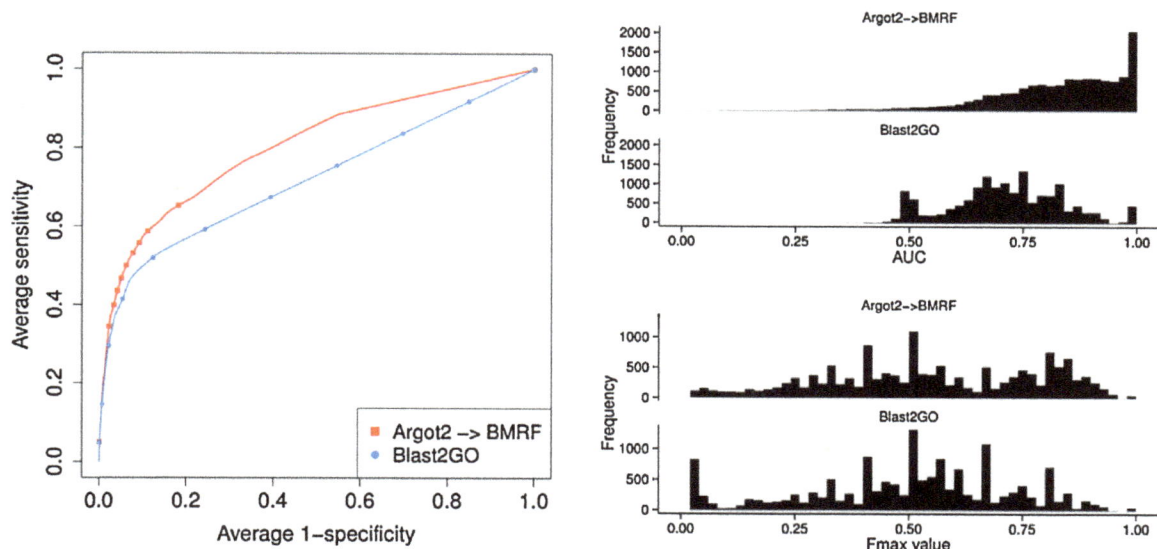

Fig. 2. Performance assessment of function prediction on rice proteins. (A) Receiver operator characteristic curve showing 1-specificity vs. sensitivity of the predictions of Argot2-seeded BMRF and Blast2GO. Specificity and sensitivity were averaged over all cross-validation runs. Dots indicate evenly spaced intervals of the underlying prediction score, line represents complete curve. Performance is summarized as AUC which is the area under these curves. (B) Histogram of AUC values per GO term of every cross-validation run calculated for Argot2-seeded BMRF and Blast2GO. (C) Histogram of F_{max} values per GO term of every cross-validation run calculated for Argot2 seeding BMRF and Blast2GO.

2.2. Application to crop species

Argot2-seeded BMRF using PlaNet [24] co-expression networks as input (Table S4) was applied to predict BP protein functions in a selection of model and crop plants comprising *O. sativa* (rice), *M. truncatula* (barrel clover), *G. max* (soybean), *P. trichocarpa* (poplar) and *S. lycopersicum* (tomato). The posterior probability of a protein associated with a certain GO term was estimated for all GO terms and all proteins in the network. In order to answer a question such as "does protein X perform biological process Y", a finite set of predictions is needed. To obtain such finite set, an *F*-score-based cut-off was applied to the posterior probability. As Arabidopsis has the highest coverage of experimental data, this cut-off was adjusted per GO term by comparing Arabidopsis predictions with available experimental data, as previously described [17]: for each GO term, a threshold on the posterior probability was defined that results in the maximum *F*-score for that GO term. All predictions are available online (http://www.ab.wur.nl/bmrf/). The online resource can be queried for predictions of proteins or for GO terms of interest, and the results can be downloaded in bulk. Queries can be based on protein identifiers, biological process GO identifiers, or text descriptors of biological processes (Fig. 3). By default, only the most detailed Gene Ontology terms (leave terms in the GO structure) are displayed, in order to focus on the most relevant predictions.

The fraction of proteins out of the complete proteome annotated with at least one biological process (annotation coverage) varies considerably between the species: rice shows the highest annotation coverage (99%), followed by poplar (77%). Soybean (43%) and

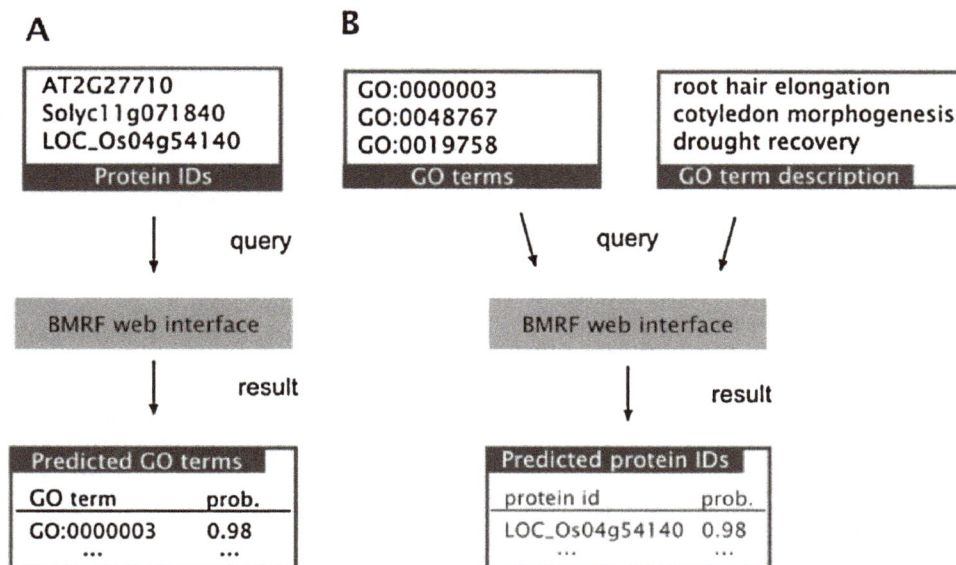

Fig. 3. Use case scenarios for the web interface. Argot2-seeded BMRF results can be queried in two ways. (A) Protein identifiers as query input. The result consists of predicted GO terms for each protein. (B) GO terms (or GO term descriptions) as query input. The result consists of predicted protein identifiers for the relevant GO term(s) and associated posterior probabilities (prob).

barrel clover (39%) show lower coverage. Tomato has the lowest coverage (12%). Such differences in annotation coverage can have at least two reasons. First, although for every biological process every protein in the input network will have an associated posterior probability, these probabilities can be below the F-score-based cut-off. This means that not necessarily every protein in the input network will be annotated. In addition, because BMRF only predicts functions for proteins in the input network, the maximum possible annotation coverage is limited by the number of proteins in the respective network. This limit is reflected by the tomato annotation coverage, as the tomato network is the smallest with 4355 proteins. With exception of soybean, the annotation coverage correlates with the number of proteins in the respective network (Table S4).

To investigate differences between available gene function annotation data and Argot2-seeded BMRF, we compared the results with existing protein function predictions from the reference genomes of barrel clover [25], poplar [26], tomato [27], rice [28] and soybean [29]. Except for tomato, the existing annotations have a much lower coverage than the above mentioned coverage obtained by Argot2-seeded BMRF (Table S5). The increase of percentage of number of proteins with at least one biological process predicted by our approach varied per species. The percentage increase ranged from ~60% for rice (24,160 in existing annotation vs. 38,998 in our annotation) to over 100% for poplar (13,682 vs. 32,119).

To complement the above presented results on coverage, which focused on the question how many proteins obtain at least one annotation, we also compared the number of predicted functions per protein. The average number of GO terms per protein in the available experimental annotation data for Arabidopsis is 4.4. As additional experimental evidence is supposed to accumulate, this number should be regarded as a lower bound of the average real number of GO terms a protein should be annotated with. Existing sets of predicted annotations for the plant species included here are considerably below this bound, whereas our set of predictions is relatively close to this bound (Table S5). Note that in this assessment, only the most granular level of the Gene Ontology is taken into account (i.e. only leaf-node terms are considered, and not more general parent terms). For those proteins for which existing annotations are available, these annotations are to a large extent a subset of what we predict (~80% of the existing annotations is also predicted by Argot2-seeded BMRF; data not shown). The higher annotation coverage in combination with the good prediction performance demonstrates the appreciable added value of the Argot2-seeded BMRF strategy for obtaining gene function annotations.

2.3. Predicted protein functions: showcases

To illustrate the potential of the functions predicted, we screened all predictions for newly annotated biological processes that are considered particularly relevant for the individual species (Table S6). Biological processes considered comprise: seed development for rice and soybean; nitrogen fixation for barrel clover; fruit development for tomato; and lignin related processes for poplar. Inspection of the selected predictions shows that the functions of proteins tend to become more specific: broadly defined functions are replaced by or augmented with more specific biological processes. For example, the rice protein *LOC_Os10g38080*, was previously annotated with anatomical structure morphogenesis, and is annotated by Argot2-seeded BMRF with seed (coat) development. *LOC_Os10g38080* is a subtilisin homologue which according to available RNAseq data is expressed in amongst other reproductive organs and seeds [28]. As additional evidence for the Argot2-seeded BMRF prediction, in Arabidopsis subtilisin and related proteases are involved in seed coat development [30]. An example for an annotation for a previously completely unannotated protein is *LOC_Os05g02520*, a cupin domain containing protein,

which was annotated by Argot2-seeded BMRF with seed maturation.

2.4. Divergence and conservation of biological processes in ortholog groups

The set of function predictions delivered above allows to compare function annotation between different plants, a task which is much less easily performed with existing annotations that are derived from various methods and that have a much lower coverage than our approach. Such comparison between orthologous genes in different plants allows to assess the limits of orthology-based function prediction, and to analyze gene function evolution.

To characterize ortholog groups with functional predictions that differ from expectations based on sequence similarity, orthologs and paralogs were identified with orthoMCL [31], resulting in 25,347 groups (Table S7). Group members for which no functions were predicted were removed. To assess the similarity of function predictions within ortholog groups, the mean functional distance within each ortholog group (dubbed 'inner group distance') was calculated (see Section 4). In case the predicted biological processes in such a group are different despite high sequence similarity, this would be indicative of evolutionary divergence by, e.g., neofunctionalization. To identify such cases, groups with at least four different organisms (6073) were ranked by their largest inner group distance and the most divergent groups ($n = 100$) were selected. In those groups, biological processes that were significantly overrepresented (more present than randomly expected) were obtained. A variety of biological processes was found (Supp. Figure S1), indicating the widespread occurrence of changes in biological processes proteins are involved in. Most prominent are processes related to cell development, regulation, and response to chemical stimulus. For the latter group, biological processes involved are shown in Fig. 4A. Among the top ranking groups (with highest 'inner group distance') involved in those processes, we chose as example a phosphatase with existing experimental annotation in Arabidopsis, PURPLE ACID PHOSPHATASE 26 (PAP26). PAP26 plays a role in the phosphate metabolism [32] and phosphate starvation [32] in Arabidopsis. The majority of the proteins with function predictions in the orthologous group (five out of seven) are indeed predicted by Argot2-seeded BMRF to be involved in phosphate metabolism or the response to phosphate starvation. However, additional function predictions differ. Populus and soybean proteins are predominantly annotated with cell death related terms; Arabidopsis with pollination and pollen germination processes; tomato with DNA repair and rice with microtubule cytoskeleton organization. This diversity in function is not reflected by orthology predictions and phylogenetic relationships of the group members (Fig. 4B and C). Independent expression data indicates that Arabidopsis PAP26 is expressed in a housekeeping-like manner, but the expression pattern varies between paralogs in other species, e.g. soybean, and to a lesser extent orthologs, e.g. between tomato and soybean (Fig. 4D). The different expression patterns give credibility to the variation in function predictions of Argot2-seeded BMRF. This indicates that PAP26, although its molecular function presumably is invariant, is involved in various biological processes in various plant species. More generally, the analysis of functional divergence presented here highlights the potential of using our set of predicted gene functions for large scale comparisons between various plant species.

3. Discussion

Finding associations between proteins and biological processes is a major challenge in non-model plants. Most experimental studies are aimed towards model organisms; hence

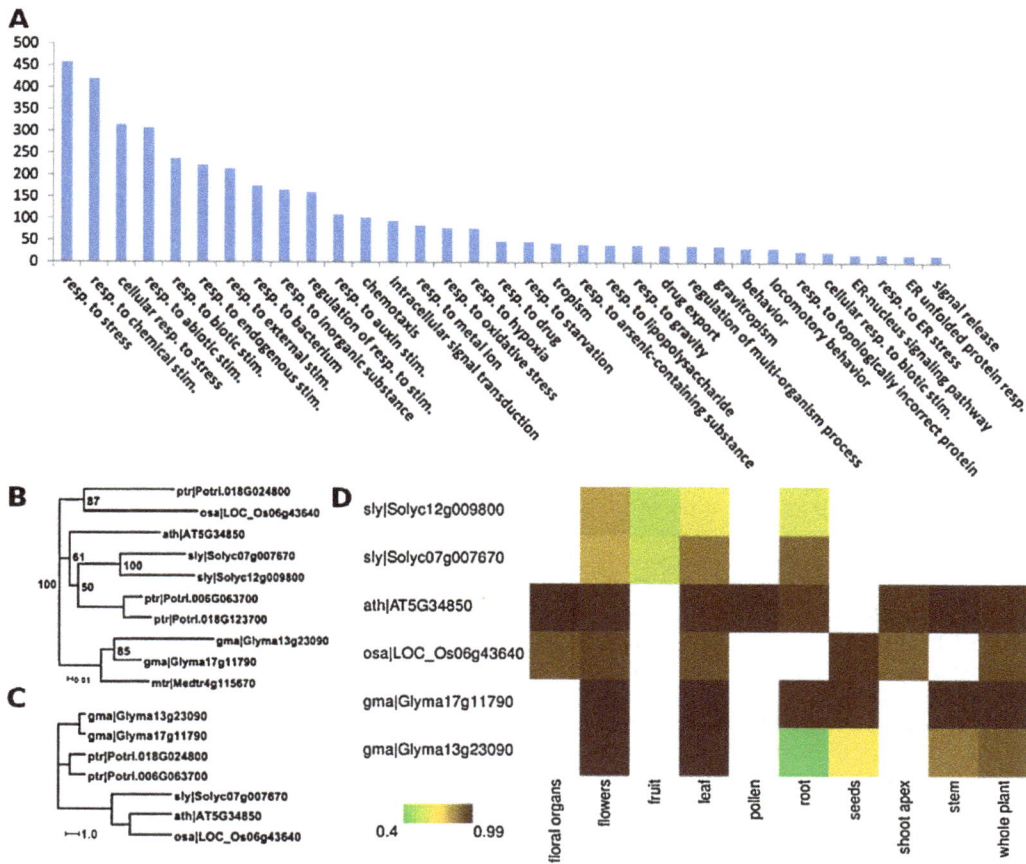

Fig. 4. Comparison between sequence divergence and functional divergence. (A) Overview of the most frequent GO terms in the top 100 most functionally divergent ortholog groups that are represented by "response to chemical stimulus" (Figure S1). (B and C) Phylogenetic relations of Arabidopsis PURPLE ACID PHOSPHATASE 26 orthologs. Trees contain Arabidopsis (ath), soybean (gma), tomato (sly), Populus (ptr) and rice (osa) PAP26 orthologs. (B) Unrooted phylogenetic tree based on sequence data. The tree was calculated with 1000 bootstraps. Confidence values are indicated at the branches in percent. (C) Distance tree based on our function predictions. Missing identifiers were not part of the co-expression network and are therefore not part of the functional distance tree. (D) Expression ranking of PURPLE ACID PHOSPHATASE 26 orthologs and paralogs in different tissue clusters. The heatmap color represents a mean percentile rank of normalized expression studies aggregated by averaging to ten tissue clusters (Table S8). Missing data is indicated in white. An overview of the aggregated studies is available in Table S8. (For interpretation of the references to color in this figure legend, the reader is referred to the web version of this article.)

experiment-based function annotation is sparse in the remainder of sequenced plant genomes. High-throughput experiments to define protein functions are overall less informative than those provided by low-throughput experiments [33]. Moreover, the experimental setup in large-scale approaches might restrict the type of function annotation that can be obtained. An example is the characterization of overexpressed rice genes in Arabidopsis [34] to infer function. Here, the problem is that the biological process of a protein is often bound to the local environment or a specific condition and a different (plant) environment might change the outcome. Another large scale analysis of gene families in Arabidopsis used prokaryotic gene information to predict function [35]. This semi-manual approach yielded good results for conserved gene families; however, gene families with low conservation were not covered.

Several computational approaches to protein function annotation exist, albeit mostly not targeted to plants, or to model plant species only [36]. An integrated platform such as Phytozome [2] provides a consistent set of Gene Ontology annotations for various plant species and hence overcomes the above-mentioned problem that annotations associated with genomes are obtained by various methods. However, Phytozome only provides sequence-based predictions. The recently published MORPH algorithm ranked genes for their membership of Arabidopsis and tomato pathways, based on a set of known genes from the target pathway, a collection of expression profiles, and interaction and metabolic networks [37].

Approaches such as PlaNet construct networks based on expression data [24] but such networks do not directly lead to gene function annotation. Similarly, a recently presented text mining approach generated networks in Arabidopsis and not gene function annotations [38]. Here we provide a structured approach to extract gene function information from networks and combine that with sequence-based information.

The combination of sequence- and network-based function prediction obtained by seeding BMRF with Argot2, offers a significant benefit over applying these methods separately. We validated the method in rice and demonstrated greatly improved performance compared to each of the methods separately and compared to Blast2GO. This performance assessment was performed using two complementary indicators, AUC and F-score, which both gave consistent results. Existing annotations provided for the plant genomes to which we applied our method have been obtained by various, mostly sequence-based approaches. A clear description of the methods and input data is often lacking, leading to the risk of error propagation and circular reasoning [3,39]. Our approach has the benefit of applying a standard method to the various genomes. Moreover, for many proteins which so far were not associated with any biological process, we now provide predictions of biological processes. Nevertheless, the combination of Argot2 and BMRF is indirectly constrained by the experimental data in databases such as UniProt [40] or PFAM [41], and by the proteins covered in available networks. It will however be straightforward to integrate newly available datasets such as additional co-expression networks or novel gene function annotations in the framework presented. Depending on availability of novel network or annotation data, we indeed plan to update our resource. An additional limitation of our current approach is that the structure of the Gene Ontology is not taken into account in the prediction process. Most existing computational methods for gene function prediction suffer from this drawback. It is feasible to make a set of GO term predictions consistent with the GO-structure [42] and we plan to apply this method to Argot2-seeded BMRF predictions in the future.

BMRF output consists of a list of probabilities for each gene to be associated with each biological process. This allows to rank proteins in order of their likelihood of association with a biological process of interest. However, it can also be important to have a finite set of predictions. To provide that, we applied a cut-off to the probabilities, based on Arabidopsis, the only species from which enough data was available. It is difficult to assess how valid the application of this cut-off in other plant species is. However, the average number of predictions per protein that we obtain in each of the species based on the cut-off that was applied is close to the observed average for Arabidopsis, giving some credibility to this cut-off. For one species, tomato, the number of predicted BP terms per protein is somewhat higher than the experimentally observed number for Arabidopsis. Hence, argot2-seeded BMRF possibly suffers from overprediction in this case. This could possibly be caused by the higher density (number of interactions compared to number of proteins) of the tomato network. However, in any case, the probabilities associated with the predictions allow narrowing down the prediction results to the most reliable ones, if so desired.

With the consistent annotation of multiple plant genomes that we performed, the relation between homology and biological process predictions can be analyzed. Ortholog groups with divergent functions indicated cases where conclusions based on sequence similarity might be inappropriate. Such inappropriate conclusions may be more common than generally acknowledged. Indeed, as recently noted in the context of comparing putative orthologs between species, relying on sequence similarity alone might identify an ortholog with the correct molecular function, but will more often than not fail to identify an ortholog that participates in the correct biological process [43]. In a comparison of gene expression patterns between different plant species, the number of times for which the homolog with the most similar pattern of expression ("expressolog") was not also the most similar at the sequence level, ranged between 15% and 50% [44]. Similarly, about half of a collection of Arabidopsis loss-of-function mutants had only low or moderate phenotypic similarity with mutants of putative orthologs in tomato, rice or maize [45]. Large scale evolutionary comparisons between plant species, for example aimed at identifying patterns in retention of duplicated genes [46,47] or functional biases in single-copy genes [7], are currently performed based on function annotations obtained using sequence similarity. Such studies will benefit from the gene annotations presented here, which overcome the limitations of purely sequence-based annotation of gene functions.

In the example of PAP26 homologs, homology captures the molecular function, but at the biological process level there is divergence. Our integrated sequence- and network-based function annotation method allows to predict such divergent biological processes. Differences in expression between the different PAP26 homologs in different species provide additional evidence for our function predictions. More generally, the results on biological process divergence are in line with the concept that evolution acts in particular by "tinkering" with genes, coopting available components of a genome for new processes.

The combination of sequence-based and network-based predictions is a huge improvement for sparsely annotated plant genomes. With the advent of RNA-seq [48] coexpression network-based protein function prediction can become a preferred method. Combined with additional analysis, such as genome-wide association studies (GWAS), potential candidate genes for traits-of-interest could be identified more reliably. Such candidate genes will be of great help in applications related to plant breeding. The ability to associate unannotated proteins to particular biological processes will spark experimental work and be essential for the advancement of understanding of gene function in plant genome evolution.

4. Materials and methods

4.1. Function prediction methods and their integration

BMRF uses network data as input. Each protein is represented as a node in the network, and connections in the network indicate functional relationships between proteins. A statistical model (Markov Random Field) describes how involvement of a protein in a particular BP influences the probability that its neighbors in the network are also involved in that BP. The parameters in the statistical model describe for each BP how strongly neighbors influence each other. Parameter values are trained using a Bayesian approach by performing simultaneous estimation of the model parameters and prediction of protein functions. This strategy needs a set of known protein functions as initial labelling of the network. Argot2 is a purely sequence-based prediction method, using searches of the UniProt and Pfam databases as input. To combine these two methods, two strategies were applied. In the first integration method, for each biological process, ranks for the different proteins were obtained from both BMRF and Argot2, by ordering the proteins based on their score for that process. These ranks were added to obtain a final ranking, which was used as the prediction score for that biological process. In a second integration strategy, initial predictions were generated with Argot2. These were supplied to BMRF as training data, meaning that the initial labelling of the nodes in the network was based on the Argot2 predictions.

4.2. Sequence and domain data

Sequence data for Arabidopsis, rice, soybean and *M. truncatula* were obtained from the Phytozome database v8.0 [2]. Poplar sequence data was downloaded from the JGI (ftp://ftp.jgi-psf.org/pub/JGI_data/Poplar/annotation/v1.1), annotation version 1.1. Tomato sequence v2.4 and annotation v2.3 data [27] were retrieved from the SGN network (http://www.solgenomics.net). Arabidopsis Interpro domains were retrieved from TAIR10 [49]. Domains of transcript isoforms were merged into one set per gene.

4.3. Function annotation data

Annotations from the Gene Ontology project, version 1.1418 [9], and from Gramene [50], were used as input for training and cross-validation. Annotations from Oryzabase version 4 [23] were used as an independent validation set. Only genes for which no annotation was available in the data from the Gene Ontology project were used for validation. In all cases, only Biological Process (BP) terms with evidence codes IDA (inferred from direct assay), IGI (genetic interaction) and IMP (mutant phenotype) were used.

4.4. Network data

Co-expression networks based on microarray data for Arabidopsis, rice, *G. max*, *M. truncatula* and poplar were obtained from PlaNet [24]. For tomato, a recently published microarray-based co-expression network [51] was used. The probe ids of the tomato co-expression network were obtained from Affymetrix (http://www.affymetrix.com) and mapped with BLAST v2.2.26 [11] to the tomato protein sequences. Further network data for Arabidopsis and rice was obtained from FunctionalNet (http://www.functionalnet.org/) [19] and STRING [18]. Arabidopsis yeast-two-hybrid data were acquired from literature [52]. The rice-Arabidopsis interspecies network was generated by using BLAST (cut-off on E-value of 1e−4). BMRF requires all proteins to be part of the input network. Thus, proteins not contained in the input network were removed. In all cases, the longest isoform of alternatively spliced variants was used.

4.5. Validation setup

Performance assessment was performed with rice. HMMER v3 (http://hmmer.org/) search against PFAM [41] and BLAST [11] alignment against UniProt [53] were used to generate the input for Argot2 [16]. In the context of the validation setup, all rice proteins were removed from the UniProt database to avoid Argot2 using information from those proteins.

For comparison, sequence similarity-based annotation was carried out with Blast2GO [21]. Rice protein sequences were queried against the non-redundant part of GenBank (NR) [54], using an E-value cut-off of 1e−4. In the context of the validation setup, hits to monocot proteins in NR were removed from the BLAST results before supplying them to Blast2GO.

Prediction runs of different method and network combinations were assessed with 100 cross-validation runs. In each run, randomly, a subset ($n = 200$) of proteins was chosen and the annotation was removed (masked). For every run, predicted functions were compared with the masked ones. Only biological process terms with at least three masked proteins were used in the performance assessment in order to allow for sufficient statistics. In the performance assessment, negative cases consisted of gene-BP associations which were not annotated as such in the experimental data.

Performance was assessed by the area under the receiver operating characteristic curve (AUC) and the F-score. The AUC is the area under the curve of 1-specificity vs. sensitivity, and is equal to the probability that a classifier will rank a randomly chosen positive instance higher than a randomly chosen negative one [20]. Specificity is the fraction of proteins experimentally known not to perform a given function which are indeed not predicted to do so, whereas sensitivity (or recall) is the fraction of proteins experimentally known to perform a given function which are indeed predicted to do so. F-score is based on the precision–recall (precision vs. sensitivity) curve. Precision is the fraction of proteins predicted to perform a given function which are indeed experimentally known to do so. The F-score is equal to the harmonic mean of precision and recall, and the maximum value of the F-score (F_{max}-score) was used for each biological process.

To obtain a finite set of predictions, functions of a protein were assigned by using an F-score-based cut-off. The F-score was calculated per GO term and its maximum (F_{max}-score), calculated with Arabidopsis data as previously described [17], was used to set a cut-off on the posterior probability. The threshold obtained with Arabidopsis data was used in the other species, because in those species, too few annotations are available to obtain a species-specific threshold. All performance measures were calculated with the R-package ROCR [55] and custom R-scripts.

4.6. Application setup

Function annotations predicted for barrel clover, poplar, rice, soybean and tomato were compared with existing predictions in terms of coverage of proteins and number of predicted functions per protein. Barrel clover, poplar and rice biological process predictions were obtained from the official genome annotations version Mt3.5v5 [25], v1.1 [26] and v7.0 [28], respectively. Soybean annotation was obtained from Phytozome [2]. Tomato function annotation data was extracted from the ITAG annotation v2.3 [27].

To determine the total number of proteins and total number of GO terms for which annotations were obtained, the annotation of each protein was expanded by including the parent GO terms of all assigned GO terms. For the calculation of the number of annotations per protein, only the leaf-terms of the Gene Ontology were included.

4.7. Evolutionary and functional distance calculation

Groups of orthologs were predicted with OrthoMCL [31]. To calculate functional divergence, BMRF posterior probabilities for each protein were interpreted as vector. The Euclidian distance for each combination of proteins within a group of orthologs was calculated. The mean of distances within a group (inner group distance) was used to rank groups of orthologs. For the PAP26 example, only groups with existing experimental annotation in Arabidopsis were taken in to account. The PAP26 tree was estimated with RaxML version 7.2.8-ALPHA [56] using the PROTGAMMA-JJTF substitution model and 1000 bootstraps. Expression data for PAP26 was obtained from the AtGenExpress developmental set [57]; publicly available RNA-seq datasets from tomato (*S. lycopersicum* cv. Heinz 1706; data SRA049915) were retrieved from the SRA database (http://www.ncbi.nlm.nih.gov/sra). Reads were mapped with GSNAP [58] against the tomato reference genome (v. 2.40, Sato et al. [27]) and the expression was determined with cufflinks [59] with default parameters. Soybean expression data was obtained from SoyBase [60]. Rice expression data was obtained from the Rice Genome Annotation Project (http://rice.plantbiology.msu.edu/). All expression experiment data were z-score normalized and percentile ranked to facilitate comparison. Replicates were merged by averaging over the expression for each gene.

Acknowledgements

We thank Dr. Yiannis Kourmpetis for helpful discussions. This work was supported by the FP7 "Infrastructures" project Trans-PLANT Award 283496 and by the BioRange program of the Netherlands Bioinformatics Centre (NBIC), which is supported by a BSIK grant through the Netherlands Genomics Initiative (NGI).

References

[1] M.C. Schatz, J. Witkowski, W.R. McCombie, Current challenges in de novo plant genome sequencing and assembly, Genome Biol. 13 (2012) 243, http://dx.doi.org/10.1186/gb4015.

[2] D.M. Goodstein, S. Shu, R. Howson, R. Neupane, R.D. Hayes, et al., Phytozome: a comparative platform for green plant genomics, Nucleic Acids Res. 40 (2012) D1178–D1186, http://dx.doi.org/10.1093/nar/gkr944.

[3] L. Du Plessis, N. Skunca, C. Dessimoz, The what, where, how and why of gene ontology – a primer for bioinformaticians, Brief Bioinform. 12 (2011) 723–735, http://dx.doi.org/10.1093/bib/bbr002.

[4] S. De Bodt, J. Hollunder, H. Nelissen, N. Meulemeester, D. Inzé, CORNET 2.0: integrating plant coexpression, protein–protein interactions, regulatory interactions, gene associations and functional annotations, New Phytol. 195 (2012) 707–720, http://dx.doi.org/10.1111/j.1469-8137.2012.04184.x.

[5] M. Van Bel, S. Proost, E. Wischnitzki, S. Movahedi, C. Scheerlinck, et al., Dissecting plant genomes with the PLAZA comparative genomics platform, Plant Physiol. 158 (2012) 590–600, http://dx.doi.org/10.1104/pp.111.189514.

[6] R. Monclus, J.-C. Leplé, C. Bastien, P.-F. Bert, M. Villar, et al., Integrating genome annotation and QTL position to identify candidate genes for productivity, architecture and water-use efficiency in *Populus* spp., BMC Plant Biol. 12 (2012) 173, http://dx.doi.org/10.1186/1471-2229-12-173.

[7] R. De Smet, K.L. Adams, K. Vandepoele, M.C.E. Van Montagu, S. Maere, et al., Convergent gene loss following gene and genome duplications creates single-copy families in flowering plants, Proc. Natl. Acad. Sci. U. S. A. 110 (2013) 2898–2903,

[8] S.Y. Rhee, M. Mutwil, Towards revealing the functions of all genes in plants, Trends Plant Sci. (2013), http://dx.doi.org/10.1016/j.tplants.2013.10.006.

[9] Gene Ontology Consortium, Gene Ontology: tool for the unification of biology, Nat. Genet. 25 (2000) 25–29, http://dx.doi.org/10.1038/75556.

[10] P. Radivojac, W.T. Clark, T.R. Oron, A.M. Schnoes, T. Wittkop, et al., A large-scale evaluation of computational protein function prediction, Nat. Methods 10 (2013) 221–227, http://dx.doi.org/10.1038/nmeth.2340.

[11] S.F. Altschul, W. Gish, W. Miller, E.W. Myers, D.J. Lipman, Basic local alignment search tool, J. Mol. Biol. 215 (1990) 403–410, http://dx.doi.org/10.1016/S0022-2836(05)80360-2.

[12] D.M.A. Martin, M. Berriman, G.J. Barton, GOtcha: a new method for prediction of protein function assessed by the annotation of seven genomes, BMC Bioinform. 5 (2004) 178, http://dx.doi.org/10.1186/1471-2105-5-178.

[13] W.T. Clark, P. Radivojac, Analysis of protein function and its prediction from amino acid sequence, Proteins 79 (2011) 2086–2096, http://dx.doi.org/10.1002/prot.23029.

[14] A. Vazquez, A. Flammini, A. Maritan, A. Vespignani, Global protein function prediction from protein–protein interaction networks, Nat. Biotechnol. 21 (2003) 697–700, http://dx.doi.org/10.1038/nbt825.

[15] Y.A.I. Kourmpetis, A.D.J. van Dijk, M.C.A.M. Bink, R.C.H.J. van Ham, C.J.F. ter Braak, Bayesian Markov Random Field analysis for protein function prediction based on network data, PLoS ONE 5 (2010) e9293, http://dx.doi.org/10.1371/journal.pone.0009293.

[16] M. Falda, S. Toppo, A. Pescarolo, E. Lavezzo, B. Di Camillo, et al., Argot2: a large scale function prediction tool relying on semantic similarity of weighted Gene Ontology terms, BMC Bioinform. 13 (Suppl. 4) (2012) S14, http://dx.doi.org/10.1186/1471-2105-13-S4-S14.

[17] Y.A.I. Kourmpetis, A.D.J. van Dijk, R.C.H.J. van Ham, C.J.F. ter Braak, Genome-wide computational function prediction of Arabidopsis proteins by integration of multiple data sources, Plant Physiol. 155 (2011) 271–281, http://dx.doi.org/10.1104/pp.110.162164.

[18] D. Szklarczyk, A. Franceschini, M. Kuhn, M. Simonovic, A. Roth, et al., The STRING database in 2011: functional interaction networks of proteins, globally integrated and scored, Nucleic Acids Res. 39 (2011) D561–D568, http://dx.doi.org/10.1093/nar/gkq973.

[19] I. Lee, B. Ambaru, P. Thakkar, E.M. Marcotte, S.Y. Rhee, Rational association of genes with traits using a genome-scale gene network for *Arabidopsis thaliana*, Nat. Biotechnol. 28 (2010) 149–156, http://dx.doi.org/10.1038/nbt.1603.

[20] J.A. Hanley, B.J. McNeil, The meaning and use of the area under a receiver operating characteristic (ROC) curve, Radiology 143 (1982) 29–36.

[21] A. Conesa, S. Götz, J.M. García-Gómez, J. Terol, M. Talón, et al., Blast2GO: a universal tool for annotation, visualization and analysis in functional genomics research, Bioinformatics 21 (2005) 3674–3676, http://dx.doi.org/10.1093/bioinformatics/bti610.

[22] J. Davis, M. Goadrich, The relationship between precision–recall and ROC curves, in: Proceedings of the 23rd International Conference on Machine Learning – ICML'06, ACM Press, New York, NY, USA, 2006, pp. 233–240, http://dx.doi.org/10.1145/1143844.1143874.

[23] N. Kurata, Y. Yamazaki, Oryzabase. An integrated biological and genome information database for rice, Plant Physiol. 140 (2006) 12–17, http://dx.doi.org/10.1104/pp.105.063008.

[24] M. Mutwil, S. Klie, T. Tohge, F.M. Giorgi, O. Wilkins, et al., PlaNet: combined sequence and expression comparisons across plant networks derived from seven species, Plant Cell 23 (2011) 895–910, http://dx.doi.org/10.1105/tpc.111.083667.

[25] N.D. Young, F. Debellé, G.E.D. Oldroyd, R. Geurts, S.B. Cannon, et al., The Medicago genome provides insight into the evolution of rhizobial symbioses, Nature 480 (2011) 520–524, http://dx.doi.org/10.1038/nature10625.

[26] G.A. Tuskan, S. Difazio, S. Jansson, J. Bohlmann, I. Grigoriev, et al., The genome of black cottonwood, *Populus trichocarpa* (Torr. & Gray), Science 313 (2006) 1596–1604, http://dx.doi.org/10.1126/science.1128691.

[27] S. Sato, S. Tabata, H. Hirakawa, E. Asamizu, K. Shirasawa, et al., The tomato genome sequence provides insights into fleshy fruit evolution, Nature 485 (2012) 635–641, http://dx.doi.org/10.1038/nature11119.

[28] S. Ouyang, W. Zhu, J. Hamilton, H. Lin, M. Campbell, et al., The TIGR rice genome annotation resource: improvements and new features, Nucleic Acids Res. 35 (2007) D883–D887, http://dx.doi.org/10.1093/nar/gkl976.

[29] J. Schmutz, S.B. Cannon, J. Schlueter, J. Ma, T. Mitros, et al., Genome sequence of the palaeopolyploid soybean, Nature 463 (2010) 178–183, http://dx.doi.org/10.1038/nature08670.

[30] C. Rautengarten, B. Usadel, L. Neumetzler, J. Hartmann, D. Büssis, et al., A subtilisin-like serine protease essential for mucilage release from Arabidopsis seed coats, Plant J. 54 (2008) 466–480, http://dx.doi.org/10.1111/j.1365-313X.2008.03437.x.

[31] L. Li, C.J. Stoeckert, D.S. Roos, OrthoMCL: identification of ortholog groups for eukaryotic genomes, Genome Res. 13 (2003) 2178–2189, http://dx.doi.org/10.1101/gr.1224503.

[32] B.A. Hurley, H.T. Tran, N.J. Marty, J. Park, W.A. Snedden, et al., The dual-targeted purple acid phosphatase isozyme AtPAP26 is essential for efficient acclimation of Arabidopsis to nutritional phosphate deprivation, Plant Physiol. 153 (2010) 1112–1122, http://dx.doi.org/10.1104/pp.110.153270.

[33] A.M. Schnoes, D.C. Ream, A.W. Thorman, P.C. Babbitt, I. Friedberg, Biases in the experimental annotations of protein function and their effect on our understanding of protein function space, PLoS Comput. Biol. 9 (2013) e1003063, http://dx.doi.org/10.1371/journal.pcbi.1003063.

[34] T. Sakurai, Y. Kondou, K. Akiyama, A. Kurotani, M. Higuchi, et al., RiceFOX: a database of Arabidopsis mutant lines overexpressing rice full-length cDNA that contains a wide range of trait information to facilitate analysis of gene function, Plant Cell Physiol. 52 (2011) 265–273, http://dx.doi.org/10.1093/pcp/pcq190.

[35] S. Gerdes, B. El Yacoubi, M. Bailly, I.K. Blaby, C.E. Blaby-Haas, et al., Synergistic use of plant–prokaryote comparative genomics for functional annotations, BMC Genomics 12 (Suppl. 1) (2011) S2, http://dx.doi.org/10.1186/1471-2164-12-S1-S2.

[36] I. Lee, Y.-S. Seo, D. Coltrane, S. Hwang, T. Oh, et al., Genetic dissection of the biotic stress response using a genome-scale gene network for rice, Proc. Natl. Acad. Sci. U. S. A. 108 (2011) 18548–18553, http://dx.doi.org/10.1073/pnas.1110384108.

[37] O. Tzfadia, D. Amar, L.M.T. Bradbury, E.T. Wurtzel, R. Shamir, The MORPH algorithm: ranking candidate genes for membership in Arabidopsis and tomato pathways, Plant Cell 24 (2012) 4389–4406, http://dx.doi.org/10.1105/tpc.112.104513.

[38] G. Blanc, K.H. Wolfe, Functional divergence of duplicated genes formed by polyploidy during Arabidopsis evolution, Plant Cell 16 (2004) 1679–1691, http://dx.doi.org/10.1105/tpc.021410.

[39] B.E. Engelhardt, M.I. Jordan, J.R. Srouji, S.E. Brenner, Genome-scale phylogenetic function annotation of large and diverse protein families, Genome Res. 21 (2011) 1969–1980, http://dx.doi.org/10.1101/gr.104687.109.

[40] E.C. Dimmer, R.P. Huntley, Y. Alam-Faruque, T. Sawford, C. O'Donovan, et al., The UniProt-GO annotation database in 2011, Nucleic Acids Res. 40 (2012) D565–D570, http://dx.doi.org/10.1093/nar/gkr1048.

[41] R.D. Finn, J. Mistry, J. Tate, P. Coggill, A. Heger, et al., The Pfam protein families database, Nucleic Acids Res. 38 (2010) D211–D222, http://dx.doi.org/10.1093/nar/gkp985.

[42] Y.A. Kourmpetis, A.D. van Dijk, C.J. Ter Braak, Gene Ontology consistent protein function prediction: the FALCON algorithm applied to six eukaryotic genomes, Algorithms Mol. Biol. 8 (2013) 10, http://dx.doi.org/10.1186/1748-7188-8-10.

[43] S. Netotea, D. Sundell, N.R. Street, T.R. Hvidsten, ComPlEx: conservation and divergence of co-expression networks in *A. thaliana*, *Populus* and *O. sativa*, BMC Genomics 15 (2014) 106, http://dx.doi.org/10.1186/1471-2164-15-106.

[44] R.V. Patel, H.K. Nahal, R. Breit, N.J. Provart, BAR expressolog identification: expression profile similarity ranking of homologous genes in plant species, Plant J. 71 (2012) 1038–1050, http://dx.doi.org/10.1111/j.1365-313X.2012.05055.x.

[45] J. Lloyd, D. Meinke, A comprehensive dataset of genes with a loss-of-function mutant phenotype in Arabidopsis, Plant Physiol. 158 (2012) 1115–1129, http://dx.doi.org/10.1104/pp.111.192393.

[46] W. Jiang, Y. Liu, E. Xia, L. Gao, Prevalent role of gene features in determining evolutionary fates of whole-genome duplication duplicated genes in flowering plants, Plant Physiol. 161 (2013) 1844–1861, http://dx.doi.org/10.1104/pp.112.200147.

[47] H. Guo, T.-H. Lee, X. Wang, A.H. Paterson, Function relaxation followed by diversifying selection after whole genome duplication in flowering plants, Plant Physiol. (2013), http://dx.doi.org/10.1104/pp.112.213447.

[48] S. Marguerat, J. Bähler, RNA-seq: from technology to biology, Cell. Mol. Life Sci. 67 (2010) 569–579, http://dx.doi.org/10.1007/s00018-009-0180-6.

[49] P. Lamesch, T.Z. Berardini, D. Li, D. Swarbreck, C. Wilks, et al., The Arabidopsis Information Resource (TAIR): improved gene annotation and new tools, Nucleic Acids Res. 40 (2012) D1202–D1210, http://dx.doi.org/10.1093/nar/gkr1090.

[50] K. Youens-Clark, E. Buckler, T. Casstevens, C. Chen, G. Declerck, et al., Gramene database in 2010: updates and extensions, Nucleic Acids Res. 39 (2011) D1085–D1094, http://dx.doi.org/10.1093/nar/gkq1148.

[51] A. Fukushima, T. Nishizawa, M. Hayakumo, S. Hikosaka, K. Saito, et al., Exploring tomato gene functions based on coexpression modules using graph clustering and differential coexpression approaches, Plant Physiol. 158 (2012) 1487–1502, http://dx.doi.org/10.1104/pp.111.188367.

[52] Arabidopsis Interactome Mapping Consortium, Evidence for network evolution in an Arabidopsis interactome map, Science 333 (2011) 601–607, http://dx.doi.org/10.1126/science.1203877.

[53] The UniProt Consortium, Reorganizing the protein space at the Universal Protein Resource (UniProt), Nucleic Acids Res. 40 (2012) D71–D75, http://dx.doi.org/10.1093/nar/gkr981.

[54] D.A. Benson, M. Cavanaugh, K. Clark, I. Karsch-Mizrachi, D.J. Lipman, et al., GenBank, Nucleic Acids Res. 41 (2013) D36–D42, http://dx.doi.org/10.1093/nar/gks1195.

[55] T. Sing, O. Sander, N. Beerenwinkel, T. Lengauer, ROCR: visualizing classifier performance in R, Bioinformatics 21 (2005) 3940–3941, http://dx.doi.org/10.1093/bioinformatics/bti623.

[56] A. Stamatakis, RAxML-VI-HPC: maximum likelihood-based phylogenetic analyses with thousands of taxa and mixed models, Bioinformatics 22 (2006) 2688–2690, http://dx.doi.org/10.1093/bioinformatics/btl446.

[57] M. Schmid, T.S. Davison, S.R. Henz, U.J. Pape, M. Demar, et al., A gene expression map of *Arabidopsis thaliana* development, Nat. Genet. 37 (2005) 501–506, http://dx.doi.org/10.1038/ng1543.

[58] T.D. Wu, S. Nacu, Fast and SNP-tolerant detection of complex variants and splicing in short reads, Bioinformatics 26 (2010) 873–881, http://dx.doi.org/10.1093/bioinformatics/btq057.

[59] C. Trapnell, B.A. Williams, G. Pertea, A. Mortazavi, G. Kwan, et al., Transcript assembly and quantification by RNA-seq reveals unannotated transcripts and isoform switching during cell differentiation, Nat. Biotechnol. 28 (2010) 511–515, http://dx.doi.org/10.1038/nbt.1621.

20

Expression of a transferred nuclear gene in a mitochondrial genome

Yichun Qiu, Samuel J. Filipenko, Aude Darracq, Keith L. Adams*

Department of Botany, University of British Columbia, 6270 University Blvd, Vancouver, BC, Canada V6T 1Z4

ARTICLE INFO

Keywords:
Mitochondria
Intracellular gene transfer
Genome evolution
Brassicaceae

ABSTRACT

Transfer of mitochondrial genes to the nucleus, and subsequent gain of regulatory elements for expression, is an ongoing evolutionary process in plants. Many examples have been characterized, which in some cases have revealed sources of mitochondrial targeting sequences and cis-regulatory elements. In contrast, there have been no reports of a nuclear gene that has undergone intracellular transfer to the mitochondrial genome and become expressed. Here we show that the *orf164* gene in the mitochondrial genome of several Brassicaceae species, including *Arabidopsis*, is derived from the nuclear *ARF17* gene that codes for an auxin responsive protein and is present across flowering plants. *Orf164* corresponds to a portion of *ARF17*, and the nucleotide and amino acid sequences are 79% and 81% identical, respectively. *Orf164* is transcribed in several organ types of *Arabidopsis thaliana*, as detected by RT-PCR. In addition, *orf164* is transcribed in five other Brassicaceae within the tribes Camelineae, Erysimeae and Cardamineae, but the gene is not present in *Brassica* or *Raphanus*. This study shows that nuclear genes can be transferred to the mitochondrial genome and become expressed, providing a new perspective on the movement of genes between the genomes of subcellular compartments.

1. Introduction

Since the origins of mitochondria and plastids by endosymbiosis, three genomes have been coexisting in plant cells. There has been a tendency for DNA from the organelle genomes to be transferred to the nuclear genome, creating many nuclear mitochondrial (numt) sequences and nuclear plastid (nupt) sequences. Numerous pseudogenes of mitochondrial or chloroplast origin are present in nuclear genomes of a wide variety of eukaryotes (reviewed in Refs. [1,2]). In some cases, large regions of mitochondrial and chloroplast DNA have been transferred to the nuclear genome (e.g., [3–5]). Some mitochondrial and plastid genes were transferred to nuclear genome and then became expressed by acquiring existing nuclear cis-regulatory elements, as well as mitochondrial or chloroplast targeting sequences, then often replacing the functions of their counterparts in the organellar genomes (reviewed in Refs. [6–8]).

Angiosperm mitochondrial genomes contain DNA derived from the nuclear genome, although amounts vary among species. A large amount of the nuclear-derived DNA is from transposable elements, although sequences derived from exons of nuclear genes also are present in some mitochondrial genomes [9–13,37]. It has been inferred that the sequences derived from nuclear genes in mitochondrial genomes are pseudogenes. No nuclear-derived sequences have yet been reported as expressed. Here we show a case of a mitochondrial gene transferred from the nuclear genome that has become expressed.

2. Methods and materials

Sequences of *orf164* and *ARF17* from *Arabidopsis thaliana* were obtained from TAIR (v.10). Sequences of *orf164* and *ARF17* from *Arabidopsis lyrata* were obtained from the PLAZA v3.0 Dicots database (http://bioinformatics.psb.ugent.be/plaza/; [14]). BLAST searches of GenBank were used to search for sequences homologous to *orf164* and *ARF17* in other species. The nucleotide and amino acid alignments were generated by MUSCLE and followed by manual adjustments [15].

To analyze sequence rate evolution of *orf164*, sequences of *ARF17* were obtained from several eurosid species including *Carica papaya*, *Citrus sinensis*, *Eucalyptus grandis*, *Populus trichocarpa* and *Prunus persica* from PLAZA v3.0 Dicots [14], and *Tarenaya hassleriana* from GenBank's wgs database (gb|AOUI01012032.1), and aligned with *orf164* and *ARF17* using MUSCLE with the default settings [15]. The dN/dS ratio along each branch was determined using a phylogeny-based free-ratio test using Codeml in PAML [16].

Total RNA was extracted from multiple organ types of *A. thaliana* (ecotype col-0) and from seedlings of *A. arenosa* using the Ambion RNAqueous Kit following the manufacturer's protocol.

* Corresponding author.
  E-mail address: keitha@mail.ubc.ca (K.L. Adams).

Leaves from *Capsella bursa-pastoris*, *Turritis glabra*, *Erysimum pulchellum*, *Cardamine flexuosa* and *Armoracia rusticana* were used for RNA extraction as above. Nucleic acid concentration and purity were determined using a spectrophotometer and quality was visualized through gel electrophoresis on 2% agarose gel. RNA was treated with DNase-I (New England Biolabs) as outlined by the manufacturer's instructions. Reverse transcription was carried out using M-MLV reverse transcriptase (Invitrogen) following the manufacturer's instructions along with random hexamer primers (IDT). Then PCR reactions were performed with cDNAs as templates. Two pairs of *orf164*-specific primers were: forward-1, 5'-ATTGACGGCTGAAGCTGTCTCTGA-3'; reverse-1, 5'-ACGCCATGGACCAGTTTCCTGATA-3'; forward-2, 5'-TGTAGTTATTATCAGAGCAATGGAGGCG-3'; reverse-2, 5'-ATAGTGAAGGGGATCTTATACCTGAAGC-3'. Primers for other genes included: *orfX* forward (5'-TGGAGAACAAAGGACGAAATACA-3') and reverse (5'-TATCCGGAGGTGTGGAAAGA-3'); *ccb203* forward (5'-GACCACTACTTCGCCTCTTTG-3') and reverse (5'-CTATGAACGGGAGCTAGCAATC-3'); *matR* forward (5'-TTAAGGACAGGTCGTCGTATTG-3') and reverse (5'-GGTCTCTCATGGCCCAATTAT-3'); *cox2* forward (5'-CGATGAGCAGTCACTCACTTT-3') and reverse (5'-ATTGGATACCCGAGAACCATAATC-3'). The PCR cycling program for *orf164* amplification was 94° for 3 min; 20–35 cycles of 94° for 30 s, 55° for 30 s, 72° for 30 s; and 72° for 7 min. PCR cycling conditions for the other genes were the same except that 52° was used as the annealing temperature. PCR products were visualized on 1.2% agarose gels, the bands were cut out of the gels, DNA was eluted and then sequenced to confirm that the amplified sequences were the correct targets.

To identify other mitochondrial open reading frames of nuclear origin, all sequences of nuclear genes in *A. thaliana* were obtained from TAIR (v.10) and aligned against the *A. thaliana* mitochondrial genome (GI:26556996) using YASS software [17] with default parameters. We identified genes having an *e*-value <1.0E−10 and not located in the chr2:3247243-3509307 region (corresponding to the mitochondrial genome insertion into chromosome 2). The resulting list was filtered to remove transposable element-related sequences, mitochondrial sequences transferred to the nucleus, short open reading frames (less than 300 bp), nuclear intron-derived sequences, and mitochondrial-nuclear sequence pairs with less than 60% identity (Table S1).

Supplementary Table S1 related to this article can be found, in the online version, at doi:10.1016/j.cpb.2014.08.002.

3. Results and discussion

3.1. Orf164 in the mitochondrial genome of A. thaliana

Orf164 is a predicted gene in the *A. thaliana* mitochondrial genome [36], located between the tRNA gene *trnQ* (tRNA-Gln) and a pseudo-tRNA gene *ψtrnW* for tRNA-Trp (Fig. 1). *Orf164*, which has a locus number ATMG00940, contains an intact open reading frame of 495 nucleotides corresponding to 164 amino acids, according to TAIR (v.10) database (http://www.arabidopsis.org/). However, when we analyzed the genomic DNA and cDNA of *orf164* by Sanger sequencing following PCR, we detected a sequencing error close to the 3' end of the predicted coding sequence, where an additional A should be present after the 446th nucleotide A, causing subsequent frame shift, and introducing a new stop codon that ends the coding region earlier. We also checked the recently sequenced and assembled complete mitochondrial genomes from three different *A. thaliana* ecotypes (C24, Ler and Col-0) from Davila et al. [18] and we found the same additional A. The corrected *orf164* open reading frame should be 462 nucleotides and 153 amino acids.

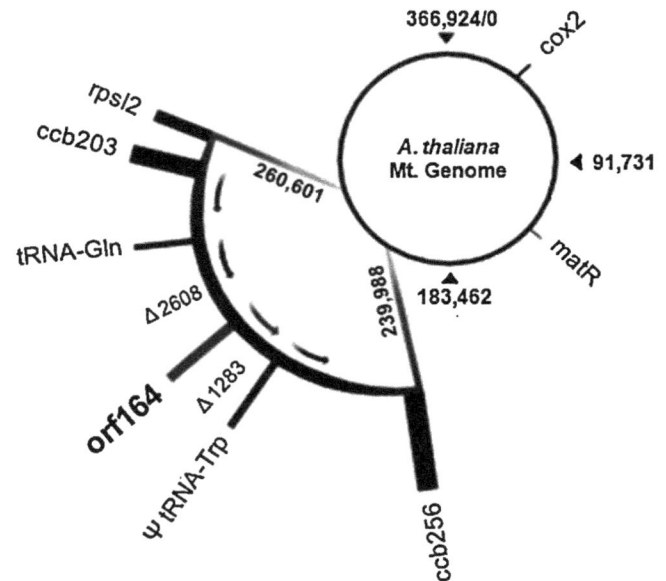

Fig. 1. Diagram of the *Arabidopsis thaliana* mitochondrial genome, with the region around *orf164* shown in detail. Arrows indicate the direction of transcription of the genes. Triangles followed by numbers indicate the sizes of the intergenic regions upstream and downstream of *orf164*.

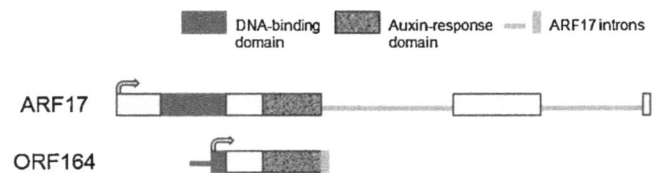

Fig. 2. Structures of *orf164* and *ARF17*. Arrows indicate the transcription start sites. Boxes indicate exons and bars indicate introns.

3.2. Orf164 is similar to nuclear ARF17 and derived from nuclear to mitochondrial intracellular gene transfer

Orf164 has high sequence similarity to a nuclear gene, *ARF17* (*AUXIN RESPONSE FACTOR 17*, AT1G77850). Comparing the sequences, *orf164* and *ARF17* share 79% nucleotide sequence identity and 81% amino acid identity. *ARF17* has two exons, and the first exon contains a DNA-binding domain and a domain regulating auxin-response gene expression (Fig. 2). *Orf164* starts at the position corresponding to the 206th codon within exon 1 of *ARF17*, using an ATG start codon that corresponds to an internal methionine codon in *ARF17*. At the 3' end of the *orf164* coding region, there are eight out of nine consecutive nucleotides that are identical to the intron at the exon–intron junction within *ARF17* (Fig. 3). The nucleotide in this region that is not identical to *ARF17* was a mutation that created the *orf164* stop codon. Eighty-four bp of the 5'UTR of *orf164* is derived from *ARF17* (Fig. 3). A mitochondrial sequence with similarity to *ARF17* was noticed by Hagen and Guilfoyle [19] and Liscum and Reed [20] in articles on *ARF* genes, but neither report identified the mitochondrial sequence as being *orf164* nor did any further characterization.

Using BLAST searches, we found many *ARF17* orthologous genes in a variety of angiosperm species. However, *orf164* has no homologous sequence in any sequenced mitochondrial genomes other than in *Arabidopsis*. We found a sequence from *A. lyrata*, AL3G32400, which is almost identical to *orf164* but annotated as a nuclear gene. However, a block of ten thousand base pairs surrounding AL3G32400 is about 99% identical to the *A. thaliana* mitochondrial genome, indicating that the sequence in *A. lyrata* is actually

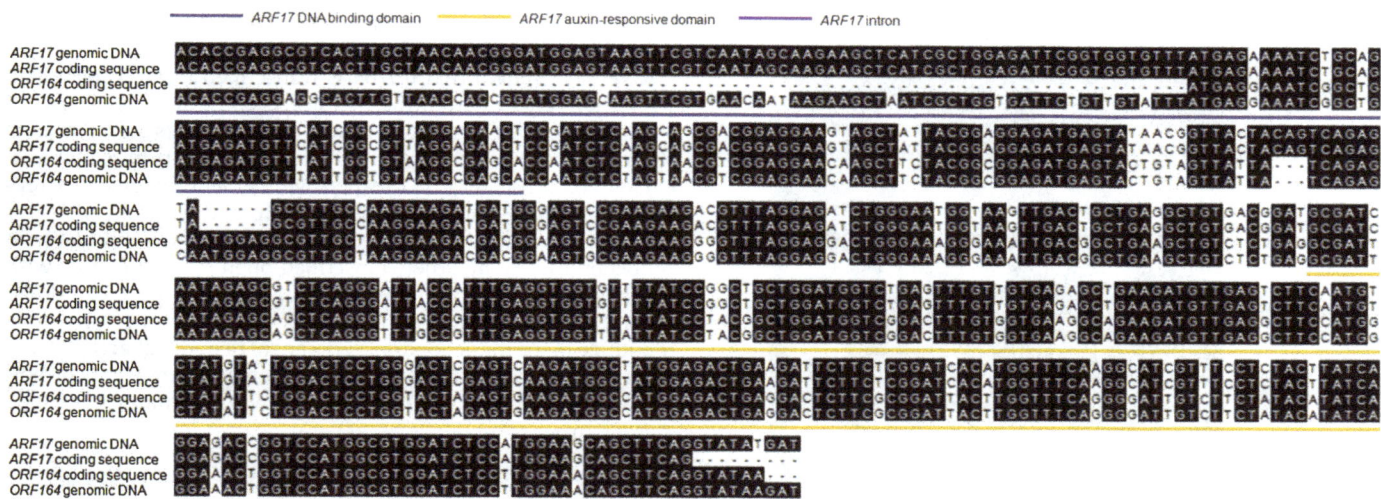

Fig. 3. Alignment of *ARF17* and *orf164* in the region of *orf164* corresponding to *ARF17*. Functional domains and the intron sequence are indicated by lines below the corresponding alignment region.

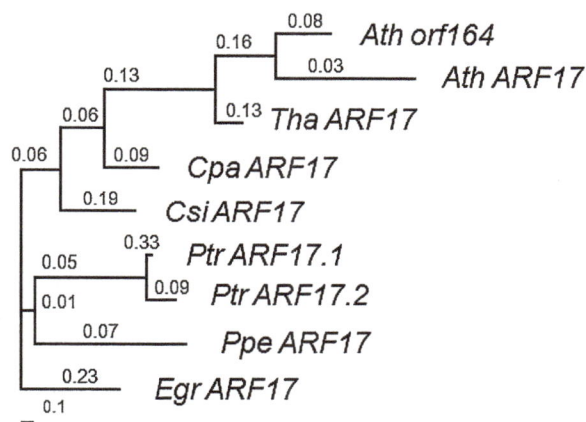

Fig. 4. Sequence evolution analysis of *orf164* and *ARF17* sequences. Numbers above the branches indicate dN/dS values and the scale bar indicates substitutions per codon. Taxa abbreviations include: Ath – *Arabidopsis thaliana*, Tha – *Tarenaya hassleriana*, Cpa – *Carica papaya*, Csi – *Citrus sinensis*, Egr – *Eucalyptus grandis*, Ptr – *Populus trichocarpa* and Ppe – *Prunus persica*.

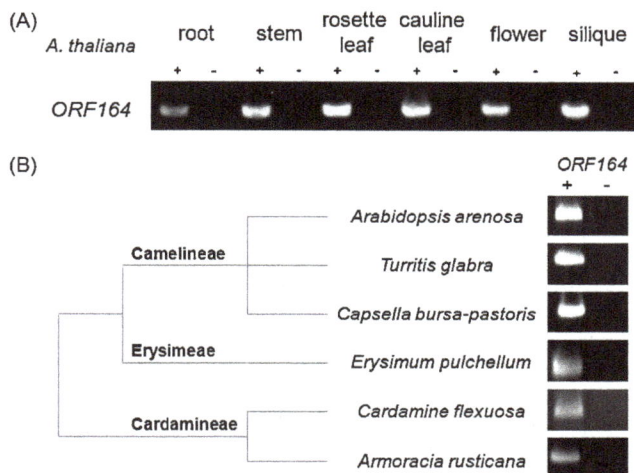

Fig. 5. Expression of *orf164*. (A) RT-PCR products of *orf164* in multiple organ types of *Arabidopsis thaliana*. Plus signs indicate the presence of reverse transcriptase and minus signs indicate absence of reverse transcriptase. (B) RT-PCR products of *orf164* in Brassicaceae species.

mitochondrial. *ARF17* in *A. thaliana* and *A. lyrata* are 90% identical, whereas *orf164* and AL3G32400 have an identity of over 99%, with only two nucleotides substituted out of 462 base pairs. We suspect this error regarding annotation of the *orf164* sequence in *A. lyrata* is due to the insertion of the whole mitochondrial genome into the centromere of chromosome 2 in *A. thaliana* [3,21]. When this region was used as a reference to assemble and annotate the *A. lyrata* genome [22], the mitochondrial genome of *A. lyrata* was annotated as being in the nucleus.

Collectively the comparative analyses presented above indicate that *orf164* is derived from *ARF17* through duplicative intracellular gene transfer, from the nuclear genome to the mitochondrial genome. We hypothesize that the transfer was DNA-mediated, and not RNA-mediated, because at the 3′end of *orf164* there are eight out of nine consecutive nucleotides that are identical to the first intron of *ARF17* (Figs. 2 and 3).

To analyze *orf164* for possible purifying selection, we performed a branch-wise dN/dS test on *A. thaliana orf164* and *ARF17* in the phylogeny with several outgroup *ARF17*s across eurosids, using a free-ratio model in PAML (Fig. 4). The dN/dS ratios of *orf164* and *ARF17* are 0.08 and 0.03, respectively, which are statistically not significantly different. Although the dN/dS ratio of *orf164* suggests

purifying selection, it may due to the very low sequence evolution rate in plant mitochondria instead of evolutionary constraints on the sequence.

3.3. Orf164 is expressed in several organ types and in five other genera within the Brassicaceae

We used RT-PCR to determine if *orf164* is transcribed. Our results show that *orf164* is transcribed in roots, rosette leaves, stems, cauline leaves, flowers and siliques of *A. thaliana*, indicating a broad expression pattern (Fig. 5A). We sequenced the *orf164* RT-PCR products to confirm their identity. To verify that the expressed *orf164* is the mitochondrial copy and not the identical copy present in nuclear chromosome 2, derived from transfer of a mitochondrial genome to the nucleus [3,21], we assayed *orf164* expression in *A. arenosa*. *A. thaliana* and *A. arenosa* are estimated to have diverged about 5 million years ago [23], whereas the timing of the whole mitochondrial genome transfer to nuclear chromosome 2 in *A. thaliana* was estimated at 44,000–176,000 years ago [24]. We detected expression of *orf164* in *A. arenosa* (Fig. 5B).

orf164 orfX ccb203 matR cox2

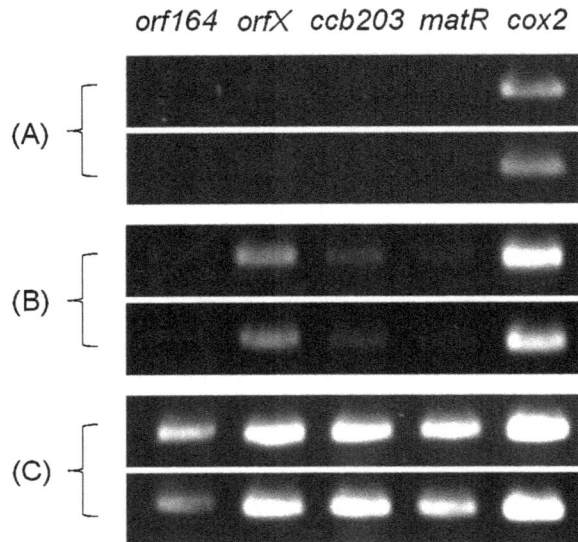

Fig. 6. Expression of *orf164* in comparison to four other mitochondrial genes in *A. thaliana*. PCR products amplified with the same amount of cDNA were generated using 20 cycles (A), 25 cycles (B), and 30 cycles (C). Two lanes in each section represent two replicates. Each column represents a mitochondrial gene labeled at the top.

To determine if *orf164* is present and expressed in other Brassicaceae genera, we performed RT-PCR using RNAs from *Capsella bursa-pastoris*, and *Turritis glabra*, both of which are in the tribe Camelineae along with *Arabidopsis*, as well as *Erysimum pulchellum* in the tribe Erysimeae, *Cardamine flexuosa* and *Armoracia rusticana* in the tribe Cardamineae which are close sister tribes to Camelineae [25]. We focused on these tribes because *orf164* is not present in the published mitochondrial genomes of *Brassica* or *Raphanus* [26,27], which are in the tribe Brassiceae. We detected expression of *orf164* in all five species (Fig. 5B) and sequenced the RT-PCR products which confirmed their identity.

To compare expression levels of *orf164* to other mitochondrial genes, we performed RT-PCR with *orf164* along with *cox2*, *matR*, *ccb203*, and *orfX* using varying numbers of PCR cycles (20, 25, and 30). Although not a quantitative assessment of transcript levels, the assay allows for rough comparisons of transcript levels among the different genes. *Cox2* transcripts were easily detectable at 20 cycles and were most abundant among the five genes, whereas *orf164* transcripts were detectable only with 30 cycles and appeared to be the least abundant (Fig. 6). These results suggest that *orf164* transcripts are less abundant than those of several other mitochondrial genes in *A. thaliana*.

How might *orf164* have acquired regulatory elements for expression? One possibility is that the transferred copy inserted near existing *cis*-regulatory elements, similar to the mechanism by which many mitochondrion-derived genes gained expression after being transferred to the nucleus. *Orf164* is located upstream of the tRNA-Trp pseudogene $\psi trnW$ which was derived from the chloroplast *trnW* (Fig. 1; [28]). The chloroplast-derived *trnW* genes are present and expressed in the mitochondrial genomes of several other angiosperm species, including potato [29], wheat [30], sunflower [35], maize [31], and beet [32]. Thus it is possible that, after transfer from the nucleus, *orf164* in the Brassicaceae inserted upstream of *trnW* and seized its *cis*-regulatory elements, acquiring expression while abolishing expression of *trnW*. It is also possible that pseudogenization of *trnW* was not caused by the insertion of *orf164* and instead by mutations in its cis-regulatory elements that abolished transcription. Although *orf164* in *A. thaliana* is 1283 bp upstream of *trnW*, the actual insertion site of *orf164* in an ancestral

Brassicaceae species could have been closer to *trnW*, followed by expansion of the intergenic region.

We have shown that *orf164* is transcribed in *Arabidopsis* and five other genera within the Brassicaceae, but it is not known if the transcripts are translated. Even if the transcripts are translated, the resulting proteins might not be functional in mitochondria. *Orf164* contains the auxin responsive element, involved in regulating auxin-response gene expression, derived from *ARF17*. It is not clear what type of function such a protein would have in mitochondria. Many other transcribed open reading frames with no obvious functions in mitochondria, and typically not conserved among species nor derived from the nuclear genome, have been identified in the mitochondrial genomes of rice and tobacco [33,34]. Thus plant mitochondria may contain numerous transcribed ORFs that do not code for functional proteins, with the number and type varying by species.

3.4. Search for other nuclear-derived open reading frames in the A. thaliana mitochondrial genome

To search for other sequences in the mitochondrial genome of *A. thaliana* that are derived from a nuclear protein-coding gene, we aligned the sequences of all annotated nuclear genes to the mitochondrial genome (see Section 2). We found only one other gene of possible interest, *orf160* which is partly derived from *MMD1* (AT1G66170), but the sequence has many indels relative to *MMD1* that disrupt the reading frame; thus we did not study it further.

3.5. Conclusions

This study shows a case of transfer of a nuclear gene to the mitochondrial genome and expression of the transferred gene, which is a phenomenon that has not been previously reported. The transfer appears to be DNA-mediated, rather than RNA-mediated. It is possible that *orf164* gained transcriptional regulatory elements from the *trnW* gene for tRNA-Trp. *Orf164* is present and expressed in several genera of the Brassicaceae, but not in *Brassica* or *Raphanus*, and thus the transfer may be a relatively recent evolutionary event. Other angiosperm mitochondrial genomes may contain genes that were transferred from the nucleus and gained regulatory elements to become expressed. This study provides a novel perspective on the movement of genes between the genomes of subcellular compartments.

Acknowledgements

We thank the UBC Botanical Garden for providing *Turritis glabra* (042231-6774-2013), *Erysimum pulchellum* (039120-0167-2008), and *Armoracia rusticana* (037622-0075-2005), and Jamie Fenneman for identifying *Capsella bursa-pastoris* and *Cardamine flexuosa*. This research was funded by a grant from the Natural Science and Engineering Research Council of Canada.

References

[1] D. Bensasson, D. Zhang, D.L. Hartl, G.M. Hewitt, Mitochondrial pseudogenes: evolution's misplaced witnesses, Trends Ecol. Evol. 16 (2001) 314–321.
[2] T. Kleine, U.G. Maier, D. Leister, DNA transfer from organelles to the nucleus: the idiosyncratic genetics of endosymbiosis, Annu. Rev. Plant Biol. 60 (2009) 115–138.
[3] X. Lin, S. Kaul, S. Rounsley, T.P. Shea, M.I. Benito, C.D. Town, C.Y. Fujii, T. Mason, C.L. Bowman, M. Barnstead, et al., Sequence and analysis of chromosome 2 of the plant Arabidopsis thaliana, Nature 402 (1999) 761–768.
[4] A.N. Lough, L.M. Roark, A. Kato, T.S. Ream, J.C. Lamb, J.A. Birchler, K.J. Newton, Mitochondrial DNA transfer to the nucleus generates extensive insertion site variation in maize, Genetics 178 (2008) 47–55.
[5] L.M. Roark, A.Y. Hui, L. Donnelly, J.A. Birchler, K.J. Newton, Recent and frequent insertions of chloroplast DNA into maize nuclear chromosomes, Cytogenet. Genome Res. 129 (2010) 17–23.

[6] K.L. Adams, J.D. Palmer, Evolution of mitochondrial gene content: gene loss and transfer to the nucleus, Mol. Phylogenet. Evol. 29 (2003) 380–395.

[7] L. Bonen, S. Calixte, Comparative analysis of bacterial-origin genes for plant mitochondrial ribosomal proteins, Mol. Biol. Evol. 23 (2006) 701–712.

[8] S.L. Liu, Y. Zhuang, P. Zhang, K.L. Adams, Comparative analysis of structural diversity and sequence evolution in plant mitochondrial genes transferred to the nucleus, Mol. Biol. Evol. 26 (2009) 875–891.

[9] T. Kubo, S. Nishizawa, A. Sugawara, N. Itchoda, A. Estiati, T. Mikami, The complete nucleotide sequence of the mitochondrial genome of sugar beet (Beta vulgaris L.) reveals a novel gene for tRNA(Cys)(GCA), Nucleic Acids Res. 28 (2000) 2571–2576.

[10] Y. Notsu, S. Masood, T. Nishikawa, N. Kubo, G. Akiduki, M. Nakazono, A. Hirai, K. Kadowaki, The complete sequence of the rice (Oryza sativa L.) mitochondrial genome: frequent DNA sequence acquisition and loss during the evolution of flowering plants, Mol. Genet. Genomics 268 (2002) 434–445.

[11] A.J. Alverson, X. Wei, D.W. Rice, D.B. Stern, K. Barry, J.D. Palmer, Insights into the evolution of mitochondrial genome size from complete sequences of Citrullus lanatus and Cucurbita pepo (Cucurbitaceae), Mol. Biol. Evol. 27 (2010) 1436–1448.

[12] A.J. Alverson, D.W. Rice, S. Dickinson, K. Barry, J.D. Palmer, Origins and recombination of the bacterial-sized multichromosomal mitochondrial genome of cucumber, Plant Cell 23 (2011) 2499–2513.

[13] V.V. Goremykin, P.J. Lockhart, R. Viola, R. Velasco, The mitochondrial genome of Malus domestica and the import-driven hypothesis of mitochondrial genome expansion in seed plants, Plant J. 71 (2012) 615–626.

[14] M. Van Bel, S. Proost, E. Wischnitzki, S. Movahedi, C. Scheerlinck, Y. Van de Peer, K. Vandepoele, Dissecting plant genomes with the PLAZA comparative genomics platform, Plant Physiol. 158 (2012) 590–600.

[15] R.C. Edgar, MUSCLE: multiple sequence alignment with high accuracy and high throughput, Nucleic Acids Res. 32 (2004) 1792–1797.

[16] Z. Yang, PAML 4: phylogenetic analysis by maximum likelihood, Mol. Biol. Evol. 24 (2007) 1586–1591.

[17] L. Noe, G. Kucherov, YASS: enhancing the sensitivity of DNA similarity search, Nucleic Acids Res. 33 (2005) W540–W543.

[18] J.I. Davila, M.P. Arrieta-Montiel, Y. Wamboldt, J. Cao, J. Hagmann, V. Shedge, Y.Z. Xu, D. Weigel, S.A. Mackenzie, Double-strand break repair processes drive evolution of the mitochondrial genome in Arabidopsis, BMC Biol. 9 (2011) 64.

[19] G. Hagen, T. Guilfoyle, Auxin-responsive gene expression: genes, promoters and regulatory factors, Plant Mol. Biol. 49 (2002) 373–385.

[20] E. Liscum, J.W. Reed, Genetics of Aux/IAA and ARF action in plant growth and development, Plant Mol. Biol. 49 (2002) 387–400.

[21] R.M. Stupar, J.W. Lilly, C.D. Town, Z. Cheng, S. Kaul, C.R. Buell, J. Jiang, Complex mtDNA constitutes an approximate 620-kb insertion on Arabidopsis thaliana chromosome 2: implication of potential sequencing errors caused by large-unit repeats, Proc. Natl. Acad. Sci. U. S. A. 98 (2001) 5099–5103.

[22] T.T. Hu, P. Pattyn, E.G. Bakker, J. Cao, J.F. Cheng, R.M. Clark, N. Fahlgren, J.A. Fawcett, J. Grimwood, H. Gundlach, et al., The Arabidopsis lyrata genome sequence and the basis of rapid genome size change, Nat. Genet. 43 (2011) 476–481.

[23] M. Jakobsson, J. Hagenblad, S. Tavare, T. Sall, C. Hallden, C. Lind-Hallden, M. Nordborg, A unique recent origin of the allotetraploid species Arabidopsis suecica: evidence from nuclear DNA markers, Mol. Biol. Evol. 23 (2006) 1217–1231.

[24] C.Y. Huang, N. Grunheit, N. Ahmadinejad, J.N. Timmis, W. Martin, Mutational decay and age of chloroplast and mitochondrial genomes transferred recently to angiosperm nuclear chromosomes, Plant Physiol. 138 (2005) 1723–1733.

[25] T.L. Couvreur, A. Franzke, I.A. Al-Shehbaz, F.T. Bakker, M.A. Koch, K. Mummenhoff, Molecular phylogenetics, temporal diversification, and principles of evolution in the mustard family (Brassicaceae), Mol. Biol. Evol. 27 (2010) 55–71.

[26] S. Chang, T. Yang, T. Du, Y. Huang, J. Chen, J. Yan, J. He, R. Guan, Mitochondrial genome sequencing helps show the evolutionary mechanism of mitochondrial genome formation in Brassica, BMC Genomics 12 (2011) 497.

[27] Y. Tanaka, M. Tsuda, K. Yasumoto, H. Yamagishi, T. Terachi, A complete mitochondrial genome sequence of Ogura-type male-sterile cytoplasm and its comparative analysis with that of normal cytoplasm in radish (Raphanus sativus L.), BMC Genomics 13 (2012) 352.

[28] A.M. Duchene, L. Marechal-Drouard, The chloroplast-derived trnW and trnM-e genes are not expressed in Arabidopsis mitochondria, Biochem. Biophys. Res. Commun. 285 (2001) 1213–1216.

[29] L. Marechal-Drouard, P. Guillemaut, A. Cosset, M. Arbogast, F. Weber, J.H. Weil, A. Dietrich, Transfer RNAs of potato (Solanum tuberosum) mitochondria have different genetic origins, Nucleic Acids Res. 18 (1990) 3689–3696.

[30] P.B. Joyce, M.W. Gray, Chloroplast-like transfer RNA genes expressed in wheat mitochondria, Nucleic Acids Res. 17 (1989) 5461–5476.

[31] P. Leon, V. Walbot, P. Bedinger, Molecular analysis of the linear 2.3 kb plasmid of maize mitochondria: apparent capture of tRNA genes, Nucleic Acids Res. 17 (1989) 4089–4099.

[32] T. Kubo, Y. Yanai, T. Kinoshita, T. Mikami, The chloroplast trnP-trnW-petG gene cluster in the mitochondrial genomes of Beta vulgaris, B. trigyna and B. webbiana: evolutionary aspects, Curr. Genet. 27 (1995) 285–289.

[33] S. Fujii, T. Toda, S. Kikuchi, R. Suzuki, K. Yokoyama, H. Tsuchida, K. Yano, K. Toriyama, Transcriptome map of plant mitochondria reveals islands of unexpected transcribed regions, BMC Genomics 12 (2011) 279.

[34] B.T. Grimes, A.K. Sisay, H.D. Carroll, A.B. Cahoon, Deep sequencing of the tobacco mitochondrial transcriptome reveals expressed ORFs and numerous editing sites outside coding regions, BMC Genomics 15 (2014) 31.

[35] L.R. Ceci, P. Veronico, R. Gallerani, Identification and mapping of tRNA genes on the Helianthus annuus mitochondrial genome, DNA Seq. 6 (1996) 159–166.

[36] J. Marienfeld, M. Unseld, A. Brennicke, The mitochondrial genome of Arabidopsis is composed of both native and immigrant information, Trends Plant Sci. 4 (1999) 495–502.

[37] D.W. Rice, A.J. Alverson, A.O. Richardson, G.J. Young, M.V. Sanchez-Puerta, J. Munzinger, K. Barry, J.L. Boore, Y. Zhang, C.W. dePamphilis, et al., Horizontal transfer of entire genomes via mitochondrial fusion in the angiosperm Amborella, Science 342 (2013) 1468–1473.

21

Gene and genome duplications and the origin of C4 photosynthesis: Birth of a trait in the Cleomaceae

Erik van den Bergh[a], Canan Külahoglu[b], Andrea Bräutigam[b], Julian M. Hibberd[c], Andreas P.M. Weber[b], Xin-Guang Zhu[d], M. Eric Schranz[a,*]

[a] Biosystematics, Wageningen University and Research, Droevendaalsesteeg 1, 6708 PB Wageningen, The Netherlands
[b] Institute of Plant Biochemistry, Center of Excellence on Plant Sciences(CEPLAS), Heinrich-Heine-University, D-40225 Düsseldorf, Germany
[c] Department of Plant Sciences, University of Cambridge, Cambridge CB2 3EA, United Kingdom
[d] Plant Systems Biology Group, Partner Institute of Computational Biology, Chinese Academy of Sciences/Max Planck Society, Shanghai 200031, China

ARTICLE INFO

Keywords:
Plant genome evolution
Synteny
Cleomaceae
Brassicaceae
Bioinformatics
Whole genome duplication
Paleopolyploidy
C4 photosynthesis

ABSTRACT

C_4 photosynthesis is a trait that has evolved in 66 independent plant lineages and increases the efficiency of carbon fixation. The shift from C_3 to C_4 photosynthesis requires substantial changes to genes and gene functions effecting phenotypic, physiological and enzymatic changes. We investigate the role of ancient whole genome duplications (WGD) as a source of new genes in the development of this trait and compare expression between paralog copies. We compare *Gynandropsis gynandra*, the closest relative of Arabidopsis that uses C_4 photosynthesis, with its C_3 relative *Tarenaya hassleriana* that underwent a WGD named Th-α. We establish through comparison of paralog synonymous substitution rate that both species share this paleohexaploidy. Homologous clusters of photosynthetic gene families show that gene copy numbers are similar to what would be expected given their duplication history and that no significant difference between the C_3 and C_4 species exists in terms of gene copy number. This is further confirmed by syntenic analysis of *T. hassleriana*, *Arabidopsis thaliana* and *Aethionema arabicum*, where syntenic region copy number ratios lie close to what could be theoretically expected. Expression levels of C_4 photosynthesis orthologs show that regulation of transcript abundance in *T. hassleriana* is much less strictly controlled than in *G. gynandra*, where orthologs have extremely similar expression patterns in different organs, seedlings and seeds. We conclude that the Th-α and older paleopolyploidy events have had a significant influence on the specific genetic makeup of Cleomaceae versus Brassicaceae. Because the copy number of various essential genes involved in C_4 photosynthesis is not significantly influenced by polyploidy combined with the fact that transcript abundance in *G. gynandra* is more strictly controlled, we also conclude that recruitment of existing genes through regulatory changes is more likely to have played a role in the shift to C_4 than the neofunctionalization of duplicated genes.
DATA: The data deposited at NCBI represents raw RNA reads for each data series mentioned: 5 leaf stages, root, stem, stamen, petal, carpel, sepal, 3 seedling stages and 3 seed stages of Tarenaya hassleriana and Gynandropsis gynandra. The assembled reads were used for all analyses of this paper where RNA was used. http://www.ncbi.nlm.nih.gov/Traces/sra/?study=SRP036637, http://www.ncbi.nlm.nih.gov/Traces/sra/?study=SRP036837

* Corresponding author.
E-mail addresses: erik.vandenbergh@wur.nl (E. van den Bergh), canan.kuelahoglu@uni-duesseldorf.de (C. Külahoglu), andrea.braeutigam@uni-duesseldorf.de (A. Bräutigam), jmh65@cam.ac.uk (J.M. Hibberd), andreas.weber@uni-duesseldorf.de (A.P.M. Weber), zhuxinguang@picb.ac.cn (X.-G. Zhu), eric.schranz@wur.nl (M. Eric Schranz).

1. Introduction

Over sixty lineages of both monocot and eudicot angiosperms have evolved a remarkable solution to maximize photosynthesis efficiency under low CO_2 levels, high temperatures and/or drought: C_4 photosynthesis [1]. The evolution of this modified photosynthetic pathway represents a wonderful example of convergent evolution. While the changes necessary for the transition from C_3 to C_4 photosynthesis are numerous, the trait has a wide phylogenetic distribution across angiosperms, with 19 different plant families across the globe known to contain one or multiple members capable of C_4 photosynthesis [2]. Much research on eudicot C_4 has focused on *Flaveria* species (Asteraceae), which contains not only C_4 species but also a number of C_3/C_4 intermediates [3]. With the emergence of genomics and the choice of *Arabidopsis thaliana* as the genomics standard model organism, species in the Cleomaceae, a sister-family to the Brassicaceae (containing Arabidopsis and Brassica crops) have been proposed for genetic studies of C_4 [4,5].

C_4 plants spatially separate the fixation of carbon away from the RuBisCO active site by using phospho*enol*pyruvate carboxylase, an alternate carboxylase that does not react with oxygen. As a consequence they are more efficient under permissive conditions [6]. The typical C_4 system is characterized by a morphological change: so-called Kranz anatomy [7]. In this anatomy, specialized mesophyll (M) cells surround enlarged bundle sheath (BS) cells, with the leaf veins internal to the BS. Generally, the veination in C_4 leaves is increased [8]. This internal leaf architecture physically partitions the biochemical events of the C_4 pathway into two main phases. In the first phase, dissolved HCO_3^- is assimilated into C_4 acids by phospho*enol*pyruvate carboxylase (PEPC) in the mesophyll cells. In the second phase, these acids diffuse into the chloroplast loaded bundle sheath (BS) cells, where they are decarboxylated and the released CO_2 is fixed by RuBisCO. The increased CO_2 concentration in the BS cells allows carbon fixation by RuBisCO to be much more efficient by reducing photorespiration. Two subtypes of the C_4 biochemical pathway are defined, based on the most active C_4 acid decarboxylase that liberates CO_2 from C_4 acids in the bundle sheath: NADP-malic enzyme (NADP-ME), NAD-malic enzyme (NAD-ME); a facultative addition of phosphoenolpyruvate carboxykinase (PEPCK) activity can be present in either subtype [9]. The subtypes are used as a classification scheme for C_4.

The process of carboxylation and decarboxylation costs more energy than the simpler C_3 form of photosynthesis, but it diminishes photorespiration. In conditions of low atmospheric CO_2 pressure, photorespiration causes a major loss in photosynthetic output and the elaborate concentrating mechanisms of C_4 photosynthesis circumvent this [10].

All genes important for the C_4 pathway are expressed at relatively low levels in C_3 leaves [11]. The mechanism for recruitment of these genes into the C_4 pathway remains to be elucidated. For some ancestral C_3 genes changes in *cis*-regulatory elements, while in others changes in *trans* generate M and BS cell specificity [12–14], indicating variation in the mechanisms underlying gene recruitment into the C_4 pathway. It has been proposed that gene duplication and subsequent neofunctionalization of one gene copy has facilitated the alterations in gene expression that underlie the evolution of C_4 photosynthesis [15,16]. Gene duplication is proposed to be a (pre)condition for the evolution of C_4 because it allows the organism to maintain the original gene while a duplicate version can acquire beneficial changes. This can lead to significant changes in metabolism without the deleterious effect of modifications to essential genes. A recent study that compared convergent evolution of photosynthetic pathways with parallel evolution concluded that duplications are not essential for the development of C_4 biochemistry, but rather changes in expression and localization of specific genes [11,17]. However, this study highlighted just the number of C_4 genes and did not take into account the age and mechanism of gene duplications.

The modifications necessary for the anatomical changes from C_3 to C_4 photosynthesis are not well established. Recent work has shown that the SCARECROW (SCR) gene that is responsible for vein formation in roots, can produce proliferated bundle sheath cells as well as other changes that can be coupled to the shift to the Kranz anatomy [18]. Further work supports this relation by describing the role that the upstream interacting partner of SCR, SHORT-ROOT

(SHR) plays in the variations in anatomy seen in various C_4 species [19,20].

Gene duplicates must be further refined by the mechanism by which they arise; either as single gene tandem duplication or whole genome duplication (WGD). Tandem duplications occur frequently, but the duplicates are often lost again resulting in a constant birth–death cycle of duplicate genes [21]. Second, there is whole genome duplication (WGD) or polyploidy, where all genes are simultaneously duplicated. After duplication there are often dramatic changes in the plant genomic structure, a process referred to as diploidization in which most genes return to single copy. However, the genes that are maintained in duplicate after WGD often have important functions in enzyme complexes (e.g. to maintain proper gene balance [22]) or can diversify and evolve new gene functions (e.g. neo-functionalization).

The contribution of WGD to photosynthesis-related genes has been studied in soybean, barrel-medic, Arabidopsis, and sorghum [23,24]. The polyploid and non-polyploid duplicated gene retention in *Glycine max*, *Medicago truncatula* and Arabidopsis for four classes of photosynthesis-related genes was compared: the Calvin–Benson–Bassham-cycle (CBBC), the light-harvesting complex (LHC), photosystem I (PSI) and photosystem II (PSII). It was found that photosystem genes were more dosage sensitive, with more duplicates derived only from WGD whereas CC gene families were often larger with more non-polyploid duplicates retained. In *Sorghum bicolor*, a recent WGD was reported to be an important origin of C_4 specific genes. Several key C_4 genes of this crop were found to be collinear with genes that function in C_3 photosynthesis when compared to maize and rice. Here, we combine the approaches of these two studies to examine the evolution of photosynthesis and C_4-related genes in C_3 and C_4 Cleomaceae species.

Gynandropsis gynandra (Fig. 1, blue clade) belongs to the NAD-ME C_4 photosynthesis sub-type [25,26] and is an important South-East Asian and African dry-season leafy vegetable (sometimes referred to as Phak-sian or African cabbage), and is closely related to horticultural C_3 species *Tarenaya hassleriana* (Fig. 1, pink clade). Both species are easily cultivated in the greenhouse, and a robust phylogenetic framework for Cleomaceae species is emerging [4,5,27]. There are two other independent origins of the C_4 within the Cleomaceae, *Cleome angustifolia* and *Cleome oxalidea* (Fig. 1, yellow clade), identified by carbon isotope discrimination [5,25]. Because of the economic importance and ease of growth, the C_4–C_3 contrast between *G. gynandra* and *T. hassleriana* makes this system most attractive and tractable. Both species also have relatively small genome sizes (*T. hassleriana* = 292 Mb and *G. gynandra* $\approx$ 1 Gb). *T. hassleriana* underwent a WGD named Th-α [28] but it is not yet known whether this event is shared with all or a subset of other Cleomaceae.

In this study we compare C_3 *T. hassleriana* of the Cleomaceae with C_4 *G. gynandra* of the same family. We use the knowledge of Brassicaceae gene functions to identify the important photosynthetic genes in both species and address the following questions: Does *G. gynandra* share the Th-α event? What is contribution of duplicate genes to photosynthesis and C_4-related gene families? And finally, what is the role of gene duplicates from WGD compared to continuous small-scale duplications?

2. Methods

2.1. Transcriptome sequencing and assembly

All transcriptome data was used directly from the Cleomaceae transcript atlas [17]. In the atlas, *T. hassleriana* genes were used as a reference to map transcripts from both species to Cleomaceae "unigenes" indicated by the gene name coined in the published *T.*

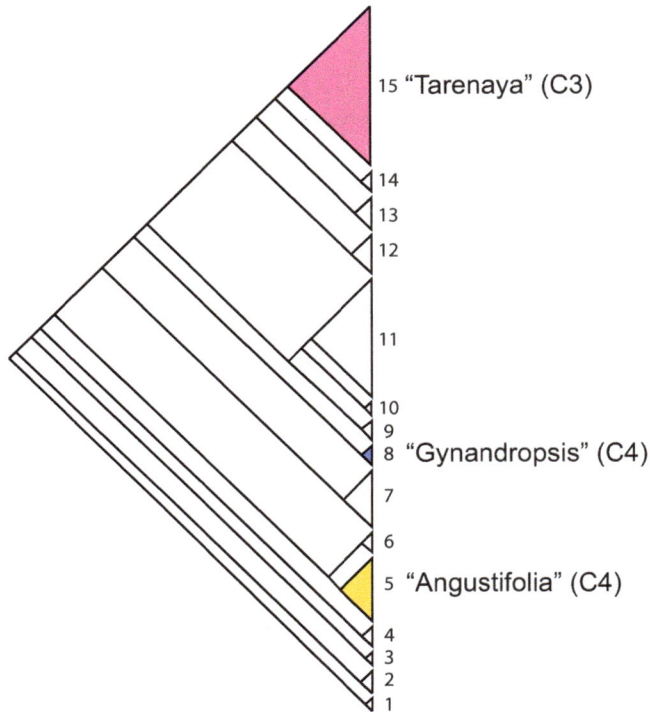

Fig. 1. Simplified phylogeny of Cleomaceae. Clades are numbered following the most recently published Maximum Likelihood phylogeny of Cleomaceae [25]. Clade 15 containing *T. hassleriana* is marked in pink. Clade 8 containing *G. gynandra* is marked in blue. Clade 5 (Yellow) contains the other origin of C_4 in Cleomaceae, with *C. angustifolia* and C_4/C_3 intermediate *C. paradoxa*. (For interpretation of the references to color in this figure legend, the reader is referred to the web version of the article.)

hassleriana genome [29]. For gene quantification we used default BlatV35 parameters [30] in protein space for mapping, counting the best matched hit based on e-value for each read uniquely.

2.2. Homolog selection

A TBlastX [31] search of transcriptomes of *T. hassleriana* and *G. gynandra* was performed with default parameters (no evalue cutoff) to have a maximum number of hits for subsequent filtering. To filter paralogs and orthologs from these results, CIP/CALP filtering was used [32]. Cumulative Identity Percentage (CIP) is defined as the sum of the number of matching nucleotides for each high-scoring segment pair (HSP) of a pair of genes divided by the total lengths of those HSPs. Cumulative alignment length percentage (CALP) is defined as the sum of the alignment lengths of all HSPs of a matching gene pair divided by the total length of the query sequence. Both of these values give a reliable estimation of the similarity of two genes and is a more accurate method than evalue or bit score threshold filtering. A CIP/CALP threshold of 50/50 was chosen as a suitable cutoff point for orthology and/or paralogy.

2.3. Ks/4dtv calculation of paralog pairs

Paralogs identified with CIP/CALP filtering were aligned using Exonerate [33] with the coding2coding model parameter, using a custom output format through the "roll your own" parameter. The exact command line was: "exonerate -m c2c seq1.fasta seq2.fasta –ryo \"%Pqs %Pts\\n" –showalignment false –verbose 0". The output from this command was fed into CodeML from the PAML package using standard parameters (Codonfreq = 2, kappa = 2, omega = 0.4). Output from PAML [34] was parsed using

custom Perl scripts to read the synonymous substitution rate (*Ks*) and the fourfold transversion rate (4dtv). This workflow is identical to the established paralog identification pipeline Duppipe [35] using updated tools and more stringent selection using CIP/CALP.

2.4. Homolog clustering

Photosynthesis genes were selected from known functionally annotated Arabidopsis genes. Gene identifiers used for each family are listed hereafter and in Table 2. βCA: AT1G23730, AT1G58180, AT1G70410, AT3G01500, AT4G33580, AT5G14740. MDH (cytosolic): AT1G04410, AT5G43330, AT5G56720. MDH (mitochondrial): AT1G53240, AT2G22780, AT3G15020, AT3G47520, AT5G09660. MDH (peroxisomal): AT1G53240, AT2G22780, AT3G15020, AT3G47520, AT5G09660. MDH (plastidic): AT1G53240, AT2G22780, AT3G15020, AT3G47520, AT5G09660. NAD-ME: AT2G13560, AT4G00570. NADP-ME: AT1G79750, AT2G19900, AT5G11670, AT5G25880. PEPC: AT1G21440, AT1G53310, AT2G42600, AT3G14940. PPCK: AT1G08650, AT3G04530, AT3G04550, AT4G37870, AT5G28500, AT5G65690. These genes were then used as a BLAST database and queried with *T. hassleriana* and *G. gynandra* atlas unigenes. Hits were then filtered using a 50/50 CIP/CALP cutoff. Using custom Perl scripts, the hits of these hits were picked up, iterating recursively until convergence (no new hits found). All unique genes resulting from this process form a family cluster.

2.5. Synteny analyses

T. hassleriana genes were used as a query in the CoGe Synfind [36] program using the following parameters: Comparison algorithm: Last, Gene window size: 40, Minimum number of genes: 4, Scoring Function: Collinear, Syntenic depth: unlimited. As query genomes, the following were used: *A. arabicum* VEGI unmasked v2.5, *A. thaliana* Col-0 TAIR unmasked v10.02 and *T. hassleriana* BGI; Eric Scranz Lab; Weber lab unmasked v5.

3. Results

3.1. Evidence of WGD in both species confirming a shared event

Using the transcript sets of *G. gynandra* and *T. hassleriana*, paralogs were matched to each other by BLAST search and CIP/CALP filtering. In total, 55,014 paralogs were found: 26,883 in *T. hassleriana* covering 49% of transcript space and 28,131 in *G. gynandra* covering 48% of transcript space. Of all paralog pairs, *Ks* and fourfold transversion substitutions (4dtv) were determined and binned to establish an evolutionary time distribution (Fig. 2). In both species a large gene birth event has taken place around *Ks* = 0.4 (Fig. 2 between *Ks* = 0.25 and *Ks* = 0.5), which corresponds to the *Ks* window established earlier for the Th-α hexaploidy event [28]. The same analysis was performed using 4dtv values and results were extremely similar. Enumerating the paralogs that fall within the Th-α peak, we see that 15,785 gene pairs in *T. hassleriana* are retained from the Th-α paleohexaploidy, or ~29% of the total transcriptome. For *G. gynandra*, 16,096 gene pairs fall within the Th-α window, or around 27% of all transcripts.

3.2. Duplicate loss and retention in essential C_4 families

We examined six gene families that are essential in C_4 photosynthesis in detail: NAD malic enzyme (NAD-ME), NADP malic enzyme (NADP-ME), β carbonic anhydrase (βCA), malate dehydrogenase (MDH), phospho*enol*pyruvate carboxylase (PEPC) and phospho*enol*pyruvate carboxykinase (PPCK). Using Arabidopsis genes as a reference, homologous clusters were created using a

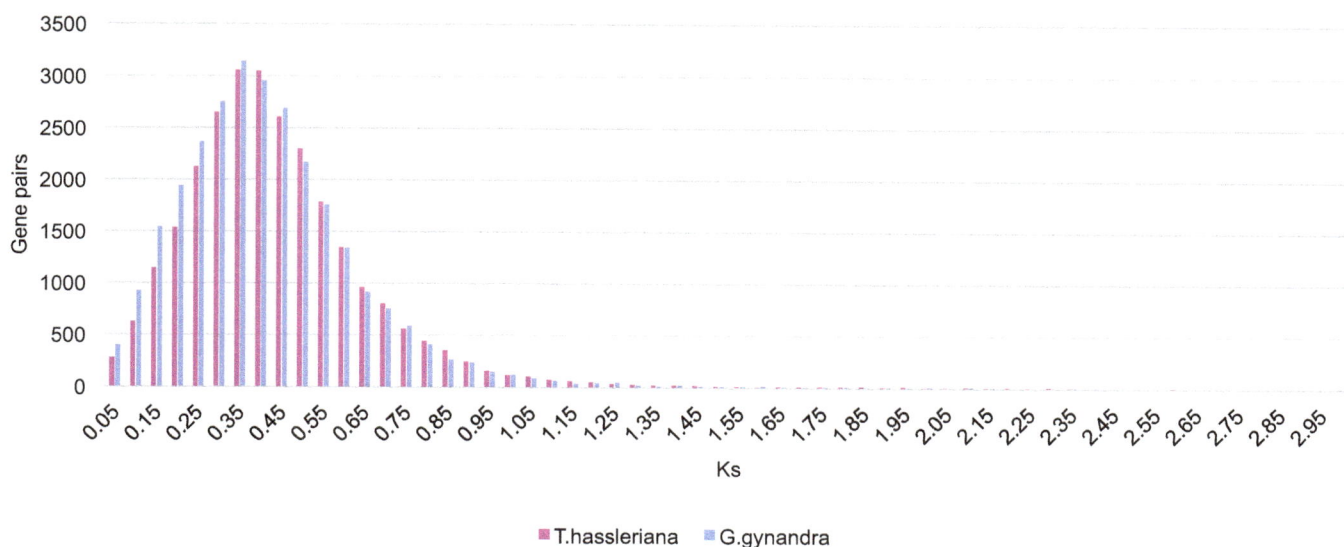

Fig. 2. Histogram showing the amount of gene pairs per Ks bin for *T. hassleriana* (pink) and *G. gynandra* (blue). The peak at around $Ks = 0.45$ is an indication of a massive gene birth event and is considered evidence of paleopolyploidy. Both species have an extremely similar peak, indicating that this is a shared polyploidy event. The Ks values of these peaks corresponds with Ks values found earlier for the Th-α hexaploidy event, indicating that this event has occurred before divergence of *T. hassleriana* and *G. gynandra*. (For interpretation of the references to color in this figure legend, the reader is referred to the web version of the article.)

CIP/CALP cutoff of 50/50. 146 homologous pairs could be placed in a cluster across the three species comprising 105 unique genes (Table 1); 40 in *A. thaliana*, 57 in *T. hassleriana* and 49 in *G. gynandra*. In most cases both Cleomaceae species have around 1.5 times the number of genes of *A. thaliana* except, interestingly, the NADP-ME family where numbers are almost the same in all species. Also of note is that *T. hassleriana* has 16% more C4 related genes in total than *G. gynandra* (57 over 49).

All genes of one species in a cluster were then aligned to each other and the Ks value of each pairing was established and subsequently binned with a stepsize of $Ks = 0.15$ (Fig. 3). At the Ks corresponding to the Th-α hexaploidy, both *T. hassleriana* and *G. gynandra* show a relative increase of gene pairs with this amount of synonymous substitutions. *A. thaliana* at the Ks of its older At-α event shows a similar, if slightly lower increase. Even longer ago in

Table 1

C4 photosynthesis homolog cluster sizes in *A. thaliana*, *T. hassleriana* and *G. gynandra*. Both Cleomaceae species have around 1.5 times the number of genes of *A. thaliana* except the NADP-ME and NAD-ME families where numbers are lower than average in the Cleomaceae species resulting in a similar amount of homologs in each species for these two gene groups.

	A. thaliana	*T. Hassleriana*	*G. gynandra*
βCA	6	10	7
MDH (cyt.)	3	6	6
MDH (mit.)	5	6	6
MDH (per.)	5	8	6
MDH (plast.)	5	6	6
NAD-ME	2	3	3
NADP-ME	4	4	3
PEPC	4	8	6
PPCK	6	6	6
Total	40	57	49

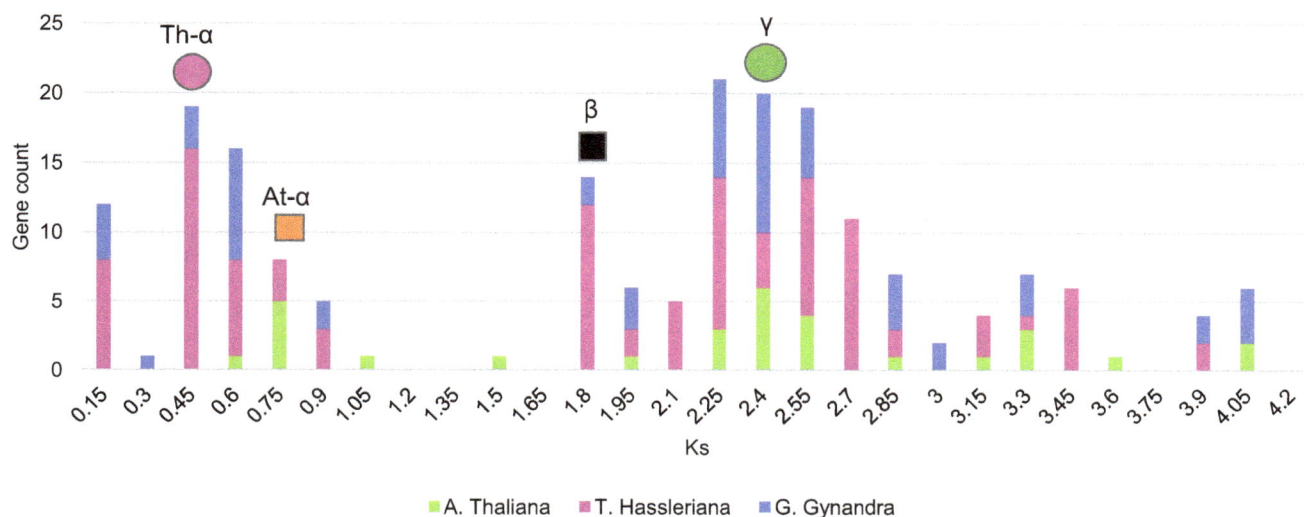

Fig. 3. Histogram showing Ks values of homolog gene clusters associated with C4 photosynthesis: MDH, NAD-ME, NADP-ME, PEPC and βCA. Gene duplication events are marked at their associated Ks value and colored according to earlier publication [28]; a square indicates a duplication (tetraploidy), a circle indicates a triplication (hexaploidy). The contribution of the Th-α (pink circle) and the At-α (orange square) on photosynthesis related gene copy number can be seen at $Ks = 0.45$ and $Ks = 0.6$ respectively. The β event at $Ks = 1.8$ (blue square) has contributed substantially to the expansion of gene copy number in *T. hassleriana*. Further in evolutionary time, around $Ks = 2.4$, the γ event (green circle) that is also shared by all three species has contributed equally to the polyploid presence in photosyntenic orthologs. (For interpretation of the references to color in this figure legend, the reader is referred to the web version of the article.)

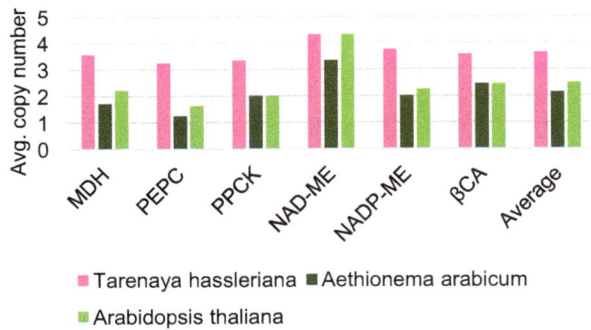

Fig. 4. Histogram showing average syntenic region copy number for *T. hassleriana*, *A. thaliana* and *Aethionema arabicum*. Because *A. arabicum* and *A. thaliana* both share a paleotetraploidy, the expected ratio of syntenic regions for *T. hassleriana*: *A. thaliana*: *Aethionema arabicum* is 3:2:2. In most cases, syntenic regions follow this distribution which is also reflected in the average ratio of all families being 3.6:2.1:2.5 (rightmost bars). The exception is NAD-ME, where the average region number in both *A. arabicum* and *A. thaliana* is as high as *T. hassleriana*.

evolutionary time at the *Ks* corresponding to the β event *T. hassleriana* has retained ~20% of C_4 related genes, where the other species show 2% and 0% retention for *G. gynandra* and *A. thaliana*, respectively. The final confirmed paleohexaploidy that all three species share, the ancient γ event at *Ks* = 2.4, has contributed substantially to the genetic makeup of all three species. In *A. thaliana* the number of relations that stem from the γ paleohexaploidy is 23%, with both Cleomaceae at 15% and 21% for *T. hassleriana* and *G. gynandra*, respectively.

3.3. Syntenic copy number variation

Syntenic analyses of the previously mentioned gene families was performed using CoGe Synfind [36]. Each *T. hassleriana* c4 related ortholog was used as a query with *T. hassleriana*, *Arabidopsis thaliana*, *A. arabicum* [37] as a basal representative of Brassicaceae. Thus for the *T. hassleriana*: *A. thaliana*: *A. arabicum* ortholog ratio we would theoretically expect 3 (Th-α):2 (At-α):2. Query results were enumerated and the average number of regions per family was determined (Fig. 4). For many families, the average is comparable to the 3:2:2 ratio, which is also represented by the average ratio (Fig. 4, rightmost set of bars) being 3.6:2.1:2.5. The exception is the NAD-ME family, which has seen more than expected retention with an orthologs ratio 4.3:3.3:4.3. The PEPC family also seems slightly under-retained in Brassicaceae, with a ratio of 3.3:1.3:1.6. Unfortunately, syntenic data is impossible to obtain without a sequenced genome so data syntenic regions of *G. gynandra* will have to be obtained in future work.

3.4. Regulation of photosynthetic homolog expression

Both Cleomaceae have substantially more copies of photosynthetic genes (Fig. 4). Using the Cleomaceae expression atlases [17], the expression of separate copies was compared in the C_3 and the C_4 species. In the expression atlas, the *T. hassleriana* coding sequence was used as a reference to map expression in both *T. hassleriana* and *G. gynandra* to a single Cleomaceae 'unigene'. Expression was quantified in nine different tissues including three developmental series: development from young to mature leaf (six stages), root, stem, stamen, petal, carpel, sepal, a seedling developmental series (three stages) and a seed time series (three stages).

For the photosynthetic gene families (NAD-ME, NADP-ME, PEPCK, PEPC, MDH, CA), homolog selection resulted in a data set of 43 unigenes with expression data for both Cleomaceae species.

Table 2
List of Arabidopsis genes used as representatives of C4 photosynthesis families. ATG identifiers correspond to identifier following the ATG system from the Arabidopsis Information Resource [43].

Gene family	ATG identifiers
βCA	AT1G23730
	AT1G58180
	AT1G70410
	AT3G01500
	AT4G33580
	AT5G14740
MDH (cytosolic)	AT1G04410
	AT5G43330
	AT5G56720
MDH (mitochondrial)	AT1G53240
	AT2G22780
	AT3G15020
	AT3G47520
	AT5G09660
MDH (peroxisomal)	AT1G53240
	AT2G22780
	AT3G15020
	AT3G47520
	AT5G09660
MDH (plastidic)	AT1G53240
	AT2G22780
	AT3G15020
	AT3G47520
	AT5G09660
NAD-ME	AT2G13560
	AT4G00570
NADP-ME	AT1G79750
	AT2G19900
	AT5G11670
	AT5G25880
PEPC	AT1G21440
	AT1G53310
	AT2G42600
	AT3G14940
PPCK	AT1G08650
	AT3G04530
	AT3G04550
	AT4G37870
	AT5G28500
	AT5G65690

Expression levels were normalized and compared amongst photosynthetic gene families, examples of which are plotted for NAD-ME and βCA (Fig. 5). Immediately noticeable is the highly similar expression profiles of *G. gynandra* when compared to the more chaotic profiles of *T. hassleriana*. This is observed in all except one gene family. *G. gynandra* has 176 expressed unigenes with a highly correlated expression pattern (Pearson correlation > 0.95) whereas in *T. hassleriana* 87 unigenes share a highly correlated expression pattern (Pearson correlation > 0.95).

The expression pattern that is observed in *G. gynandra* in the β-CA family also correspond to their *A. thaliana* highest ranking match (Table 2). The cluster consisting of C.spinosa_00253, C.spinosa_13896, C.spinosa_18526 and C.spinosa_10164 for example all match highest to *A. thaliana* gene β carbonic anhydrase 4 (AT1G70410). The cluster consisting of C.spinosa_07642 and C.spinosa_13410 both map to carbonic anhydrase 1 (AT3G01500). A similar pattern is present in NAD-ME where the cluster of C.spinosa_03046 and C.spinosa_09126 both map to NAD-ME1 (AT2G13560) and the C.spinosa_12536 singleton maps to NAD-ME2 (AT4G00570).

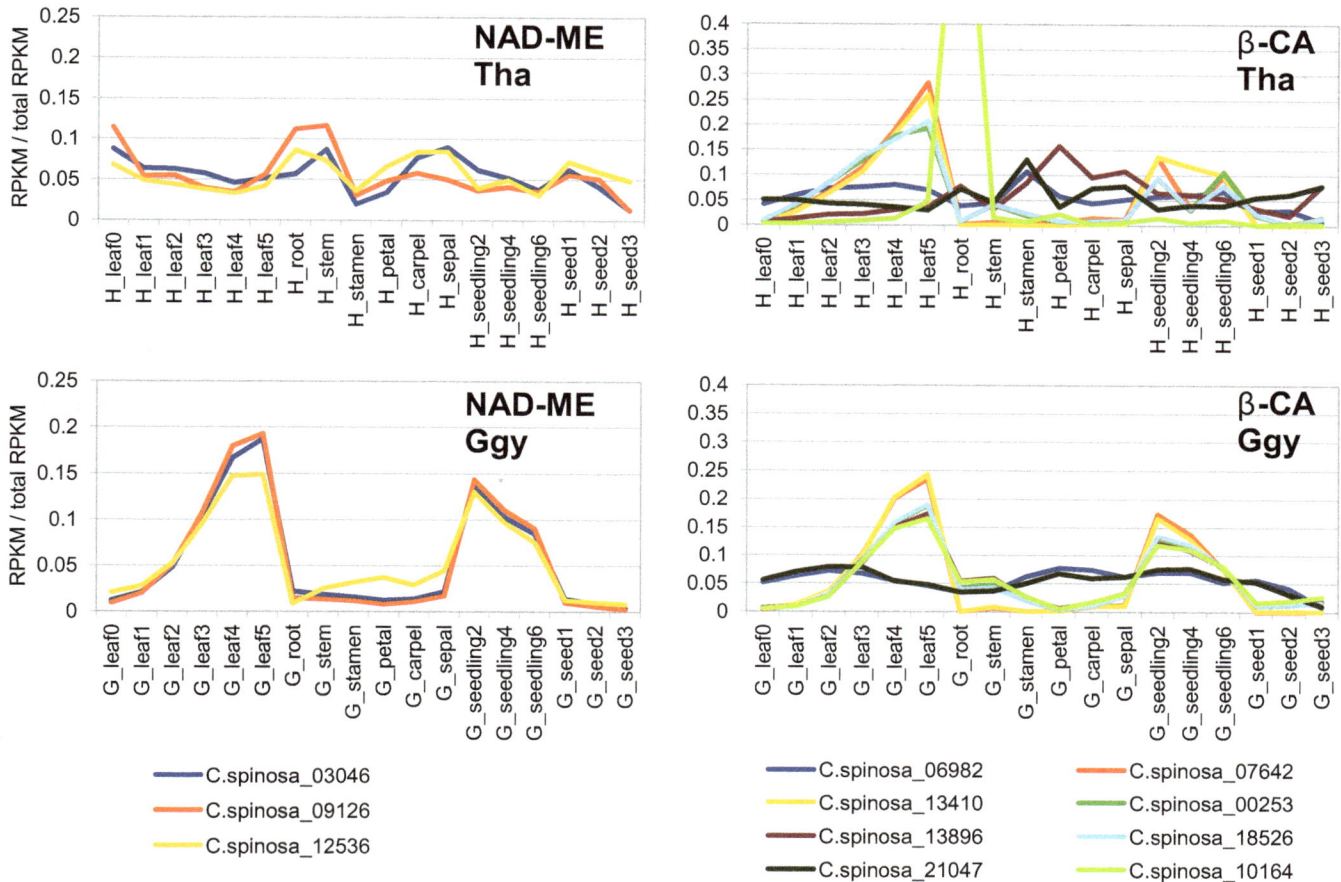

Fig. 5. Canalization in expression of NAD malic enzyme (top and bottom left) and β carbonic anhydrase (top and bottom right) homologs in *T. hassleriana* and *G. gynandra*. Top left: NAD-ME expression in *T. hassleriana*. Top right: βCA expression in *T. hassleriana*. Bottom left: NAD-ME expression in *G. gynandra*. Bottom right: βCA expression in *G. gynandra*. (Mapped) gene names and associated colors are displayed, see Materials and Methods for more details on the mapping of *G. gynandra* transcripts to *T. hassleriana* genes. Note that leaf0–leaf5 as well as seedling2–seedling6 and seed1–seed3 are time series of the same organ, with the leaf and seedling gradient being two days separated by stage. Transcription levels in *G. gynandra* (lower graphs) are more strictly regulated across organs, seeds and seedlings. The chaotic patterns in *T. hassleriana* (upper graphs) results in half the genes having a Pearson correlation > 0.95 compared to *G. gynandra*.

4. Discussion and conclusions

In this study, we have analyzed the transcriptomes of the C_3 *T. hassleriana* and C_4 *G. gynandra* to address the potential contribution of WGD and recent gene duplicates to the evolution of photosynthesis and C_4-pathway related genes. The initial comparison of *T. hassleriana* and *G. gynandra* was performed to identify the differential expression of key-genes involved in the NAD-ME C_4 biochemical pathway. However, it did not consider the role of gene duplicates. We show that very distinct patterns will occur when the duplication history is taken into account.

We could confirm the Th-α hexaploidy that has been found in *T. hassleriana* using an independent transcriptome dataset. We also find that *G. gynandra* shares this WGD with *T. hassleriana*, further establishing the occurrence of WGD in this lineage. Based on the phylogenetic position of both species in Cleomaceae, the Th-α duplication took place at least before the divergence of the two species which means that it is shared across Cleomaceae lineages 8–15 according to the latest phylogeny of the family [25]. Dating this polyploidy event in terms of absolute age is always a difficult task, however, here we find that the *Ks* rate of *G. gynandra* is extremely similar if not identical to *T. hassleriana*. Assuming then that mutation rates between these two species are the same, we can reaffirm the previous date estimation of Th-α at 13.7 mya [38].

The influence of the Th-α WGD event on photosynthetic gene composition is apparent, both in ortholog number as well as in syntenic region copy number for both species. From absolute orthologs numbers we can see that there is no increased retention between Cleomaceae species and even a slightly lower rate of retention in *G. gynandra*. This indicates that both species have experienced similar evolutionary constraints for a significant amount of time. Also we need to consider that genes sharing a similar sequence, do not necessarily have to share the same function. Even using strict CIP/CALP filtering which has been proved to be an accurate measure for the prediction of true orthologs [32], differential expression either in time, localization or regulation can substantially change the function of a gene. This is especially the case for genes in the core C_4 photosynthesis pathway, where many C_3 genes have been recruited into new functions [13,39].

When establishing *Ks* values of deeper ortholog nodes of photosynthesis genes, a large proportion of genes seem to have been retained from the γ duplication. For a trait that is likely to be highly dosage sensitive [23], we expect that gene loss will be rare and that remnants from this old paleohexaploidy are still present. However, considering the time that has passed since the γ paleohexaploidy event and on the basis of absolute gene copy numbers some gene loss has taken place predating the transition from C_3 to C_4.

The evolutionary importance of WGD events is made clear from the dominant presence of retained Th-α genes in both Cleomaceae species. However, certain questions remain: Can we couple this importance to the evolution of specific traits or in this case, C_4 photosynthesis? This is an old discussion, dating back to the works of

Ohno who was the first to suggest that the massive radiation of vertebrates was caused by a whole genome duplication in the ancestor [40]. An earlier study on the evolution of photosynthesis in soybean, showed that the Calvin–Benson–Bassham cycle (CBBC) and the light harvesting complex (LHC) gene families show a greater expansion from single gene duplications than both photosystem groups. This is explained by the increased dosage sensitivity of photosystem genes: if some subunits are expressed differently due to duplications while others are not, this is deleterious for the system as a whole [23]. This acts as a conservation mechanism for gene copy number that does not affect the more loosely connected enzyme collection of the CBBC and LHC genes.

In *G. gynandra*, where the expression of C_4 genes is tightly linked in clusters we would expect a high retention of orthologs. However, this dependency on transcriptional regulation has not lead to an increased retention of photosynthetic genes, as evidenced by lower copy numbers for all C_4 gene families when compared to *T. hassleriana*. It is not likely that neofunctionalization of genes after polyploidy has played a major role in the shift to C_4 photosynthesis. The much more stringent transcriptional regulation of C4 cycle genes in *G. gynandra* when compared to *T. hassleriana* as evidenced in this study is in accordance with the alternative hypothesis, which states that this process was mainly due to recruitment of existing genes in transcriptional space as suggested by several authors [12,14,41,42].

We still have much to learn regarding the development of C_4 photosynthesis. When studying this exceptional trait, we must always consider the genetic history of the species in question. Here, we give evidence that duplications, on a large scale and small, contribute to trait evolution. The exact mechanisms behind the recruitment of these genes into new biochemical pathways however are still largely unknown. Current sequencing efforts for *G. gynandra* will significantly aid in finding the detailed mechanisms of gene and C_4 photosynthesis evolution. The *Cleome* genus provides an excellent model system for unraveling the evolutionary origin and workings of C_4 photosynthesis and hopefully will enable us to harvest the fruits of our knowledge on this remarkable form of plant energy conversion.

Competing interests

The authors declare that they have no competing interests.

Authors' contributions

Cleome transcriptome sequencing, processing, assembly and quantification was done by CK. AB and APMW provided comments on handling highly expressed duplicates as well as proofreading the manuscript. EvdB performed the bioinformatic analyses. EvdB and MES prepared the manuscript. JMH and XZ proofread and edited the manuscript.

Acknowledgements

The work of EvB and MES was funded by NWO Vernieuwingsimpuls Vidi grant number 864.10.001. APMW acknowledges support by the Deutsche Forschungsgemeinschaft (SPP 1529; EXC 1028).

References

[1] R.F. Sage, P.-A. Christin, E.J. Edwards, The C4 plant lineages of planet Earth, J. Exp. Bot. 62 (2011) 3155–3169.

[2] R.F. Sage, The evolution of C4 photosynthesis, N. Phytol. 161 (2004) 341–370.

[3] M.S. Ku, J. Wu, Z. Dai, R.A. Scott, C. Chu, G.E. Edwards, Photosynthetic and photorespiratory characteristics of Flaveria species, Plant Physiol. 96 (1991) 518–528.

[4] N.J. Brown, K. Parsley, J.M. Hibberd, The future of C4 research – maize, Flaveria or Cleome? Trends Plant Sci. 10 (2005) 215–221.

[5] D.M. Marshall, R. Muhaidat, N.J. Brown, Z. Liu, S. Stanley, H. Griffiths, R.F. Sage, J.M. Hibberd, Cleome: a genus closely related to Arabidopsis, contains species spanning a developmental progression from C3 to C4 photosynthesis, Plant J. 51 (2007) 886–896.

[6] X.-G. Zhu, S.P. Long, D.R. Ort, Improving photosynthetic efficiency for greater yield, Annu. Rev. Plant Biol. 61 (2010) 235–261.

[7] G.E. Edwards, V.R. Franceschi, E.V. Voznesenskaya, Single-cell C4 photosynthesis versus the dual-cell (Kranz) paradigm, Annu. Rev. Plant Biol. 55 (2004) 173–196.

[8] A.D. McKown, N.G. Dengler, Vein patterning and evolution in C4 plants, Botany 88 (2010) 775–786.

[9] Y. Wang, A. Bräutigam, A.P.M. Weber, X.-G. Zhu, Three distinct biochemical subtypes of C4 photosynthesis? A modelling analysis, J. Exp. Bot. 65 (2014) 3567–3578.

[10] J.R. Ehleringer, T.E. Cerling, B.R. Helliker, C4 photosynthesis, atmospheric CO2, and climate, Oecologia 112 (1997) 285–299.

[11] A. Bräutigam, K. Kajala, J. Wullenweber, M. Sommer, D. Gagneul, K.L. Weber, K.M. Carr, U. Gowik, J. Maß, M.J. Lercher, An mRNA blueprint for C4 photosynthesis derived from comparative transcriptomics of closely related C3 and C4 species, Plant Physiol. 155 (2011) 142–156.

[12] N.J. Brown, C.A. Newell, S. Stanley, J.E. Chen, A.J. Perrin, K. Kajala, J.M. Hibberd, Independent and parallel recruitment of preexisting mechanisms underlying C4 photosynthesis, Science 331 (2011) 1436–1439.

[13] J.M. Hibberd, S. Covshoff, The regulation of gene expression required for C4 photosynthesis, Annu. Rev. Plant Biol. 61 (2010) 181–207.

[14] K. Kajala, N.J. Brown, B.P. Williams, P. Borrill, L.E. Taylor, J.M. Hibberd, Multiple Arabidopsis genes primed for recruitment into C4 photosynthesis, Plant J. 69 (2012) 47–56.

[15] K. Monson Russell, Gene duplication, neofunctionalization, and the evolution of C4 photosynthesis, Int. J. Plant Sci. 164 (2003) S43–S54.

[16] R.K.S. Monson, F. Rowan, The origins of C4 genes and evolutionary pattern in the C4 metabolic phenotype, in: Academic Press (Ed.), C4 Plant Biology, 1999, pp. 377–410, San Diego.

[17] C. Külahoglu, A.K. Denton, M. Sommer, J. Maß, S. Schliesky, T.J. Wrobel, B. Berckmans, E. Gongora-Castillo, C.R. Buell, R. Simon, et al., Comparative Transcriptome Atlases Reveal Altered Gene Expression Modules between Two Cleomaceae C3 and C4 Plant Species, The Plant Cell Online (2014), http://dx.doi.org/10.1105/tpc.114.123752, Advance online publication.

[18] T.L. Slewinski, A.A. Anderson, C. Zhang, R. Turgeon, Scarecrow plays a role in establishing Kranz anatomy in maize leaves, Plant Cell Physiol. 53 (2012) 2030–2037.

[19] T.L. Slewinski, A.A. Anderson, S. Price, J.R. Withee, K. Gallagher, R. Turgeon, Short-root1 plays a role in the development of vascular tissue and Kranz anatomy in maize leaves, Mol. Plant 7 (2014) 1388–1392.

[20] P. Wang, S. Kelly, J.P. Fouracre, J.A. Langdale, Genome-wide transcript analysis of early maize leaf development reveals gene cohorts associated with the differentiation of C4 Kranz anatomy, Plant J. 75 (2013) 656–670.

[21] S.B. Cannon, A. Mitra, A. Baumgarten, N.D. Young, G. May, The roles of segmental and tandem gene duplication in the evolution of large gene families in Arabidopsis thaliana, BMC Plant Biol. 4 (2004) 10.

[22] P. Edger, J.C. Pires, Gene and genome duplications: the impact of dosage-sensitivity on the fate of nuclear genes, Chromosome Res. 17 (2009) 699–717.

[23] J.E. Coate, J.A. Schlueter, A.M. Whaley, J.J. Doyle, Comparative evolution of photosynthetic genes in response to polyploid and nonpolyploid duplication, Plant Physiol. 155 (2011) 2081–2095.

[24] X. Wang, U. Gowik, H. Tang, J.E. Bowers, P. Westhoff, A.H. Paterson, Comparative genomic analysis of C4 photosynthetic pathway evolution in grasses, Genome Biol. 10 (2009) R68.

[25] T.A. Feodorova, E.V. Voznesenskaya, G.E. Edwards, E.H. Roalson, Biogeographic patterns of diversification and the origins of C4 in Cleome (Cleomaceae), Syst. Bot. 35 (2010) 811–826.

[26] E.V. Voznesenskaya, N.K. Koteyeva, S.D. Chuong, A.N. Ivanova, J. Barroca, L.A. Craven, G.E. Edwards, Physiological, anatomical and biochemical characterisation of photosynthetic types in genus Cleome (Cleomaceae), Funct. Plant Biol. 34 (2007) 247–267.

[27] R.D. Marquard, R. Steinback, A model plant for a biology curriculum: spider flower (Cleome hasslerana L.), Am. Biol. Teach. 71 (2009) 235–244.

[28] M.S. Barker, H. Vogel, M.E. Schranz, Paleopolyploidy in the Brassicales: analyses of the cleome transcriptome elucidate the history of genome duplications in Arabidopsis and other Brassicales, Genome Biol. Evol. 1 (2009) 391–399.

[29] S. Cheng, E. van den Bergh, P. Zeng, X. Zhong, J. Xu, X. Liu, J. Hofberger, S. de Bruijn, A.S. Bhide, C. Kuelahoglu, The Tarenaya hassleriana genome provides insight into reproductive trait and genome evolution of crucifers, Plant Cell Online 25 (2013) 2813–2830.

[30] W.J. Kent, BLAT – the BLAST-like alignment tool, Genome Res. 12 (2002) 656–664.

[31] S.F. Altschul, T.L. Madden, A.A. Schäffer, J. Zhang, Z. Zhang, W. Miller, D.J. Lipman, Gapped BLAST and PSI-BLAST: a new generation of protein database search programs, Nucleic Acids Res. 25 (1997) 3389–3402.

[32] F. Murat, Y. Van de Peer, J. Salse, Decoding plant and animal genome plasticity from differential paleo-evolutionary patterns and processes, Genome Biol. Evol. 4 (2012) 917–928.

[33] G.S. Slater, E. Birney, Automated generation of heuristics for biological sequence comparison, BMC Bioinform. 6 (2005) 31.

[34] Z. Yang, PAML: a program package for phylogenetic analysis by maximum likelihood, Comput. Appl. Biosci. 13 (1997) 555–556.
[35] M.S. Barker, K.M. Dlugosch, L. Dinh, R.S. Challa, N.C. Kane, M.G. King, L.H. Rieseberg, EvoPipes.net. Bioinformatic tools for ecological and evolutionary genomics, Evol. Bioinform. 6 (2010) 143–149.
[36] E. Lyons, M. Freeling, How to usefully compare homologous plant genes and chromosomes as DNA sequences, Plant J. 53 (2008) 661–673.
[37] A. Haudry, A.E. Platts, E. Vello, D.R. Hoen, M. Leclercq, R.J. Williamson, E. Forczek, Z. Joly-Lopez, J.G. Steffen, K.M. Hazzouri, et al., An atlas of over 90,000 conserved noncoding sequences provides insight into crucifer regulatory regions, Nat. Genet. 45 (2013) 891–898 (advance online publication).
[38] M.S. Barker, H. Vogel, M.E. Schranz, Paleopolyploidy in the Brassicales: analyses of the Cleome transcriptome elucidate the history of genome duplications in Arabidopsis and other Brassicales, Genome Biol. Evol. 1 (2009) 391.
[39] U. Gowik, J. Burscheidt, M. Akyildiz, U. Schlue, M. Koczor, M. Streubel, P. Westhoff, cis-Regulatory elements for mesophyll-specific gene expression in the C4 plant Flaveria trinervia, the promoter of the C4 phosphoenolpyruvate carboxylase gene, Plant Cell Online 16 (2004) 1077–1090.
[40] S. Ohno, U. Wolf, N.B. Atkin, Evolution from fish to mammals by gene duplication, Hereditas 59 (1968) 169–187.
[41] U. Gowik, P. Westhoff, The path from C3 to C4 photosynthesis, Plant Physiol. 155 (2011) 56–63.
[42] B.P. Williams, S. Aubry, J.M. Hibberd, Molecular evolution of genes recruited into C4 photosynthesis, Trends Plant Sci. 17 (2012) 213–220.
[43] P. Lamesch, T.Z. Berardini, D. Li, D. Swarbreck, C. Wilks, R. Sasidharan, R. Muller, K. Dreher, D.L. Alexander, M. Garcia-Hernandez, et al., The Arabidopsis Information Resource (TAIR): improved gene annotation and new tools, Nucleic Acids Res. 36 (2011) D1009–D1014.

PERMISSIONS

LIST OF CONTRIBUTORS

Fazhan Qiu, Yanli Liang, Yan Li, Yongzhong Liu, Liming Wang and Yonglian Zheng
National Key Laboratory of Crop Genetic Improvement, Huazhong Agricultural University, Wuhan 430070, China

Yuki Monden, Kentaro Yamaguchi and Makoto Tahara
Graduate School of Environmental and Life Science, Okayama University, 1-1-1 Tsushimanaka Kitaku, Okayama, Okayama 700-8530, Japan

Arthur Tavares de Oliveira Melo
University of New Hampshire, College of Life Science and Agriculture, Department of Biological Sciences, Durham, NH, USA

Saurabh Raghuvanshi, Pratibha Gour and Shaji V. Joseph
Department of Plant Molecular Biology, University of Delhi South Campus, Benito Juarez Road, New Delhi, 110021, India

Jamie Waese and Nicholas J. Provart
Cell and Systems Biology, University of Toronto, Toronto, Ontario M5S 3B2, Canada

Manish K. Vishwakarmaa, V.K. Mishraa, P.S. Yadava and H. Kumara,
Department of Genetics and Plant Breeding, Institute of Agricultural Sciences, Banaras Hindu University, Varanasi, India

P.K. Guptab
CCS University, Meerut, India

Arun K. Joshi
CIMMYT, South Asia Regional Office, P.O. Box 5186, Kathmandu, Nepal

Dinesh Kumar Jaiswal
Department of Biology, University of North Carolina, Chapel Hill, NC 27599, USA

Emily G. Werth, Evan W. McConnell and Leslie M. Hicks
Department of Chemistry, University of North Carolina, Chapel Hill, NC 27599, USA

Alan M. Jones
Department of Biology, University of North Carolina, Chapel Hill, NC 27599, USA

Department of Pharmacology, University of North Carolina, Chapel Hill, NC 27599, USA

Frances Soman
Department of Electrical Engineering and Computing Systems, College of Engineering and Applied Sciences, University of Cincinnati, Cincinnati, OH 45221,United States

Xiaoting Chen, Kevin Ernst and Mike Borowczak
Department of Electrical Engineering and Computing Systems, College of Engineering and Applied Sciences, University of Cincinnati, Cincinnati, OH 45221,United States
Center for Autoimmune Genomics and Etiology, Cincinnati Children's Hospital Medical Center, Department of Pediatrics, College of Medicine, University of Cincinnati, Cincinnati, OH 45229, United States

Matthew T. Weirauch
Center for Autoimmune Genomics and Etiology, Cincinnati Children's Hospital Medical Center, Department of Pediatrics, College of Medicine, University of Cincinnati, Cincinnati, OH 45229, United States
Division of Biomedical Informatics and Division of Developmental Biology, Cincinnati Children's Hospital Medical Center, Department of Pediatrics,College of Medicine, University of Cincinnati, Cincinnati, OH 45229, United States

Zareen Khana, Hyungmin Rhoa, Shang Han Hunga, Soo-Hyung Kima and Sharon L Dotya
School of Environmental and Forest Sciences, University of Washington, Seattle, WA 98195-2100, USA

Andrea Firrincieli
Department for Innovation in Biological, Agro-food and Forest systems, University of Tuscia, Viterbo 01100, Italy

Virginia Luna and Oscar Masciarelli
Laboratorio de Fisiología Vegetal, Departamento de Ciencias Naturales, Fac. de Cs. Exactas, Universidad Nacional de Río Cuarto, Ruta 36 Km 601, 5800 RíoCuarto, Argentina

Locedie Mansueto, Roven Rommel Fuentes, Dmytro Chebotarov, Frances Nikki Borja, Jeffrey Detras, Juan Miguel Abriol-Santos, Ruaraidh Sackville Hamilton, Kenneth L. McNally, Nickolai Alexandrov and Ramil Mauleon
International Rice Research Institute, College, Los Baños, Laguna, 4031, Philippines

Kevin Palis
International Rice Research Institute, College, Los Baños, Laguna, 4031, Philippines
Boyce Thompson Institute, Ithaca, NY 14853, USA

Alexandre Poliakov and Inna Dubchak
Lawrence Berkeley National Laboratory, Berkeley, CA 94720, USA
DOE Joint Genome Institute, Walnut Creek, CA 94598, USA

Victor Solovyev
Softberry, Inc., Mount Kisco, NY 10549, USA

Maggie E. McCormack, Jessica A. Lopez, Tabitha H. Crocker and M. Shahid Mukhtar
Department of Biology, University of Alabama at Birmingham, Birmingham, AL, USA

Mariko M. Alexander
Plant Pathology and Plant-Microbe Biology Section, School of Integrative Plant Science, Cornell University, Ithaca, NY, USA
Boyce Thompson Institute for Plant Research, Ithaca, NY, USA

Michelle Cilia
Plant Pathology and Plant-Microbe Biology Section, School of Integrative Plant Science, Cornell University, Ithaca, NY, USA
Boyce Thompson Institute for Plant Research, Ithaca, NY, USA
Robert W. Holley Center for Agriculture and Health, USDA-ARS, Ithaca, NY, USA

Madlen Nietzsche and Ramona Landgraf
Plant Metabolism Group, Leibniz-Institute of Vegetable and Ornamental Crops (IGZ), 14979 Großbeeren, Germany

Takayuki Tohge
Max-Planck-Institute of Molecular Plant Physiology, Am Mühlenberg 1, D-14476 Potsdam-Golm, Germany

Frederik Börnke
Plant Metabolism Group, Leibniz-Institute of Vegetable and Ornamental Crops (IGZ), 14979 Großbeeren, Germany

Institute of Biochemistry and Biology, University of Potsdam, 14476 Potsdam, Germany

Hans-Jörg Mai
Institute of Botany, Heinrich Heine University Düsseldorf, Universitätsstraße 1, Building 26.13, 02.36, 40225 Düsseldorf, Germany

Petra Bauer
Institute of Botany, Heinrich Heine University Düsseldorf, Universitätsstraße 1, Building 26.13, 02.36, 40225 Düsseldorf, Germany
CEPLAS Cluster of Excellence on Plant Sciences, Heinrich Heine University Düsseldorf, Düsseldorf, Germany

Dóra Szakonyi, So ie Van Landeghem, Lieven Baeyens, Jonas Blomme, Stefanie De Bodt, Fabio Fiorani, Nathalie Gonzalez and Dirk Inzé
Department of Plant Systems Biology, VIB, and Department of Plant Biotechnology and Bioinformatics, Ghent University, B-9052 Ghent, Belgium

Katja Baerenfaller,   Asuka Kuwabara, Wilhelm Gruissem and Sean Walsh
Department of Biology, ETH Zurich, CH-8093 Zurich, Switzerland

Rubén Casanova-Sáez, David Esteve-Bruna, Sara Jover-Gil, Tamara Muñoz-Nortes, David Wilson-Sánchez, José Luis Micol
Instituto de Bioingeniería, Universidad Miguel Hernández, 03202 Elche, Alicante, Spain

Jesper Grønlund and Vicky Buchanan-Wollaston
Warwick Systems Biology Centre, and School of Life Sciences, University of Warwick, Coventry CV4 7AL, United Kingdom

Richard G.H. Immink, Aalt D.J. van Dijk and Gerco C. Angenent
Plant Research International, Bioscience, 6708 PB Wageningen, The Netherlands

Yves Van de Peer
Department of Plant Systems Biology, VIB, and Department of Plant Biotechnology and Bioinformatics, Ghent University, B-9052 Ghent, Belgium
Genomics Research Institute (GRI), University of Pretoria, Private Bag X20, Pretoria 0028, South Africa

Pierre Hilson
Department of Plant Systems Biology, VIB, and Department of Plant Biotechnology and Bioinformatics, Ghent University, B-9052 Ghent, Belgium
INRA, UMR1318, and AgroParisTech, Institut Jean-Pierre Bourgin, RD10, F-78000 Versailles, France

Alexandr Koryachko, James Tuck and Cranos Williams
Electrical and Computer Engineering, North Carolina State University, Raleigh, NC, USA

Anna Matthiadis and Terri A. Long
Plant and Microbial Biology, North Carolina State University, Raleigh, NC, USA

Joel J. Ducoste
Civil, Construction, and Environmental Engineering, North Carolina State University, Raleigh, NC, USA

Allison Gaudinier
Department of Plant Biology, College of Biological Sciences, University of California, Davis One Shields Avenue, Davis, CA 95616, USA

Michelle Tang
Department of Plant Biology, College of Biological Sciences, University of California, Davis One Shields Avenue, Davis, CA 95616, USA
Department of Plant Sciences, College of Agriculture and Environmental Sciences, University of California, Davis One Shields Avenue Davis, CA 95616, USA

Daniel J. Kliebenstein
Department of Plant Sciences, College of Agriculture and Environmental Sciences, University of California, Davis One Shields Avenue Davis, CA 95616, USA
DynaMo Center of Excellence, University of Copenhagen, Thorvaldsensvej 40, DK-1871, Frederiksberg C., Denmark

Ricardo A. Chávez Montes, Humberto Herrera-Ubaldo, Joanna Serwatowska and Stefan de Folter
Unidad de Genómica Avanzada (LANGEBIO), Centro de Investigación y de Estudios Avanzados del Instituto Politécnico Nacional (CINVESTAV-IPN), Km. 9.6Libramiento Norte, Carretera Irapuato-León, CP 36821 Irapuato, Guanajuato, Mexico

Joachim W. Bargsten
Applied Bioinformatics, Bioscience, Plant Sciences Group, Wageningen University and Research Centre, Wageningen, The Netherlands
Netherlands Bioinformatics Centre (NBIC), Nijmegen, The Netherlands
Laboratory for Plant Breeding, Plant Sciences Group, Wageningen University and Research Centre, The Netherlands

Edouard I. Severing
Laboratory of Genetics, Plant Sciences Group, Wageningen University and Research Centre, Wageningen, The Netherlands

Jan-Peter Nap
Applied Bioinformatics, Bioscience, Plant Sciences Group, Wageningen University and Research Centre, Wageningen, The Netherlands
Netherlands Bioinformatics Centre (NBIC), Nijmegen, The Netherlands

Gabino F. Sanchez-Perez
Applied Bioinformatics, Bioscience, Plant Sciences Group, Wageningen University and Research Centre, Wageningen, The Netherlands
Laboratory of Bioinformatics, Plant Sciences Group, Wageningen University and Research Centre, Wageningen, The Netherlands

Aalt D.J. van Dijk
Applied Bioinformatics, Bioscience, Plant Sciences Group, Wageningen University and Research Centre, Wageningen, The Netherlands
Biometris, Wageningen University and Research Centre, Wageningen, The Netherlands

Yichun Qiu, Samuel J. Filipenko, Aude Darracq and Keith L. Adams
Department of Botany, University of British Columbia, 6270 University Blvd, Vancouver, BC, Canada V6T 1Z4

Erik van den Bergh and M. Eric Schranz
Biosystematics, Wageningen University and Research, Droevendaalsesteeg 1, 6708 PB Wageningen, The Netherlands

Canan Külahoglu, Andrea Bräutigam and Andreas P.M. Weber
Institute of Plant Biochemistry, Center of Excellence on Plant Sciences(CEPLAS), Heinrich-Heine-University, D-40225 Düsseldorf, Germany

Julian M. Hibberd
Department of Plant Sciences, University of Cambridge, Cambridge CB2 3EA, United Kingdom

Xin-Guang Zhud
Plant Systems Biology Group, Partner Institute of Computational Biology, Chinese Academy of Sciences/Max Planck Society, Shanghai 200031, China

Index

Introduction to Horticulture